Associative Remote Viewing

The Art and Science of Predicting Outcomes for Sports, Finances, Politics and the Lottery

Debra Lynne Katz, Ph.D.
Jon Knowles, M.A.

Printed in the United States of America
First Printing, August 2021

Published through Living Dreams Press, Santa Barbara, CA

ISBN: 978-1-943951-28-4 (print)
ISBN: 978-1-943951-29-1 (eBook)

Library of Congress Control Number: 2021908328

Cover Design by Debra Lynne Katz

We are adapting our book to the digital age. This book has a web page: http://www.ARVBook.com

We have put URL's that would ordinarily be referenced in the text onto a website for the book. The reason is that many URLs soon become out of date and cease working.

The web page contains links to the authors and contributors to the book as well as other relevant sites.

The print version of the book is black and white. However, the e-book version will have color graphics and tables.

Dedication

Debra: I dedicate this book to Marty Rosenblatt and to my first grandson River Conner Katz.

Jon: I dedicate this book to Jamie Knowles and Naima Hart.

Acknowledgements

We heartily thank these viewers, project managers and researchers who permitted their ARV work and/or experiments or other materials to be included in our book: Tunde Atunrase, Dr. Tom Atwater, Teresa Fendley, Michael Ferrier, Frank&Friends, Bill Gattis, Igor Grgić, John Herlosky, Sandra Hilleard, JFK, Greg Kolodziejzyk, Nicola Laurino, Lincoln Lounsbury, Joe McMoneagle, Sean McNamara, Dr. Ed May, Dr. Garret Moddel, Dr. Julia Mossbridge, Maximilian and Gisela Mueller, Alexis Poquiz, Marty Rosenblatt, Daz Smith, Dr. Don Walker.

We are grateful for permission to use sketches and extracts from sessions by Michael Ash, Dale Graff, Carl McClelland, Russell Pickering, Joffre Perreault, Dave Silverstein, Lyrysa Smith, Joyce Wahlberg, Mark White, Aala Zahrah.

Many thanks to a most generous and extremely talented copy-editor, Teresa Fendley, and to Pili Torre, Coral Carte and Aggie Hinman for valuable assistance in editing, research or proofreading.

A note about the text: Both authors contributed to each chapter, with one or the other doing the first draft and both involved in revising each chapter. As you will see, we have different writing styles and we have let these stand for the variety they offer. At times we will indicate that one or the other of us is primarily speaking in a chapter. In addition, we designed the book in hopes it will become a resource for the field – a sourcebook – and we like the authenticity of the many contributors presenting their materials in their own voices.

Contents

Introduction

This book is a celebration. It is the coming together of decades of inquiries, endeavors and experiments to understand and make use of profound intuitive aspects of human consciousness through sustained individual effort and group interactions. It is an honoring of many who have worked, both tirelessly and at times very tired, not merely for their own personal benefit, but for the advancement of human knowledge.

Our book is a collection of projects, studies and other undertakings that have three things in common: They focus on intentional precognition; they seek to understand the variables involved in generating predictions through the use of psi-based perception; and they involve an empirically based formal or informal experimental approach. Many include an applied component such as making predictions about sports events, elections, financial trades or the lottery. Additionally, excluding a few formal studies, these projects were devised and conducted by people with moderate to substantial levels of remote viewing training and experience under their belt.

Several of the informal exploratory efforts harken back to the days before science became an institutionalized, hierarchal domain in which only credentialed individuals studying certain sanctioned topics were recognized.[1,2,3] While this statement reveals disappointment with the way history has played out, it should not be taken to mean we don't respect formal research, particularly that which involves repeated trials done under rigorous protocols. In fact, we have included what is probably the most comprehensive literature review of formal and informal Associative Remote Viewing projects to date. We are hopeful the approaches and results noted in this book will be useful to all levels and types of researchers, whether homegrown or lab-oriented.

As will be noted throughout this book, remote viewing was born in research laboratories. This highly collaborative environment allowed for sharing ideas and insights by participants, in particular Ingo Swann, even if they didn't always receive recognition in the literature for their intellectual contributions.

ARV was created and utilized by such researchers more with the idea that it could be used to make predictions for money-making ventures than as a testing procedure for psychics. In this respect, ARV is one of the first examples of psi researchers taking an experimental, quantitative-based approach and applying it to real life. Likewise, much of what was brought into the labs, even before Ingo Swann and others formalized the practice of remote viewing, was inspired by a combination of spontaneous psi experiences and contemporary psychical research. In this way, information was truly bidirectional, moving back and forth between the arenas of science and public life.

The formal labs in parapsychology with official funding, on-campus locations, and a hierarchal structure of prestige and power have largely faded into the past. But the questions raised and lessons learned by past researchers live on and have been carried forward by present-day researchers, some of whom have joined forces with groups such as the Applied Precognition Project (APP), the International Remote Viewing Association (IRVA), and a plethora of social media sites such as Facebook, Reddit, Yahoo and groups formed by remote viewing teachers who have professional-level remote viewers. While there is often a spirit of egalitarianism and the recognition that in ARV there are no experts, there are, as always, power dynamics at play which sometimes foster and sometimes hinder the development of the field.

Many of those who study the sociology of scientific knowledge such as P.J. Bowler[4] have found the tide seems to be changing with regard to who has the power to conduct research and in what form and format these activities are undertaken. As K. Franczak[5] wrote, "the pressure to recognize the legitimacy of forms of knowledge previously deprived of institutional legitimacy can also be observed" (p. 21). He suggests that the "number of areas and disciplines that scientists can count on unconditional deference is drastically shrinking" (p. 19), which he attributes to "representatives" who do not have "legitimate institutional or scientific authority" but are in positions of power to voice skeptical or alternative viewpoints, such as politicians, the media, business or religious leaders, new social movements and "determined amateur enthusiasts" (p.20). These individuals are newly empowered through the growing and diverse forms of technology that make acquisition and communication of all forms of knowledge possible without reliance on authority figures sanctioned through institutions. Franczak asserts: "Many of these successfully defend themselves against labels of 'counter-knowledge' or 'pseudoscience' and seek supporters within channels not necessarily sanctioned by scientists" (p. 20).

In precognitive ARV projects, not only are we seeing highly capable people who don't have formal academic backgrounds taking them up, but in recent years, parapsychologists and academic researchers from other fields seem to be taking more interest, particularly in application-based RV activities, although numbers are still small. This interest has led to joint projects between both groups of researchers, which has then led to informal researchers moving to the adoption of more formalized research practices, while a few academic researchers have taken up the practice of remote viewing themselves. How widespread these migratory behaviors are remains to be seen. However, all one has to do is travel to the Swann archives at the University of West Georgia's Ingram library to discover this migration is not new – Ingo was a pioneer in undertaking the dual role of subject and researcher, and in the archives we found evidence of many formal researchers and academics who tried RV themselves. It is also possible the SRI/military projects, which have dominated the remote viewing narrative, wouldn't have received funding renewals if this hadn't happened.

What to expect from this book

Since we present such a mixture of projects, ranging from informal explorations to formal trials, we will endeavor to make clear in each instance what type of project is being presented. To do so, we include numerous references in the chapters and full details in the bibliography, while balancing the need to make the chapters fully readable. In some cases, positive results are reported; in others, failures; and in still others, the results are not yet in. In describing these varied projects and results, our aim is not to establish proof but to present, often for the first time, what practitioners in the remote viewing community have been doing with their time, often over many years. We hope our book sparks interest and opens minds to the rich range of possibilities available to us as a species.

Wager wisely

We understand that many readers of this book will have little interest in formal research and will not share our passion to understand every factor and variable in remote viewing, in ARV, and in precognitive-based applications projects. Rather, they may be curious about the way human consciousness, time or life itself works and whether our book sheds light on these. (We think and hope for you it will!) Some will be interested in learning how to improve their remote viewing skills, which we will cover in Chapter 25. For many, the usefulness of the book will be in what it can offer in applying psi to financial investments, business decisions, gambling, and sports betting. In other words, many will hope the book will help them make money.

As Marty Rosenblatt, the head of the largest ARV organization, is fond of saying, "Wager wisely!" We would not want readers who are struggling financially to take risks they wouldn't otherwise take after reading about the success of ARV projects. Rather, we'd like to see those who already engage in higher-risk financial activities and investments explore the role that intuition plays in all their decisions and see what happens when they supplement their present logic-based predictions with psi-based ones, whether performed by remote viewers or by themselves as budding intuitives. Highly trained and experienced remote viewers are taking part in financial and sports predictions and others are looking for opportunities to do so. One group has a dozen full- and part-time employees, some earning a living doing cryptocurrency predictions.

All remote viewers undertaking these activities deserve to be treated with respect and compensated for their work. However, it should be kept in mind that the level of success may be affected not only by their skills, but by how well a project is set up and by the "synergy" between the client and themselves.

With that preview of what's coming up, a few words about the authors.

Jon Knowles

When Debra suggested in late 2019 that we write a book together, I snapped at the chance. We had both been very active in the Applied Precognition Project for several years, had been viewers on the same projects, had met at conferences and we got along quite well. Debra had an extensive background in psi prior to becoming involved in remote viewing and she was already the author of three books about psi. The fact that she had written a half dozen articles for non- and peer-reviewed journals related to RV, ARV and parapsychology further convinced me she would be an excellent writing partner.

I have been active in the remote viewing field since 1999, which is when I began training with Pru Calabrese of TransDimensional Systems. I detailed my experiences in that successful remote viewing company in my 2017 book, *Remote Viewing from the Ground Up*.[6] After TDS closed down in 2003, I helped form and was a viewer, project manager and admin person in the Aurora Remote Viewing Group, the first multi-method RV group (viewers used different forms of remote viewing). Following that, after working solo and with a few others, in 2010 I joined Marty Rosenblatt's Physics-Intuitions-Applications group (which soon became the Applied Precognition Project). I was active in APP as a group manager, viewer, volunteer staffer and membership coordinator for five years. Since 2016, I have worked with a few other viewers, done solo viewing, and maintained 120+, a site with links to all facets of the field. To keep 120+ up to date, I kept an eye on as many developments as I could in this (finally!) fast-growing field.[7] While doing so I have met quite a few practitioners over the years, including new folks and others who keep a low profile on public media and in the remote viewing community.

I have long been interested in mainstream research on consciousness and the body and have read a great deal of mainstream and parapsychological literature. I decided early on not to attempt to publish in peer-reviewed journals and instead submitted articles to Daz Smith's *Eight Martinis* magazine and presented at APP Conferences. There were several reasons for my decision, but in brief, for someone like me who is very interested in theory, not a scientist, and practically oriented when it comes to RV, I felt I could accomplish more by avoiding the sclerotic peer-review process. (Our Bibliography includes all the articles published in *Eight Martinis*.)

My focus in remote viewing for the past ten years has been on ARV. That's in part because accessing numbers and letters is one of the hardest things to do in remote viewing, and ARV offered promise of being able to get both indirectly (through association). Also, ARV offered the possibility of use in finance and sports events – where there was already evidence of success. Making money is something society pays attention to and success in finance or sports could help propel remote viewing into the mainstream where it belongs – and eventually will be. Why it hasn't entered the mainstream sooner is another story, having to do with the history of RV, particularly after it entered the public domain in 1995 (a story that will have to wait for another day). It turned out ARV can also be used in winning lotteries, which could be a game changer, and so we devote a chapter on the significant – but almost unknown – lottery successes by remote viewers, including some of my own.

Another point of note is that in ARV you get a clear and distinct result. The prediction comes true or it does not. The team wins or it does not. Further, you can do many ARV trials in a short period of time and thus quickly acquire statistics to chart your progress, while in non-ARV projects the results are generally less clear and statistics harder to come by. If it is a client project, you may not even get feedback, and you can do far fewer trials or projects in a given time span. ARV statistics can be another tool in understanding all facets of the massive shift in understanding the world that is underway, and of which remote viewing is one augur.

An important aspect of ARV is the basic fact of precognition. If precognition is real, and there is an avalanche of evidence indicating it is, then there is something about the future we can sense and get a handle on beforehand. This goes against the dominant viewpoint in science that the future does not exist in any real sense and therefore you cannot predict it with any significant degree of accuracy using psi. (It is accepted that you can predict the future to some extent using statistical methods, including big data and the "wisdom of crowds.") The floor ARV success rate of more than 60% described in these pages achieved by many groups and individuals destroys the belief that psi is not real. As a former philosophy major in college, the subjects of time, precognition and the nature of reality have remained a strong interest of mine over the decades since graduation – and in remote viewing, we are in the midst of these mysteries.

Regarding my personal history, my degree in philosophy was from Harvard University in 1960 and I received an M.A. in English Language and Literature from U.C. Berkeley in 1966, following two years in the US Army courtesy of President Kennedy and the draft. I am retired and by occupation I was a teacher (14 years) and a medical transcriptionist (20 years). I devised a typing abbreviation system (ABCZ Software), which has been used with text expanders by many medical transcriptionists. For about 20 years I was very active on the extreme Left trying to build a revolutionary party in the United States – that was my true "occupation" at the time. That effort obviously never took wing and in 1999 I turned to the fascinating field of remote viewing, which remains my focus. When it is time, I'd like to return to composing music ('modern classical'), something I greatly enjoy doing.

I live with my wife in the Bay Area, California, and am very happy that technology has enabled me to meet and be in touch daily with so many wonderful people around the globe who have also discovered the remarkable field of remote viewing. We have a genuine community of discussion and activity in this surprising and paradigm-changing endeavor.

Debra Katz

Recently, I carried out a survey of 106 experienced remote viewers with Dr. Patrizio Tressoldi of Italy. We wanted to find out about the different kinds of remote viewing applications projects remote viewers are engaged with today. They included many of the categories addressed in this book, but also others such as crime solving, finding missing objects, medical applications, and esoteric and archeology projects. We asked the viewers, "What do you enjoy most about RV applications work?"

While I wasn't surprised to hear positive sentiments, I didn't expect so many to express that remote viewing actually makes them "high." They expressed an extreme intensity of emotion, such as "love," "thrill," "most fascinating." Many appreciated the level of personal development, insight and learning that happens in relation to remote viewing. Other comments included "helping clients realize their goals," "getting a deeper understanding of myself" and "learning about my subconscious mind." One loved "how it works – there is an inner threshold and I like to step over it and back." Through RV, one said, "we can have any understanding, knowledge or wisdom that we desire." Others appreciated that there's "always something new," "the process and learning," and "the confirming feedback." One viewer wrote: "I feel it helps me get over my Fear of Being Publicly Wrong (FOBPW)." Another said, "Being helpful to society, and totally blown away clients is fun, too."

From the above comments, I've recently started to wonder if there isn't something that gets stimulated in the pleasure centers of the brain when remote viewing, at least when one is on target. Something about seeing a great match is like a drug and produces a sense of euphoria. For example, I had a target where I saw guys sitting at computers, side by side with heads turned, and I wrote that it reminded me of the Mission Control Center at NASA. I felt a sense of exuberance. I also had written down the words, "narrow miss" and "close call." When I saw the feedback photo, I could hardly believe it – even though I've been doing intuitive work for decades. The target was NASA Mission Control during the Apollo 11 launch.

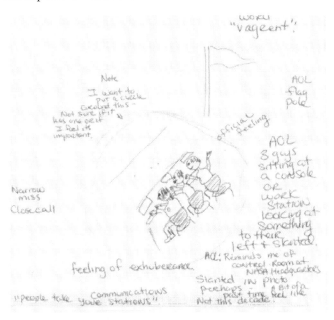

Remote Viewers Transcript (Debra)

Feedback Photo – Mission Control

The euphoria I felt in the session lasted for almost a week. At first, I thought I was really happy because I had been working so hard for so long to improve my remote viewing skills and here was evidence of the payoff. But at some point, it occurred to me I may have empathically merged with the emotions of the immense sense of accomplishment they must have felt in Mission Control when the shuttle successfully landed, after some very close calls.

Regardless, experiences like these keep me up way too late at night, not just remote viewing, but researching, writing and talking about it as if I'd just discovered it yesterday. While remote viewing is not always easy or joyous, it is always an adventure and never dull. No two sessions are the same. I've now been blind-tasked (given only a target number) on the Statue of Liberty on four separate occasions and every time the information came in differently. The most memorable of these was when I had the awareness that I was standing up in a very high tower looking out of windows from what could only be described as a human head. What a weird, eerie, exciting realization that was, which told me I had to be in some kind of human-like statue. Still, I didn't realize it was the Statue of Liberty. It really is moments like these that make not just remote viewing, but all of life so darn cool!

Just a few words about my background: Like many of my colleagues, I did not get involved in ARV for the money or wagering aspect. I jumped in purely because I was looking for free practice opportunities. I had recently joined up with some researchers who were exploring to what extent remote viewing could be used to diagnose viruses in tomato plants and to describe microscopic organisms. I was attending my very first International Remote Viewing Association conference and in the attendee folder was a flyer that said "ARV4Fun."[8]

Little did I know when I reached out to Marty Rosenblatt, who was running ARV trials with both new and experienced viewers, that I'd go on to meet so many incredible people, including my co-author here, Jon Knowles. At that time, I also never imagined I'd become the co-webmaster of the Applied Precognition Project (APP), participate in several research and remote viewing groups, attend training in almost every remote viewing modality and rarely miss a remote viewing conference. Nor did I realize that I'd attend the Ph.D. program in Psychology (with an emphasis in Consciousness and Society) at the University of West Georgia. Another surprise was discovering Ingo Swann's archives had recently been donated to the university, which led to my spending most Fridays with Ingo (posthumously of course) for two years.

I'd already known Jon for several years through hearing his talks at conferences, running into him at trainings, engaging in lively discussions with him on multiple RV online forums, and working as a viewer on some of his innovative projects involving pictograms, SUARV, and one involving searching for silver in the wilderness. As a rater/judge on several projects over the years, from time to time I'd come across an excellent remote viewing session and wonder – who is this viewer? It would turn out to be Jon. Still, it wasn't until he came out to stay with me for a week to help with cataloging the Swann archives that I really realized what an impressive guy he is. Within three days of cataloging, he had already far surpassed what had taken me two semesters to accomplish. Jon was organized, efficient, concise, and focused – not easy when you are surrounded by endlessly fascinating letters, reports, memos, articles, books, about the most intriguing subjects in the world. For our ARV book together, Jon has made use of his extensive knowledge of the subject matter, paired with a knack for staying current with just about everything going on in the field – he's been in touch with dozens of people at the forefront of what really could be considered an ARV movement to learn what they are up to and to assess their data (in several cases, data that had not been previously shared). He's essentially a walking encyclopedia of ARV and RV.

ARV is endlessly fascinating as it is truly a bridge between consciousness (all that exists within the mind and between minds) and the material, physical world. At the same time, it is a mean of utilizing one's psychic powers to make money.

With that I will just say that whatever has drawn you to join us here, we appreciate the honor and responsibility of serving as your guides. We will do our best to handle the topics in a balanced and comprehensive manner. We hope you will get a feel for the richness, depth, complexity, rewards, trials, tribulations and mega-awesomeness of remote viewing, Associative Remote Viewing and the art, practice and science of intentional psi.

PART ONE

THE NATURE AND HISTORY OF REMOTE VIEWING AND ASSOCIATIVE REMOTE VIEWING

CHAPTER 1

What is Remote Viewing?

It is recommended that future experiments focus on understanding how this phenomenon works, and on how to make it as useful as possible. There is little benefit to continuing experiments designed to offer proof, since there is little more to be offered to anyone who does not accept the current collection of data. —Jessica Utts[9]

When people speak of remote viewing, they typically use the term in one of two ways. First, it is used as a synonym for clairvoyance – an intuitive-based activity obtaining information at a distance through visual pictures.[10] The word and definition display their French origin *clair* (clear) + *voir*. Dictionaries broaden the meaning beyond the visual to "1. the power or faculty of discerning objects not present to the senses, 2. ability to perceive matters beyond the range of ordinary perception" (Merriam-Webster) or "the supposed faculty of perceiving things or events in the future or beyond normal sensory contact" (Oxford).

Second, it is a term used within a specific historical context linked to a scientific protocol, and the variety of methods within the protocol.

The origins of remote viewing at the American Society for Psychical Research

Remote viewing has a rich and colorful history. It was developed in research labs and funded and used as an information-gathering tool by numerous US intelligence and military agencies from approximately 1972 through 1995. It has been taught and practiced in many public venues in the years since. Much of the early history has been presented in volumes by former members of the Star Gate project and by outsiders. We present an abbreviated history of the early years, an account enriched by Ingo Swann's documents from the University of West Georgia archives (not available when many of the books were written).

Testing and use of psi for practical purposes in diverse cultures for centuries is well documented.[11,12,13] However, applied psi and remote viewing were developed within the historical context discussed here when Ingo Swann, recruited as an experimental subject, joined forces with Janet Mitchell, Karl Osis, and Gertrude Schmeidler at the American Society for Psychical Research (ASPR). Their work began the long series of events that led to Swann becoming widely recognized as "the father of remote viewing."

Having had exceptional spontaneous experiences as a child, Swann decided in his late thirties to seek assistance from researchers, hoping they could help him gain insight into his intuitive potential. From approximately 1970 through 1973, ASPR researchers conducted numerous experiments with Swann. Their aim was to study psi-based perceptions in relation to Out of Body Experiences (OBE). Initially, target materials were located in the lab and Swann was tasked with describing objects placed on a shelf about 10 feet above his head while hooked up to monitoring equipment.

During this time at the ASPR, researchers observed "learning curves" taking place as they varied environmental conditions. Mitchell noted that Swann was not simply a subject of these experiments, but an active participant whose input led to meaningful changes and discoveries in their lab. These included the importance of sketching,[14] the usefulness of receiving feedback after every trial so adjustments could be made by the viewer and specific changes implemented in the experimental protocols by the other researchers.[15] This type of collaborative effort would characterize the work of remote viewing not just within the ASPR lab at that time, but also in many arenas in which remote viewing would be studied and practiced.

Remote viewing comes to the Stanford Research Institute

Swann had read a paper by Dr. Harold E. Puthoff titled "The Physics of Psychoeneretic Processes, Research Proposal" and he was impressed, as he shares in this passage:

> Somewhat into his paper, Puthoff made statements which electrified me – since they had to do with my own understanding that psi phenomena were species-wide, part of our species life potentials. "When one considers basic life processes within the framework of modern scientific theory, particularly modern quantum theory, two basic viewpoints emerge…One is that quantum theory as now understood is, in principle, essentially capable of encompassing the biological and psychological principles of existence as manifested in life processes…From this viewpoint, the fact that we have not done so is due simply to the complexities of analysis presented

to the theorist by even the simplest of living processes…This viewpoint we refer to as the reductionist viewpoint. Here it is considered that even the most complex of life processes can in principle be reduced step by step through layers of complexity, to the basic principles encompassed by present quantum theory."

WOW! Here Puthoff was talking about BASIC life processes and was including psi phenomena among them. PROCESSES! And processes at the fundamental quantum level, processes which, if faster than light, might help account for the instantaneous perceptions I had noted during the ASPR experiments. Well, I had never encountered this view in parapsychology (as it existed at that time). I was not a quantum physicist, of course, but I felt from my own experiential levels that I completely understood what Puthoff was talking about. In reading his ideas, I felt I was reading a version of my own concepts. So. If you had been in my position and hunting for knowledge that might help elucidate complicated things, and all other things considered, what would you have done? I didn't need to decide to write Puthoff. I just sat down and did it with a letter dated 30 March 1972.[16]

In 1972, Swann joined forces with Hal Puthoff and Russell Targ at the Stanford Research Institute (later called SRI International) in what became known as the psycho-energetics program. Their program became the research and training arm of the US government's clandestine remote viewing program. During this time, remote viewing evolved through creating and testing experimental projects, inspired by Swann's early work and ongoing collaborative efforts by multiple parties within SRI and between other research labs.[17] They were aided by dozens of supporting parties within the US government and private sector.

Within this creative environment many interests were explored, from proving psi exists to determining how it could be most useful and to understanding the mechanisms behind it. However, no matter what the task or who was tasking it, Swann remained focused on and documented changes to his own internal mental processes and related somatic responses, and how adjustments to protocols and influences from social and environmental factors in and out of the lab affected them. These observations would form the basis of Swann's development of what eventually became known as Controlled Remote Viewing.[18]

Experiments or methods that attracted the attention of the CIA included Swann's moving the recording needle of a heavily shielded magnetometer supposedly impervious to outside influence and his viewing of the planet Jupiter prior to a flyby, which confirmed his impression that Jupiter had rings (unknown at the time).

While traditional parapsychological experiments were also carried out at SRI involving forced-choice tasks, four other types of sustained projects were instrumental in acquiring renewed funding for the development of operational remote viewing and further research: outbounder experiments, remote viewing by coordinates, the development of controlled remote viewing (CRV), and the exploration of "analytics": "My word for non-physical targets or subjects."[19]

Outbounder experiments

Outbounder experiments were a carryover from Swann's work at the ASPR.[20] They involved a viewer, a researcher sent to a distant location (the outbounder) and a third person acting as a monitor or "interviewer."[21] Sitting together, the interviewer and remote viewer would get into a relaxed state and the interviewer would direct the viewer to use their imagination to make contact with the outbounder. One technique was to invite the viewer to move around to different vantage points: 500 feet up looking down, moving to the right or the left, moving through doors, or even turning on lights.

One of the first published reports of remote viewing was Targ and Puthoff's *Information Transfer Under Conditions of Sensory Shielding*.[22] This paper reported on the outbounder approach with former police officer Pat Price as a subject. They wrote: "A study by Osis led us to determine whether a subject could describe randomly chosen geographical sites several miles from the subject's position and demarcated by some appropriate means (remote viewing)" (p. 604). The SRI co-directors constructed their target pool based on

> the theory that natural geographical places or manmade sites that have existed for a long time are more potent targets for paranormal perception experiments than are artificial targets prepared in the laboratory. This is based on subject's opinions that the use of artificial targets involves a trivialization of the ability compared to natural, pre-existing targets (p. 605).

They found:

> Pat Price's ability to describe correctly buildings, docks, roads, gardens and so on, including structural materials, color, ambience and activity, sometimes in great detail, indicated the functioning of a remote perceptual ability. But the statements contained inaccuracies as well as correct statements (p. 605).

Targ and Puthoff[23] further summarized the results of 50 experiments with subjects, both experienced and new to this sort of task, viewing remote geological locations and buildings up to several thousand kilometers away. At the locations were "buildings, roads, laboratories apparatus, and the like." They asserted:

> The development of SRI of a successful experimental procedure to elicit this capability has evolved to the point where visiting government scientists and contract monitors, with no previous exposure to such concepts, have learned to perform well; and subjects who have trained over a one-year period have performed excellently under a variety of experimental conditions (p. 330).

Locations included museums, a city hall, a miniature golf course, a nature preserve, the BART transit system and a shielded room. They found no decline in psi by increasing the distance between remote viewer and target.

In this same report, Puthoff and Targ[24] described 12 additional experiments carried out by five subjects; two of them were "visiting government officials." The target material included real objects that researchers visited and interacted with while the remote viewers tuned in. Here again, the remote viewers were "interviewed" by a researcher such as Targ, who was blind to the actual target. This interviewing process involved inviting the subject to mentally interact with the object through intent and visualization, explore it through imagined movement commands and then produce a sketch. Subjects recorded their responses verbally and in writing at this time. Targets included "a drill press, Xerox machine, video terminal, chart recorder, a random number generator and typewriter." Results were significant across both groups – the experienced remote viewers and the inexperienced visiting subjects. Some of the newer subjects' sketches were said to be "exceptional." However, newer participants' results were found to be less consistent than those of the experienced subjects (p. 345).

Controlled Remove Viewing – The birth of Operational Remote Viewing

On the heels of these positive results, a series of projects, still largely classified at the time, was carried out "to determine the utility of remote viewing under operational conditions."

Project Scanate was the brainchild of Ingo Swann. After his initial few months at SRI being run through standard parapsychological experiments using machines, Swann felt they were getting off track from the original aims of those funding the experiments, which was to discover whether remote viewing might be a viable tool for information gathering.[25,26] Based on his past experiences at ASPR, Swann questioned the then-dominant theoretical framework in parapsychology holding that a telepathic connection had to be present between sender and receiver in order for psi-based information to be accessed. He pleaded with SRI directors to conduct a series of trials that would demonstrate a viewer could be successful having nothing more than latitude/longitude coordinates. After much pushback, they agreed to put him and another viewer through such trials, and it immediately became clear to them and to their clients with the CIA that they were onto something.[27]

Whereas Project Scanate had used *National Geographic* photographs as feedback for the coordinates, now real-life targets were given to Swann and Pat Price. One such target was a vacation cabin on the East Coast. At first, the experiment appeared a failure because instead of a vacation home Price described a military installation. However, Ken Kress,[28] CIA Project Manager, sent an agent to the site and they discovered a very sensitive military site close by. Kress noted:

> The evaluation was, as usual, mixed. Pat Price, who had no military or intelligence background, provided a list of project titles associated with current and past activities including one of extreme sensitivity. Also, the codename of the site was provided. Other information concerning the physical layout of the site was accurate. Some information, such as the names of the people at the site, proved incorrect (p. 10).

In 1977, Kress declared that remote viewing was promising, but it remained to be seen how useful it would be as an intelligence gathering tool. A number of reports have revealed the types of projects undertaken by the remote viewers in those years. A classified 1983 Defense Intelligence Report on Project Grill Flame[29] wrote:

> RV is the ability of certain individuals to access and describe, by means of mental processes, information blocked from ordinary perception by distance and shielding. Targets for RV have ranged from small objects in nearby light-tight canisters to remote technical facilities at intercontinental distances, from numbers generated at random by a computer, to nuclear tests in a foreign country. Successful viewings for the DOD/Intelligence communities include: A secret NSA facility, including code word retrieval; Soviet R & D facility at Semipalatinsk, USSR, known to have ongoing operations; static tests of Minutemen and Poseidon solid-propellant missile firings in the Western United States; circumstances regarding the release of Iranian hostages (p. 11).

The authors noted that since 1976, more than a dozen "seminal papers" had been reported in the literature, "the bulk of which had been successes" (p. 15). Under the heading, "Value of data" was written:

In the FY '81-FY '83 An evaluation process in conjunction with its DOD sponsors (DIA, Army INSCOM) has investigated US capabilities in applied intelligence applications, both to determine the potential for application in the U.S. efforts, and to provide data useful in the threat potential of corresponding Soviet/East-bloc applications. To carry out this task, SRI pursued application tasks that were of interest to the intelligence community and have responded to quick-reaction requirements set by DOD representatives monitoring the progress of the work (p. 32).

They reported that investigations had shown that:

remote viewing, both at SRI and ARMY INSCOM personnel, has in many cases provided meaningful descriptions of East-bloc targets of interest to the intelligence community. Evaluation by appropriate intelligence community specialists indicates that a remote viewer is able by this process to generate useful data corroborated by other intelligence data. As is generally true with other human sources, the information is fragmentary and imperfect, and therefore should not be relied alone but is best utilized in conjunction with other resources. Although efforts to establish the exact degree of accuracy and reliability are not yet complete, the data generated by the RV process appear to exceed any reasonable bounds of chance correlation or acquisition by ordinary means and therefore constitutes an exploitable information source (p. 32).

In a follow-up report, Puthoff [30] wrote:

As a result of the material being generated by both SRI and CIA remote viewers, interest in the program in government circles, especially within the intelligence community, intensified considerably leading to an ever-increasing number of clients, contracts, and tasking, and therefore expansion of the program to a multi-client base, and eventually to a joint services program under DIA leadership (p. 10).

Establishment of training and operations at Ft. Meade

In September 1977, the US Army's remote viewing program Gondola Wish (the first of many project names) was established at Ft. Meade, an army base in Maryland. The work environment was set up per Swann's instructions that remote viewing personnel should work as part of a team in an isolated environment free from interference. They would focus only on remote viewing and maintain separation of roles – viewers, monitors, analysts and clients. The program was overseen by Lt. F. Holmes "Skip" Atwater at the direction of the Assistant Chief of Staff for Intelligence (ACSI), Maj. Gen. Edmund Thompson.

As SRI did not yet have its formal training program in place, Atwater established a training program of his own that incorporated some of the practices and lessons learned at SRI. It was also influenced by other sources, including his time spent with Robert Monroe at The Monroe Institute.[31] This methodology would later be referred to as *Extended Remote Viewing (ERV)*. Atwater believed one cannot prove that one is actually leaving their body during an OBE or RV experiment, but one can make use of imagined visualizations of extending oneself to a distant location and moving around, which is similar to what viewers were doing in the outbounder experiments at SRI.[32] In ERV, viewers are brought into a deeper, more relaxed state bordering sleep. The viewer is accompanied by a monitor/interviewer who records the session, takes notes and makes suggestions. A summary report is typed up and then analyzed and passed on to the client.

In October 1978, US Army's INSCOM was tasked by the ACSI with developing a parapsychology program of its own. In late 1978 to 1979, a few viewers were selected for project Grill Flame (the project's new name). In 1979, the first operational remote viewing session was conducted[33] and in December 1982, the US Army's RV project's name was changed to Center Lane.

During the late 1970s, Puthoff and Swann conducted research and development toward a new training method, one now called Controlled Remote Viewing (CRV). With Puthoff as his supervisor, Swann was awarded a training contract for this work. The contract stated that Swann was the proprietary owner of this methodology.[34] In 1983, with Swann as instructor, two individuals from the Ft. Meade Unit began their SRI-based CRV training; in 1984, a second group of CRV candidates began training.

Controlled Remote Viewing is a stage-based method that was in development while soldiers were being trained in some stages of it. CRV is a highly structured methodology not intended to make someone psychic but rather to train people to use their innate ability, which Swann felt was a capability of the human species. CRV was intended to serve as a replacement for viewers who might not have the luxury of being monitored by others like those at SRI. It also allowed viewers to make greater use of their unconscious somatic reflexes (through ideograms, sketching and clay modeling). Perhaps most importantly, it was designed to decrease incidents of "analytic overlay"[35] – the tendency by remote viewers to try to "name the target" – which leads to misinterpretations and can easily derail the session.

A report dated August 1984 entitled *Defense Intelligence Agency: Directorate for Scientific and Technical Intelligence*[36] noted both the value of and the differences between the two approaches (CRV and ERV) and stated:

Within the DOD, the intelligence community is the prime user of data gathered by remote viewing. Because intelligence must be gathered surreptitiously and requires access to forbidden and guarded places, remote viewing provides an excellent, and sometimes the only, means of getting the desired information (p. 33).

A formerly classified 1984 Science Panel Report[37] noted:

A considerable variety of material was presented with photographic backup in support of the validity of the perceptual method. Much of this was highly impressive. The data showed the effects of training on the success rate, which typically reached a sustained plateau at a level higher than prior to training, both for groups of subjects as well as for individual trainees (p. 4).

Still, multiple documents suggest that a formal assessment by a completely independent committee of the SRI training methods was planned, but never completed by SRI directors, partially because they had not yet developed suitable protocols for testing that could ensure the testing procedures themselves would not affect results.[38] Funding agencies also pressured them to move the training along faster so the viewers could apply their learning to more classified projects that the SRI researchers could not participate in. Swann was resistant to this and in the end, per his contract that stated CRV was his proprietary methods, kept all his students' lesson notes and transcripts. However, two of his students (Tom McNear and Paul Smith) decided to recreate his training from memory and pass it on to other viewers in the military unit.[39]

As the years passed, a large mix of viewers came into being – those who had directly trained with Swann[40] or later trained with his students in CRV methodology.[41] Over time, CRV teachers introduced modifications. Some were "exposed" to Swann's methods[42] but continued to use extended remote viewing or other approaches in conjunction with monitored sessions.

Declassified operational military projects

Since the defunding of the above programs in the mid-1990s, many documents have been released by both the Ft. Meade remote viewers and by the SRI and other researchers.[43,44,45] While many projects remain classified, the materials now available reveal the nature of some of the operational targets. These included describing a Russian military installation, which revealed the existence of the previously unknown Typhoon class Russian submarine[46] and viewing Pan Am Flight 103, which crashed in Lockerbie, Scotland in 1988.[47] Other operational targets included drug interdiction in alliance with the US Navy Air Stations Joint Task Force and collecting intelligence on foreign military leaders' plans and tracking their activities. These leaders included Muammar Gaddafi, Saddam Hussein and General Manuel Noriega.[48,49] Ft. Meade viewers also searched for hostages such as Col. Rich Higgins, who was kidnapped by Hezbollah terrorists in Lebanon; William M. Buckley, kidnapped by Shiite guerrillas; and General James Dozier, taken hostage by Red Brigade terrorists. The Chernobyl nuclear meltdown was another target.

Field work in the public domain

Swann and other early remote viewers participated in a variety of applied projects financed by independent investors and researchers outside the government's purview. Swann referred to the projects as "field work." They involved finding hidden items and helping locate natural resources.

Some of the most highly publicized projects were sponsored by the Mobius Group, established in 1977 by Stephan Schwartz, another remote viewing pioneer. The first of these was Deep Quest – a submarine RV experiment conducted with SRI personnel. From 1979 to 1981, Schwartz spearheaded the Alexandria Project, a remote viewing archaeology project conducted in Egypt searching for buried artifacts.[50] Another project of note was the Columbus Caravel Project, which was designed to locate and excavate the remains of the last two of Columbus' missing ships from St. Ann's Bay, Jamaica.[51] These were thoroughly documented and highly successful projects.

Dale Graff has been credited with having secured the very first contract between the US Air Force and SRI, and he served as the DIA Coordinator/Director of Project Star Gate. In recent communications with Debra, he confirmed that in 1976, he set up an experiment with Ingo Swann and Harold Sherman. Their assignment was to track him on the Coppermine River in the NW Territories in Canada. These details are documented in *Tracks in the Psychic Wilderness*,[52] Chapter 6, "Arctic Search." This chapter has only Sherman's data since Graff was unable to obtain Ingo's scripts at the time the book was being prepared, but it does describe the essence of the experiment. According to Graff, Sherman used a simplified form of psi/RV. This same book documented "Trouble in Reactor Bay," which was a combined RV and RV/dream communication project involving a deep ocean diving vessel in 1977.

In his second book, *River Dreams*,[53] Graff detailed several RV search projects, including the remote viewing of a rocket motor test; a Soviet airplane crash in Africa,

"Airplane Down"; "The Search for General Dozier" and the quest to locate a customs department fugitive, Charles Jordan. "On the Run," outlined in chapter 5 of *River Dreams*, describes the project in which the term "eight-martini session" was coined by Swann – it refers to a remote viewing session so unbelievably accurate you have to go out and drink eight martinis to get over the amazement of it all. (Hence the title of Daz Smith's popular remote viewing magazine.)

Other unknown or less publicized applied RV projects undertaken by Swann were the Ft. Huachuca Treasure Project and the Robert Jones Buried Treasure Project. Several projects involved oil exploration – the Halbouty Oil Exploration Project, 1976 Ghana Exploration; 1981–1985 Washburn Oil Exploration and the Ada Oil Company Sites.[54]

International Remote Viewing Association

In 1999, a few years after the defunding and declassification of the US Government programs, Angela Thompson Smith[55] co-organized the very first professional conference on remote viewing in Ruidoso, New Mexico. Smith was one of the first civilian students to be trained in remote viewing and has since published several remote viewing books. The conference was followed by a meeting in Alamogordo, New Mexico, that included Smith, former military remote viewers, and SRI researchers. At that time, they drafted the bylaws of the *International Remote Viewing Association*, which celebrated its 20th anniversary in 2019. IRVA's primary goal was to provide an unbiased approach in presenting information, training, research and education to the public. As IRVA's president, Skip Atwater[56] wrote in an early newsletter:

> Applications now involve remote viewing in areas as diverse as commercial forecasting and the development of successful business strategies, medical diagnosis, criminal investigation and forensics, financial investing, scholarly inquiries, historical explorations, and much more. The "reality" of remote viewing is no longer in question, except in the most determinedly skeptical circles (p. 2).

While formal membership is modest, IRVA has members in numerous countries. Its main activities have been sponsoring annual conferences, publishing a magazine (*Aperture*), offering remote viewing practice targets, sponsoring research-oriented awards such as the Warcollier Prize (of which Debra has twice been the co-recipient), the Gabrielle Pettingell Fund and the Lyn Buchanan Scholarship. In addition, IRVA has served as a center for discussion and interaction. In 2020, IRVA established a new research initiative, the IRVA Research Unit (IRU), spearheaded by Debra Katz and Dale Graff and assisted by the IRVA Directors. IRU is designed to encourage informal and newer researchers to participate in more formal projects. It brings more awareness about remote viewing to academic researchers while advancing all efforts in the area of research and psi.

Atwater's statement above may have been overly optimistic, considering the slow development of remote viewing in commercial applications and the lack of "scholarly inquiries" in the following years. However, a review of applications-based projects published in RV-related publications and presented at conferences spanning the past 20 years reveals a great deal of activity and gains in the areas to which Atwater refers. Some of these were reported at the IRVA Conference in 2020, while others date from an earlier period. The following includes ventures presented at IRVA, as well as others.

Business consulting. Early on (1998-2003), Prudence Calabrese's TransDimensional Systems' business with a team of viewers (core of 5, total team about 14) included a large insurance company and other paying clients. Joe McMoneagle has had many commercial clients over the decades. Probably the most successful recent example is IRIS in France with Alexis Champion, CEO of IRIS Consultancy Services reporting the company has had more than 90 clients since 2008 and has been involved in more than 120 projects and interventions. Clients have included banks, industries, museums, energy, nuclear, transportation, police & tribunal, traders, think tanks, universities, artists. Their work has been featured in more than 25 European media companies. Champion defined the different situations applications were best suited to: emergency situations, innovation, art, archaeology and history, communication, human resources, finance, entrepreneurship, industry, crime solving and judiciary.[57]

Crime solving. Pam Coronado,[58] a former IRVA President, discussed her experiences working with 50 police departments across the United States, as well as international agencies and the FBI. Calabrese's TDS team did extensive anti-terror work with the FBI after 9/11.

Healing and medical applications. Husick[59] reported on a project in which remote viewing was used to understand and help twins with autism. Calabrese[60] presented on the use of RV in remote diagnosis and healing; Klieman[61] demonstrated how human consciousness can be used for healing; and Atunrase[62] discussed a project in which viewers were tasked with a cure for cancer.

Humanitarian work. Remote viewing has been used to assist an adopted son understand the circumstances of his adoption by helping him locate his birth mother.[63] In Project Blind Awareness, blind children learned to use remote viewing to locate their parents on another part of the campus.[64] Angela T. Smith[65] reported on *Remote Viewing in Humanitarian Aid Work* in Haiti, an inter-group effort to form a team to locate missing men.

Locating downed aircraft. Angela T. Smith's Mindwise Consulting searched for the downed plane of Amelia Earhart and crew using remote viewing.[66]

Presidential elections. Katz and Bulgatz[67] designed a project to determine whether 11 remote viewers using a double-blind protocol could describe a human subject in enough detail so raters could choose between two potential candidates in order to predict the outcome of the 2012 United States Presidential Election. For the 2016 election, Katz, et al.[68] conducted another double-blind Associative Remote Viewing project, in which 41 experienced remote viewers were tasked with describing a feedback photo they would see at a future date after the election.

Scientific discovery. In 2012, Hitomi Akamatsu went to Hawaii for on-site, intensive training with the Hawaii Remote Viewers' Guild. She demonstrated a remarkable ability to see, sketch and describe things that were physically distant, without any foreknowledge of the target. During her advanced training she was given a blind tasking, the creation of the Higgs-Boson subatomic reaction, the so-called "God Particle." Hitomi went into a room alone and worked for hours, assembling more than 40 pages of sketches and descriptions.[69] In 2019-2020, she viewed for the CryptoViewing team, which has been successful.

Morse, et al.[70] completed hundreds of binary trials to see if they could consistently determine whether tomato plants were healthy/unhealthy or contained/did not contain a virus.

Katz and Beem[71] reported on a double-blind, free-response, exploratory experiment, in which 39 remote viewers used their intuitive skills and training to describe a bacteriophage, which is a virus that attacks bacteria.

Other projects have attempted to use remote viewing to explore remote viewing itself. For example, several have attempted to demonstrate whether viewers are directly tuning in to an existing target or rather to the "tasker's intent."[72,73]

Training and project development. IRIS trained staff of client companies to develop creative solutions and innovations. For example, in the Watch Project the customer (a major French bank, La Société Générale) was seeking to design and build a prototype for a watch that could do micropayments. Facilitators trained staff to remote view and then tasked them to describe a mystery object as it would be on December 31, 2015. In another collaborative project with the cultural administration of the city of Bourges, IRIS facilitators taught artists to use their intuition to create works of art related to an archaeological site.[74]

Music composition. Nancy Smith was a remote viewing group manager and her husband Sam Smith is Associate Principal Cellist of the Boise Philharmonic Orchestra and a music professor at the College of Idaho. With the assistance of APP's Marty Rosenblatt, they created *Music from the Fringe*. This was a three-day collaboration of composers, artistic directors, remote viewers, analysts, and four cellists, who were taught remote viewing skills as part of a creative process to compose music.[75] The music was featured in the documentary *Third Eye Spies*.[76]

Planetary targets. Many remote viewers have been tasked with planetary targets. While Sherman and Swann's 1973 experiment described unknown aspects of the planet Jupiter, many others have involved blind tasking in which the viewers received only a target number and the knowledge that the target was a location. Angela T. Smith[77] reported viewing the ring anomalies of Saturn. Brown[78] conducted a study that explored the creation of the asteroid belt. McNear[79] presented a compilation of 18 remote viewers' transcripts describing Mars. Viewers had been tasked for different projects relating to Mars by different managers spanning the past 40 years, yet displayed remarkable correspondence.

Esoteric targets. An esoteric target is one that is oriented toward the mysterious, the spiritual, the unexplained. Atwater[80] and Smith[81] noted that occasionally esoteric targets would be given to Ft. Meade viewers to break up the tediousness and seriousness of their operational targets. Sometimes this was done out of personal interest by the tasker. At IRVA 2020, Williams[82] noted esoteric targets can be controversial and should only be given to viewers once they have established a track record for accuracy with verifiable targets. Recent projects have involved exploring the possibility of life in the Sirius star system and UFO sightings by a 747 freighter flight crew.[83,84] Brown[85] has done extensive work in this area investigating subjects like Area 51 in Nevada and the appearance of UFOs in Phoenix.

There is a significant connection between remote viewing and UFOs (now called UAPs). Many people prominent in the field have reported UFO experiences or observations, e.g., Swann, Pat Price, McMoneagle, Lyn Buchanan, Edward Dames, Courtney Brown, Pru Calabrese, and Angela T. Smith. Practices of a few of these remote viewers have resulted in some high-profile controversies.

IRVA has not been the only organization at the center of remote viewing activity. A widespread ecosystem of grassroots viewers, taskers and project managers are undertaking informal research, practice projects and work for clients. Regarding magazines/newsletters, in addition to IRVA's *Aperture*, there is Daz Smith's *Eight Martinis*. APP is the largest organization in the field (membership about 1,200), does extensive ARV viewing in groups, sponsors two conferences a year, and has posted many videos relating to psi as well as presentations by leading figures in the field. Further, there has been active discussion and the practice of remote viewing in forums like TenThousandRoads, Yahoo groups, Facebook, and reddit.

In our next chapter, we will discuss projects that typically fall under the category of ARV, including: stock market and Forex trading predictions, sporting event predictions, lottery predictions, etc.

Historical resources

A number of books specifically focusing on this rich history have been published by the first generation of researchers, viewers and managers. Some books have been written by journalists who interviewed the early key players (e.g., Marrs,[86] Schnabel,[87]). In recent years, the second generation of remote viewers has written their own books, many focusing on training methods.

Also, several television shows and films have focused on remote viewing. One we highly recommend is *Third Eye Spies*,[88] co-produced by our friend Lance Mungia and directed by Russell Targ.

In addition, there is an expanding body of archival collections, such as Ingo Swann's at the University of West Georgia (which Debra and Jon have both spent extensive time examining) and Ed May's archives of SRI materials housed at Baylor University. Many viewers, project managers, and researchers discuss and display remote viewing in all its facets on numerous social media sites.

Given the abundance of resources, we will not focus further on the history except as it applies to the main focus of our book – the practice of Associative Remote Viewing. However, in our chapter on "How to Remote View for ARV Projects" (Chapter 25) we share methods for accessing psi-based information and credit the individuals and groups that first conceived of these techniques.

All histories are embodied accounts lacking an all-knowing "God's eye" perspective

As with any historical account, every presentation of human experience is embedded in a highly personalized and, hence, narrow perspective. This is particularly true of the history of remote viewing. Many projects were classified, participants were trained in keeping secrets and operating on a need-to-know basis, and remote viewers and researchers operated in small units with a division of roles. A score or more US government agencies became involved during the 20-year duration of the Star Gate projects. An extremely comprehensive and detailed summation of this remote viewing history, six years in the making, is presented in Ed May and Sonali Bhatt Marwaha's *Star Gate Archives*.[89]

The larger social context within which the programs were embedded matters too. In particular, given his importance in remote viewing, Ingo Swann was both a psi subject and researcher, yet because of the customary assignment of roles and the power dynamics involved, he rarely received credit for his role as a researcher. His name was omitted from papers and many of his interesting observations about what he was learning and experiencing about his internal mental processes went unmentioned in the formal literature of parapsychology and were only partly acknowledged in private training manuals and books.

Therefore most of the credit for ideas, formulations of projects and experiments went to the credentialed scientists, while contributors such as Ingo got to enjoy (or lament) the status of being a very talented subject. Meanwhile, as it turns out, based on recent admissions in personal communications and in conference presentations, many of the scientists themselves had extraordinary psi-based personal experiences that also went largely unmentioned since it was not part of their "proper" official role. If direct first-person experiences by a researcher or other participant are not fairly represented, then whoever attempts to portray the histories of each of the individual players, the advancements, successes and failures, is at a severe disadvantage.

Throughout this book, we have attempted to flatten out this hierarchal approach, bringing forward the voices and work of those who may not have advanced degrees or impressive research credentials but who have nevertheless made contributions to the field. They are doing innovative and important work, demonstrating success, offering theories and new ways of thinking about these topics, and have served as our mentors, colleagues, partners and friends.

CHAPTER 2

What is Associative Remote Viewing?

In this chapter we provide an overview of ARV and discuss its origins and early experiments.

Associative Remote Viewing is a form of remote viewing for making predictions about future events involving more than one possible outcome. It's "associative" because the viewer is not tasked with getting information directly about the event, but rather about independent subject matter that is associated by the tasker with the event's outcome. This independent subject matter can be simple – like a color, taste or smell – or it could be a photo, location, object. It could also be an emotion having a connection with the event. From the beginning, ARV projects have primarily used photographs of locations or objects as potential targets.

ARV only works for events in which all possible alternative outcomes are known in advance, allowing for specific predictions to be made and acted upon by making a decision and, if warranted, placing a wager, making a purchase or analogous action. Prior to the event, no one knows which of two or more possible outcomes will become "actualized" – that is, will happen. ARV is designed to make a prediction about the outcome prior to the event taking place.

Examples of such events include sporting events – Which team will win or Will the total be over or under the points scored by both teams? Casino games – Will the roulette ball fall into a black, red or green slot? Stock trades – Will a stock rise or fall in a designated time period? And presidential elections – Who will win?

ARV is not used to look into the future in an open-ended manner to see what might transpire. For example, if I wanted to know how my life will shape up over the coming year, I would not use an ARV protocol. I might visit a clairvoyant reader or as a clairvoyant reader myself (Debra speaking), I might just close my eyes and see what images come up.

If I wanted to know where a hostage is located, I would use remote viewing protocols but not Associative Remote Viewing.

If I wanted to predict the winner of a horse race, I would use ARV and get my ducks in a row. I'd have to become familiar with all the factors involved: when the race will take place, the appropriate target material and the photo to pair with each horse in the race. After I did my remote viewing session, I would compare it with the options, apply a method of judging and make a prediction.

ARV therefore is a system that uses psychic functioning, but this is only one of many cogs in a complex wheel with multiple components, factors, materials, roles, timing and decisions. ARV is often performed in a group setting with people assigned specific roles and responsibilities. While it has been successfully carried out by many individuals working solo, those individuals had to make use of materials and technologies created by others or by themselves at an earlier date in order to remain blind to (uninformed about specifics of) the target material involved.

The main components of successful ARV trials include choosing the right project manager, viewers and other participants (e.g., a separate bettor and photo selector), good target material and tasking, sound judging and sound scoring methods. Other important aspects are making the prediction, wagering, tracking the outcome of the event, providing feedback for the viewers, and the timing of all these communications and client interactions. These activities, tasks, responsibilities and roles will be outlined below, with many described in detail using real-world examples along with similar examples in upcoming chapters.

ARV trials are often carried out in a manner similar to psi-based experimental trials, with pre-planning, implementation of a structured methodology, tracking of data and careful analysis. In real world applications, many ARV trials are carefully planned in advance, put into place, effectively carried out and assessed to see what worked and what could be improved on.

Why use ARV instead of another psi-based approach?

ARV is helpful when multiple predictions will be needed for similar events or when an event has little unique data. For example, if the weekly goal is to predict the winner of a football game, a remote viewer tuning in to the game might perceive large men wearing uniforms and helmets battling over an egg-shaped ball. While rather descriptive, such details could obviously describe both teams. If the remote viewer describes the winning uniform as blue rather than red or describes the logo of a bird rather than a bear, the project manager might be able to predict the winning team. However, many teams have the same color and similar logos. Viewers often have difficulty distinguishing between team mascots. Also, over time, the remote viewer or the judge may learn to recognize team colors and logos, and then get tripped up by their own preferences or biases.

ARV protocols were developed to circumvent these challenges to precognitive work and to allow for repeated blind trials over time, essentially turning what could become forced-choice tasks between binary outcomes into open-response ones for the viewer. This is done, for example, by pairing photos that are different in all possible ways with two possible outcomes, and then having a remote viewer tune in to the single photograph they will see in the future (called the feedback photo). In many ARV setups, the judge compares the remote viewer's written transcript with each photo and scores each transcript against the photos. Each photo is associated with one outcome of the event and the association is often made randomly after the viewing has been done, while some determine the associations of the photos prior to the sessions. Theoretically in ARV, the highest-scored photo should be the photo the remote viewer sees in the future as feedback, and the project manager accordingly issues a prediction and may wager on it or pass (not make a bet), depending on the scores given for each photo.

Understanding a typical ARV trial

In ARV trials, viewers don't need to know anything about the overall project itself, nor do they need to be aware of the individual events or possible outcomes. All they need to know is their task – to use their psi abilities to describe a photo (or other type of target) they will see at a specified future time and date. Viewers may approach this task using a variety of techniques. They do, however, have to provide the judge a written response prior to the event, usually consisting of both words and sketches. The judge will compare the transcript to one photo at a time and score it according to a predetermined scale. This enables the judge to predict which photo the viewer is likely to see in the future.

If prior to an event the viewer turns in a description that strongly matches one photo, with little correspondence to the other, the judge can be more confident the outcome linked to that particular photo will occur. In such a case, the judge or project manager often issues a prediction.

If there isn't a strong match to one photo, this may indicate the viewer was either not able to tune in at all or was unable to describe their feedback photo well enough (e.g., both scores are low), or they tuned in to more than one photo. In these instances, a pass will be called.[90]

Process breakdowns that lead to passes may also occur, such as when a game is cancelled. In stock trades, a pass may be warranted if the trader makes an error setting up the trial, fails to enter a trade into the online system in a timely way or if the trading system goes down. Passes may also be called if the judge/project manager makes and recognizes an error.

In informal applied ARV projects, passes are recorded but typically not factored into the overall statistics.[91] Only outcomes of actual predictions are used to calculate the hit/miss ratio. In such projects, the end goal is often to use the prediction to make money and with a pass no money is won or lost. While passing can prolong a project that has a preset number of predictions (which is needed for formal research projects measuring statistical strength), in informal applied ARV projects passes are useful since they prevent misses.

Whether a formal prediction or a pass is issued, following the event the manager needs to do as promised and show the viewer the target photo, sometimes at a pre-specified time. The feedback photo is always the photo associated with the winning outcome. This completes the "feedback" or hypothesized "retrocausal loop" and closes out that trial.

But how important is the feedback photo? Müller, Müller and Wittmann[92] found that feedback didn't make a difference when they compared two groups of viewers. They found their viewers did equally well when they simply tasked themselves to tune in to the photo attached to the viewing outcome that they would not see. That is, they would get no feedback. However, it's not known whether viewers were told up front they wouldn't see the photo, nor whether they were told to direct their focus, or if they thought they would receive feedback but then did not.

If a viewer intentionally tunes in to a target by directing their attention to the future date, time and place when they will receive the feedback, and then does not get the feedback, this creates a breakdown in the process and compromises the result. Through observation of the process, they may have found they did poorly or accidentally tuned in to the wrong photo in cases when they didn't receive their feedback photo. Accidentally tuning in to the wrong photo is called displacement. This happens so often and is such a problem that we have devoted two chapters to it.

Essentially, the ARV process serves to overcome the inherent problems of forced-choice repetitive tasks by pairing limited choices (the outcome of an event) with unlimited options (the target can be anything in the world).

While ARV protocols and purposes vary, one commonality is that viewers use intuitive processes to attempt to accurately describe and draw sketches of a photo, video clip (or other media) that is paired with a potential future outcome. In ARV setups which use photos as targets (the most common type of ARV), viewers try to describe the associated image they will see in the future, rather than directly describe the outcome or event itself. This enables viewers to remain blind to the subject matter (which could be one of millions of images), even if they have foreknowledge of the target event and its limited number of possible outcomes. Depending on a project's goals, successful ARV predictions may result in financial gain, may demonstrate evidence of psi and precognition, and may assist those seeking information about the future, such as predicting which candidate will win the upcoming presidential election.[93]

While applied ARV protocols vary, they usually include the nine phases outlined in the table below, although not always in that order.[94]

Nine Phases of a Standard Binary ARV Protocols with Two Photographs Process

1. An event is selected.
2. A pair of unlike photos is selected.
3. Remote viewers are given a random alphanumeric "tag".
4. Viewers do a remote viewing session.
5. A judge scores the session against the pair of photos.
6. A tasker or manager issues a prediction or a pass.
7. Someone on the team places a wager or makes a purchase.
8. The event occurs.
9. The feedback photo is shown to the viewer.

Figure 2.1

Depending on the setup, the tasker, manager or researcher carries out most of these steps in the process, often with the aid of specialized computer software. Viewers view and receive feedback. Judging is done by the tasker, by a third party or by the viewers themselves. Wagering is done by the manager, the viewers or by a designated buyer or a client. Some protocols depart from the above order and include other particulars such as more blinding of participants about a) the nature of the event, b) the specific event, c) the photos or other potential targets.

Phase One — *Event selection*: Let's imagine a sports enthusiast has recruited a project manager to help predict which team will win a game. They decide the sport they are most interested in is football and specifically the February 2021 Super Bowl between the Kansas City Chiefs and the Tampa Bay Buccaneers.

Phase Two — *Photo pairing*: To set up the trial, the manager can choose any two photos, provided the photos are as different from each other as possible in terms of color, shape, size, and overall visual and conceptual content.

Furthermore, the pictures will not be related to the teams or the game; one will simply be randomly paired with the Kansas City Chiefs and the other will be paired with the Tampa Bay Buccaneers. For example, a photo of an office building in a city environment may be paired with Kansas City to win (Photo A) and a photo of black horses running through a green field may be paired with Buccaneers to win (Photo B).

Phase Three — *Tasking*: The manager sends a random target number to the remote viewers, who are blind to the photo options. The remote viewers are told the target number is representative of the feedback photo they will receive at a future date. They are given a deadline by which to submit their written remote viewing session transcript. The tasker writes his intention that Photo A stands for a win by Kansas City while Photo B stands for a win by Tampa Bay. This intention is not shown to the viewers. In some setups, the association is made by a random process after the viewer has done the session. One form the cue could take is "Super Bowl February 2021 / photo associated with winning team." However, some prefer full sentences for the cue.

Phase Four — *Remote viewing:* This is the only phase in which the participants' intuitive faculties are utilized. Here remote viewers attempt to access the feedback photo they will see in the future and record their impressions on paper. The transcript will include information about the viewer, the time of the session, cooldown, etc., and will include words, graphic impressions and/or sketches. Some of these impressions may be based on visual images received by the viewer, while some may be internal auditory, gustatory, rhinal (nose) or other somatic impressions. Viewers then submit their transcripts to the manager.

Phases Five and Six — *Judging and issuing prediction*: Suppose the manager notices the remote viewers' transcripts include words such as "movement," "animals," "several," "open spaces" and "green fields," and there is even a sketch resembling a four-legged animal. There is no mention of a manmade structure or anything to do with an urban environment. The manager would rate how well each transcript matched each picture, in this example assigning a high score to matches with the photo of the horses (Photo B) and a low score to matches with the city building photo (Photo A). Because the remote viewers will only be shown the photo associated with the winning outcome and Photo B has been paired/associated with Tampa Bay, the manager can have a strong degree of confidence that Tampa Bay will be the winner. Thus, the Tampa Bay Buccaneers are recorded as the predicted winner.

However, if the manager was not confident about either photo being a strong match – perhaps both photos had an equal number of matching elements or one remote viewer's transcript strongly matched one photo and another's matched the other photo – instead of issuing a prediction, the manager might issue a pass. This is a non-prediction that is not factored into the hit/miss rates in applied ARV projects since there is no financial loss or gain.[95]

Phase Seven — *Wagering:* A bet is made. For example, the manager contacts the client and lets them know that the prediction is for Tampa Bay Bucs to win. It is best if the client is not a viewer on the project.

Phase Eight — *The event:* The game is played and which team wins becomes known.

Phase Nine — *Feedback:* In order to complete the process and in keeping with the theoretical retrocausal assumptions underlying the ARV process, soon after the game is completed the remote viewers are shown the photo associated with the winning team, regardless of whether the prediction was correct. This is considered essential since the target assigned to the viewers was the feedback photo attached to the winning team. If the correct photo is not seen by the viewers, there is no feedback, and hence there was never actually a target for the remote viewers to describe in the first place.

Origins of Associative Remote Viewing

In 2020 Stephan Schwartz[96] revealed the origins of ARV[97] in an article published in *Mindfield,* a newsletter published by the Parapsychological Association.

Schwartz was at a dinner party in Beverly Hills in 1971 and a "very arrogant plastic surgeon" questioned the reality of psychic phenomena as Stephan enthusiastically explained his Deep Quest experiment. The surgeon opined that "I don't think there is a shred of evidence to support it; it is just statistical mumbo-jumbo. It's all coincidence and self-delusion." Schwartz asked what it would take to convince him and the surgeon said, "Win a horse race. Everybody knows no one can predict the outcome of horse races." Schwartz agreed to take the surgeon up on it. While driving home, Schwartz thought about the fact that it was hard for viewers to get names or numbers so it would be difficult to psychically get a horse's name or number to predict a winner. Perhaps ARV would be able to circumvent that problem.

Schwartz was doing research as Special Assistant for Research and Analysis for Admiral Elmo Zumwalt, Chief of Naval Operations and came across an account of the Battle of Aboukir Bay (also known as the Battle of the Nile), during which the British Vice-Admiral Horatio Nelson was tasked with finding and destroying Napoleon's fleet. Schwartz writes:

One of the central problems Nelson faced was how to communicate with his ships once the battle commenced. The ships of the line were spread out across miles of the Mediterranean Sea and wreathed in clouds of black powder smoke spewed out by the cannons. Nelson realized he would not be able to see his entire line of ships, so he used his frigates, smaller ships that he ordered to cruise up and down the line during a battle. He sent the frigates out with a message to relay as they sailed, and the frigates returned with messages sent by the battle-line captains, passing them up the line to Nelson. Each message was that of a few colored flags, each having a previously agreed-upon meaning, specifying the ship's number and command. Nelson's 18th century communication technology allowed Nelson to command his ships and prevail. As the result of this epic battle, Napoleon's and France's sea power was broken (p. 6).

Schwartz continues:

I began to think of his basic idea in terms of my own nonlocal consciousness research. It was already clear to me that certain kinds of linear information, such as numbers, were hard to get. Many sessions had taught me that sense impressions and a sense of knowingness were what came through best. Nelson's solution, I saw might be a way to get analytical information. I could use a previously agreed-upon association to substitute complex messages made up of abstractions, with the kind of things that produced the best results in my own remote viewings and those of my remote viewers: objects, individuals, other living beings, and places. By linking, for example, an apple with a name or number, or scissors with a specific analytical outcome, I could acquire information it would otherwise be hard to obtain nonlocally (p. 6).

Along with SRI researcher Ed May, Schwartz reminisces:

We created two target sets of Los Angeles locations and ran each session independently. The assigned task was "at 4:30 PM tomorrow, we will be standing somewhere. Please describe, using all sense impressions, where you are." The session data were judged in a blind, rank-ordered assessment of the data against the target images. Both women unequivocally selected the target associated with the 6th horse in the 6th race at Hollywood Park. That night, we went to the racetrack and placed a two-dollar bet. We won $14 and jumped around, clapping each other on the back, as if it were a hundred times that amount. ARV had made money (p. 13).

The first public mention of ARV appears to have been by Schwartz in August 1977 at the Philosophical Research Society Conference on Extraordinary Human Functioning.[98]

In his 2020 article discussing the origins of remote viewing, Schwartz stated he had turned $5,000 into $150,000. In an interview on June 8, 2020 with one of the present authors (Debra), he explained this was accomplished over the course of 42 weeks, using the S&P 500 by making a call on Thursdays for the next close of the market for Friday. This occurred sometime prior to the published studies of Targ and Puthoff, although Schwartz could not recall the exact time period.

Schwartz brought five remote viewers into his Mobius lab, some of whom were prominent people; for example, bestselling novelist Michael Crichton and Judith Orloff (psychologist and psychic). Viewers submitted sessions on Thursday each week and Schwartz would get the information to the stockbroker, who would make the prediction on a Friday. Schwartz created a photo pool from National geographic photos, consisting of approximately five different kinds of targets, with various categories such as buildings and bridges, with the photos being orthogonal enough to be able to determine which photo the viewers were describing.

He said in the end they had to stop after 42 weeks "because it was eating the lab alive." People were just getting too excited and wrapped up in money. It wasn't because the money was going to the individual viewers – it was going to the lab for future projects, but when they would get together to talk about other projects – such as finding missing persons – all anyone wanted to talk about was how well they were doing for this project. This is one reason he believes ARV is best done for a shorter duration and not on an ongoing basis.

Schwartz said the high fees paid to the stockbroker were another challenge. These fees could skew the numbers and final statistics regarding profits earned.

During the 2020 interview, Schwartz reminisced about his friend and colleague, statistician James Spottiswoode, the first person to do a complex project involving lotto numbers using Associative Remote Viewing and algorithms. When James went to print out the numbers just before the deadline, the printer failed. The numbers were correct, but he wasn't able to buy a ticket in time. (See Chapter 21 for more details about this experiment.)

Beginning in 1985 Edwin May served as Director of SRI and the SRI-Consciousness Laboratory (SRI-C), which was considered the research arm of the US intelligence community remote viewing projects. May advised the present authors in email correspondence that the military programs used ARV as one of their information-gathering and decision-making tools. He wrote, "instead of remote viewing yes/no questions with its low effect sizes, we could get the answers using free response in an ARV protocol."[99]

May explained there was use of ARV even while they were working on a parallel project (Deep Quest, 1976) with Stephan Schwartz in which viewers were attempting to use their psi abilities while submerged in a submarine. SRI was testing whether remote viewers (Hella Hammid) could pick up information nonlocally (such as describing the location of other researchers) while doing their sessions underwater in a pressurized sphere providing an electromagnetically isolated environment. According to Schwartz[100] and Targ,[101] who further documented Schwartz's efforts, the viewers were successful.

Ed May referred to another method which he called "the computer analogue of Associational Remote Viewing (ARV)." The viewer is shown a square on a computer display. The viewer is told that answers to the "question of the day" are sliding under the square at a rate of one per millisecond and the viewer should press a mouse button when the "correct" answer is under the square. The technique was used to try to locate a government official kidnapped in Lebanon.[102]

Further evidence of the use of ARV use in Star Gate projects is found in a document about the proposed exploration of "Mental Communication." Five stages were proposed, the first being the use of abstract symbols, "emotion transmitting," "colors, objects, scenes, etc." The second stage was to "develop viewer proficiency." The third stage was to "develop message traffic." "Each symbol will have a build-in word assigned to it." Fourth stage: "Wean the team from the computer." Fifth stage: "Wean the team from the laboratory situation." The document refers to "three already-proven methodologies," one of which is "Associative Remote Viewing (ARV)."[103]

The statement that ARV was an "already-proven" methodology is supported in a February 1981 letter from SRI Director Hal Puthoff to a corporation in which he writes about "forging a link" between choosing a computer-generated random number (which can be done "more often than expected by chance") and "a problem of interest (e.g., which oil well will be the highest producer)." Puthoff writes that the "procedure has been used by us successfully in an exploratory pilot series to determine the locations of hidden persons, hidden ammunition, and hidden radioactive material."[104]

Additional confirmation of the official use of ARV by SRI involved one of many government proposals to develop the very expensive MX-Missile system. The idea was to foil any Russian attack by deploying 200 missiles on tracks and randomly shuffling them among 4600 silos over 1500 square miles. Hal Puthoff told authorities about SRI's use of "RV-generated data to show rather forcefully that the application of a sophisticated average technique (sequential sampling) could in principle permit an adversary to defeat the system."[105] This information was given to Department of Defense Secretary Weinberger in March 1981 and this version of the "Peacekeeper" system was cancelled. However, it's unknown what role the RV results played in the decision. Russell Targ offered the opinion that this was indeed a form of ARV.[106]

These early uses of ARV were apparently not widely known. For example, Dale Graff was delegated control of the DIA RV project in 1990, including the SRI external contract and all functions for management of the Star Gate Ft. Meade unit. He advised Debra via email that he was not aware of any use of ARV for official government business.

There was informal use of (non-binary) ARV at Ft. Meade to try to win the lottery. We have not been able to obtain details of these attempts but apparently they were not successful.[107]

While Stephan Schwartz is rightly credited with originating the practice of Associative Remote Viewing, an earlier effort should be mentioned. Czech parapsychologist Milan Ryzl conducted a proof-of-principle experiment in the early 1960s to determine if it was possible to communicate information with psi. He chose the task of conveying five three-digit numbers using a complex and extremely time-consuming format with green and white colors on cards with the numbers coded (converted from decimal to base 2) by a sequence of colors. Hence a correct binary choice of colors would be able to identify a three-digit number, which is a form of ARV. An elaborate error-correcting method was used. The result was successful communication of the five numbers without a single error, demonstrating that an association of a sequence of colors on cards with numbers could be used to transmit information. However, the experiment took 19,500 color calls at a rate of 400 per hour, with two participants.[108]

In addition, Ryzl later trained viewers in hypnosis and meditation and used associative methods to win substantial amounts in Czech lottery draws. (We discuss Ryzl and his methods in Chapter 21.)

In the next chapter we will outline subsequent formal experiments, right up to the present day.

CHAPTER 3

A Closer Look: Selected Research and Practical Applications of ARV

In this chapter, we provide a review of formal as well as informal research and practical applications of ARV. The goal is to give an idea of the rich history of ARV and show that ARV is not only alive and well today, but is blossoming. One of our main motivations for writing this book is that ARV is probably the most active area of remote viewing activity. While we can only speculate why ARV is now so prominent, we believe it is a combination of factors.

First, the decades-long contributions and efforts of Marty Rosenblatt and those involved in his P-I-A (Physics-Intuitions-Applications) and APP (Applied Precognition Project) groups cannot be overestimated. At all times Marty has sought to be inclusive of people, moving forward with unmatched dedication and drive despite a few conflicts and controversies. While not a formal research organization, APP has done a tremendous job in actually doing ARV, keeping statistics, getting the word out, and educating thousands of people through workshops and conferences. As examples of the latter, APP has posted a library of free educational videos on remote viewing. Marty has sponsored the "Talk With" series featuring remote viewing researchers and viewers, as well as wide range of experts in psi, psychology, physics, energy, ufology and health. Both authors worked intensively with Marty in APP and learned much from the experience.

As we have noted, ARV is a process originally conceived and designed to make predictions for wagering. It offers the potential to make money, while regular remote viewing can boast few such examples. This potential may be the biggest reason for the intense interest in ARV, which accelerated during the pandemic of 2020. Sharing information about some of the little-known ARV financial success stories is one reason we decided to write this book.

Even for those not expressly interested in its money-making potential, ARV inspires the imagination, calling into question the physics of time and space and the nature of reality. It speaks to the questions: Can we really know the future? Can the future influence the past? What is the effect of multiple conscious beings working together? What is the interface between consciousness and technology whether that technology is simply pen and paper or complex computer programs? We also wonder how the potential for making money affects viewers and results. Finally, ARV lends itself very well to a research protocol and the generation of data that can be easily analyzed statistically. You make a prediction of a specific event and you get a clear and definite result. You know whether your prediction was correct or not.

For these reasons, ARV continues to be of interest to parapsychological and psychological researchers, to traders and investors, and to many people who are interested in discovering how one's intuitive perceptual abilities can become useful in everyday life.

The following presentation of selected ARV research and ventures, while not exhaustive, is fairly comprehensive. We make that assertion with some confidence, given our extensive literature reviews, interviews and our joint participation in many applied efforts and formal research projects. Our presentation is organized chronologically. It is a mixture of projects, from which we hope a useful picture emerges of the themes and areas of interest that will be fleshed out in other chapters of this book. We include peer-reviewed as well as informal ARV research. Some of the themes have already been mentioned because they are integral components of Associative Remote Viewing projects, such as those related to judging and scoring considerations, the use of individual vs. teams of remote viewers, feedback, timing, retrocausation, displacement and target selection. In addition to what is included here, on the website for this book, we include a comprehensive chart of documented ARV projects (http://www.arvbook.com).

In one of the earliest and best-known efforts, Keith Harary and Russell Targ used ARV in 1982 to forecast changes in closing prices of the silver futures market. They made nine consecutive correct forecasts, which yielded earnings of $120,000.[109]

Also in 1982, Dr. Harold E. Puthoff used ARV to predict the daily outcome of silver futures contracts for 30 consecutive days. Seven remote viewers conducted from 12 to 36 trials per person over the entire series. Each day, predictions were made using consensus judging. Twenty-one of the 30 trades were profitable, yielding profits of $250,000 to the investor.[110]

In 1994, Russell Targ, Jane Katra, Dean Brown, and Wenden Wiegand conducted an ARV experiment in which remote viewers had time to receive feedback before starting another trial. In this nine-week series, objects were associated with the two possible outcomes, "Up" or "Down," of the weekly silver futures contract. Using an error-detecting protocol, a judge compared the remote viewers' descriptions to the targets and rated the accuracy of the description on a scale of 0 to 7. If the session scored 4 or higher, a prediction was made. Results yielded seven trades and two passes. This was a simulation, so no purchases were made, and capital was not risked. Six of the seven trade predictions were correct.[111]

In 2000, Marty Rosenblatt, operating under Physics-Intuition-Applications (P-I-A), reported results from an ARV experiment ("the AVM project") that predicted stock market closing points. Seven viewers were paid to do 500 sessions each, for a total of 3,500 predictions with 700 investment targets. According to Rosenblatt's report the overall performance was about what you would expect based on chance.[112]

Two of these viewers were CRV student Lori Williams (now an instructor) and her then-husband. In private correspondence with Debra in 2016, Williams explained a few things about the project. When asked whether they had found the high number of trials to be stressful or tedious (500 for any participant in any project is almost unheard of), she explained these targets were really more forced-choice. All they had to do was a quick ideogram (reflexive mark the subconscious makes) for each trial, and then state whether this looked and felt more like their pre-learned ideograms for animal, vegetable or mineral. With this method they could work quickly and were not required to provide in-depth, psi-based information. When asked to speculate what happened to halt the progress after the first 100 predictions, she said she couldn't be sure. Initially the participants were not aware money was being wagered. Once this was revealed, she and her husband started to feel pressure about the potential of losing money for other people. She therefore believes it could have been the stress related to wagering that affected results.

Greg Kolodziejzyk is an accomplished triathlon athlete, entrepreneur, and algorithmic trading strategy developer and trader who a conducted a 13-year study using a unique computer-based approach to the ARV protocol that allowed a single operator (himself) to conduct 5,677 trials. Of these, 52.65% correctly predicted the outcome of their respective future events, yielding a statistically significant score of $z = 4.0$. These 5,677 trials addressed 285 project questions intended to predict the outcome of a given futures market. Multiple ARV trials answered a single question. Of these project questions, 60.3% were answered correctly, resulting in a statistically significant $z = 3.49$. One hundred eighty-one project questions resulted in actual futures trades where capital was risked. Of those, 60% of the trades were profitable, yielding a profit of $146,587.30.[113] Greg K reported that he went for quantity, rather than quality, in his remote viewing sessions. (See Chapter 15 for a longer account of Greg K's work.)

University of Colorado ARV project

Garret Moddel is a professor in the Department of Electrical, Computer, & Energy Engineering at the University of Colorado. He conducts research in quantum engineering, device technology in optics, nanostructures and bioengineering, is former president of the Society for Psychical Research, and has published more than 200 papers.

In 2012, Moddel joined two people in a course he taught at the University of Colorado, D. Laham and Christopher Smith (son of former military remote viewer Paul Smith) to conduct an experiment using ARV with ten inexperienced remote viewers. The goal was to predict the outcome of the Dow Jones Industrial Average (DJIA). One of the project's unique aspects was that participants conducted their viewing sessions together in the same room, as opposed to by themselves or in the presence of a single interviewer, as ARV experiments are usually done. They also used a novel rating system that we discuss in Chapter 4, the U of C 3-Point Rating Scale, which takes into consideration the strength of the matches to each photo. In aggregate, the participants described the correct images and successfully predicted the outcome of the DJIA in all seven attempts (binomial probability test, p < .01). An initial investment of $10,000 yielded a gain of about $16,000, with a total of $26,000 at the end of Trial 5.[114]

Samuelson's conceptual replication of the University of Colorado Project

We met Mark Samuelson years ago when working on Applied Precognition projects. Mark is not only a talented remote viewer but has the ability to analyze large amounts of data. He has made some major contributions to the way we think about variables and approaches to ARV. From August 2014 to August 2015, Samuelson, serving as project manager and rater, recruited several APP members to perform a conceptual replication of Smith, Laham, and Moddel's project.[115] Some viewers were quite new and some very experienced, including the authors. One notable difference from the U of C project was that the group met via GoToMeeting's synchronous online technology rather than in person. These meetings were held a couple of times each month for about a year. Katz and Michelle Bulgatz took turns leading the group in an opening meditation, then the participants, who were at home, would be given a target number and would do their remote viewing session while the webinar continued. After the viewers emailed their transcripts, the meeting ended. A day or two later, viewers received the feedback photo paired with the side that actualized.

Group predictions were rated using a simple judging method. Samuelson's group predicted professional sporting events rather than stock market fluctuations. The goal of exceeding their 65% hit rate also differed from the University of Colorado group's goal of making money. After 26 trials, Samuelson's group had 13 hits, 7 misses, 4 passes, and 2 pushes—maintaining, but not exceeding, their 65% accuracy rate. Samuelson later explained in email correspondence with Katz[116] that despite the hit rate being impressive in terms of statistics, he was disappointed in the results. He cited a single viewer as having derailed many of the predictions because he did an excellent job describing the wrong photo in the set. Samuelson blamed himself because he placed more weight on this one viewer's sessions. Despite usually having between 8 and 20 viewers, this viewer had a reputation as an excellent remote viewer for non-ARV projects. Displacement once again seemed to be the culprit, although the other theme to emerge was the manager's admission that he placed more weight on one participant's remote viewing response.[117]

A few notes from Debra here: I really enjoyed working on this project, both as a viewer and in occasionally leading meditations. I learned a few things, too. At this point, I was used to leading groups in highly detailed, visually oriented meditations, using an unscripted, intuitive approach for my guided meditations. While this works quite well in other venues, I found I could not stop imagery from upcoming targets from flowing into my subconscious. Too often I'd find myself telling the participants to visualize an image and I'd find out later it had been part of the target. I also began to worry that even if the visuals were not pictures from the future, they might still distract or plant ideas into the viewers' minds. While these meditations are designed to stimulate visuals, the last thing we'd want right before an RV session is to stimulate distracting ones! Therefore, I decided as time went on to either forgo the meditation altogether or simply lead the viewers through simple breathing exercises focused on relaxing the body. I don't think this had any influence on the trials overall, but I wanted to mention it for other managers who would like to add a cooldown or meditation prior to a group RV session. My pal Michelle Bulgatz also had similar concerns.

The other thing I noticed was my sessions sometimes weren't as good when I did them in the meeting compared to when I did them logged out and alone. For one thing, I preferred to have more time to do my sessions than the 10 minutes typically allocated, so it's hard to say if my performance was improved by working independently or simply due to having more time. (Most viewers find 10 minutes is long enough for ARV.)

Finally, I learned one very important lesson that is echoed throughout this book. I expected Samuelson would do a formal report on this project, especially since I viewed it as a loose replication of the University of Colorado study. I have to say, a bit ashamed, I gave him a really hard time about his decision, quoting words of skeptics that this could be construed as being evidence of the *file drawer effect* in parapsychology.[118] Mark wrote me back explaining he did not have a formal academic background, didn't have any idea how to do formal write up for publication, and reminded me that he had been forthright in making his data available to everyone on our ARV discussion lists as it became available. In retrospect, I realized that because this project was conducted in such a careful, methodological way, I had thought of it as a formal project. I also realized I saw Mark as someone who was much more capable of collecting, organizing, analyzing and interpreting data than most graduate students I knew (including myself), so I was quite surprised to hear he didn't have an advanced degree.

This put me on the path of doing all I could to connect ARV and RV project managers with more formal researchers. Yet the question remains and debates continue about how important it is for informal projects to move into a more formal arena by publishing in scholarly journals. I personally like the idea because it establishes a formal record other researchers can build on, even decades later. Social media posts simply will not be easy to find in 50 years. Journals like the *JSPR* have been in existence for 130 years, and there is a good chance it and similar publications will continue. I suspect only time will tell as more open collaborative online technologies emerge.

Moddel and Wimberger's ARV stock options project

Lisa Wimberger was hit by lighting on her fifteenth birthday and underwent a series of blackouts, seizures and what she refers to as "freezing." These traumatic experiences took her on a journey in search of healing. Along the way, she stumbled upon teachers who taught her how to access her own intuitive abilities. She eventually formulated her own meditation-based modality, "Neurosculpting," which she teaches through the institute of the same name and discusses in her books, *Neurosculpting: A Whole-Brain Approach to Heal Trauma, Rewrite Limiting Beliefs, and Find Wholeness.*[119]

Lisa made the acquaintance of former Star Gate Director Dale Graff. He was interested in her approach to intuitive work, which she describes as a mixture of clairvoyance and OBE. Graff referred her to Dr. Garret Moddel, who wanted to conduct further exploratory ARV trials to make stock market predictions. His aim was not to publish another formal study, but rather to learn from each trial what worked, while seeing if he could effectively use ARV for trading. He was willing to share all his data and notes with the present authors provided we made clear the findings have not been peer-reviewed nor structured in a way he would have done for a formalized study. From his copious progress notes and email correspondence with both Moddel and Wimberger, we briefly summarize their project here.

Starting in November 2012, Moddel conducted a series of 25 informal trials with Lisa Wimberger. Of these, 19 trials resulted in predictions (6 trials either without strong enough correspondence with either photo or where a trial could not be completed due to scheduling or procedural issues). Of the 19 predictions, there were 13 hits and 6 misses (68.4% hits).

When asked about his trading approach, Moddel explained:

I didn't just trade shares, which would go up or down slightly, but instead traded stock options, which tend to greatly amplify shifts in the market. I did that because I was buying and selling within about 24 hours, when there would not be much market movement. The trading costs can be quite high with options. I don't remember the specific amounts, but if one is right 50% of the time, there is significant loss.

Many of the questions discussed in his notes illustrate the complexities and challenges of placing trades in a timely manner. For example, he planned to start with $5,000 for the first trial but the money didn't end up in his account on time. The prediction was correct but they had risked only a small amount and earned $74.65. The result of the next trial was also correct and yielded much larger earnings - $1,999. The third trial was also correct, but they sustained a financial loss of almost that amount:

Because of delays in placing the trade it came out a loss, but the actual prediction was technically correct. Looked at another way, if I had made the opposite choice and bought a call instead of a put, it would have lost more. Sold 100 puts @0.29 at 10:15 am. $2,900 – 23.50 commission, for a loss of $1,947.

Moddel's notes show he clearly gave more attention to his losses than to the larger number of successes. He wrote, "Note: Lisa was away on August 17 and could not see the image that I sent her until later. I think the lesson here for me is when Lisa's description does not provide an obvious description of an image, I should not bet, and last-day trading is risky, period. It will be interesting to see if Lisa's descriptions are less accurate when the option choice is more ambiguous, i.e., does her perception depend not only on the picture I send her, but also on the decision process I had to go through to decide on the picture?"

The following week there was a miss and on August 28, Moddel wrote:

There were clearly elements of image 2 in Lisa's prediction. I see the following possibilities for the incorrect prediction:

1. I made a bad judgment. Looking at the reading again, I can see how Lisa's description would lead to image 1.
2. This was erroneous, due to leakage of the image into her mind, because she's perceiving not just the image that she'll be sent but also the alternate.

3. There was some sort of fundamental uncertainty in the market outcome, so that it was not possible to predict accurately. This again brings up the question as to whether Lisa sees what she will be sent, or what seems to be correct. In both this prediction and the Aug. 17, there were some hints of both images in Lisa's description. I can test for this by not giving Lisa any feedback, i.e., not ever sending her the correct image. I'd like to do that after I've established a success rate with the current process.
4. My wavering belief. This time and Aug. 17 I felt doubt as I opened up the stock options website. I need to get over those doubts.
5. The degree of drama in the images may affect the outcome. In the most recent case once might argue that image two was more dramatic.

Improved protocol: Based on these observations I will not invest unless the prediction is very clear, and then I will do it with full confidence of its being correct. I will also try to make the images have equal emotional value, and sufficient detail for Lisa to pick out several features. Another observation: the days on which the coin flip was the same as the investment did better than when the coin flip was earlier. I will flip the coin just before I open Lisa email with the descriptions. (If this is really a factor, then it implicates the coin flip as an important factor in the prediction.)

Again, these are notes and not intended to be definitive findings, but we wanted to include them because they raise a number of considerations such as displacement or "leakage" to the alternative, uncertainty of market outcomes, the project manager's own "wavering beliefs," the content within target materials, and wagering decisions based on confidence levels.

Additionally, in looking over this data, we noticed that the viewer was particularly adept at describing the colors of the correct photo and often named aspects of the photos. For example, when there was a vehicle with wheels, she mentioned wheels and did not do so in any of the other trials. When the target was a penguin on a completely flat, light-colored surface facing to the left, she described toes pointed to the left and thought she was looking at a turkey on top of a flat table.

As is typical in RV trials, correct information with specific details was mixed with distortion of the information. The viewer didn't include sketches but only words. This seemed to force her into overuse of unrelated symbols to match the shapes she was trying to describe. She often went to great lengths to provide detailed descriptions of complex shapes, and many of the photos had structural elements of a moderate to high level of complexity. Without the viewer providing a sketch, the judge had to create a mental image and then compare the descriptions to that image.

It's actually quite challenging to verbally describe shapes someone else needs to comprehend and replicate. Lyn Buchanan, a former Star Gate project viewer and well-known instructor, runs new students through group exercises in which one student looks at a complex photo and is only allowed to describe shapes or dimensionals, while the other students attempt to produce a sketch of the photo from the verbal description alone. It is usually clear that much is lost in translation.

Lisa said she had not been trained in formal remote viewing methods. She was unaware of the usefulness of sketching but will try it in the future. She likely has a promising future in ARV.

Moddel's project clearly demonstrates the complexities of using psi to place stock market trades. Project managers interested in using RV for trading clearly need to have a solid knowledge of trading platforms and an understanding of how to best to translate and integrate remote viewing data into a structured trading protocol. In this regard, as mentioned by Dr. Moddel, financial loss can occur even when there is a correct prediction. Others in the field have experienced this for various reasons, such as trading fees relative to gains made, mismatches between the high and low prices predicted and those that actually happen, and failure of the trailing stops selected. As we will discuss in Chapter 18, similar complexities apply when using ARV for horse racing predictions.

A background in trading, statistics, modeling and data analysis is probably the last thing people think of when they get excited about the idea of making money with ARV. Most people assume finding talented psi participants is the issue. Instead, finding those who are skilled at working with and analyzing the data, particularly large amounts of it, are in shorter supply, at least in ARV/RV circles. Still, there are workarounds for those who are challenged in these areas or just don't have the desire to put in the time to become proficient. For instance, a growing number of experienced traders offer "For entertainment purposes only" membership services to traders interested in receiving predictions without having to manage their own projects (see Chapter 15).

Project Firefly: A yearlong endeavor to create wealth by predicting Forex currency moves with Associative Remote Viewing

A few words on Project Firefly (PFF) here – we expand on this greatly in the Appendix. In this innovative project, more than 60 remote viewers contributed 177 intuitive-based ARV predictions over a 14-month period. These viewers had been active in pre-established, self-organized groups. PFF was supervised by experienced ARV group managers operating under the umbrella of the Applied Precognition Project (APP), an organization exploring precognition and leveraging ARV methodology as an investment enhancement tool.

Based on predictions from the ARV sessions, PFF used a Kelly wagering strategy to guide trading on Foreign Exchange (Forex) currency markets. While not a scientific experiment, viewers performed under some of the typical scientific protocols including double-blind conditions, randomization, etc., using a variety of ARV methodologies. Investors, nearly all of whom were also participants (viewers and judges), pooled investment funds totaling $56,300 with the stated goal of "creating wealth aggressively." Rather than meeting that goal, however, most of the funds were lost over the course of the project.

Adapting a form of ethnographic study, Katz, Grgić and Fendley, with the aid of many other APP members including Knowles, not only referred to the statistical results produced by the PFF effort, but also employed a mixed-methods qualitative approach to share the information and insights contributed by the many participants about what happened, what worked and what didn't. This work created a reference that the authors felt was useful for those conducting future applied precognition projects involving multiple participants or groups. We felt the insights gleaned from this study could improve both ARV experimental design and execution of research protocol, benefiting professional and amateur researchers alike in their future ARV experimentation.

As noted above, the published article by Katz, Grgić and Fendley[120] is an Appendix in this book. Findings in the article are summed up as follows:

First, predictions based on aggregate groups on a single trade day did not fare as well as single entities (groups or solos). Instead, the data generally support using the best viewers and teams, as per their hit rates listed in Table 8, and keeping the protocol simple. An exception to this was seen in Phase Two, Runs 2 and 3, when the top solo viewers' hit rates dropped from around 70% to roughly 50%. Those data were not statistically significant, however, because no solo viewer did more than 11 non-passing predictions during those runs.

Second, the goal of having 240 trades in a single year placed a great deal of stress on the trading team. Of 249 predictions, 72 were passes. This may be an example of too many predictions in too short a time span, as seen in the Targ/Harary study.[121]

Third, an independent Oversight Committee could provide valuable support for the trading team by serving as a check and balance on trading activity, monitoring protocol, and implementing a process to make changes with greater transparency for the viewer/investors. This could be critical if an aggressive wagering method is being used and early losses are incurred.

Fourth, the Kelly wagering method should be used only after verifying the hit rate for the specific viewers and a specific protocol. In this instance, subsequent examination of the pre-Firefly data showed many of the entities used in Firefly had hit rates below chance for similar financial predictions. In such cases, a more conservative approach than investing 20% of all monies should be applied. Further study on the hit rates of different protocols is needed (p. 44).

ARV dream project

In recent years, Katz and friends and colleagues Nancy Smith, Dale Graff (former DIA Director of the Star Gate project), Michelle Bulgatz and Duke University Professor Emeritus Dr. James Lane[122] conducted a yearlong, double-blind study using dreaming as a precognitive tool developed by Graff, within an ARV protocol. A cohesive group of seven experienced remote viewers (the APP Sublime group, which included Katz and Bulgatz, along with David Silverstein, Chris Georges and Marty Rosenblatt) participated in 56 trials in which they attempted to have precognitive dreams of a future feedback photo. Their protocols, mirroring RV protocols, required them to produce a written transcript upon awakening that included descriptor words and sketches.

A single judge (Nancy Smith) served as project manager. She rated the transcripts using the SRI 7-point scale, pooled the transcripts for each trial to make a group prediction and wagered on a sports event. Five of the seven remote viewers/dreamers consistently produced dreams at will, resulting in 278 transcripts. Two dreamers had high individual hit rates (76% on 17 trials and 64% on 25 trials). With 56 trials, 28 group predictions yielded 17 hits and 11 misses, which a binomial test showed to be at chance levels.

The overall monetary gain was almost 400% of the initial stake. (For more about this project including examples from the viewers' transcripts please see Chapter 17.)

German Stock Index Project (DAX) –
Predicting the stock market: An Associative Remote Viewing study

In 2017, a team of researchers from Germany won the IRVA Warcollier Prize for an ARV-related proposal. This provided them with $3,000, which helped finance their wagering attempts. The main research objectives were to determine the hit rate for predictions of the German stock index DAX (*Deutscher Aktienindex*) using Associative Remote Viewing, to test whether feedback is necessary for ARV predictions and to explore factors that might influence the quality of the viewers' perceptions in ARV sessions. In addition, they wanted to "identify a design for subsequent studies in the sense of a proof-of-principle study" (p. 2).

One of the variables they wanted to test was the importance of feedback. They postulated that intention rather than feedback was important. To test this, they split the trials in half so each viewer would receive feedback for only half the trials.

The hit-miss ratio was quite impressive. The researchers indicated "that the ARV method used in our study predicted the near future of a stock index above chance level" with 38 of 48 correct predictions amounting to

> a highly significant result ($p = 2.3$ x $10-5$, binomial distribution, B48 (1/2); $z = 3.897$), reflecting the hit ratio of 79.16%. The z-score divided through the square root of n = 48 trials corresponds to an effect size (ES) of 0.56. In contrast, a true random number generator (RNG: random.org) was not able to predict the stock index significantly (24 out of 48, binomial distribution, B48 (1/2), is $p = 0.11$; $z = 0$) (p. 335).

However, wagering did not fare as well, earning them only (237€), which was not significantly higher than the profit the RNG would have produced. The average profit per trial for the ARV predictions was 4.93€ and for the RNG predictions was 1.60€ ($t = 0.722$, $p = 0.472$).

The researchers determined it was not necessary for a viewer to receive feedback in the near future in order to do well, at least for this particular protocol involving a short turnaround. They wrote:

> 24 out of 48 trials were sessions with a feedback for the viewers, the other half was without feedback. Both conditions were independently significant: In the feedback condition, the viewers succeeded 20 times and failed only 4 times ($\chi2 = 10.667$, $p = 0.001$). In the non-feedback condition, the viewers succeeded 18 times and failed only 6 times ($\chi2 = 6.000$, $p = 0.014$). A Chi-Square test for the frequency of hits and misses shows that there is no significant difference between both conditions ($\chi2 = 0.505$, $p = 0.477$).

For future studies attempting to replicate their project, they suggested that a

> new hypothesis would be that the hit rate of ARV with binary outcomes for targets existing at the present moment is significantly higher than the hit rate of ARV with binary outcomes in the future. If the results were positive according to this hypothesis, the probabilistic future would be an additional factor for predictions with ARV leading to more misses.

They also suggested future studies should continue to attempt to predict the stock market on an hourly basis, noting that

> this is even more difficult by conventional means because of the high volatility of the market across a given day. Generally, if the ARV method is properly conducted, it has the potential to become a probed and tested paradigm for the research field and can convincingly prove that Psi effects are robust and replicable.

Further, they found:

> The overall ARV hit rate for future predictions is primarily influenced by target selection, data collection and judging. These factors are mainly controllable and it would be simple to conduct a replicable ARV experiment, if the necessary experience and human resources were available (p. 340).
>
> (For a more detailed account of this experiment, see Chapter 15.)

CHAPTER 4

Scoring, Judging and Prediction Methods

A variety of methods have been used to judge and score sessions in remote viewing, ARV and experimental parapsychology experiments. Here we present a representative sampling of these methods, including those most commonly used in ARV. Throughout the book we refer to these methods and go into some of them in detail in later chapters – this chapter will serve as a reference for all of these methods. We also offer criticisms of some of these methods.

Methods include the following: Whately Carington's; simple matching; matching with decoy photos and a sum of ranks statistical approach; SRI/Targ 7-point scale; Greg Kolodziejzyk 4-point scale; University of Colorado (Garret Moddel) 3-point scale; Joe McMoneagle's Gestalt method; the Military Scoring Method (Bruce Miller, Lyn Buchanan); Alexis Poquiz method (aka "Dung Beetle"); and Applied Precognition Project's HAG (Harsh Analysis with Gestalts).

APP has worked extensively on scoring methods starting with the SRI/Targ scale. (See Chapter 12.) Ed May's Computer Assisted Scoring (CAS) and Julia Mossbridge's Positive Precognition are two additional methods. These innovative approaches are based on "profiles" of photos and the use of categories to score them in order to reduce "feedback loops" by preventing the viewer and analyst from seeing both photos in a binary ARV trial. We discuss both these methods in Chapter 9.

Whately Carington's method of scoring drawings

English researcher Whately Carington (1892–1947) used a very complex method of scoring drawings by percipients against other drawings (which were the targets) concentrated upon by another person. 741 people made drawings, which were used to form a "Catalogue of Frequencies of Objects Drawn or Mentioned in Experiments I to VII."

The frequency with which a Tomato, a Cauldron, a Mirror, Corkscrew (Gimlet), Medusa or hundreds of other objects were drawn by the percipient in each series (I-IV, VI, and VII). Whether the object was accompanied by another was also tabulated in the Catalogue. The number of "hits" was calculated taking several factors into consideration. One was whether the object drawn was rare or frequent in the catalog. Some hits were considered better than others. Another factor was a comparison of "all other correspondences between *percipient's drawing* and *drawing made by the agent on some other occasion*."[123] Further: "We made it a practice to grade our resemblances as "a," "b," or "y" according to the degree of confidence we had in them, which very much approximately corresponded to the degree of resemblance discernable…"

Hence there was a subjective three-level scale. Although Carington wanted to devise a sound repeatable method for scoring such drawings (which had not been achieved previously), the method clearly had a large subjective component, as can be noted in the above quote.

The most important facet of Carington's pamphlet for present-day ARV is the mention and description of displacement, a term he coined. We devote two chapters to this important concept. Carington observes that displacement is "that which Dr. Thouless has also well named temporal dislocation of response" (p. 13).

Simple matching task – Choose best match, no scoring

In this method, the judge simply compares the viewer's transcript to the photo options and chooses the best match. No scores are assigned, and the judge must select one of the photos. Once results are known, statistics are calculated based on whether the photo chosen as the best match was actually the target. If so, a hit is registered. If not, a miss is registered.

This approach has been widely used in many experimental parapsychology projects and by many informal or casual practitioners. Criticisms of this approach include a tendency for judges to notice or place greater importance on some matching aspects while missing others. Most problematic is the inability to assess how strong a match is – a transcript with very minor correspondence receives equal weight to one with very strong matches. When there is little matching data, judges are forced to simply guess (or use their own intuition) to make a determination. Still, some people argue this method should be sufficient if the remote viewing is really good.

Matching task, with decoy photos, ranking (sum of ranks)

This method builds on the one above. It also compares the remote viewing transcript to the photo option without use of additional scoring. However, if a set includes more than one photo option, the judge selects first best match, second best match, third best match, etc. In calculating the statistics, the second and third best matches receive some level of credit, but not as much as a first-place choice. This method has been very widely used in parapsychology experiments.

The methodology is described in detail by Milton in her article, *A Meta-Analytic Comparison of the Sensitivity of Direct Hits and Sums of Ranks,*[124] and by Solvin, Kelly & Burdick in their 1978 article, *Some new methods of analysis for preferential-ranking data.*[125] Recently it was used by Katz, Lane & Bulgatz[126] in their study comparing objects set in different background conditions.

SRI 7-point Confidence Ranking (CR) scale (sometimes called the Targ scale)

The SRI scale, developed in the early 1970s, is the most widely used method in scoring ARV transcripts. The first discussion of the scale in a formal paper appears to have been in *Viewing the future: A pilot study with an error-detecting protocol* by Targ, Kantra, Brown and Wiegand.[127] The remote viewing transcript is compared with a photo and assigned a score between 0 and 7. The judge repeats this process with each photo in the set. Each transcript is judged independently against the photos.

This scale takes into consideration that some data will occur by chance alone. It also incorporates the idea that viewers will often have a mixture of correct and incorrect elements and presupposes that while incorrect data should not negate correct data, a transcript containing both should not be scored as high as a transcript with all correct data. The term "confidence ranking" (CR) underscores the idea that the scale is not intended to simply assign a grade, but to give the project manager a degree of confidence that the viewer described the target and not something else.

Over time, the Applied Precognition Project has developed standards for applying the 7-point scale in group ARV trials to decide whether to make a prediction or pass. Once all scores are arrived at and recorded, the project manager compares the scores, and the photo with the higher average score will be selected as the group prediction, subject to other factors.

One such factor is that the high score must reach a minimum value. Russell Targ used a cutoff score of 4.0, while Marty Rosenblatt and APP have often used a CR of 3.5 as the minimum to make a prediction. Katz, Smith, Graff, Bulgatz and Lane[128] found in their ARV dream study that they would have had greater success if they had set the threshold at 5 or higher, since scores even as high 4.5 did not always result in a hit.

While a CR of 3 for one photo is higher than a score of 2 for the other photo, this difference doesn't offer a high degree of confidence that the viewer has strongly accessed the target. A 3 on this scale does, however, indicate at least some contact was made. Still, it is just too easy for some matches to occur because of chance alone. One rule of thumb that should be kept in mind is that about a third of a response may match any target – a figure cited by Ed May and others from their extensive lab experiments.

A second factor is the amount of the spread between the scores for each photo. If a CR of 7 is given to one photo and a CR of 6 to another, the spread of 1 point is too narrow to have confidence that the higher-ranked photo is the feedback target. The minimum spread APP relies on is 2 points. As an example, if scores are 3 and 1, no prediction is made because the high score is below 3.5. If the scores are 4 and 2 or 5 and 1, a prediction is made – namely, that the photo corresponding to the higher score will be the one shown as feedback to the viewers because the associated event has occurred.

A formal procedure has not been established for group trials in which multiple viewers have higher scores for one photo while others have higher scores for the other photo. A conservative approach calls for a pass in this case. A less conservative approach adds the scores for each photo and selects the highest total as determining the prediction. This latter procedure, with variations, has been used in betting on sports events at all the APP conferences – particularly Joe McMoneagle's gestalt method and more recently the HAG method (both described below).

When Katz, Smith, Graff, Bulgatz and Lane[129] analyzed their ARV dream study results, they found the manager/judge sometimes took the following approach. If one viewer had a CR score of 4 and another viewer had a 5 for the photo but three viewers received equal or higher scores for the other photo, a prediction would be made for the side with the higher total scores. Researchers studied whether such calls resulted in more misses or hits and found an even divide – meaning, had passes been called when some high scores pointed to one photo and other high scores to the other photo, some trials would not have been misses, but some hits would have been missed, as well.

Another factor to consider is a viewer's past record. A project manager might make use of a viewer's past hit rate and rely on their sense of the viewer's displacement rate (which has rarely been quantified) to determine whether their CR scores should be given more consideration than those of other viewers. Formal studies are clearly needed of multiple viewers with high CR scores for different photos.

One of the criticisms of the SRI/Targ method is the wide differences in scores judges assign. This could be due to the judges' personalities. Some are more lenient and "permissive," trying to discern what the viewer intended or experienced (such as when they say they felt it was very warm at a seaside location), while other judges are conservative, giving credit only for perceptions clearly visible within the frame of the photo. Another example would be if a photo has red in it, a permissive judge might allow pink or mauve to be considered a match, while a conservative judge might not.

ARV expert Igor Grgić noted this lack of consistency in an assessment of the gap between the lowest and highest scores assigned by judges for the same photo target. The maximum score difference was 4.0 and a gap of 2.5 was taken as a rule of thumb.[130] Igor found judges were by no means 100% in agreement in group predictions. Generally, most judges agreed and called a prediction for the same side, while some called a pass. In one trial, there was a call for the opposite side. On an individual transcript level, most of the time all the judges chose the side that the transcript scored higher on. If not in agreement, in most cases the judges would pass on a particular individual pick. There was also an example of one judge picking the opposite side from the other judges.

Still, even with these scoring discrepancies, the 7-point scale is fairly easy to apply and is practical to use when there are many trials to rate and time is an issue. The SRI scale has been used in more formal projects and by more judges with more testing than other methods that assign scores, so these issues should not be construed to mean the scale is more problematic than others.

The SRI (Targ) Seven-Point Confidence Ranking (CR) Scale

7	Excellent correspondence, including good analytical detail (e.g., naming the target), and with essentially no incorrect information.
6	Good correspondence with good analytical information (e.g., naming the function of the target), and relatively little incorrect information.
5	Good correspondence with unambiguous unique matchable elements, but some incorrect information.
4	Good correspondence with several matchable elements intermixed with some incorrect information.
3	Mixture of correct and incorrect elements, but enough of the former to indicate that the viewer has made contact with the target.
2	Some correct elements, but not sufficient to suggest results beyond chance expectation.
1	Little correspondence.
0	No correspondence.

Figure 4.1

Greg Kolodziejzyk's 4-point scale (.1 to 4.0)

Greg K conducted a successful pioneering 13-year ARV experiment. His method assigns a confidence score from .1 to 4 to each transcript. The low score (.1) indicates there were no similarities between the sketch and either image while the high score (4) indicates high confidence that there were similarities between the image selected and the sketch. Greg's experiment is discussed in Chapter 15.

University of Colorado 3-Point confidence ranking scale

This scale was developed as part of a successful ARV experiment conducted by University of Colorado professor Garret Moddel. The experiment was first discussed in an article titled *Stock market prediction using associative remote viewing by inexperienced viewers.*[131] To date, the scale has been used formally only in one other experiment, which compared judging methods and rater reliability.[132]

This 3-point scale incorporates two measures: judges score the transcripts as having a "low," "medium" or "high" match for each photo option. While this method isn't as sensitive as the 7-point scale since there are only three levels of grading, its advantage over the 7-point scale is that a comparison between *both* photos is a component part of the score, as can be seen in the table below. The score rates the individual transcript for each photo.

University of Colorado Three-Point Confidence Ranking Scale	
A-1	Low similarity to the "Side A" target and with little similarity to the "Side B" target.
A-2	Good (medium) similarity to the "Side A" target and with usually low or no similarity to the "Side B" target.
A-3	Excellent (high) similarity to the "Side A" target, with usually low or no similarity to the "Side B" Target.
B-1	Low similarity to the "Side B" target, with little similarity to the "Side A" target.
B-2	Good (medium) similarity to the "Side B" target and with usually low or no similarity to the "Side A" target.
B-3	Excellent (high) similarity to the "Side B" target, with usually low or no similarity to the "Side A" target.
M-0	M-0 Means that it can't be judged between the two and the transcript has low or not similar to both targets.
M-1	Means that it can't be judged between the two and the transcript has medium similarity to both targets.
M-2	Means it can't be judged between the two and the transcript has high similarity to both targets.

Figure 4.2

McMoneagle method

At the APP conferences, Joe McMoneagle introduced a method to score binary ARV sessions. The following account is taken from notes by attendees at the 2014 APP conference.

Marty Rosenblatt credits Joe with the idea of giving the "gestalt" of the target 50% of the total score. What is a gestalt? This German word has no distinct English equivalent, but refers to the overall impact of significant elements. In this usage of the term, it is what stands out the most, contains other smaller elements, or what is repeated. An ocean target contains the gestalt of water. A skyscraper's gestalt is a manmade building. A waterfall has gestalts of water, downward movement and possibly mountain. Joe felt the gestalt of a photo can be represented either by words or graphics (drawings). Even a color can be a gestalt if it dominates the scene. In a photo with many different colored objects, color would not be a gestalt, but if everything in the photo was different shades of blue, the color blue could be considered a gestalt.

Quantifying the Data in the McMoneagle Method

1. Indicate whether or not a transcript identifies a clear gestalt. It could be 0=None, ½= Weak, or 1=Clearly defined. Joe indicated this by writing "Gestalt" or "Gestalt ½."

2. Count the total number of descriptors or attributes on each transcript. Joe recorded this number as the dividend (numerator) in his fraction.

3. Count the total number of relevant descriptors on each transcript. Joe recorded this as the divisor (denominator). (Note: contrary to conventional formula math, Joe acknowledged that he records his fractions reversing the dividend and divisor positions).

4. Identify if the transcript is in contact and most relevant to either Target A, Target B, or neither.

Figure 4.3

Doing the Math in the McMoneagle Method

The example uses the numbers obtained on a particular score sheet at the conference:

1. Total the number of gestalts per target and multiply by 10 and convert to a % number. 6½ gestalts for Target A = 65%. 5½ gestalts for Target B = 55%.

2. Take the total number of relevant descriptors per target and divide by the overall total number of descriptors. Target A: 50/104 = 48%, Target B: 38/78 = 49%.

3. Add the percentage of relevant descriptors to the percentage of gestalts (per target) as whole numbers. Target A: 48% + 65% or 48 + 65 = 113.

4. For Target B. 49% + 55% or 49 + 55 = 104

5. Divide this sum by 2. 113/2 = 56.5%

6. This quotient is the score for Target A. 56.5%

7. The score for Target B. 104/2 = 52%

8. This particular trial was a Pass. (The notetaker did not recall what the established threshold was for scores or how differences played into it.)

Figure 4.4

Notetaker comments: The method could have resulted in a figure over 100%. Joe was asked if the scoring was basically from experience, and he said it was.

Military CRV Method: Scoring data by categories

This system of scoring was first reported in the US Military remote viewing manual, in which the image below was published.[133] Bruce Miller developed the method during the 1980s after training was moved from SRI laboratories to the Ft. Meade Unit. Lyn Buchanan served as the unit's database manager and later reported on this system in Appendix 5 of his book, *The Seventh Sense*.[134] Buchanan has collected more than 7,000 score sheets of remote viewers using this method. Angela Thompson-Smith used the method in a CRV experiment described in IRVA's very first *Aperture* newsletter.[135] The study explored variables such as gender, age, type and length of training, the trait of absorption, and range of handedness, distance from target, time, local sidereal time, and presence or absence of solar storms

When introduced, the approach was unique in asking remote viewers to assign each perception and sketch to a pre-listed category such as color, pattern, relationship, shape, size, meaning, etc. The viewer then enters the number of correct perceptions per category on the data sheet. This information is entered manually into a computer database. This method helps project managers know which viewers are best matched for different types of projects.

According to Alexis Poquiz, who carefully reviewed this method prior to developing his scoring method (see next section), this approach requires the viewer to indicate how often each correct perception and sketch applies to one of 28 categories. Nowhere can the word itself be entered on the sheet, which can result in what he refers to as "massive data losses." Further, it involves *manual* worksheets, judging, categorization, calculations and database entry, which are very time-consuming.

Remote Viewing Scoring Sheets

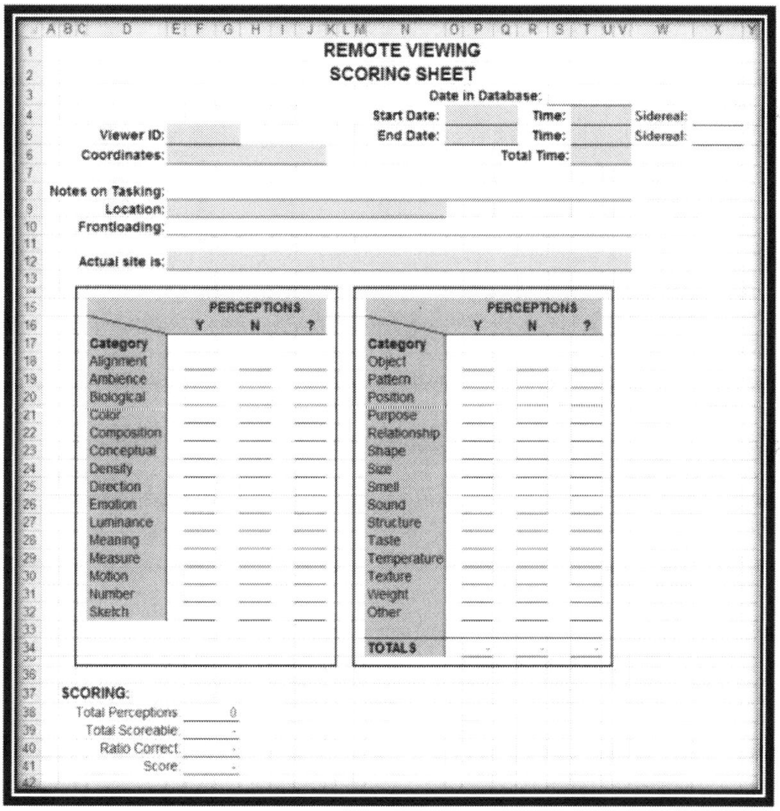

REMOTE VIEWING
SCORING SHEET

Date in Database:
Start Date: Time: Sidereal:
Viewer ID: End Date: Time: Sidereal:
Coordinates: Total Time:

Notes on Tasking:
Location:
Frontloading:

Actual site is:

Category	PERCEPTIONS Y	N	?
Alignment			
Ambience			
Biological			
Color			
Composition			
Conceptual			
Density			
Direction			
Emotion			
Luminance			
Meaning			
Measure			
Motion			
Number			
Sketch			

Category	PERCEPTIONS Y	N	?
Object			
Pattern			
Position			
Purpose			
Relationship			
Shape			
Size			
Smell			
Sound			
Structure			
Taste			
Temperature			
Texture			
Weight			
Other			
TOTALS	-	-	-

SCORING:
Total Perceptions 0
Total Scoreable -
Ratio Correct -
Score -

Figure 4.5

Remote Viewer		Date	
Target No.		StartTime	
Viewer location		EndTime	
Feedback		Lst (local Sidereal)	

Notes:

CATEGORY	DATA			AOL				DATA			AOL		
	Y	N	?	Y	N	?		Y	N	?	Y	N	?
Alignment							Position						
Shapes							Energies						
Colours							Relationship						
Smells							Composition						
Sounds							Sizes						
Taste							Mass/Density						
Texture							Dimensions						
Temperature							Structure						
Life form/s							Tangibles						
Luminescence							Emotions						
Measure							Ambience						
Movement							Intangibles						
Objects							Other						
	Y	N	?	Y	N	?		Y	N	?	Y	N	?
TOTAL 1							TOTAL 2						
COMBINED 1 & 2 TOTAL													

1. Total qty perceptions (Y+N+?)	Do not include AOLs.	
2. Total qty of no-feedback items (?)	No data points without feedback.	
3. Subtract Line 2 from Line 1.	Total scorable for this session.	
4. Total correct perceptions (Y)	As determined by feedback.	
5. Divide Line 4 by Line 3 (Y/total)	% of accurate scorable data.	
6. Multiple Line 5 by 100, add a %.	Overall 'general' session profile.	

Figure 4.6

Dung Beetle (aka Poquiz method of scoring ARV transcripts)

This memorably named system could belong to none other than ARV expert and maverick, Alexis Poquiz, founder of the largest and longest running Facebook RV group. Alexis dubbed this system "the Dung Beetle" to reflect the sometimes-maddening challenges ARV presents. He also chose the name to counter the dryness and ultra-seriousness characteristic of the formal scientific domain. However, as anticipated, once Debra and co-researchers started using the name in formal projects, peer reviewers struggled with its informality. Poquiz allowed it to be referred to as the "Poquiz method of scoring." The methodology is intended to be more refined and sensitive than the 7-point SRI scale or the military method. Every word and sketch on a transcript is entered into an Excel sheet, and scored as correct, incorrect or "questionative" for each photo option. A granular overall hit/miss ratio is then automatically calculated using Excel scripting.

When asked by the authors about his ranking scale, Poquiz[136] explained:

The Poquiz Methodology has developed into a computational approach to qualitatively and quantitatively evaluate a remote viewing session. At its very core, judging remote viewing sessions is subjective because judges may differ in their evaluation of a given perception. Arriving at a true score is not possible; we can only approximate the score of a session. The Poquiz Methodology acknowledges this subjective nature by borrowing on the concepts of variance, standard deviation, and uncertainty. Rather than providing a definitive score, it produces a base score and establishes a range that attempts to isolate the true score, between a defined minimum and maximum.

The Poquiz Methodology was first posted on the Internet, social media and at various remote viewing conferences, and then formally used in a project conducted by Katz, Beem, and Fendley.[137] Remote viewers were tasked with describing a microscopic organism, specifically a bacteriophage. Three biologists were recruited to rate the sessions using the Poquiz system of scoring.[138] This required them to individually assess every word and sketch, and then to subtract the number of correct responses from incorrect ones to derive an overall hit rate. The data sheets were inadvertently lost and about a month later the raters were asked to repeat their rating tasks. Soon after these were completed, the original data sheets were found. The researchers compared the two and found all the raters had changed some of their responses. As many as 50% of one rater's responses on the two sheets were different.

A revised Poquiz system was used in a remote viewing project designed to predict the outcome of a US presidential election.[139] While the project's sample size was too small to be statistically significant, it revealed specific challenges in judging remote viewers' perceptions.

The method is quite time-consuming since it requires either the viewers or an independent analyst to input all the impressions into the spreadsheet. One good feature is the ability to keep the scores hidden. Once the words and sketches are input, the spreadsheet can be passed to multiple judges who can remain blind to each other's ratings. The score sheet can also be easily modified to fit the needs of different projects. For example, a modified version was used by Katz, Lane and Bulgatz[140] in a project designed to test the effect of different background conditions on the perception of objects within photographic targets.

Poquiz has always made the Dung Beetle system available to all who wish to use it. He has graciously written up in-depth explanations of his system and the several iterations it has undergone. (This is presented in Chapter 11.)

Tom McNear / Marty Rosenblatt – HAG (Harsh Analysis with Gestalts)

Tom McNear, the person Ingo Swann considered his best viewer, joined Marty Rosenblatt and APP in 2019. He and Marty introduced a modification of Joe's method, which they call Harsh Analysis with Gestalts (HAG).

The gestalt accounts for 50% of the score, as with the McMoneagle method. The other half of the score is figured differently, however. The FIG (First Impression Gestalt) is assigned a value of 0 to 1.0 based on the scorer's impression of how much the dominant "gestalt" (words or graphic) matches each photo. According to Tom McNear, the summary in the transcript provides good guidance in assigning the FIG.

For an example, let's assume the transcript is given a score of .5 for Photo A, meaning there is moderate correspondence between the transcript's "gestalt" and Photo A. To get the remaining half of the score, all the elements in the transcript are numbered (for our example, the transcript has 12 elements). AOLs and words used to describe the ideograms are not included. The "APPI Analysis Review Tool" is used. If an element in the photo is significantly present in the session, it receives a score of 1. If it is present in a minor way, the score is .5. These scores are totaled; let's say the total is 6. This is divided by the total number of elements in the transcript. Example: 6/12 = .5.

The FIG (.5) and the elements (.5) are added. The total in this case is 1.0. This number is multiplied by 3.5. Why 3.5? Because 3.5 is the score used by APP over the last decade in thousands of trials as a cutoff on whether to risk money (e.g., make a bet). Using 3.5 is convenient for comparison because it put the HAG scores in the same range as all the previous trials.

Here the resulting score for Photo A is 1.0 x 3.5 = 3.5. If the score for Photo B was two points lower (1.5 or below), the tasker might consider making a bet. However, 3.5 is the minimal passing score, so confidence would not be high for this trial.

APPI Analysis Review Tool

	Name:	
	Date:	
HAG: Harsh Analysis with Gestalt	APPI/ Coord:	
	Session Time:	
	Reviewer:	

	Photo Site A	Elements	Photo Site B				
FIG_A = 0.7		1		FIG_B =			
# of A 2 x 1.0 = 2		1		# of B 1 x 1.0 = 1			
# of a 2 x 0.5 = 1	A	2	B	# of b 1 x 0.5 = 0.5			
Total Matches: 3		3		Total Matches: 1.5			
	a	4					
$(FIGA + \dfrac{\text{Num Match}}{\text{Total Factors}}) \times 3.5 = 4.55$	a	5		$(FIGB + \dfrac{\text{Num Match}}{\text{Total Factors}}) \times 3.5 = 1.05$			
		6					
		7					
HAG Confidence Ranking: 4.55		8		HAG Confidence Ranking: 1.05			
		9					
		10	b				
	A	11					
		12					
		13					
		14					
		15					
		16					
		17					
		18					
		19					
		20					
		21					
		22					
		23					
		24					
		25					
		26					
		27					
		28					
		29					
		30					
		31					
		32					
		33					

	34	
	35	
	36	
	37	
	38	
	39	
	40	
	41	
	42	
	43	
	44	
	45	

Figure 4.7 – Harsh Analysis with Gestalts

Associative Remote Viewing projects:
Assessing rater reliability and factors affecting successful predictions

Judging capability, decision-making by predictors and scoring methods are crucial factors in the success or failure of remote viewing trials. Even before Grgić and Poquiz ran their informal series of tests, this had been noted by parapsychologists such as Honorton;[141] May, Utts, Humphries, Luke, Frivold & Trask;[142] Humphries, May, Trask, and Thomson;[143] Humphries, May and Utts;[144] Jahn, Dunne and Jahn;[145] and Targ, Puthoff, and May.[146] May, et al.[147] concluded, "If multiple analysts are used, additional problems arise concerning inter-analyst reliability. If an individual analyst judges a number of responses in a series, within-analyst consistency becomes an individual problem" (p. 194).

These researchers observed inconsistencies in judging, particularly when matching session data to the target and "decoy" photos. However, it doesn't appear any of these researchers conducted a formal series of tests to demonstrate the extent of this problem, which may be a factor in why more attention is not paid by researchers and managers to these issues, as evidenced by continued recycling of the method of a single target and decoys.

As a result, Igor, Debra and T.W. Fendley, who had jointly authored the paper on project Firefly, gathered again to undertake what became known as the ARV rejudging project. They invited the esteemed Dr. Patrizio Tressoldi of the Science of Conscious Research Group, Dipartimento di Psicologia Generale, Università di Padova, Italy, to serve as their statistician. Dr. Tressoldi a few years before had heard of the extensive efforts being undertaken by the Applied Precognition Project, many of which were not being formally documented. He offered to assist the APP group managers and, true to his word, joined our project as adviser and statistician. The project won a $3,000 research grant from the Parapsychology Association's 2018 PEAR award.

The overall aim of the ARV rejudging project was to understand which factors contribute to successful or unsuccessful predictions in ARV trials. For this purpose, a sample of 86 completed ARV trials for sport or financial events was collected. The study examined factors related to generally accepted protocols in applied ARV projects, specifically the rating of remote viewing transcripts and making predictions. It also sought to test inter-rater reliability.

Three teams of independent judges operating under blind conditions – some working independently, some working as teams – repeated the judging, scoring and predicting, while keeping all other variables unchanged. Some judges used the original 7-point confidence ranking (CR) scale, while others used a 3-point CR scale. These new scores and predictions were compared to the original scores and predictions, as well as to each other. Judges were in 100% agreement in only six of 86 trials (6.9%). In seventeen trials (19.7%), eight of nine judges agreed with each other. Only five trials (5.8%) resulted in a hit for every judge, while in 10 trials (11.6%), eight of nine judges had hits.

The original judges did better than all of the new judges. Small differences between ranking scales were found. Trials using multiple remote viewers were slightly more successful than trials using single viewers. Criteria for making predictions – setting minimum scores, differentials between scores and options for passing – helped minimize misses. Given these factors, the project set forth new guidelines for ARV and other parapsychology projects.

The data indicated those with more experience judging other remote viewers' ARV sessions had higher hit rates than those who had little to no experience beyond judging their own sessions. (All judges had some remote viewing and self-judging experience.) As noted, the original judges outperformed new judges. Whether this was due to experience, or to knowing the identity of the viewers (new judges were blind to their identity) or to the fact they were the original photo selectors is unknown. It's possible that timing was a factor as original judges often only set up and judged one to two trials per week whereas some of the newer judges "binge judged," rating all the sessions in one sitting. This happened despite researchers' instructions to restrict this type of behavior to prevent rushing or burn out.

The sample sizes of the groups of judges were not large enough to make statements that could be confidently extrapolated to other studies. Given the extremely low correspondence between even experienced judges when working with sets of only two photos, this project suggests researchers in all types of projects involving photosets may do well to include a pre-trial phase with training and practice to establish some inter-rater reliability before proceeding to the formal trials. This may be especially important if a design involves a large number of photos in a set, where pairing of photos so they are orthogonally balanced becomes even more challenging.

Generally, a higher pass rate resulted in a lower miss rate. Passes are a feature unique to applied remote viewing projects. Most experimental psychology projects do not issue passes because their goal is solely to determine whether there is a psi effect. In an applied remote viewing project in which one tries to determine on which side to wager, a pass means no wager is placed. The judges who for the most part followed the criterion of making a prediction only if the spread was two or more points between each score generally did better than other judges, but only in trials with multiple viewers. About 20% more hits than misses occurred in trials with multiple viewers when judges issued a CR score of 5 or higher (using the 7-point scale).

Another variable studied was the scoring systems. The SRI 7-point scale produced only slightly more hit rates than the University of Colorado 3-point scale. This difference could have been due to researchers being initially more familiar with the 7-point scale.

Another variable was consensus team judging. This involved using two raters to work as a team. They were instructed to rate all sessions together and arrive at an agreed-upon score, through debate, if necessary. Researchers wanted to understand if working in this manner would lead to stronger results. They did not find a difference. Still, it was felt that the consensus judges may have rushed their choices, may not have debated each other as rigorously as hoped, and at least one lacked the same level of experience as the others. It was determined this is an area that warrants further exploration.

The question of whether to base predictions on scores from a single viewer or the aggregate score from groups of viewers is considered in designing applied ARV projects. As noted, overall, the use of groups with three to four viewers as opposed to a single viewer yielded more successful predictions. However, for some trials, single remote viewers outperformed all other remote viewers in the group and predictions that missed would have been successful if only their prediction had been used. In other instances, had a single viewer's prediction been used instead of all the others, the predictions would have led to misses. Some, but not all, remote viewing projects have been able to identify which viewers consistently did better or worse than others.

The authors also noted that "several of the researchers we approached prior to the commencement of this project, either no longer had access to their original data in an organized manner, or were unwilling to share it." Given the amount of time and resources that go into any project, it was recommended that if data collection methods were initially solid and initial scores were high enough to suggest that a psi effect might be present, researchers may want to reevaluate the data using a new scoring/prediction method, a new team of raters or other parameters to understand how such changes could affect results.[148]

How to improve rater reliability in ARV projects

The experiment above tested rater reliability. This is of concern within many scientific disciplines that use human observers to rate some kind of response. It has been defined as "the extent to which two or more raters (or observers, coders, examiners) agree. It addresses the issue of consistency of the implementation of a rating system…High inter-rater reliability values refer to a high degree of agreement between two examiners. Low inter-rater reliability values refer to a low degree of agreement between two examiners."[149] A variety of statistical methods have been developed to help evaluate reliability. There are two kinds of reliability – intra-rater, which has to do with consistency by the same rater, and inter-rater – consistency between two or more. While it may not be possible to achieve perfect consistency for ARV remote viewing managers and judges, some simple exercises and steps can be taken in preparation of the trials that really matter. The following suggestions could be useful with most scoring systems.

Running tests with self

From Debra: We'd like to thank Alexis Poquiz for inviting us, many years ago now, to participate in testing his Dung Beetle methodology as he developed it. He sent us a list of 10 words and a photograph via email, and asked us to indicate whether each word matched or not. He wanted us to send these to him so he could compare between viewers. I completed my word list, but then couldn't find it, so redid it. I subsequently found my original scoring and happened to notice several of my responses were not the same as before. This was quite disturbing!! Part of me agreed with my first set of responses, part with the second. In the end, I had a third set of scores different from the first two. What was up here? This is an example of a problem with intra-rater reliability. I would later learn this is not unusual.

Alexis next called an online meeting with the other judges (all of whom already had judging experience with ARV sessions). This was an eye-opener as well. As a group, we were all over the place in our responses. The discussions that ensued were quite interesting. Some of this had to do with the different ways we defined words, even though we all spoke the same language.

One word I still recall because Alexis and I got into quite a debate over it. This word was "structure." The photo was of a mountain and the viewer had the words "natural" and "structure." Alexis thought I should not receive credit for the word "structure," since he defined a structure as something manmade. I adamantly disagreed, feeling many things in nature are structural, such as rock formations, cliffs, mountains, etc. As a viewer, I often refer to these myself as "natural structures." He maintained that structure tends to be angular and more organized, unlike most things in nature. He also reminded us that Joe McMoneagle, famed military viewer, had also stated "structures" should refer to what is manmade, not natural.

While that might be a good idea, I replied, every viewer has their own perceptions of words and these are subjective. Not long before then, I lived in Sedona, Arizona, where several times a day I drove past highly structured, square- and rectangular-shaped red rock formations. I happened to be returning to Sedona to lead a workshop. While there, I took some photos to send him. One showed what looked like chiseled steps in the side of a mountain.

Alexis said he was starting to see my point, but was not quite there. We also discussed how maybe it wasn't so important for a judge to rate a session per their own definitions, but rather to understand what the viewer meant by them. Often judges are not given the opportunity to ask the viewer questions, especially in more formalized trials, where this type of communication might break formal protocols.

Sometime during this period, I had another eye-opening experience regarding this issue. Researcher Lance Beem and I were in the throes of our first IRVA Warcollier remote viewing project in which 39 remote viewers had been tasked with describing a microscopic organism, a bacteriophage. We had finally recruited three biologists/virologists to evaluate the top-scored transcripts that had already gone through a pre-selection process. (This is a bit unusual, but the aim of the project wasn't simply to prove remote viewing, it was to see if the virologists could learn from the transcripts what triggered replication.) Each scientist/rater had about a dozen transcripts to rate that had already been deemed to have strong matches to known elements. We were using a modified version of the Dung Beetle/Poquiz approach, and this required the scientists to rate every word and sketch as correct or incorrect. Volunteer assistants entered the viewers' words onto score sheets so the scientists wouldn't have to do that part.

Somehow their score sheets were lost and, much to our embarrassment, we had to ask them to redo their ratings. Soon afterward, however, we found them. We did a comparison and found extensive differences between the two. The widest differences came from our expert scientist, who had the most knowledge about phages, which he studied for his doctoral work. Since we were paying him from our very small budget, we felt justified to ask him to reconcile the differences. His third score sheet seemed to be a compilation of his first two. So here again, was an example of problems with intra-rater consistency. One has to ask: If a single person can't agree with himself/herself over a period of time, how the heck can other raters agree? What is going on here?

Probably several things. From this experiment, we concluded that it doesn't work to ask the same people to learn from information they are being asked to rate as correct or incorrect. For example, we decided to do word analysis to find out the top repeating words in all the remote viewing transcripts, which theoretically should speak to the undisclosed tasking question – the trigger for replication of the bacteriophage. We found the word "heat" was the top repeating word. However, this was not a word any of the raters could say with certainty was correct. (How could they? They didn't yet know the trigger for replication.) This is a paradox. The rater is being asked to learn something from what they don't know, while being asked to judge the sessions as correct or incorrect.

This doesn't have to be the case. For a project like this where we wish to learn something, it makes sense to avoid rating and judging altogether. When raters are asked to act as experts/authorities of what's right or wrong, they will dismiss any word that lies outside their current experience as being applicable. However, if they are only reading over the transcripts with the idea in mind, they might relate in an entirely different manner.

Is a judge an authority on words or a student of words? What does a remote viewing session reveal about human perception, language, psychology and communication, in addition to whether or not it's a good match to a photograph or some other target?

Additionally, I felt viewers came up with many words that cannot be confirmed as accurate or inaccurate. Alexis Poquiz concluded that this factor, which he called the "questionative" factor, is not a small thing. Some of the above rating scales take it into consideration more than others; in all cases, judges should have a plan for how to deal with this factor.

For example, a conservative judging approach would be to set aside impressions that can't be clearly decided. That's what Michelle Bulgatz and I did while doing consensus judging of sessions about winners of presidential elections. Our disagreement about a single word would lead to heated debate, requiring extensive online research. At some point, we realized how silly this was. That's when we set up the rule that if we argued for more than five minutes, we would disregard the word. Of course, we could have asked the viewer to explain what they meant, but that didn't occur to us, even though we had access to their emails and knew them all.

Recommendation – Practice judging the sessions!

What is the solution? While complete agreement among judges is highly unlikely, we advocate doing many more pre-trial practice sessions than are typically done for ARV projects. This is especially important before embarking on trials where the results really matter, such as those that will be wagered on or where one is attempting to establish high hit rates. This judging practice should occur at both the "macro level" (eyeballing the transcripts and choosing best matches or applying scores from one of the scales) and also at the "micro level" – evaluating every single word, as is done in the Dung Beetle/ Poquiz method and the McNear-Rosenblatt HAG method.

Raters may work individually and then come together to compare their responses (online meetings work fine with screen share). It can also be done through consensus team judging in meetings where the raters discuss/debate every word. This latter approach would be more time-consuming, but in real time one gets to learn the differing perspectives, so resolving differences may be quicker. We suggest using this procedure more than once, including while the trials of a project are ongoing or with examples from other projects.

Further, if one is concerned about preparing well for future high-stake trials, we recommend practicing with the same target pool and, if possible, with the viewers who will take part in the trial. This will help familiarize judges with the types of verbal and graphic impressions the viewers get. If a series of events is involved and you opt to replace photos used after each event, which is a viable option, the images in the pool will not be constant.[150]

Working with viewers to assist with the judging process

Project managers/judges may not realize they can get viewers to adjust what they are doing. For example, when I was a viewer in a project in which Ed May was a coder (using his CAS program), he noted that I provided way too many words, which made his job harder. I adjusted to this, even though initially I was worried it might lead me to leave out important words.

As an instructor myself, I've certainly asked my students to better organize their responses on paper and to write more neatly or type out summaries when their handwriting was not legible. I've also required viewers to provide summaries, which can be extremely valuable. Viewers can be flexible, and they often appreciate learning new ways of doing their sessions. They may need to be told to provide fewer words or more sketches, or to better define or vary their in-session movements (as when they tell themselves to move above a target or below). They may also need to be told not to use terminology that is idiosyncratic, as in some forms of CRV or its offshoots. Viewers trained with a particular instructor may not realize that someone looking at their sessions who is not familiar with CRV may not be able to tell the difference between the terms (e.g., how stages are labeled) and the actual impressions. So a notation like P2, AOL, Break or Matrix may need explanation ahead of time.

Bottom line, it is easy to build into any protocol ways for project managers to request or require viewers to make adjustments to aid in the judging process. While viewers may not be able to comply, it won't hurt to ask. For example, in one project we had 12 remote viewers conducting 30 remote viewing trials. The viewer was a highly experienced psychic detective who had worked primarily with law enforcement personnel leading her through visualizations. When I noticed her sessions were devoid of sketches, I asked my co-researcher, Michelle Bulgatz, who was managing viewer communications, to find out what we could do. The psychic felt she could not sketch but Michelle was able to train her over the phone. The detective ended up producing sketches that matched the photo quite nicely. This took time, but with 25 trials left to go on the project, it was worth it, and the psychic-turned-remote-viewer appreciated the learning experience.

Criticisms of the entire experimental set up and a few words on the statistics

Jessica Utts[151] has authored remote viewing research reports assessing the efficacy of the US governmental RV research programs. She has also served as chair of the Statistics Department at UC Irvine and is a former President of the American Statistical Association. In a 1991 paper describing the use of metastudies in parapsychological research, she summed up the state of affairs regarding statistics and parapsychology:

> Parapsychology, as this field is called, has been a source of controversy throughout its history. Strong beliefs tend to be resistant to change even in the face of data, and many people, scientists included, seem to have made up their minds on the question without examining any empirical data at all. A critic of parapsychology recently acknowledged that "the level of the debate during the past 130 years has been an embarrassment for anyone who would like to believe that scholars and scientists adhere to standards of rationality and fair play." (Hyman, 1985a, p 89). While much of the controversy has focused on poor experimental design and potential fraud, there have been attacks and defenses of the statistical methods as well, sometimes calling into question the very foundations of probability and statistical inference (p. 363).

Many psi researchers express frustration over the expectation that experimental designs must include matching tasks that will include unnecessary photos that serve no purpose except to aid in the analysis phase of the project (so the viewer's transcript can be compared with the various photos). Still, many invariably cycle back to the argument that "this is the best we have that allows for statistical analysis and rejection of the null hypothesis, so we have to just go with it." Stephan Schwartz (in personal communication with Debra) explained this is why he will no longer do such experiments.

For statisticians, the more photos there are in a set, the more confidence when a correct match is made that it was due to the quality of the remote viewing as opposed to chance. For this reason, even in ARV studies some researchers will add extra photos to be judged, not because they are all associated with different potential outcomes, but because it will impact their statistical calculations. For example, for a football game with only two teams, they might add four photos. As challenging as it can be to find two photos that are different from each other in every way, the difficulty is far greater with a set of six or more photos. The more sets that are needed, the more difficult this challenge becomes, not because the remote viewing doesn't hold up but because it's too hard to come up with photos that are "orthogonal" (extremely different) from each other in every way, especially in an experiment with many trials.

We see three main problems with methods that link the statistics to the methodological setup. As noted above, they add complex tasks (e.g., decoy photos) that if not set up correctly can derail the process, they are reductionist and they introduce noise into the system.

Reductionism in ARV trials

Perhaps the most unusual target I was ever assigned was while serving as a viewer in an RV project by Stanley Krippner, a highly esteemed researcher. During the session, for which I had received nothing more than a target number, I received and recorded one image only – that of a half man/half bug cartoon image. (The session lasted only a few minutes as I was already past the deadline for turning in my transcript.) When I saw the photos in the set, it was clear I had received a perfect match to a highly unusual image – a half man/half bug cartoon image.

What are the statistical odds that of every possible image that could be chosen of all the images in the universe, that I would have guessed the target would be a half man/half bug? I don't have any way of calculating that likelihood, but to be conservative, let's say there was a one in 500 chance. Even if it was closer to one in a million, it doesn't matter whatsoever in the statistical calculations. If a judge scores it a 7 on a 7-point scale or a 3 on a 3-point scale, that simply helps the rater/project manager issue a prediction or determination as to whether I was describing the target or another photo in the pool. For statistical calculations, this is a single hit, and a single hit equals one point only.

Likewise, let's say for another trial I hardly have any matching data and have a miss. This also equals one point. If all we have are these two trials, my hit rate is 50%, which is at chance level and translated as "no psi present" by statisticians. Even if we have 1,000 trials and for 500 of them I and other viewers have fantastic correspondence to the target leading to hits, but on the other half our sessions are off track and lead to misses, then this is still not statistically significant and therefore all the statisticians of the world will state based on these numbers "no psi was present." The project was a failure, remote viewing doesn't work. Case closed. I say, "No, this stuff is real!"

Parapsychology in the US blossomed under a forced-choice protocol involving card "guessing" tasks.[152,153] The idea was that in guessing quickly, the subconscious would kick in and access the information. It made sense that each one of these trials would count as 1. To assess if psi was present required many more hits than misses. Remote viewing changed all that. Now there are many more than five choices, although the possibilities are not endless. Most targets are not as unique as a half man/half bug in cartoon form.

In fact, after one has been assigned hundreds of targets, particularly location targets, one realizes there is actually less diversity than one might think, at least in the overall gestalts of targets. For example, water is a very common feature in a potential remote viewing target. If I was assigned 10 location targets, water would likely be in at least one of these and probably more. Same thing with people or vehicles. It is true that some matches could be considered guesses or coincidences, at least to an observer. But tell this to someone who wrote down the word "water" after having a sensation of liquid being splashed onto him, or smelling ocean salt air, or feeling the motion of waves, or sensing sunlight glaring off what seemed to be a water's surface. Tell them water was just a "lucky" guess and see what kind of reaction you get. (I won't, but you can!). Their reaction is due to somatic experiences during their session. Something notable happened in their mental and bodily processes, such as a vision, a feeling, a sensation – possibly all three at once – that conveyed "water." This would be like asking someone to do a very complex math problem and when they come up with the answer, you say, "Hey, that was just a guess." No – they went through a productive process in their mind and body. Remote viewing isn't "guessing."

Still, not every word that emerges in an RV session is going to produce or accompany a notable experience or sensation. Many RV sessions are mediocre. This is why, as noted earlier in this chapter, it's very useful to employ a rating scale along with matching tasks. Sometimes a remote viewing transcript will be given more credit than it actually deserves. Underestimated sessions and overestimated trial results balance each other out in the statistics. Still, it's very frustrating to most everyone directly involved. It's as if someone from the outside has a rule book and is not only applying the rules, but gets to make assertions about what the rules mean for the subject overall.

This is what is meant by reductionist (I'm not speaking of materialism here). It's hiding what has really taken place on many levels; it hides the quality (or lack thereof) of the remote viewing, and it hides all that happened in the trial.

Professor Courtney Brown, after decades of intensive RV investigations, made no bones about dismissing this entire matching task design in the introduction to his book, *Remote Viewing, The Science and Theory of Nonphysical Perception:*[154]

Another (and more important) example of how a commonly applied element of the scientific method conflicts with characteristics of the remote-viewing phenomenon itself is with what has become a standard experimental setup for many researchers. In this set-up, comparisons are made between remote-viewing data and a set of potential targets, one real and the others decoys. The setup normally uses five potential targets, and judges are used to make the comparison between the possible targets and the remote-viewing data to see how well the remote viewer describes the real target...

Much of this volume is dedicated to explaining a nearly obsessive series of experiments that we ran at The Farsight Institute to understand what is in fact going on with this procedure, and our conclusion finds that the procedure itself deeply conflicts with the psychic targeting process of remote viewing, leading to the near total corruption of the data type and gathering process. The reason underlying this conclusion is not simple, and I ask readers to hold off on their judgment of these statements until they have read the remaining chapters in this volume. Our initial suspicions regarding the "pick the correct target out of the bunch" idea came because of our own personal experiences with this procedure. Certain phenomena occur when this procedure is used, and the phenomena are so repeatable that we concluded that the fault was not with our remote-viewing capabilities, but rather with the experimental setup. Speaking from a personal level, I am a scientist who has invested years learning how to remote view, and repeated remote-viewing experiences have taught me to trust that these experiences are real. Something was happening with the "pick the correct target out of a bunch" experimental setup that made my own remote-viewing experiences experience go awry, and it was because of my long experience as a remote viewer that I decided to question how elements of this routinely used experimental design might influence the remote-viewing experience in an unexpected fashion (p. 14–15).

In terms of "something" that was happening and going "awry," we explore every aspect of this in our next two chapters, which are on displaced psi – Displacement.

CHAPTER 5

Displacement: Its Nature and History

Even if one regards displacement as a curse, an understanding of its functioning should still be welcome, since it seems likely that displacement can be better prevented if its causes are known. In addition, like any nuisance, displacement may be an interesting phenomenon in its own right, and may provide some insights into the workings of psi. —Julie Milton[155]

In this chapter, we will examine one of ARV's and perhaps even parapsychology's greatest nemesis – displacement. We will offer examples of the problem, which afflicts both applied and formal experimental projects. In the chapter after this, we will examine theories of displacement, offer ways to minimize it and suggest how displacement may offer insights into the very origins of psi itself.

Displacement takes place when a psi percipient who is attempting to obtain information about a target or subject matter instead accesses information that is spatially or temporally removed from the designated target. They are still using their psychic perceptual abilities but toward something other than what they were assigned. This can occur in varying ways and degrees.

Sometimes a viewer will totally displace to something other than the target. At other times, the viewer will obtain information about the target but also about something else. Since it is widely believed in the field, based on tens of thousands of examples, that viewers sometimes miss the target completely, it should not be assumed they missed it due to displacement, as we will explain.

As we have noted earlier, in traditional binary ARV there are two possible outcomes, each associated with one of two photos (although real objects, tastes, colors and smells have also been used). Let's take a fictional example, but one that is illustrative of a far-too-common situation in ARV trials.

Let's say the project manager or a tasker has chosen the photo of the Taj Mahal to be associated with the Boston Celtics winning a game, and she associates a photo of the Queen Mary docked in harbor with the Miami Heat winning. These are arbitrary associations of each team with one object. The goal is for the remote viewers to use their psi abilities to describe only one of the photos – the one that will be associated with the winning team. The viewer will see the photo at feedback time – the Taj Mahal if Boston wins, Queen Mary if Miami wins. This setup allows for a prediction to be made in advance of the event.

Let's say everything is going beautifully: the remote viewer noted she had a vision of a large ocean liner, with a woman's name on it. The liner is positioned in water, but not moving. People are eating at fancy tables. The viewer is reminded of a vacation she took to a sunny, warm place and has scribbled the words "California" and "haunted" on her pages. The drawings and words are not at all a match for the Taj Mahal. Wow! With no wrong data, the judge gives a score of 7 for the Queen Mary photo on the 0-7 SRI scale, and a score of 1 for the Taj Mahal. It seems very clear which one the viewer saw.

Everyone involved in the project gets excited and wagers on the Heat. The game is played and ... what? No way?!!! The Heat lose? Are you sure? Yep, the score was close, but the big screens in the casino don't lie (well most of the time – we were at one ARV conference where they did). Still, in this scenario, yep, the Celtics won and the Heat lost.

But how could this be? The viewer will not see the photo of the Queen Mary. In this setup, the viewers did not self-judge and the ARV protocol allows for just one photo to be shown to the viewers - the photo associated with the winning outcome, which is the Taj Mahal. Something clearly has gone wrong. Not only are the project manager and friends who wagered on the Heat out the money, but the viewer is very disappointed when she sees the feedback photo. She knows it has nothing to do with the target she saw and described. There is not even any water. She was very off! So what happened? Dastardly displacement!

Some practitioners say displacement is most commonly observed when self-judging is involved, since in comparing transcripts to both photos what could be considered premature or "false feedback" is possible. Unfortunately, displacement can and frequently does occur even when the viewer is not exposed to the "judging photos" (shorthand term for the photos to be judged) and an independent judge has looked at the photos and scored the transcript.

How can this happen? While theories will be offered later in this chapter, let us introduce a few ideas here.

Many remote viewers, especially those new to ARV, feel a sense of pride if they describe either photo well. They are still in the process of growing their wings, so to speak, and they appear to have flown. Such evidence of psi for them is a morale booster, especially if it's a very close match. One explanation is that to experience success with psi, a remote viewer will default to the easier photo to describe rather than the actual feedback photo.

Another suggested explanation is that displacement may happen if one target emotionally stimulates the tasker or viewer more than another. Another conjecture is the protocol design opens the door for displacement to the alternative target when there is not a single target. (This has been Jon's perspective for a long time and led to his preference for Strict Unitary ARV, which is discussed in Chapter 14.)

Whatever the reason, doing a great job describing the wrong photo is highly problematic to the project overall, because it leads whoever is judging to assign a high score to the photo. That can lead to an incorrect prediction, which lowers the success rate and can lead to losing money, as well. Displacing to the wrong photo or option is therefore much worse than simply failing to get correct or enough data, because when the latter happens and a low score is issued for both photos, it is clear a prediction should not be made. Correctly describing aspects of both photos creates a dilemma for the judge, who may wisely decide to pass (make no bet or trade).

Forms of displacement

There are several kinds of displacement. "Backward displacement" takes place when the potential target cognized precedes the intended target by one, two or more steps (designated as -1, -2, etc.). "Forward displacement" takes place when the data matches a target that takes place later in the series than the intended target by one, two, or more steps (designated as +1, +2, etc.).[156]

Expanding on the above types (Forward and Backward), we note a further division into two types of displacement:

1. **Out of sequence (aka "displacement in time")** – When a viewer is doing a series of trials, whether all on the same day or spaced out over days or weeks, the participant may view a target that is presented before or after the designated target. The participant views a past or future target rather than the present one. Displacement often occurs to the very next trial in a series, but it can also be to a trial much later, as well. This form of displacement has been observed since the earliest psi experiments.

2. **Single trial displacement: displacement to an alternative potential target provided in the trial** – While many ARV trials are binary (two options), some projects have three or more options. In experimental parapsychology projects in which the aim is to show whether there is a statistical effect, it is possible to establish above-chance levels with fewer trials when more than two options are judged. When offering multiple alternatives (usually photos), it is hard to find

ones sufficiently different from each other so a judge can reliably determine which photo the viewer was viewing. This increases the potential for displacement since there are more photos to displace to. The procedure with one target and one or more alternatives or decoys has been employed in many psi experiments and in numerous attempts to use ARV to predict outcomes of games, lotteries, elections and horse races. In these types of projects, the alternatives to the designated target have been dubbed decoys, as we discussed previously. These are also called "control photos" or (Debra's term) "judging photos."

As an example, when I (Debra) first began doing ARV, I started with a group that was predicting the outcome of horse races. I really wanted to do a good job describing my feedback photo so I spent a long time on the session, producing about a dozen pages of data and sketches. For this project, I was required to self-judge. I used Marty Rosenblatt's computer program, which required me to log in, receive the target number, upload a transcript and then click on a button that showed the photos (eight in this instance), allowing me to scroll through each one in order to choose the best match.

I was very pleased with the first photo because of strong matches between my words and sketches and the photo, but I also had a lot of data that didn't match. As I continued to look through the rest of the photos, I couldn't believe what I saw. Not only did I have matches of words and sketches in every single photo option, but in a few cases, I had named the main object or feature in the photo.

From this and subsequent experiences, I've begun to suspect that staying in session for a long time may not be the best practice for ARV trials. Once enough data is provided, the subconscious may get bored and decide to move on to something else. I am also beginning to have an awareness when something has switched and I may be describing another photo. At a certain point, the information seems to not only look and sound different, but perhaps more significantly, it has a different feel. When I have this sense, I never fail to see matches to more than one photo when engaging in self-judging. I now only self-judge if I'm not concerned about the results such as when practicing what I call "speed-viewing" with online apps like RV Tournament.

Suggested causes of displacement

Let us return to several possible causes of displacement.

Displacement from the environment. A common example of this type of displacement is when a remote viewer describes a photo that was emailed to them or that they happened to see online immediately before they looked at their feedback photo.

Interpersonal Displacement. This happens when a viewer displaces to something related to another person involved in the trial, such as a judge, manager, experimenter or another participant. Often the viewer will describe something regarding the person tasking them. The viewer may be too focused on that person, or that person too focused on the viewer, leading to what has been called "telepathic overlay." An intermingling of thoughts shows up in the session.

Recently Debra chatted with former SRI researcher Russell Targ. He gave an example of displacement that occurred at a workshop he was teaching. He had taken several objects from home as potential targets, but chose only one. His daughter, Elisabeth, participated in the exercise and was disappointed to see her transcript contained no matches to the chosen target. Instead, she had drawn a pair of glasses with a crack in them. What had happened didn't even occur to Russell until he arrived back home and began emptying his pockets of the potential targets. One item was the pair of glasses with a cracked lens he had worn several decades ago. They had been hiding in the back of his desk drawer for years until he had pulled them out for this demonstration. He realized how much sense this made. Even though they weren't what he ultimately chose as the target, as a baby, Elisabeth had seen the glasses on his face hundreds of times and had touched them with her little hands. Why wouldn't she connect with these instead of a meaningless target? Also, he was personally connected to the glasses and did not have such a connection to any of the other potential targets. This could be considered an example of both interpersonal displacement and displacement to another item in a set.

Example of interpersonal displacement

Years ago, I (Debra) was managing a complex remote viewing project involving 39 remote viewers who were tuning in to a microscopic organism. One morning a friend emailed a photograph of the Eye of Horus – an Egyptian symbol she had a dream about and wanted help interpreting. I had the image open on my desktop and was closely studying it when an email came in from one of the viewers (whom I did not know well). The transcript in the email contained an image very strongly suggestive of the Eye of Horus! I wrote her back and asked if during the session she felt she had been particularly focused on me for any reason (a strange, but necessary question). The viewer admitted she had been wondering what I was like and whether I would approve of her work. In this case, it was clear the viewer was focused on me and not the other way around as I didn't know her and had no idea she was about to email me her transcript.

Figure 5.1 – Pictures that were on Debra's Desktop

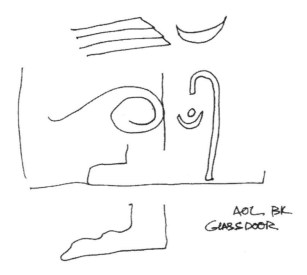

Figure 5.2 –Remote Viewer's Transcript

Interpersonal displacement example from bacteriophage project: Bottom: Viewer's sketch. Top: Page that was open on Debra's desktop when viewer sent the session.

Another example from Debra:

This type of displacement happened in a remote viewing webinar I was teaching from home. The students had 10 minutes to finish their session. The undisclosed target was a location showing elephants frolicking in the mud at a park in Kenya. While the students were working, I muted the audio (no one was on camera for the duration of the class) and my significant other, Danny, surprised me with an enormous plate of fruit slices and a fork. Usually, I'd tell him this wasn't a good idea – for the reason I'm about to share – but I couldn't turn him away after he had gone to all that trouble. We finished most of the snack just in time to return to the class, and he removed the plate. As usual, the students were very excited to discover what the target was.

Many did quite well, but one expressed disappointment. She was really off and didn't understand what had happened because her imagery had been so clear. I asked her to describe her impressions. She finally blurted out, "I wasn't getting anything! The only thing I kept seeing was two people with forks eating a large plate of fruit!" Again, I feel the viewer was more focused on me than I was on her. Since I had several other students, I wasn't focused on any particular one. Still, I had an affinity for this student, having worked with her for a while. Also, because she felt anxious about publicly sharing her work, the student had gotten into the habit of emailing her transcripts to me instead of uploading them to the online class page. The previous classes this student took with me had to do with using clairvoyance to read and heal people.

Later, in Chapter 17 on ARV and Dreaming, we present examples in which dreamers displaced to something that occurred the following day instead of the intended feedback photo or to something very specific occurring in the project manager's life.

Displacement vs. confirmation bias

Many factors can cause poor results in ARV or psi-based trials. Sometimes the viewer just never makes it to the target. They may describe some unrelated thing or the contents of their own musings. They may distort or misinterpret impressions, which snowball into all sorts of images and scenarios, or they may be so vague it's hard to know what is going on. It's really important to not make more of a session than is there.

Unfortunately, as psychologists and skeptics point out, some psi participants who don't do well propose explanations to support the notion that they are psychic. They look for matches outside the target. Viewers and even judges or managers have done this many times. One could write an entire chapter on this alone; however, since the skeptics have done the job, we will be brief.[157]

In most cases, no one is trying to lie or deceive. Their egos are simply getting the best of them in their determination to be right. This type of "fishing for confirmation" has been referred to as "confirmation bias."[158]

I (Debra) have seen students engage in confirmation bias plenty of times. For example, in the football prediction described previously, imagine a viewer saw the image of a pigeon and nothing else. Upon receiving the Taj Mahal feedback photo, they do not see any pigeons in the photo. They do their own online research, googling the terms "Taj Mahal" and "pigeons." Eventually they find a photo of one bird – not even a pigeon – flying above the Taj Mahal. They say, "Oh look, I was right, there are birds there!" Although true, a bird is so insignificant to the target we could not say they were correct. This could happen with a sketch, as well – a sketch might be a circle with a slight protrusion, looking vaguely like a bird. The viewer says this must be a bird nesting on the roof of the Taj Mahal since that's what birds are known to do. It's too much of a stretch.

How can we distinguish between displacement due to psi and displacement due to another cause?

Even when we suspect displacement due to psi has occurred, we can't say with certainty an element of coincidence wasn't involved. For example, a viewer writes "there is a dog" in their transcript. Perhaps while they were completing the session, a dog was barking loudly and that noise made it into the session. It just happened to coincide with a dog being in the feedback photo or another photo in the set.

One might look at the following three factors as pointing to a higher likelihood that it is psi displacement as opposed to another explanation:

1. When a viewer has a very detailed match to a photo in the set, perhaps even naming one or more central or unique elements, and thus a high score is assigned to the photo.
2. When the match is close in time or in a trial sequence.
3. When a matching element is unique or highly specific. For example, it's one thing to say there is an animal in the photo when there is a dog in the non-actualizing photo – that may or may not represent psi displacement. It's another thing to say there is a dog, and it's positioned to the far left of the paper, situated above a yellow triangle, with its face pointing to the right, and this matches the non-actualizing photo in the same trial or the content of the photo in the very next trial. In the latter case, we'd be more open to the idea that the viewer perceived the wrong photo on an intuitive level.

Here is an example from a recent study that used objects set in different backgrounds.[159] After the study, researchers were startled when they noticed Chris Georges' session in trial #6, which contained the words "Bellows-like, Expanding and Contracting." The actual target for that trial was the photo of a house, but about ten targets later, the target was an accordion. The sketch and photo are undeniably extremely close. But is the sketch of an accordion distinctive enough (especially since a piano was the target in another trial) to say for sure this was an example of psi displacement? (Debra says "Maybe, maybe not," and Jon says "Definitely!")

Figure 5.3 – Remote Viewer's Sketch (Chris Georges)

Figure 5.4 – Target Photo in Later Trial

Why does it matter if something went wrong due to psi displacement or for some other reason? It doesn't make any difference regarding the outcome of an ARV trial. It is about understanding what may have happened in order to mitigate the effect in the future and ultimately to understand more about psi itself. That being said, some have suggested that while displacement is a real thing, it's not something people should spend too much time thinking about. Our friend Marty Rosenblatt held this position when we both were active in APP. His view at the time was that whatever you focus on, you create or enhance. Instead of trying to understand displacement, he wanted viewers to move on and focus on getting better by setting it aside. His view was, and to some extent still is, "It's all psychological."

Others suggest those who try to ignore displacement still encounter it. If the issue is not addressed, a viewer may become discouraged, not understanding what is happening. They may think they are somehow the only one, when displacement is something all viewers in ARV projects encounter. We can pretty much guarantee a new viewer will notice displacement, especially when self-judging, within a couple months of beginning ARV trials, and often sooner than that. The person may be quite perplexed by this while others are saying, "Geez, how many more times do we have to address this?" That is one reason we felt the need to write this chapter.

Will working on psychological factors be enough to lessen or resolve the issue or are adjustments to the methodological design required? These factors could include composition of the target pool, the pairing of photos, the extent of "blindness" in the protocol for all participants, reducing feedback loops, and making use of the extensive research that has been done on best ARV practices. These points are further explored in the next chapter, as well as in Chapter 9, which discusses Ed May's Computer Assisted Scoring and Julia Mossbridge's Positive Precognition protocol.

Let us now hark back to the early days of psi research to see what was going on and then move forward to present-day research and informal trials.

Displacement a century ago

As far back as 1884, French physiologist Charles Richet published an article on "mental suggestion" in the *Revue philosophique de la France et de l'étranger*, which according to Carlos Alvarado[160] of the Parapsychology Foundation "is an early classic of experimental parapsychology" (p. 543). Richet noted features of mental telepathy, finding it was "very capricious, wandering, uncertain." It manifested "in different degrees with different individuals" (p. 616). Richet referred both to target displacement and declines in the subject's performance. The displacements seemed to be related to "consistent confusions of one target for another" (p. 546).

In 1925, Dr. Carl Bruck[161] provided one of the first detailed descriptions of displacement, remarking upon it in experiments he performed with hypnotized subjects.

His work was published in German in 1925 and was not known at the time to English-speaking researchers. The targets were drawings and photographs and the set up was as follows:

> These experiments were done simultaneously with R. and Z., using the drawing of a *ladder* as original. The two subjects were placed about ten feet apart and it was not possible that they could have seen each other's drawings. They were hypnotized simultaneously and were instructed to aim at the original (unknown to both of them, and to the witnesses of the experiment) which had been placed in the portfolio. These simultaneous attempts succeeded extremely well under conditions which forestall all objections.[162]

The narrative by Dr. Bruck continues:

> At the start of the experiment [taking place on September 11, 1922, with R. in hypnosis], one drawing was lying on top of the portfolio and the others [three of them] were contained within it. I had no special methodological purpose in doing this, but to this procedure a surprising and theoretically significant result is due. The picture lying on the portfolio represented a medicine bottle with a stopper, and I looked at it fixedly from time to time during the experiment.
>
> R. now proceeded to draw what was, for me at least, knowing the original, a completely correct response, namely, the upper rim of the bottleneck with the stopper in place; as can be seen in the reproductions of the original and the copy, he even sketched in the shading that appears on the right side of the stopper. Then he ceased drawing and let the pencil drop from his hand as an indication that he had done all he could. Since I, however, had to regard the experiment as incomplete, I insisted that R. should continue to draw if anything further came to him. I admit this is a questionable procedure which, since the bottle had failed to appear in complete form, might end in a suggestive misleading of the subject into a spurious result. To my surprise, however, R. did not follow this lead; he left the first picture unfinished and beneath the neck of the bottle drew a completely new picture. With energetic strokes, indicating a new attitude to the task, he drew, beginning at the top, a sketch of a streetlamp. A drawing of a streetlamp was in fact the second of the three other drawings which lay in the portfolio. They were entirely out of sight of all those present, including several medical men who had come to witness the experiments…

The experiment yields, so far as we know, a novel result. Instead of the *postponing* of a correct telepathic impression, it presents, under the guise of an apparent partial failure, the solution of a task which had been arranged for a *later* experiment.

In other experiments the subject drew material which had been in a preceding experiment which the subject was unaware of.

A few years later in his seminal book *Mental Radio*,[163] Upton Sinclair termed the same effect "anticipations," referring to out-of-sequence drawings done by his wife Mary Craig:

A displacement effect was observed, where Craig's reproduction attempt failed to match the target drawing but was strikingly like another one in the same batch. Attempting to read the first of a series of eight drawings given to her in sealed envelopes, she described it as 'some sort of grinning monster – see only the face and a vague idea of deformed neck and shoulders…the face of the creature is broad and weird'. This did not at all match the target drawing, a leg wearing a roller skate, but was a more-or-less exact description of the seventh drawing in the batch: In this instance, six of the eight drawings had been made by Upton's secretary, while the other two (without her knowledge) had been made by the secretary's brother-in-law, who had happened to be visiting, and were the two in question, suggesting some kind of personal influence. In another instance, Upton's drawing of an elevated railway upside down was represented by Craig as a steamboat. These clearly failed to match, but she repeated elements of the same image for Upton's next drawing in the series, which this time was actually of a steamboat.

Whately Carington

The person who named the phenomenon "displacement" was British researcher Whately Carington. Carington was badly injured landing his Royal Air Force plane during World War I and for the rest of his life devoted himself to studying psychics. He conducted extensive research but he was a shy and modest man. "His services to psychical research have never been fully recognized."[164] Indeed, Carington is basically unknown to the world of practical remote viewing. This reflects the unfortunate separation of remote viewers who are undertaking practical work, sometimes with clients, from parapsychological researchers who are primarily observing and testing the phenomenon and constructing explanatory models.

Carington first observed displacement in the following circumstances. He would post a drawing on the door of his study from 7:00 p.m. to 9:30 a.m. the next morning. Percipients

(251 of them) were asked to draw what they thought was posted during those hours on a particular day. Most of the percipients lived far away and none saw the posted drawings. The percipients were drawing the targets "blind." Carington used random numbers from tables and then a dictionary to select the targets, choosing the first drawable word based on the random numbers. He or his wife would then draw the target.

Over the course of numerous trials, he found matches between the target and the percipient's drawings seldom occurred. Eventually he began to wonder if telepathy happened on its own schedule, not on a linear one: "To put it another way, it became clear to me that a 'hit' might be *displaced* to some extent from what one would commonly regard as its natural position in the series of trials."[165]

He observed that drawings were hits on the night after or the night before the designated target drawing. This was the same phenomenon that Dr. Bruck had noticed. As another researcher put it:

> In other words, there was time displacement which might be either forwards or backwards. As a result of this displacement, the scoring card took on the 'scatter' appearance to be seen on a target around the central bull's eye, but in this case it was a 'scatter' in *time* and not in *space*.[166]

Carington noted:[167]

> The other point that bears mentioning is the remarkable tendency we noticed for the best and most convincing resemblances to occur on the wrong occurrences. We made it a practice to grade our resemblances as "a," "b," or "y" according to the degree of confidence we had in them, which very much approximately corresponded to the degree of resemblance discernable, and the table below shows the number of a's given in each of the three experiments, and how many were displaced early, late, or not at all.

Carington Table 3
Displacement

	Early	Zero	Late	Total
Expt. I	9	2	28	39
Expt. II	11	3	17	31
Expt. III	9	1	9	19
Total	29	6	54	89

Figure 5.5

Of Table 3, Carington[168] wrote:

We should of course expect about 9 zero displacements out of 89 awards if chance only were operative, and the difference is not significant. On the other hand, the excess of Late over Early displacements is quite definitely so, both for the first experiment (P < .01) and for the Totals (P < .02).

Carington went on to conduct extensive statistical tests to see whether chance alone could account for displacement effects. While they could account for some, the results were statistically way beyond chance.

It was because of these experiences that Carington, too, disapproved of the popular parapsychological design of matching participants' sketches/verbal impressions to decoy photos:

On the whole I think there is very little to be said about forced matching for this purpose, or any other. Apart from the fact it is liable to be completely wrecked from the phenomenon of displacement, it does not seem to yield any information that would not be given by a suitable system of marking, while it automatically precludes the possibility of giving recognition to the influence of more than the original on the same drawing, and is hopeless with dealing with multiple or composite drawings (p. 74).

Carington suggested to fellow researcher S.G. Soal, who also had few same-day results, that Soal re-examine his data to see if time displacement occurred. Soal did so and acknowledged that indeed the odds jumped to many millions to one. According to Milton, Soal's 1940 paper was influential in establishing that displacement exists.[169]

The Rhines

Pioneer researcher J.B. Rhine confirmed the existence of the effect:

> Unconsciousness, too, has produced an effect called displacement (the term introduced by the late Whately Carington), the tendency of the subject to hit the targets before and after the one at which he is aiming.[170]

Rhine noted that percipients would correctly predict the next Zener card in a long series of successive tries. Rhine went on to describe yet another sort of displacement:

> Recently at the Duke Laboratory still another effect of unconsciousness has been encountered by Cadoret and Pratt. They found that some subjects consistently mistake one symbol for another and produce a reliable type of missing even while they are consistently hitting certain other symbols.
>
> Using mechanical aid in carrying out extensive analysis, Pratt and Foster of the Duke Laboratory have recently found that the subject's displacement and consistent missing combined may produce a highly complex pattern of significant effects. Among other curious results they find that the subject displaces differently before and after hits (i.e., successful calls) than before and after misses. Working with very long series as they are, the findings are based on data that possess a high order of significance. This systematic wandering up and down the run of target cards and off on to the wrong symbol is quite understandable as a product of the introspective blindness that seems to be a universal characteristic of the process. Presumably if any clear introspective awareness of the occurrence of ESP were to be achieved, these peculiar effects would be eliminated and a high order of conscious control, as well as utilization, of the process would be possible. The range of possible consequences of such utilization would be very great indeed.[171]

In 1953 Louise Rhine turned away from her husband's fixation on forced-choice "guessing tasks" and explained her rationale[172,173] as follows:

> Over the years research with psi ability, though it made progress, has been slow and difficult. Over the years, subjects in tests have seldom made perfect scores, and such frustrating effects such as displacement and psi missing have frequently been encountered. For these reasons, experimenters have come to expect…a few points over chance expectations…in laboratory tests, the response is either a

hit or a miss...the opportunity of experimenters to get ideas of the way ESP is mediated into consciousness has thus been limited (p. 90).

She did not leave the field of parapsychology. Instead, she embraced qualitative research, specifically what could be considered a "case collective approach." She went on to collect survey responses from more than 7,000 spontaneous experiencers who reported intuitive impressions that she felt revealed more in-depth psi-based and psychological processes involving movement from the unconscious to the conscious.[174,175]

Displacement – Yesteryear (c. 1980–2000)
Tart and Hastings

In a report to the Parapsychological Research Group, C. Tart and A. Hastings state:

> We wish to report an apparent displacement of ESP effect in a remote viewing study from the physical characteristics of the remote site to the psychological processes of the agent (traveler). Such displacement is of methodological significance in studying this phenomenon.

They explain that Puthoff and Targ[176] created the basic remote viewing paradigm, in which an agent travels to a randomly selected target site while a percipient (viewer) attempts to describe what the target site looks like in the absence of sensory cues or previous knowledge about the site. These authors reported excellent results in a large number of experiments. Hastings and Hurt[177] adapted this basic procedure to a group-viewing situation with very good results. Twenty of the thirty-six group members correctly chose the remote site from six possible alternatives, a result that would occur by chance less than one in ten million times.

Tart and Hastings modified this protocol during a weekend workshop in 1976. In one experiment, the "outbounders" were given several preselected numbered envelopes containing a photo of a location they could potentially visit that weekend. Using a random number generator, they chose one envelope, but soon discovered this location was not open or accessible, meaning they would have to choose another. One researcher was very upset about not being able to visit the first location. When they learned the second location, another researcher became upset about not getting to go to another choice that was in one of the preselected envelopes. After the viewers described the researchers' location, analysis showed the viewers did not do well compared to the stellar results they had demonstrated in other tasks that weekend. The transcripts were judged to match the researchers' desired locations and not the actual target.

Tart and Hastings wrote: "This apparent displacement to the psychological processes of the principal agent is, of course, a post-hoc finding, and there may be alternative explanations for our results. The viewers might have been astonished enough with their success on the previous slide experiment to be frightened and so have their ESP turned off, there might be inherent differences in doing remote viewing versus a slide GESP test, or fatigue might have set in. We are inclined to report the displacement explanation given its ubiquity in other types of ESP experiments, and the primary purpose of this report is to alert others with the remote viewing procedures to it" (p. 5).

Tart[178] explored this topic further in a paper titled, "Are we interested in making Psi function strongly and reliably?" According to researcher Julie Milton, this paper reinvigorated interest in the topic of displacement. She felt the study did not control well for certain effects, but it prompted a paper by Crandall and Hite[179] "in which the relationship between target and displaced scoring has been handled more carefully."[180]

Milton's dissertation on displacement

The Rhines' work formed the basis of a dissertation entirely devoted to displacement, Julie Milton's *Displacement Effects, Role of the Agent, and Mentation Categories in Relation to ESP Performance.*[181] Her thesis covered primarily forced-choice studies from the 1930s to 1985, although she did mention a few free-response studies.

Milton justified the need for her project by stating

Although Carpenter (1977) and Palmer (1978) included discussions of displacement in various sections of reviews of various aspects of ESP, no detailed, fully comprehensive review of displacement literature had ever appeared, despite the fact that research on displacement spans the last half-century, and that over a hundred papers have been published which deal with some aspect of displacement (p. 35).

Milton also noted she was unlike past researchers, who viewed displacement as a subject of "curiosity." Most seemed to take a "pest control approach" to displacement, meaning they regarded it as a "nuisance rather than as a phenomenon of interest."[182]

This assessment was based on Milton's finding that twice as many papers included an "examination of displacement in the 1940s and '50s as in the following two decades." She felt this could be due to switching from forced-choice (such as Zener – ESP card guessing) experiments to open-response designs (Ganzfeld, dream and remote viewing experiments).

To even attempt to evaluate whether displacement was taking place on a statistical level requires very time-consuming efforts to compare every transcript or "mentation" (what the psi participant mentioned) with every photo option within a trial or between trials. This is not usually feasible.

Milton's examination of more than 60 studies that investigated displacement concluded with a rather skeptical statement. She opined that her research did not find a strong statistical basis for the realty of a displacement effect, calling into question whether or not it actually existed. Still, she felt this lack of statistical evidence could either be because displacement is less common than suggested by the numerous researchers who have independently observed it or that there have not yet been sufficient experimental procedures established to test for it.

In her overall conclusion, she criticized studies for "the post-hoc nature," which she asserted could lead to the danger of seemingly positive evidence of displacement being merely the result of analyzing only "chance fluctuations in scoring noticeable enough to yield a significant effect" and "the apparent tendency to start analyzing for displacement only after significant effects have failed to show up on the intended target." It is possible researchers have tested for displacement, failed to find it, but then never mentioned that in their write ups, thus making it seem as if the problem was greater than it was. She also noted "the lack of consistency among the results of studies examining the relationship between displacement and other variables could be taken as an indication that there is no real effect for other variables to relate to, although there are other possible explanations for this absence of solid findings" (p. 138).

Still, some relevant findings came out of her assessment. She pointed out that given Rogo's observation that displacement seemed to occur at the very end of the session, if the subjects began their trials well, and only began to lose interest after some time in the ganzfeld, then they might be expected to begin by hitting the target and progress to the controls (decoys) toward the end of the session. It would be easy for future researchers to incorporate an analysis for this effect once they had analyzed for displacement, and some interesting results might be found.

Further, as part of this same project Milton conducted experiments of her own, looking at effects for "the role of the agent; performance on specific mentation categories; the effects on correspondence judgement of picture preference; the application of information theory to ESP; and displacement." While she didn't feel there was strong evidence of psi across all experiments, she did find "a post-hoc result in Experiment One, that percipients' correspondence judgments seemed to be heavily influenced by their liking for the pictures (p. (one-tailed)" and that "the judges had been influenced by a pattern of preference similar to that of the percipients" (p. 241).

an indication that displacement may have occurred on those trials on which strong correspondences occurred to two or more pictures in the target set; scoring was lower ($0.10 > p$ (one-tailed) > 0.05) on those trials on which the judge suspected that displacement had occurred than on the remaining trials, with slightly below-chance scoring on the 'displacement' trials, as would be expected if displacement was not accompanied by some target-related scoring. If this result really was an indication that displacement occurred on these trials, rather than a weak statistical fluctuation, the apparent effect size again is quite strong and further research may prove useful. It should be borne in mind, however, that even a significant result would not have meant that displacement only occurs with displacement to two or more pictures in the target set, with no scoring on the target; it is quite possible that displacement could also occur to one single control picture, or to several pictures and include correspondences to the target (indeed, the latter possibility could have contributed to the lack of significance in this study, since only the degree of psi-missing, and not of the mentation's correspondence to control pictures, was used as an indicator of displacement (p. 237).

She concluded:

The most useful information to emerge from the review was that displacement seems to be by no means the established phenomenon it is assumed to be in the literature, but that the failings of previous research to yield clear-cut findings nevertheless point very clearly to a concrete research strategy which should give displacement its best chance to prove itself. As suspected, it may be that displacement is more prevalent than has been thought but remains undetected in most free-response studies because analyses applied to all of the trials in the experiment are too crude to allow detection of displacement. However, until replication at an acceptable level of significance of this finding has been made, this remains as speculation (p. 243).

SRI experiments

As we noted in our review of ARV literature in Chapter 3, in 1982 Keith Harary and Russell Targ used ARV to forecast changes in closing prices of the silver futures market.

They made nine consecutive correct forecasts, which yielded earnings of $120,000.[183] They repeated the experiment but were unsuccessful on all nine trials. According to Targ:

> The following year we attempted to replicate these results. In the first two trials for the new March silver contract, we had two successive misses. Particularly disturbing to us was the fact that the descriptions given by the viewer were exceptionally detailed, but of the wrong objects in the pool. We have spent much time trying to analyze the source of this problem, which in psi research is called displacement. However, these misses did not in any way negate the statistical significance of the string of successes in the first series, though they certainly motivated us to do further research before returning to the market place.[184]

In a later experiment after the failure of the second try at silver futures, Targ, Katra, Brown and Wiegand[185] tried a different approach:

> We actually used a different trading strategy, to give more trading days, based on the idea that misses (30%) are half displacement to the wrong target (15%), and half random output, with no psi associated with any target (15%). If that is true, then we can trade either when both people see targets of the same direction, or when one sees a target direction with a score of 5 or greater, and the other passes (no target is seen). In this case, we will get a miss when both people see the wrong target, (.15 x .15 = 2.25% of the time,) or when one person sees nothing and the other displaces (2 x 2.25% = 4.5% of the time.) This assumption gives a 6.75% miss rate. We trade when both agree, which will likely be 49% of the time, as stated before, plus when either viewer sees a target and the other passes (0.7 x 0.15 = 10% of the time). For the two people, this gives 2 x 10% = 20% additional trading. With these assumptions, we trade 75% of the time we have a trial, and have a 9% error rate on those trials. It is as though every trial is a "confidence call" by the judges. If they don't like the quality of a viewer's description, in their blind matching, they declare it a pass (p. 374).
>
> It appears these adjustments worked: Of the 12 viewings that were not rated pass by the judges, 11 correctly described the object that the viewer was shown at a later time $p = 0.003$). The objects shown to each viewer corresponded to the direction of the one-day change in the price of May silver futures. Of the nine trials carried out, two were passed for various reasons, and seven were recorded as traded in the market, although no purchases were actually made. Six of the seven trade forecasts were correct (p. 367).

This experiment demonstrates that multiple viewers, each with their own target pool, can be used in an associative remote viewing protocol to overcome the problems of displacement that has plagued researchers in this area. We, of course, do not know if this is a universal solution, but it is clearly a step in the right direction (p. 379).

The experiment led them to offer these guidelines:

- Use selected viewers with a proven track record.
- Pay attention to each viewer by giving consideration to his or her mental state at the time of the experiment.
- Provide trial-by-trial feedback of only the correct target, and do it as soon as feasible.
- Create trust by full disclosure, and no hidden agendas.
- Psi is a partnership, not a master/slave relationship.
- Seriousness of purpose provides motivation to both the viewer and the experimenter.
- Targets should be attractive and uniquely different: No tarantulas for viewers who don't want to experience them.
- Do not create large target pools, 2 to 4 items at most.
- Take enough time to achieve rapport, plus 10 to 30 minutes for a trial. One trial per day is plenty.
- Practice allows viewers to recognize mental noise and separate it from the psi signal.

Jack Houck on displacement

Jack Houck is best known for his enthusiasm for PK (psychokinesis) and for his spoon-bending parties. Less known is his involvement in ARV and his interest in displacement. In a paper published in 1986, Houck wrote:

The judges all became very excited because they were sure that they would be successful in their investment based on this information. When the report came back from the market the next day, however, they found that they were wrong. This phenomenon has been called *"displacement."* It was probably caused by the diversion of the viewers' minds to the time of judging rather than to the time of the feedback event. The excitement at the time of judging was probably much greater than the emotion felt at the feedback event. This problem was compounded by the fact that the judging occurred at the same place as the feedback location and that one of the judges provided the feedback to the remote viewers. When a computer is used to perform the target selection and to carry out the judging functions, and it provides information only about the correct target scene, many of the potential problems in ARV experiments that could lead to displacement are eliminated.[186]

Houck continued with a point we have already noted and will come across again, especially in our chapter on time (Chapter 7):

One of the things that became apparent to me was that a number of examples of people's experiences seemed to include time displacement (i.e., they would observe something that contained information that was not at the observed remote site at the time of the experience, but it was there either in the past or the future). Because there seemed to be so many reports, I took the attitude that if this information is "real," then how could it be explained? This is quite a different attitude than the skeptical one of "these people are lying" or their experiment is fraudulent.

Big Data ARV anthropological project – Evaluating the "Big" RV Facebook group

Within the formal scientific arena, simply stating one believes something to be a large problem is not going to garner much attention or respect without formal data to back it up. Unfortunately, much of what's happened in the field of RV and ARV over the past decade has taken place outside the formal arena and therefore has not been well-documented in the scientific literature.

For this reason, Debra and co-researcher Carl Anthony McLelland embarked on a mission to provide such documentation. They analyzed all available posts and associated comments from one of the largest Facebook groups. The group founded and overseen by Alexis Poquiz had more than 10,000 members at the time we assessed it. The data analyzed went back as far as 2007, covering a span of 13 years. Our main goal was to discover the most frequently used words during the period, speculating that what was most talked about and of most concern would be most representative of today's remote viewing culture. In this respect, it was an anthropological investigation – much as if we visited a tribe in an isolated place and attempted to learn about them by studying their tools and artifacts. For our project, the artifacts were members of the RV group and the tools were their words. Of course, we had hypotheses about what would emerge. These were confirmed, along with some surprises.

Of the top 30 words, "RV" came in first, mentioned 203,072 times in original posts and 2,085,800 times in original posts plus replies; "ARV" came in fourth place, having been mentioned 18,121 times in original posts and 218,435 times in original posts plus replies. The word "science" (interesting for a psi-based group) came in at 11th place, and was repeated 7,724 times in original posts and 218,435 in replies/comments.

"Displacement" beat this, coming in 10th place, having occurred 8,254 times in original posts and 236,198 times in all posts and replies/comments to these posts. "Wagering" came in as our 27th word, with 225 original posts and 373,973 in original/comments. Several other words related to ARV, such as "Applied Precognition Project" and "lottery," were among the top repeating words, as well. It appears we were successful in demonstrating the relevance of these words to the field of remote viewing, with displacement being in the top ten in importance. This project is still being formally assessed at the time of this writing.

Please join us for the next chapter as we consider a number of theories about displacement and offer many suggestions for how to mitigate it.

CHAPTER 6

Displacement Theories and Proposed Solutions

Over the years, along with many of our remote viewing colleagues, we have been exploring theories that attempt to explain why remote viewers experience displacement. While it may be easier to explain why newer viewers who don't understand the ARV setup would displace, it is a mystery why even experienced viewers frequently displace to the wrong photo (or photos) in a set, or even to elsewhere (such as to a photo displayed on the manager's desktop). This happens with self-judging and independent judging. In this chapter, we provide an overview of theories of displacement, both our own and others', to help raise awareness of the issues involved and inspire creative solutions.

The Skeptic – "It's not displacement"

For our first theory, we will explore a common assertion by skeptics and skeptical researchers that whatever has gone wrong is not because of displacement. Skeptical researchers will often deny that displacement is a real thing – pointing out two arguments that could be categorized as the "random bits of data" argument or the "confirmation bias" claim.

Some skeptics argue one can never prove for certain that displacement occurred because a viewer could have come up with "random bits of data" that coincide with the target or one of the alternative photos. We agree this could be, and often is, the case – but this is not what is generally in play. Former SRI-SAIC Director Ed May has often noted that a rule of thumb in the lab is that 30% of the data in any remote viewing session is likely to be found in any RV photo target. If this percentage is correct (which we have not verified, but it sounds reasonable based on our own experience and observations), this means some impressions will match a certain percentage of all photos without psi being involved.

For this and other reasons, when it comes to displacement, it's prudent to follow the motto "extraordinary claims require extraordinary evidence." In other words, if we are going to assess whether displacement has occurred, we need to establish high standards. We suggest looking to see if three elements are present – quality, quantity and uniqueness. If they are, the "random bits of data" argument falls apart.

1. The viewer's transcript contains specific details that resulted in a high score if a rating scale was used (quality).
2. The transcript contains a "sufficient number" of details to indicate their presence is not random (quantity).
3. The target consists of distinctive, unusual or less common elements (uniqueness of the target).

When all three of these are present, we can have more confidence that displaced psi may be the culprit.

For example, imagine if the target for a particular trial was a photo of a living room. The alternative photo (the one that did not correspond to the event's outcome) was an ocean scene with seals playing in the water, and the seals are positioned at the bottom left of the photo. If the viewer only stated "water" but didn't mention any other characteristics, this really shouldn't be seen as proof of displacement since water is not uncommon in targets and sessions. However, if the transcript included "water," "sea creatures, probably mammals," "leaping through the waves," "positioned on the bottom left," and "smells like salt," we could have much greater confidence that displacement was indeed a factor.

To help determine whether displacement in an ARV trial may have occurred, we have developed a verbal flow chart. This chart could be used if it's suspected displacement might have occurred to another photo or object in the set.

1. Did the prediction fail; did the event of the outcome turn out to be something other than predicted? If yes, continue.
2. Did the transcript originally receive a high score? (Such as a 5 or above on a 7-point scale?) If yes, continue.
3. Check the judging – ask the person issuing the prediction to rejudge the transcript against both photos (we refer to standard binary ARV for this example). Also ask an outside person to judge the trial without knowing the outcome. Tell the two judges

to be conservative and not give credit for common words. If both judges give a high score to the photo that did not correspond with the outcome, continue.

4. Check to make sure the photos are different from each other in every way. Ask an independent person to examine them for similarities. If notable similarities are found, then the judging should discount responses by the viewer to these aspects. For example, if both photos contain an image of a shoe and the viewer has stated the word "shoe" or has a sketch of a shoe, then there should be no credit for that with regard to either photo, and the transcripts should be rejudged. If the score is diminished for the non-winning photo but higher for the winning photo, then it's likely the culprit was the earlier scoring of the similarities. It wasn't displacement.

This flow chart would also work if it is suspected that displacement occurred between trials – such as if a match was noticed to the next target in the series. However, "suspected" is the key word. We don't recommend searching for displacement in every practical trial and the idea should not be introduced unless it is really obvious or it is part of a research effort. We don't want the manager or viewer expecting displacement will take place, since expectation might have an influence on the remote viewing itself.

Just as importantly, we want to minimize the likelihood that *confirmation bias* enters the picture. Confirmation bias is another common reason skeptics will dismiss any suggestion that displacement occurred. Usually this dismissal is by those who are not actually assessing the data directly, but simply on a theoretical basis. Still, to the credit of the skeptics, over the years we have seen many viewers, especially new viewers, exhibit this bias when they first hear about the possibility of displacement.

Confirmation bias can be defined as finding whatever one is looking for in an environment packed with information. Meaning is superimposed by the observer. Because of the tendency for humans to exhibit such bias,[187] it is imperative that viewers not look too hard, long or far for ways in which their questionable data might match something. Not only is it likely they will find a match somewhere, but the effort could well establish sloppy practices on both psychic and cognitive levels.

Instead, take it off the table unless it's highly obvious – as in Debra's plate of fruit example in the last chapter. As you recall, two of us sat at a table sharing a big plate of fruit while the viewers were doing their sessions at a distance. One viewer perceived two people sitting together eating a big plate of fruit. This meets the above criteria – very specific data and a number of matching characteristics. This is very different from me finding excuses for why the student saw fruit. Maybe I am sitting near my kitchen and have fruit in my refrigerator, or perhaps I have some curved yellow shapes on my shirt that resemble bananas, or maybe I had fruit for breakfast…maybe that was it? No. If we are going to get wobbly and sloppy like that, the whole process will start to crumble.

What to do with the knowledge that displacement has occurred?

If displacement occurs, often there's nothing to be done in the moment. For example, if a financial or sports prediction is involved, passing may be the best option. However, future project designs could ensure at least a qualitative assessment is made of instances when displacement is suspected.

Some researchers have done post-hoc assessments of displacement, as we noted with Carington and Soal, and as Senior Scientist Dean Radin has done with IONS (Institute of Noetic Sciences) online psi games and apps.

One type of assessment examines whether the session data matches the next target in the series or if there is an underlying pattern to the misses. This type of assessment gives promise of providing insights into the complexities of displacement. However, it may be a good idea to keep viewers uninformed about this planned assessment to avoid cuing them to start speculating about future trials.

The remainder of this chapter assumes that displacement effects do occur in ARV, based on our own extensive experience with them as well as what many other remote viewers and taskers report. Of note, it appears virtually all people new to ARV sooner or later (and usually sooner) come to the same conclusion about the reality of displacement without ever having heard of it. This includes experienced academic researchers from other fields who start to explore ARV. Also, as we indicated in the previous chapter, parapsychological researchers have noted and documented displacement effects going back more than a century.

False feedback theory[188]

What I (Debra) refer to as "false feedback" is not the only cause of displacement, but it may be one of the biggest culprits. The false feedback theory can be thought of as an overarching theoretical framework made up of smaller subordinate theories. I will explore a number of these "sub theories" to explain how something that is not feedback inadvertently becomes "false feedback" or what could also be called "surrogate feedback." The subconscious inclination to swap out what will later become the actual feedback for something else that is available earlier is grounded in the viewer's psychological makeup. This includes internal motivations, emotions, desires and orientations, along with physical acts, behaviors and usages, including semantic and linguistic ones. These processes, behaviors and usages are then connected to the way the participants in an ARV trial or series of trials relate and communicate with each other. False feedback may be exacerbated or minimized by the project's design, which includes all procedures, tools, materials and technologies.

Sub Theory #1 – Sloppy semantics

Far too often, even those who have been doing ARV for a long time may refer to all photos involved as "targets." For example, an ARV project manager might say, "Now it's time to compare your transcript with *both targets* to see which is a better match." A remote viewer might say, "I had to call a pass because my transcript seemed to match all the targets." Even my writing partner Jon (until now – hopefully) has often referred to both photos as "the targets." However, my position is that both of those photos *are not* targets and *must not* be referred to as *targets*. There is only one target – the photo associated with the winning outcome, also known as the feedback photo if it is predetermined that this will be "the target."

When pressing this issue, I typically get two responses: One is agreement that this is a poor choice of language. Another is, "Well yes, but everyone knows what I really mean is the potential targets." While it may be true everyone in the group knows what is meant on a subconscious level, the person calling both photos "targets" may be thinking of them as such.

My contention is this: Since the aim of all remote viewing and ARV is to tune into a specific target, the distinction between what is and is not a target needs to be stated very clearly and understood by all participants – viewers, judges, project managers, etc. If this differentiation is not clear, it may very well lead to greater instances of displacement.

Obviously, many people are inclined to call both photos "targets" in binary ARV because either photo could end up as the target once the outcome of the associated event is known. Whichever photo was paired with the winning outcome is the target; the outcome of the event drives the selection of the target. However, focus on the connection between the photos to be judged and the feedback photo really needs to be minimized. A large wall within one's conscious and unconscious mind needs to be erected between the concept of the set of potential targets and the actual target. Anything less will lead to the phenomenon I am referring to as "false feedback." Without this distinction, the judging session may become a premature feedback session. The problem can't easily be remedied by intent and positive thinking alone, as many failed attempts by experienced viewers have demonstrated. I don't say these words lightly – I've written a book on the power of positive thinking and I'm a strong advocate of it in other circumstances. I will offer some reasons for this failure and then make suggestions for solutions.

Jon's interjection:

There is much I agree with in this perspective. I agree with Debra's point that it is better not to refer to each potential target as a target without the qualifying word "potential." Further, the most often used form of ARV, binary ARV, does work – to an extent – and we should use all the means we have to understand the displacement that occurs and to lessen it. However, after doing, witnessing and studying binary ARV for over a decade, particularly in APP, which has had hundreds of viewers doing thousands of sessions, it struck me that the very fact of positing two potential targets, with one actualizing (that is, it is associated with the winning outcome) and another not actualizing (is associated with the losing outcome) means you will very often get information about both of them. You are "baking in" displacement the moment your setup includes more than one potential target – whether it's two or many (as in the five or six photo "decoy" sets). This kind of displacement is all but certain to occur when binary ARV is carried out over days, weeks, months.

This viewpoint was not voiced by others when I raised it and it was resisted, including in APP. And it is still true that nearly all ARV nowadays is binary ARV. While psychological and other factors need to be addressed and acted on, I believe the fundamental cause of this kind of displacement is the fact of having more than one potential target. (I say "this kind" because, as we have seen, there is also displacement in time, a subject that also has been given very little attention in remote viewing, even though Carington noted it 80 years ago.) We explore the alternative of having just one target in ARV, called unitary ARV and Strict Unitary ARV, in later chapters.

Returning to Debra's account

False feedback can be conceptualized in spatial and temporal terms. First, here is the most common timeline of a binary ARV trial:

Timeline of ARV Steps	
Step 1	Manager chooses an event.
Step 2	Manager pairs photos with possible outcome (this could happen between step 5 & 6 if the manager doesn't see viewer's transcript).
Step 3	Manager assigns viewer a target number, advises when feedback will come.
Step 4	Viewer does session. Submits transcript to manager.
Step 5	Judging happens (transcript compared with photo pairs).
Step 6	Judge records a prediction (some share openly, some keep private).
Step 7	Judge makes a wager (optional).
Step 8	Event happens, leading to outcome.
Step 9	Manager provides feedback to viewer based on outcome.
Step 10	Viewer does feedback session; manager assesses how things went.

Figure 6.1

Figure 6.2 – Leaping Across a Gap

False feedback is like a person leaping from a ledge across a large gap. The first ledge represents the time of doing the remote viewing session. The far ledge is the feedback photo attached to the winning outcome. What is required of the viewer is to leap over the space between them. Any of the intervening factors could plunge the viewer into the abyss. These distractors can occur during the judging phase (whether self-judging or independently judging), the prediction phase, the event itself, or during wagering. The things that the viewer might displace to are potential displacement triggers or distractors. I feel the metaphor of needing to jump over them is useful because if one falls into them, one will not be able to reach the intended side (the target). Also, this metaphor offers a warning that hopefully will create a feeling of trepidation or at least wariness. When we get to possible solutions for displacement, we'll discuss a popular behavioral approach called "adverse conditioning." Use of this metaphor is a start in that direction.

Using ARV to simply prove to oneself one is psychic

One reason remote viewers, especially newer ones, may displace to the judging photos (the "decoys") instead of to the target is simply because they either don't understand what is involved in the overall ARV design or they don't care too much. Their primary aim is to experience being psychic. Sometimes that is more easily achieved by tuning in to both of the judging photos or perhaps one of the photos is easier to comprehend or describe, is more numinous or more relatable to the viewer.

One exercise I've done many times with different groups of remote viewing students is to give them an abstract object as a target, one that could not be named even by someone staring at it with their eyes open. Students tend to do more poorly on these types of targets than ones they could name or recognize with their eyes open. Further, if I give them one of these targets first, followed by a more recognizable or familiar image right afterward, most of the class will skip over the first target and describe the next one. I've also noticed this happen with newer students and highly emotionally charged targets (which I no longer give to new students). While these exercises did not involve ARV, I've observed similar patterns of displacement when one photo is easier to name than another or one is more emotionally charged.

In the introduction to this book, I mentioned how I was motivated to join ARV projects (mostly those led by Marty Rosenblatt) not because I cared about wagering or winning anything, but purely because it provided free practice opportunities in a supportive environment and a fun group atmosphere. For the first couple years of practicing in ARV groups, I also really didn't care if I displaced to the wrong photo. All I cared about was seeing if I had really strong matches to one of the photos. Over time, though, it became apparent how this attitude was unfair to the other group members, who were working hard to achieve higher group stats and earnings, whether in applied or research projects.

As an advancing student (in all psi endeavors), I began to understand that if one isn't able to hone in on the proper target, one really isn't exhibiting a high level of skill. Simply put, a person who is consistently displacing to a "decoy" photo is not going to have a very long ARV-related career any more than someone who bowls would have a future if he continuously knocked down all the pins in his neighbor's lane. Imagine the reaction he would get!

Unfortunately, this is often not communicated for ARV projects. Will remote viewers be somehow damaged if you shake their confidence by telling them they need to shape up? Could it disturb the highly sensitive nature of their concentration, perhaps even leading to psi inhibition? While that could well be the case if you traumatize them, if the project isn't a training exercise, they are already assured displacement isn't from a lack of psi. It's actually a sign of too much psi focused on the wrong thing. They may be able to improve if their displacement is simply about defaulting to what is easiest or most interesting, emotionally charged or benign. (Some viewers like emotionally charged targets, and others avoid them at all costs.)

Attentional theory of displacement and false feedback

This theory is predicated on the observation by many parapsychologists and remote viewing professionals that psi-based perception follows many of the same principles of regular attention. This theory speaks not only to why remote viewers might displace within ARV or within a parapsychological experiment using matching tasks and photo sets, but to why viewers tend to gravitate toward some information even when focused on the correct target.

To understand how rules that govern regular human attention and perception may impact psi-based perception, a review of the psychology of attention may be useful. Modern books on the history of attention tend to start their accounts in the early 1950s with Broadbent's dichotic listening tasks, made possible by the newly developed audio recording technologies. These tasks brought the study of attention into a laboratory setting with repeated and formal experiments that focused primarily on auditory attention and inattention.[189,190]

However, decades before this, psychologists were making internal observations of their own and others' attentional processes without the aid of formal technology. These include structural psychologists such as Tichener[191] and functionalists such as William James,[192] followed by the phenomenologist Merleau-Ponty.[193] I feel these earlier psychologists may have the most to offer remote viewers who are not processing information through use of advanced technology and laboratory setups, but are directly experiencing the internal canvas of their own minds and taking part in simple project designs with materials accessible to all.

T.H. Ribot's theories of voluntary vs. involuntary attention

As far back as 1903, T. H. Ribot observed in a small book entitled *The Psychology of Attention*[194] that there are two kinds of attention – voluntary and involuntary. He noted voluntary is the kind in which a person sets their intention to focus on something in particular, while involuntary is what happens to the person's attention apart from what they have intended. In the latter, one's attention is being moved away from whatever one was already paying attention to.

Like many of the early psychologists, Ribot was first a physiologist. Therefore, he was very focused on the reflexes of the physical body. He referred to an element in attention as movement and noted that attention is always moving. In voluntary attention, we move our attention with intent, whereas with involuntary attention, it is moved for us, against our will and often so fast we don't realize it's happening until it has happened.

Within involuntary attention, there are two different kinds of distractions. One is internal – as when we are distracted by something within ourselves. Maybe we become tired or hungry or curious about something other than what we are trying to focus on – such as writing a chapter about displacement – so our attention automatically moves in a new direction.

A distraction then could be defined as something that moves our attention from where we wanted it to be or where it was previously fixed. An external distraction would be our cat meowing or a fly buzzing or something falling off the wall. External distractions often involve movement, noise or someone else's intent. Along with movement there is change. Something changes before us. Another feature is contrast. If it's light in a room and a bright light shines, we may not notice it, but if it is dark and are exposed to the same degree of luminosity, we will notice it.

While people tend to interpret distractions as irritating and wish they were not so easily distracted, Ribot pointed out these characteristics of attention are absolutely necessary to our safety and survival. They help us become aware of dangers (the bear about to eat us) as well as opportunities (the bear we may be able to eat). People often wish they could focus their mind better or for longer periods, as being able to do so is often equated with being successful in school, career, and accomplishing goals. Yet if our minds didn't constantly waver but remained stationary on one thing for too long, this could create a number of problems. He suspected this is what is happening when someone suffers from obsessive disorders; their minds are too fixated.

Ribot proposed that a person successful in school, profession, trade or skill is one who has the ability to focus and refocus attention even on those things they don't always wish to focus on. What gives some the advantage is when the thing they have to focus on is the same as what they truly, inherently are passionate about focusing on. When the internal desire is paired with an external demand, this makes it much easier to pay attention. While much of this today seems common sense, Ribot was laying the foundation for explaining a variety of experiences and challenges we have in remote viewing and associative remote viewing, even though he was not aware of these activities and did not seem to condone William James' explorations into mediumship or other parapsychological topics.

Gestalt theory

According to the American Psychological Association's online Dictionary of Psychology, Gestalt principles of organization are principles of perception derived by the Gestalt psychologists such as Max Wertheimer[195] that "describe the tendency to perceive and interpret certain configurations at the level of the whole rather than in terms of their component features" (p. 1).

These "laws" include closure, common fate, good continuation, proximity, similarity, symmetry, figure/ground and Prägnanz. While a detailed discussion of each of these lies beyond the scope of this book, I will present the ones that seem to be applicable to remote viewing and then return to the topic of displacement.

While the Gestalt psychologists were referring to ordinary visual perception rather than psi-based perceptions, chemical engineer and parapsychologist Rene Warcollier[196] felt certain principles emerging from the Gestalt schools of psychology were applicable to emerging data from his own telepathy experiments. He stated, "We can look to the psychology of perception for other principles that reveal themselves and paranormal behavior…the Gestalt theory is the school of psychology that seeks to understand the wholeness quality or structure of each perception or reaction" (p. 26).

Warcollier brought Gestalt concepts such as the relationship between figure and ground into his early experiments involving telepathic "senders" staring at images in one room while a "receiver" attempted to sketch these in another. He provided examples, concluding that "sharp contrast of figure and ground, in this case equally divided and repeated, seems favorable for telepathic communication." He added,

> What we may conclude from our knowledge of the normal psychology of perception is that there is always a dynamic core in telepathic perception, not unlike that in normal perception, whether the impression is global or detailed. There is a tendency towards organization, toward a wholeness character of the impression, in order that perception be as simple, as symmetrical, as regular and as meaningful as possible (p. 27).

The Gestalt law of *closure* is defined as "the act, achievement, or sense of completing or resolving something. This principle states that people tend to perceive incomplete forms (e.g., images, sounds) as complete, synthesizing the missing units so as to perceive the image or sound as a whole—in effect closing the gap in the incomplete forms to create complete forms."[197] Rutledge[198] adds "When looking at a complex arrangement of individual elements, humans tend to first look for a single, recognizable pattern" (p. 1).

Warcollier[199] was perhaps the earliest psi researcher to emphasize the importance of movement portrayed in target material. He theorized that even simple drawings that conveyed a sense of movement (such as a bird or a train) would be more perceivable than drawings with stationary subject matter. He wrote, "The idea of movement in the telepathic impression involves the Gestalt theoretical principle of *Prägnanz* – which is the German word for 'good figure'" (p. 28). This is also sometimes referred to as the law of simplicity, which holds that humans tend to interpret ambiguous, partial or complex images as simple and complete.

A few of the Gestalt principles seem to speak very well to an issue that doesn't seem to have much to do with displacement but very much with why remote viewers tend to get derailed in sessions: Analytic Overlay (AOL). Analytic Overlay happens unconsciously and spontaneously when aspects of the target appear as an image or a concept that is not correct as a whole but there is some correspondence. An example would be if a viewer drew a tomato when the target was really a red ball. Aspects of round, red and small came in not as a ball but as a tomato. Analytic Overlay can also happen through deduction. In this case, the viewer consciously reviews the elements and says, "I saw something red, I was inclined to draw a circle, and it's small. Maybe it is a tomato."

Gestalt theories of perception have also been applied to understanding people on a broader psychological level,[200] particularly in relation to their "neuroses." People are primed to seek closure to a stressful event. Humans are always trying to figure out an answer in advance in order to have closure even before closure is really appropriate. We are obsessed with understanding in advance what will happen, as if thinking will bring us closer to knowing and ultimately completion. This sheds light on the tendency by many, if not all, viewers to take just a few bits of information and arrive at conclusions that turn out to be unwarranted. This could be what is happening when displacement occurs. The viewer may be psychically reaching for the quickest and easiest data that allows them to feel like they know what the target is, even before that can truly and accurately be known. Whether the viewer is intuitively (and unconsciously) "grabbing" information from the two photos or the project manager's mind, whatever can help them form a conclusion first will lead to the desire for closure. It doesn't take psychotherapy to see how we do this constantly.

Deconstructing the relationship between two or more photos

Perhaps the biggest problem in standard ARV is the very intimate relationship that exists between the photo options. This relationship is created at the start of a typical ARV trial. For judging to be effective, the photos must be carefully selected so they are of equal potential interest to viewers, but as different from each other as possible.

Everything (with one exception we will discuss) needs to be different – shapes, sizes, colors, positions, patterns, luminosity, contexts and contents. Both co-authors have been involved in projects in which we spent hours creating photo pairs and sets, thinking they were sufficiently different from each other. When we got together with other judges, they immediately – and correctly – found one or more similar aspects, causing us to discard at least one of the photos. Even with teams of judges working together to create pairs sufficiently different from each other, remote viewers subsequently reported impressions or sketches in their transcripts that unveiled similarities.

The exception we refer to involves what some have called "numinosity." Some photos are "intrinsically" more interesting than others. This varies, of course, with viewers' backgrounds and interests but in some pairs, there would be almost universal agreement that one photo is more interesting than the other. Examples: Albert Einstein and a matchstick. A nuclear bomb and a plain piece of paper. With the second pair there is a factor which will be discussed in a later chapter: viewers in the Star Gate/SRI program nearly always were able to get good data whenever the target had high entropic change, such as a nuclear explosion. While potential targets should be orthogonal in many respects, they should be of equal interest or "numinosity" to the extent possible.

Further, in order to ensure differentiation, those creating the set pay attention to the relationship between the two photos. This very act may create a linkage between the two photos. Even though the end goal is for the viewers to see only one photo (the one associated with the winning outcome), perhaps viewers can't help but be exposed to the other part of the pair or set due to unconscious factors. Where there is effort, there can be intense human concentration, focus, attention and energy. This effort enhances linkage. The greater the linkage, the more likely it would be for viewers to access both photos during their remote viewing session.

We also have to pay attention to the language used. The words photo "pairs" and "sets" denote a natural and close, if not inextricable relationship between the members. These words are used in just about every parapsychological experiment in which judges compare participants' transcripts to more than one photo. Therefore, asking the viewers to see only one photo and not another in "a pair" during their remote viewing sessions is like telling them to look at a person's foot or leg and not the other. You can't look at one sock without at least wondering about the other – usually if we see a salt container, we will think of pepper, especially if these tend to sit together in a salt and pepper container on the table. If you know a set of twins and hang out with one, it's very hard to not think of the other.

Again, none of this is happening because we want the viewers to see these pairs or sets – this is all being done for the sake of the experiment or project setup. In fact, we want the viewers to do just the opposite – to avoid making this linkage. Research suggests that one person's attention on something causes another person's attention to be more focused on that same thing – even if they are in two different rooms. For example, in one study, people reading passages of books reported an easier time comprehending the passage when someone in another room linked to them relationally was reading it at the same time, even though they were not aware of this.

In addition to the attention and emphasis on connecting the photos in an ARV trial setup, additional perceptual tendencies may be at play here.

Another Gestalt psychology principle is *common fate*. According to Rutledge[201] this concept is perhaps more tied to human survival than any other. This principle states that objects functioning in the same place, and especially when moving together in the same direction, appear to belong together. When this happens, they are perceived as a single unit, such as a flock of birds. This means humans are prone to look for, and notice, groupings of objects, especially those moving through space and time. The Gestalt laws of *Proximity* - things close to one another are perceived to be more related than things that are spaced farther apart – and of *Similarity* – things that are similar are perceived to be more related than things that are dissimilar – would also apply.

Rutledge Shows this Example (the Two X's Close to Each other) When He Moves Them

X X	X	XX
Acquaintances	Single guy	Together & in love

Figure 6.3

In this example, even though these are only simple Xs, which are identical, we immediately have an impression of their relationships to each other. We have an impression of the XXs on the right as two, "a pair." They are "not related," so to speak, to the X in the middle, which is by itself. They are also "not related" to the Xs on the left, which are near each other but "not close." If any of the Xs were moved closer together, that would change our perception of them. If the two on the left moved in the same direction but kept the same spacing, we'd think they may be together but not as closely related as the two on the right. We may project relationships among people onto these simple symbols. Most people are entirely unconscious of how groupings, spacing and movement impact perceptions.

In regard to our discussion of displacement, this concept makes more sense if we look at a timeline from the point of view of the viewer in a solo ARV process.						
Tag is given	*Viewer does session*	*Transcript judged*	*Prediction*	*Wager*	*Event*	*Feedback*
1	*2*	*3 A & B* *These seem together as one*	*4*	*5*	*6*	*7*

Figure 6.4 – Discussion of Displacement

It's a continuum in time and the viewer moves along the line from one entry to the next. A few things stand out. With judging (3A&B), the photo pair is in play; both exist at the same time and both are addressed before the viewer can move on. On the timeline, the pair is closer to the viewer doing the session than to the feedback time. In a real-life practical example, this is even more likely to be true since the event often takes place a day or more (even months) before the feedback, whereas judging is usually done soon after the session has been completed.

Preliminary studies[202,203] and anecdotal examples show RV may be more accurate when there is a short timeline. But even then – with session, judging and event close together – the feedback phase is still farther from the viewing phase than the judging phase. It is always separated by three or more other phases (judging, prediction, wager/pass, event, feedback).

The Gestalt psychology *Law of Good Continuation* is also applicable here. This law says elements arranged on a line or curve are perceived to be more related than elements not on the line or curve. Some people may have the sense that all of these elements on the "line" of ARV are linked.

As will be noted in our later chapter about what makes a good remote viewing target (Chapter 8), individual differences and preferences come into play in psi-based tasks, as they do elsewhere in life. Some individuals may think more linearly or structurally, with a tendency to organize bits or chunks of information more quickly and in different ways than others do. If so, some of the Gestalt perceptual principles may be operating even more strongly for these folks, whether in relation to groupings involved in ARV or in relation to aspects of a target in the session itself.

I have observed this in newer remote viewing students who have jobs requiring the ability to be organized such as administrators and accountants. While this is just speculation, it's very possible that someone who would never leave a kitchen cabinet without first ensuring that all the cans were lined up with labels facing forward or even grouped by subject matter (God forbid the green beans are mixed in with the cans of tuna!) might be prone to seek out groupings and relationships within their own perceptions or within an ARV trial, even when these are not intended.

"Whatever is compared is paired." Often those with some knowledge of ARV history will suggest computer judging might help alleviate these connections and reduce "feedback loops" between them. While we discuss this approach in detail in Chapter 9, there is still no way in binary ARV to avoid the concept of pairing – two things are being related and compared. Even if the process is based on the computer using "profiles" of each photo and matching human scoring of the session with the profiles, somewhere in the depths of cyberspace one thing is still being compared to the other. The relationship between photos in a trial still exists.

The bottom line: it's hard to look at one's feet without seeing the ground. It also may be hard to see only one photo if it has been linked by intention with another in time and space.

Debra's arousal theory of displacement

This brings us to another theory about displacement, which I refer to as the Motivational-emotional theory of displacement. The remote viewer and their feedback are like two lovers. Think Romeo and Juliet. Nothing and no one can keep them apart. They don't want to wait years or even months to be brought together. They want each other now, and whether that has to happen in life or death doesn't matter. Likewise, the remote viewer WANTS and NEEDS the feedback NOW, no matter what the cost. This may not be true for experienced viewers, but it does seem to be for some beginners. While wanting and needing feedback is present to some degree for all viewers, some really "have it bad." They seek relief wherever and however they can get it. This premise at the core of my arousal theory of displacement is predicated upon the motivational theories of Clark L. Hull.

Hull's Drive Reduction Theory

Hull[204] attempted to create a universal theory that could describe all behavior. He based his theory on the concept of homeostasis, the idea that the body actively works to maintain a certain state of balance or equilibrium. *Hull's Drive Reduction Theory* offers insight into displacement. Hull was a psychologist who wrote "when survival is in jeopardy, the organism is in a state of need. So the organism behaves in a fashion to reduce that need…In a stimulus response relationship, when the stimulus and response are followed by a reduction in the need, it increases the likelihood that the same stimulus will elicit the same response again in the future."[205] Hull used the term *drive* to refer to the state of tension or arousal caused by biological or physiological needs. By arousal, Hull was not talking about sexual arousal, but rather a situation when the emotions and physiological responses are elevated.[206]

Again, while Hull was not referring to remote viewing or any parapsychological concepts, his theory would explain displacement effects in ARV and other experimental setups.

During or immediately before a remote viewing session, many things are happening within the remote viewer's mind and body – both emotionally and physiologically. We write, "I'm feeling somewhat nervous about starting this session" or "anxious" or "looking forward to it" or "dreading it" – whatever it is, the feeling isn't always entirely neutral. We can declare the feeling and, in fact, that is built into the CRV methodology – whether it's called declaring *a personal inclemency* or *setting aside* what might be distracting us.[207] At different times during the session, we move into higher states of arousal – again not using this word to mean "sexual," as is often the meaning – but our senses or emotions are aroused, and a feeling accompanies this in the body.

At different times prior to, during or following a remote viewing session, we feel more stress. This happens because there are so many unknowns, the biggest one regarding the question, "Will I actually be able to access the information?" Even for the most experienced viewer, there is always the possibility that no information will come or that our "monkey mind" will get in the way. Even if the viewer is just doing a practice session, it's human nature to feel the weight of "What if I don't get anything or this time I can't do this?" "What if failure here means I won't be able to get a hit next time? What will that mean for my identity, reputation, career, etc. as a remote viewer?!" If one tunes in to their own bodies as these thoughts are bubbling up, one might just start to feel as if they are about to have a heart attack!

However, if information – especially unexpected information – is flowing smoothly into one's consciousness or onto the paper, the viewer will become relaxed and the arousal level will decrease. Following a session, "How did I do?" always raises its head. Some viewers care very much about this, while others learn to let go of this to a degree. Many have a sense of waiting, lack of resolution and anxiety. One major thing can immediately decrease this state of arousal – getting feedback that shows us we did well. Everything is OK now. Even if we learn we didn't do well (which could propel us into another state of arousal), just having the opportunity to decrease this arousal can't be ignored. For some viewers, feedback is like piece of chocolate cake sitting on the kitchen counter, beckoning to us. The closer we get to it, the more our mouth waters and the need to lunge for it takes over.

This is where the photos come in. One of the two photos will be the feedback photo, and if we access both of them in our transcript, we are attempting to relieve the tension of the moment.

Look again at this linear process of ARV.

In regard to our discussion of displacement, this concept makes more sense if we look at a timeline from the point of view of the viewer in a solo ARV process.						
Tag is given	*Viewer does session*	*Transcript judged*	*Prediction*	*Wager*	*Event*	*Feedback*
1	2	*3 A & B* *These seem together as one*	4	5	6	7

Figure 6.5 Discussion of Displacement

The set of photos is like a mom holding out a spoon with batter on it to taste before the cookies are done – the photos are beckoning: "You don't have to wait for the feedback – I'll be your feedback. I can give you relief and make all your dreams come true now. Why wait?"

So again, bottom line – the photos, as a pair, provide relief.

This relief might speak to a stronger attractor in remote viewing – the very reason why some people like remote viewing so much and why others find it too anxiety-provoking. The adrenaline rush we get when we see our feedback and discover we did, in fact, do a good job can be compared to that of an adrenaline junkie who does death-defying physical acts to feel invigorated. That's not to say it's the only reason a person climbs to the top of Mount Everest or dives to the deepest depths of the sea, but there is something invigorating about the highs and lows involved and all the physical and mental challenges that have to be overcome while having very cool experiences and interactions with the natural world. Remote viewers may experience feelings similar to those of extreme sports enthusiasts, except their sport can be done from the safety of their home or office with no equipment needed except paper and pen. Of course, they don't get to see the size of their physical muscles increase (just the opposite, unfortunately), but they do get to experience what it's like to have their internal psychic faculties blossom.

Social theory of displacement – Telepathic spread

Moving to other theories of displacement, in addition to the connections between the photos, there are connections between participants in a trial.

Personal relationships are obviously very important to all of us. As mentioned in the previous chapter, there are many kinds of displacement. Viewers can displace to the wrong photo, but they can also tune in to what the project manager is doing at the moment or later that day, or to what they have on their computer screen. Hundreds of parapsychology experiments in telepathy demonstrate both physiological and intuitive connections between individuals.

This is typically referred to as "the experimenter effect"[208,209, 210,211] Much of the focus of these studies has been on whether the experimenter is "psi conductive" (usually a believer) or "psi retardant or resistant" (usually a skeptic). These studies operate from the premise that an experiment run by a psi-conductive experimenter will be more successful. The main problem, often pointed out, is if the experimenter is psychic, it's hard to isolate the psi effect. However, rarely is it considered that the more intuitive the experimenter is, the more displacement effects could take place.

It might not even be possible for remote viewing to occur if these connections weren't there from the start. Those brand new to RV often find it surprising – even doubtful – that remote viewing works by simply having a manager choose a photo or location and assign a target number for the remote viewer, with no other information being given. Yet the viewer is able to describe the target. This is happening…how? Through the intent of the tasker? Through a telepathic connection?

Until Ingo Swann came on the scene in the early 1970s and convinced researchers to conduct an experiment involving a longitude and latitude coordinate that would be randomly selected by a computer for a remote viewer to describe, it was a widely believed in parapsychology that a telepathic connection between a sender and receiver of psi information was vital for psi-based functions to occur. In other words, they thought one person needed to focus on the target for another to successfully perceive it. Over time, less emphasis was placed on this connection. However, even when a computer randomly selects a longitude-latitude coordinate, human intentions determined what such coordinates symbolized, and human intention made the decision to describe what was at the coordinates. Human intention and consciousness are very much a part of all human activities involving the acquisition of information. Displacement may, therefore, be malfunctioning telepathy.

The following examples of potential "telepathic contagion" were originally provided in a paper published in *The Journal of Nonlocality* entitled, "The tip of the iceberg: placebo, experimenter expectation and interference phenomena in subconscious information flow."[212] Bengston and Moga[213] wrote about mice that were cured with the use of a visualization-oriented healing approach referred to as "cycling." In addition to the mice in the experimental group, 80.5% of the control mice, which did not receive the healing-with-intent cure but were located in the same room, also demonstrated remission. Mice in the control groups that were located at a different facility did not survive. The authors hypothesized that under certain conditions "resonant bonds" can form. Essentially this shows an example of displacement in healing.

According to Warcollier,[214] "Our 1923 experiences…have revealed very numerous cases of analogous perception between the percipients, which were altogether independent from those that the agent intended to transmit. It was a true mental contagion of errors…probably…the transmission of a…fragmentary, subconscious thought of one of the percipients." Further, dream researchers noticed that participants in dream studies who shared the same doctors but had no other connection with each other reported the same dream mentation.[215,216] The body of research that has looked in depth at this malfunctioning more than any other is focused on the "experimenter effect."[217]

In recent years, members of the Hawaii Remote Viewer's Guild and Daz Smith, editor of *Eight Martinis* magazine, along with others, have been looking into "tasker's intent." This explores whether viewers are accessing not what they have been asked to describe but instead what the project manager was thinking about the project or their conceptualizations of the target.

Displacement to the project manager is not much different from what happens to a child who is supposed to be paying attention to the math calculations on the white board but instead is more interested in the teacher – what she looks like, where she bought her dress, what her boyfriend might be like. None of this is going to help the student pass the test, but for some it's far more intriguing than what is happening with the chalk-drawn mess of symbols and numbers on the board.

Likewise, why would we want to look at a photograph of a location or a person when we could pay attention to a nearby, real-life person (our project manager)? As mentioned previously, as viewers, we tend to be very much dialed into our project manager. On a subconscious or conscious level, we want them to like our session. We shouldn't minimize that our subconscious may like it if we turn in a session and the project manager looks at our transcript and thinks "Wow! that viewer did a great job!" We didn't want to disappoint them and now we are impressing them. This can serve as a point of false feedback – that is, when the viewer's attention is on anything other than the target itself. It's as if a viewer's attention to something in the future attracts the viewer's attention during the remote viewing session. So much is going on at this midway point.

This may be why even in independent judging the viewer gravitates to the two (or more) photos. The arousal-motivation theory posits that the more we care about impressing someone, the more we may displace. Even if we don't care about impressing them, we may spend a lot of time thinking about others involved in the project – how we can communicate with them or what they are going to do or think – because much of the project's success depends on what others do.

Psi experiments in presentiment show a physiological connection between individuals. One person in a room looks at a video while their companion's vital signs are tested in a different room. This can create an effect even between strangers paired up at the start of a trial, but it is much more likely to happen with those who are linked through marriage or genetics.[218,219] This, too, could be related to displacement. I've observed that students who first took another class with me are more likely to displace to me than new students.

Relationships between participants in a trial may also be highly relevant. Anecdotal evidence shows if there is a "star" remote viewer on a team who is held in high regard by the project manager and other viewers, the manager may place greater trust and emphasis on their session. When there is a miss, the other viewers' sessions may go in the direction of the "star's" missed session. This is on par to what might happen in a classroom filled with students where one has the reputation of being the smartest and always getting the best grade. If that one student has a response that is different from everyone else's, the others may be likely to abandon their own response and copy that student's, if given the opportunity. It's possible that psychic participants are doing just that – dialing into the remote viewer who has the best reputation. And this might be exacerbated by the project manager's inclination to put more stock in that particular viewer's transcripts. In this way, one viewer could lead everyone else off a ledge if they are really off target or displacing to the wrong photo.

Former Ft. Meade viewer turned instructor Lyn Buchanan[220] made these observations regarding participant telepathic contamination:

> The conditions which I have experienced as providing the most contamination are those where: a) multiple viewers are assigned the same target, (even in operational work, I will assign each viewer a different aspect of the task in order to prevent "telepathic overlay"). b) viewers are told that another person will be working the same target. For example, if I tell a viewer that this question will also be worked by a viewer they highly respect, he/she has almost always produced data which is very much like that produced by the other viewer. If the viewer is told that a person they don't like or don't respect will also be working the same target, the information is most always different and even opposing to what the other viewer finds. When two viewers work the same target and each is kept ignorant of the other's participation, they are not as prone to produce as much contamination (p. 9).

Displacement as a function of wanting to avoid passes

Making a pass rather than a prediction can protect a project but can feel disappointing. Without a prediction, there will be no wager. The game won't be nearly as much fun to watch as when a prediction and wager are made – especially when a group is involved. It's a party killer, a downer. Who wants that? Furthermore, passes can prolong a project or trial that some participants hope is ready to end. Therefore, one might be inclined to psychically choose a photo just so a prediction and wager can be made.

SOLUTIONS

How to test if the photos in the set were a factor

Tell participants their task is to remote view the feedback photo they will see at a later date. Then show them two photos that are both different from the target/feedback photo. Ask them to compare their transcripts to each photo and come up with a score, using an agreed-upon scale such as the SRI 7-point scale. Make sure they understand how to apply the scale. Ask them also to write down any words that are striking matches and remind them that having only a few or minor matches is all right, too. Be sure to show the target photo as feedback at a later and predetermined date. Ask them to apply the same scale. At this point, only after they provide a score, you can reveal to them that the earlier "judging photos" were not related to the target at all. As we described earlier, an alternative to this procedure would be to have independent judges score all the photos. If scores are significantly higher for the judging photos, this would support the hypothesis that the viewers were describing the judging photos and not the feedback photo.

Ways to avoid false feedback

Semantic adjustments: While it may be impossible to ever fully extinguish the concept that the photos are pairs due to how our attention works, we can at least try to minimize this. The first step is to adjust our semantics. Never refer to the judging photos as "targets." Allow the word "target" to be used only in reference to what will be paired with the winning outcome and shown to the viewers at feedback time. Use the term "judging photos" or "photos to be judged" or "photosites."

Further, the photos should not be called "pairs" or "sets." This will, of course, be hard to avoid after almost 100 years of precedents for this in parapsychology experiments – but it doesn't mean we can't try. Again, the main issue is to make sure the photos are different from each other. Project managers may even want to keep the photos in different folders or otherwise separate them spatially or physically.

Solutions for motivational and emotional issues related to false feedback: Can thinking about displacement cause it? Displacement happens to people who don't even know what it is, so we know it's not purely a function of people expecting it to happen. But can focus on displacement make it worse? We can't say. However, trying to not think about it doesn't seem to help either.

What we have noticed and want to encourage people NOT to do is to use the idea of displacement as the excuse for why they didn't do well in a session. Perhaps you've seen posts on social media by someone who wants to show an example of possible displacement. Often, they are looking for a reason to feel it wasn't a total miss. We suggest it's better in the long run to write the session off – you just never reached the target or had too many misinterpretations. Seeking reasons to explain a miss easily leads to confirmation bias, as noted in our discussion of skeptics. They have enough to give us a hard time about – there's no reason to add fuel to the fire.

Hopefully after reading this chapter viewers who already engage in ARV projects will be more aware of their own inclinations to use judging photos to get a quick "fix" of feedback.

Following are two approaches, again borrowed from psychological theory. The first is to use positive conditioning – to replace one thing that feels positive with something else that feels equally positive or even more so.

Positive conditioning – Replacing a positive with an equal or stronger positive: As noted above, when it comes to ARV projects involving self-judging, any remote viewer will tell you they experience a sensation immediately before they see a judging photo. It is a mixture of relief and excitement, often accompanied by a sense that now we are finally going to get to know more about the feedback. We won't know for sure which photo is correct, but we will soon have a lot more information than we did. We are also about to find out if we did well, at least toward matching one photo, and we anticipate a thrill of excitement. This boost seems to occur even for those of us who realize they are about to enter into a danger zone where they are going to be exposed to one photo that is not correct. The stimulated feeling results from two factors – we have been operating at a level of stress from all the unknowns and now the stress is about to be relieved. So the act of self-judging allows for stress relief and increased arousal.

How, then, can ARVers who self-judge because it provides a kind of premature feedback still get some kind of satisfaction if they are supposed to avoid this pleasurable feeling?

One might think the financial incentive to do well through only describing the feedback photo and nothing else would be enough. Some projects pay viewers and some do reasonably well in describing only the correct target. However, displacement still happens in these projects, too. As Lyn Buchanan once shared with his CRV students, money appeals to the ego and logical mind, but the subconscious (maybe most peoples') doesn't respond to financial incentives (money) in the same way it does to things that make the body feel good. He keeps a bowl of candy nearby and rewards viewers who do well in their session – he treats viewers with M&Ms so they don't gain too much weight!

The above is purely anecdotal, but it may be important even for paid projects to provide incentives to connect only with the feedback. To do this, we may benefit by sweetening the deal at feedback, such as online feedback parties held via Zoom where everyone does "show and tell" about the photo that actualized. Or the project manager could assemble and show a group compilation from the winning transcripts. What other positive things can be done so viewers will feel more arousal at feedback time than during judging? We hope our readers will let us know of creative solutions they discover.

Adverse conditioning or aversion therapy: Adverse conditioning was a concept popularized by the behavioral psychologists. This technique reduces the appeal of behaviors one wants to eliminate by associating them with physical or psychological discomfort. For example, a person with alcohol addiction could be prescribed a drug that causes nausea if they drink or have their driver's license taken away for driving while intoxicated. A version of this is "covert sensitization." Instead of subjecting the person to the actual thing, the person imagines the undesirable behavior and either imagines or is actually exposed to an unpleasant stimulus.

While we don't recommend that project managers punish their viewers for displacement, we suggest they either minimize or withdraw attention when displacement occurs. They should also make it clear to newer viewers that focusing on displacement can harm a trial, with negative consequences for the group as a whole. If displacement happens, the project manager should not give the viewer positive attention or kudos for doing a good job at describing the wrong target. Instead, they should express eagerness at seeing how well the viewer can describe the actual feedback photo attached to the winning outcome of the next trial. Everyone is different, but this approach should work for some. Viewers are sensitive and must always be handled with compassion. If a project manager notices a viewer is continuously displacing, they might gently suggest the viewer move from doing ARV trials to other forms of remote viewing that are focused on real life tasks instead of future feedback photos. This is quite often a better path for some who suffer from chronic displacement.

In addition, we know of a few professional viewers who experienced severe displacement in ARV who now focus on their remote viewing businesses and practices using regular RV instead.

Passes are your friend, not your enemy

Recently I (Debra) moved to the state of Oregon and took my driver's exam. As part of the exam, they allow you to click "pass" if you don't know an answer so you can avoid getting too many answers wrong. If you click pass, it doesn't work against your score and you get a new question to answer. However, I wanted to show I could pass the test without passes even though I hadn't carefully studied the rules of the road book. At one point, I realized I was dangerously close to failing the test and began to pass, but by then it was too late. I had to wait three months to retake the test, but even worse, I continue to have to endure my spouse's jokes about it.

In the same vein, some ARV project managers fail to make use of passing as a valuable protection against decreased stats and loss from wagers. Project managers have told us they dislike passing because they feel it lowers morale and extends the length of projects. However, Joe McMoneagle has suggested if large amounts of money are involved, those engaged in ARV should be prepared to pass more often than not. He recommends setting the threshold very high – perhaps as high as 90% correct information for one photo and none for the other. For anything less, a pass should be called. While this could seriously prolong a project, it would protect against financial loss where larger sums of money are involved.[221]

Another solution is to set firm rules regarding when predictions can be made and when passes will be issued. For example, there must be a difference of two points between scores for the photos and one score must be at least CR 3.5 or 4 for a prediction to be made. For pre-registered formal ARV research projects, one could mandate a minimum number of predictions rather than a minimum number of trials.

On a psychological level, become cognizant of how you personally feel about passes so you can work through these feelings. If your project manager passes, whether for a project in which you are the sole viewer or if multiple viewers are involved, do you feel a sense of disappointment? If you have ever been a judge, did the idea of an impending pass affect the score you gave to a transcript (either lowering it or elevating it with the idea you wanted to avoid a pass)? Have you ever felt you or your team couldn't deal with more passes because it was prolonging the length of your project? If you answered yes to any of these, you may have an aversion to passing that is affecting your decision making.

Time burps – Theories related to time

One cannot discount the possibility that something about the way time works may influence and sometimes disrupt predictions despite how the ARV protocol seems to cover time glitches. For example, if a viewer is describing the photo they will see *after* an event, if something happens immediately *before* the event to change the course of history, the feedback photo should still reflect those changes and retroactively project itself (or be projected) to the viewer at the time they are doing their session.

Some evidence shows trials involving events that are clear-cut and definitive tend to result in accurate predictions compared to events with greater last-minute fluctuation.[222] Müller, Müller & Wittmann conducted an ARV study of stock market predictions and "post-hoc analysis indicated that the session quality depended on the volatility of the stock index: The viewer's perceptions were clearer and less ambivalent when the stock index also had a larger point difference at the end of the prediction."

Other informal findings demonstrate that when a trial is completed quickly, with feedback delivered right away as opposed to far in the future, predictions are more likely to result in hits. More formal study is needed in this area, including whether feedback really matters at all. Müller and Müller[223] found that it didn't – but in their study, viewers were tasked with viewing the photo associated with the winning option, so we can't say if their results would have been the same if they had tasked their viewers specifically with describing their feedback, which they didn't receive. Anecdotally, Debra and others have observed that for trials where viewers were told the feedback was the target but then didn't get feedback due to a glitch in the system, more predictions were passes due to either lower CR scores across the board or seemingly higher instances of displacement.

Theories related to time will be discussed in Chapter 7. For now, since time is so elusive, the best suggestion we have is for project designs to focus on events that are more likely to have definite outcomes that can be known as soon as possible after the remote viewing session is conducted.

Solutions based on modifying project design and setup

Avoid self-judging as part of the project design setup: While displacement can happen even with independent judging, it is apparently most often observed with self-judging, per anecdotal accounts, including those shared at conferences and in private conversations with remote viewing researchers Russell Targ, Ed May and Joe McMoneagle. In self-judging, the remote viewer is shown the photos associated with both possible outcomes so they can determine the best match to the impressions in their transcript. In this scenario, self-judging is performed after a remote viewing session but in advance of the event's outcome being known. This allows a prediction so a wager can be made.

In dozens of public demonstrations and group activities involving self-judging tasks in which we have participated, facilitated, or observed, more often than not, at least one viewer has had excellent matches to both photos. When this happened, newer participants were perplexed and more experienced people sometimes asked why a flawed protocol would be used. It has, indeed, felt very much like the movie *Ground Hog Day*[224]– different people, different photos, but the same experience time after time after time. Such events usually turn into a demonstration and discussion about displacement more than about remote viewing itself. This is very unfortunate!

That is not to discount some advantages to self-judging. First, self-judging can be fun. Second, viewers whose transcripts are incomplete or unclear may see correspondences to the photos that an independent judge might miss. Self-judging is easy, efficient and economical compared to independent judging, which requires recruitment, training, and more planning. Also, running a group through self-judging protocols can be useful to demonstrate how the entire ARV or experimental RV set up works.

We lack formal studies of whether self-judging leads to lower hit rates than independent judging. Further, some, like experienced remote viewer Tunde Atunrase, have conducted experiments (e.g., predicting the World Cup and a horse race) with viewers using self-judging and have been successful. Tunde finds no fall-off with self-judging in his solo work either. Viewers Tom McNear and Sean McNamara have voiced similar opinions about their ARV trials.

Most of the following solutions could apply to projects involving both self-judging and independent judging.

Solutions for separating the judging photos from the feedback photos:

This solution involves making changes to the overall project design and setup.

A few years back while giving a talk at a conference, Debra suddenly had an idea of how to create a separation between judging photos and the feedback photo. She suggested changing the judging photos in some way, such as showing them in smaller sizes with a larger feedback photo. While early parapsychologists such as Pratt and Rhine[225] reported the size of targets didn't matter in terms of psi performance, the intent would be to help viewers distinguish between the judging photos and the feedback photo. Sean McNamara, who has been working with groups of viewers in Colorado, ran with this idea and recently reported preliminary positive results in his informal investigations, finding instances of displacement seemed to decrease.[226]

Debra suggests other ways to separate photos, both in conceptualization and in actual physical space. One way would be to print out the feedback photo and pin the printout on a blank wall, preferably in a room where there is little activity. (Choose a blank wall without anything else on it, with neutral colors.) Task the viewers to describe the photo they will see on the wall at a future date. During the session, they should visualize the wall at feedback time. The judging photos will be presented only digitally.

Warning – This approach has produced positive results for Debra, although she has not run formal trials using this method. One problem, however, occurred when the printer didn't work. The trials resulted in passes. She also noted a tendency to not want to print the photo if there didn't seem to be a good match. Also, activities in the house at the time of feedback seemed to be picked up in the viewing session.

Another option would be to do a simple hand drawing of the judging photos while using a photograph of the feedback photo.

We hope those reading this will come up with more creative solutions and try to integrate them into their own project designs.

Physical design solutions – Unpairing the judging photos from each other: Another way to "unpair" the judging photos from the feedback photo when independent judges are involved would be to use two judges, each receiving only one photo to score. The project manager would compare the two scores and make a decision about the prediction. However, it would be important for the judges to do several pre-trials together to ensure rater reliability. As our ARV rejudging project found,[227] when 220 remote viewing transcripts across 86 completed ARV trials were rejudged by several judges, variability was striking between judges for both individual scores and for decisions about predictions.

These ideas again are not just intended for ARV projects but for all parapsychological experiments involving judging sets. They have not yet been formally tested.

Keeping viewers blind to the overall design: This solution would happen at the project design/methodology level. If remote viewers don't know they are participating in an ARV project setup involving different photos, they may stay focused on the target itself. They could be told only that they have a remote viewing assignment, and their assignment is simply to describe the photo they will be sent on a particular date.

However, experienced remote viewers may just assume it is an ARV project if they know the manager tends to be involved in such projects. Also, complications could arise if wagering is involved. Many consider it unethical for viewers to be told a project is for one purpose if the project manager is using their work for other reasons, such as their own financial gain.

Keeping viewers blind to each other, assigning them different target numbers and different photo options: Viewers can be kept blind to each other's identities so telepathic overlay is minimized. Also, each can be assigned a different target number and photo options, although this takes much more time and effort to set up and run and might not be at all feasible for ongoing applied projects.[228]

For experimental parapsychological projects, it is always preferable to assign remote viewers different targets to avoid the "stacking effect." This occurred in forced-choice tasks in which viewers were being given a series of tasks, one after another, and began to anticipate which choices had not yet been chosen. The present authors don't believe this to be as much of an issue for remote viewing/free-response tasks. However, some statisticians will take issue with assigning viewers the same target numbers and targets, and they will cite it as a reason for denying publication. Unless it is mentioned prior to their review, this should never be a reason for a project's dismissal, according to the authors of the paper that introduced stacking effects,[229] after they realized their paper was being used to unfairly invalidate entire projects. Still, this practice of assigning each viewer their own target numbers and photos might help minimize displacement in all ARV projects.

Daz Smith's Solution – Task viewers with describing that which is different about the two photos but attached to photo paired with winning outcome

Expert viewer, author, and publisher of *Eight Martinis* magazine, Daz explained in email correspondence that he has devised a method in which, rather than trying to make the photos different, he simply tasks the viewers to come up with something that is unique to the target photo.

> I purposely picked a set of two targets/images that were very similar in: age, size/ weight, shape, form and entropy. For example: two monuments or statues. The main differences were in the Stage 2 textures, touch, tastes, shape and form. If target A were metal and curved, Target B would be stone and angular. And my main focus data-wise was then on Stage 2 impressions and the sketches to make by decision.

Respect viewers' intuitive perceptions and feelings at all times

Remote viewers are intuitive not only during their remote viewing sessions. If they share how they felt about their session or about how the prediction will go, it's important to pay attention and show respect by acknowledging this. Far too often project managers, and especially formal researchers, will place an emphasis on the remote viewer's impressions during their session in terms of the target, but then dismiss their feelings about their session, the project, or problems they intuit about the trial or project itself.

Parapsychologists will often tell participants, "Yeah, that's interesting, but not something we are assessing, so there's no place to put that" or "Well, our design doesn't allow for us to pass or to do something differently just because your intuition tells you there is a problem." This not only invalidates the participant but leads to missed opportunities for insights and for ensuring that the trial goes smoothly. For this reason, researchers may want to build into their designs the ability to call for passes or, in addition to their declared or registered hypothesis, declare in advance their exploratory intent, to allow for ad hoc assessment of unexpected phenomena or unintended effects.

Many remote viewers and other intuitives are thoughtful people who are constantly observing their own experiences and internal and external processes. Formal and informal projects need to make room for these viewers so they can excel and participate long term.

Ensuring viewers feel respected may not only help the study itself, but help encourage their ongoing participation in future projects.

Bottom line: don't underestimate viewers' ability to make contributions to the study in ways that go beyond simply describing a target.

Various projects have attempted to track how well viewers felt they did during their sessions. This is a good idea; we'd just suggest making it optional. When a project demands the viewer to describe and rate his feelings and tries to quantify it (e.g., "Choose between 1 and 10 how confident you are about this session."), it may lead to guessing. Instead, encourage viewers to write a statement at the end of their session indicating if anything stands out about how the session went – Was it easy, frustrating, confusing? Were they tuning into more than one target? Did they feel happy or stressed when the session ended? Leave it open-ended. Perhaps the judge can translate their statement into a structured scale. If the researcher wants a structured scale measuring viewer sentiments and feelings about how the trial went, have the design include an option for a "non-response." Then spontaneous feelings won't get mixed in with forced guessing, and the assessment will likely be more accurate.

Project managers need to consider how they might be contributing to displacement

For example, very little has been studied about the effects of wagering on ARV outcomes. Project managers need to take into consideration the possible emotional or retrocausal effects, even when (or especially when) viewers are unaware of their wagering behaviors. The power of remote viewers' subconscious minds should not be underestimated. Would they approve of the decisions the manager/client is making? Would they feel taken advantage of or undercompensated?

As noted above, it's far too easy for researchers and project managers to fail to take into consideration that viewers are psychic not just during a session but often in relation to whatever is happening within the project itself. Their intuition may be operating at a subliminal threshold but still affecting them emotionally. They may have agreed to certain terms of the trial, leading a manager or client to believe everyone was in agreement, but on a deeper level the viewers may not feel they are getting a fair deal, or they may have ethical concerns. We suspect these could be factors but have no way of knowing. How many projects have been derailed because of these issues? This is a unique perspective on "the experimenter effect" described earlier, which has been widely studied in parapsychology.

What to do if you just can't overcome displacement in ARV?

Many viewers who experience chronic displacement in ARV projects are on target in remote viewing projects that don't use ARV or experimental designs involving matching tasks or decoy photos. Success, or lack thereof, in ARV doesn't translate to success in other types of remote viewing projects. If you are feeling frustrated about displacement or anything related to ARV, the best thing you can do for yourself is to take a break. Start practicing with other types of remote viewing targets and protocols. Actually, we highly recommend for all viewers to supplement ARV sessions with other types of psi work.

The word on the street is that doing too much ARV results in a decline in your other RV work. In ARV, you only need enough data to differentiate one photo from a second or a set of decoys, whereas in operational remote viewing, you need a great deal more data. Regular RV sessions tend to be much longer than ARV sessions.

Based on our own extensive participation in ARV projects, we suggest taking breaks between trials. If you want to retain your skill level in remote viewing, take considerable time off from doing ARV. Viewers will have different types of experiences with different setups, goals, targets and coworkers. Sometimes after you do regular RV for a while, you may be in a much better place when you return to ARV.

CHAPTER 7

Time and Remote Viewing

The past is a chameleon that always wears a hint of the now. It fools us into thinking it is, or always was, an absolute, when, in fact, it has never been that way. —Joe McMoneagle[230]

Given that physicists are currently grappling with an understanding of time, it may be that a psychic sense exists that scans the future for major change, much as our eyes scan the environment for visual change or our ears allow us to respond to sudden changes in sound. —Ingo Swann[231]

Time has always been a great mystery to our human species. We begin this chapter with an exposition about time and since our focus is on ARV, we explore the mysteries of time by interweaving the experiences and viewpoints of four remote viewers who have been intimately involved with research on time as well as how you experience it as a remote viewer. Their views sometimes echo current scientific theories about time and sometimes strike out on their own. There used to be few such "dual threats" (researcher and viewer) but that is changing among younger parapsychological researchers at universities and as more viewers venture into formal or semi-formal research.

At Jon's suggestion, we have chosen to focus on two giants in the field, Ingo Swann and Joe McMoneagle, one emerging giant – Julia Mossbridge – and one "upstart Crow" – what 17th century academics called Shakespeare – coauthor Debra Katz, who will be receiving her Ph.D. as this book goes to press. The choice of two women is not accidental, but more to the point Julia is a top-notch and very innovative researcher and author who both remote views and teaches it, while Debra has a wealth of experience as a remote viewer and teacher and has already published half a dozen peer-reviewed articles as well as three books on the psychic realm.

What time is it?

The sun dial is the emblem of humankind's relationship with time over the ages. The sun arcs across the heavens as humans rise and shine – then sleep under the sun's reflected beams. The sun measures the pace of life and shapes our understanding of time through its majestic movement. But in the 21st century, it is atomic clocks and not the sun that measure time. One second is now said to be the "amount of time radiation would take to go through 9,192,631,770 cycles at the frequency emitted by cesium atoms making the transition from one state to another" (Wikipedia). We can measure precisely enormous velocities like the speed of light or the electron (23,250 miles per second) or the 9 billion cesium cycles. Along with these unfathomable velocities, we learn of distances travelled and expanses of time unimaginable to our ancestors and hard for us to grasp even today. One duality emerges: slow macroscopic time and ultrafast subatomic time.

When we turn to frameworks to understand time, we find another pair of perspectives – time in fixed units as measured by cesium atom and by macroscopic clocks and watches, and a second kind of time – our subjective experience of it. Julia Mossbridge expresses these two specific ways of thinking about time by contrasting the empirical and the personal (her terms) with regard to precognition.[232] One obvious personal subjective feeling is that time goes fast when we are having fun or "last an eternity" when we are bored or suffering. Remote viewers have experienced far more than changes of pace when it comes to time, however, as we shall see later in this chapter.

The place of time in understanding the universe

In recent decades, researchers have used advanced equipment to better test scientific theories about time. However, it appears that the nature of time is one of those 100-year puzzles – or perhaps 1,000 years.

As theoretical physicist Renato Renner put it: "If I look at where we have paradoxes and what problems we have, in the end they always boil down to this notion of time."[233] Another theoretical physicist observes that physics is guilty of "expelling time" by not considering it a fundamental element of reality.[234]

Today science is abuzz with new data from astronomy and subatomic physics. A quick flyover of older and current views by physicists regarding time informs us: Time slows down at great velocities. Time runs forward or backward in the equations of physics. Time may run differently in parallel universes. Time may be quantized like photons. Time travel is impossible except in wormholes. Time is an illusion: Folks, we actually live in a block universe which contains what we call past, present and future.

Theoretical physicist Stephen Hawking went so far as to claim not only is the future indefinite, but history itself does not exist! This is postmodernism invading physics with a vengeance!

> Quantum physics tells us that no matter how thorough our observation of the present, the (unobserved) past, like the future, is indefinite and exists only as a spectrum of possibilities. The universe, according to quantum physics, has no single past, or history.
>
> The fact that the past takes no definite form means that observations you make on a system in the present affect its past. (p. 71)[235]

Most people will find it difficult or impossible to accept these propositions, if that is what quantum physics truly tells us. As influential as Hawking has been among cosmologists and the general public, his interpretation of quantum physics is just one of more than a dozen competing frameworks. For our part, we venture the belief that there has indeed been a history here on earth, or put another way, definite events existed in what we call the past. We would like to think that we can influence events (and ourselves) in the past, as Hawking believes, as well as Julia Mossbridge and Marty Rosenblatt, but, with respect, for us the evidence is not yet fully in.

The imprecision and fuzziness of time

We don't usually associate time with imprecision – in fact, the watches and clocks we consult give us a number accurate to a second, which is more than we generally need in daily life, while atomic clocks are nano-accurate. However, some scientists and remote viewers depart from that perspective. One of the latter is Joe McMoneagle, universally acknowledged as one of the very best remote viewers. Joe has taken part in thousands of hours of experiments in labs and has demonstrated remote viewing for the media more often than anyone else. He has written several important books, including *The Ultimate Time Machine*.[236] What does he have to say about time?

> As I explained in my book, *The Ultimate Time Machine*, there probably isn't anything like the present …What if I told you that, in my experience, most remote viewers who target something in the present usually provide some information that is pertinent to the target in the past and future? Well, that's exactly what happens in most cases…There are numerous examples of information being provided by viewers on present-time targets where the information was slightly off in one direction or another. It's always been fascinating to me that no one ever talks about

the phenomenon, nor do they take the past/future information into account when trying to evaluate how or why remote viewing works (pp. 152-53).[237]

For some physicists, the present is only the instant of "now," and everything else is past or future. In general use, however, the present ranges from a few minutes all the way up to an era but generally meaning from minutes to a year or so. Bracketing the present, Joe refers to the "near past and near future."

One thing can be said about real-time targets: they may predominantly lie within real time, but they will usually contain near past and near future information as well. This means that fixing a target in time is critically important to the remote viewing process. The more accurately you can do that, the better…By necessity, when targeting something in present time, or for that matter, in the past or future, you should provide one of the following statements:

1.   Describe the target as it exists now.
2.   We want to know about the target, as it exists today, July 14, 1999, at 10:15 A.M.
3.   Our interest lies in present time only.

I once did a whole series of targets at SRI-International for which no specific time of interest was mentioned. Every single one failed. It was years later that buildings I had described in those remote-viewing sessions were actually built at those specific target sites.

The importance of viewing at a precise time

When helping to plan the 2016 Applied Precognition Project Conference, Debra suggested inviting Joe McMoneagle to share his approaches to finding missing people. He had spent a few years intently focused on using his RV skills to find missing people around the world while contracted with a Japanese television company. For instance, with Joe's remote viewing and the help of the Japanese government and a search team, two women who had been kidnapped in China and taken across the North Korean border were located and returned.[238]

This was a popular television show in Japan – referred to as Nippon Television's prime-time Chounouryoku Sousakan show (roughly translated, *FBI: Psychic Investigator*). Although the cases were well documented, we could not locate anything in writing documenting Joe's specific approach to finding people. We were overjoyed when he agreed to present on this topic and even more so when he laid out his approach step by step. While we are operating from memory about his talk, what stood out the most was his approach to time.

Here's a summary of the approach Joe uses when he is doing a target to locate a missing person, especially when a search party is available to act on the information and conduct a physical search for the person. At the core is the need to task himself to focus on the *exact moment in the future* where the missing person would be found by the specific search team. This was very important since the person could easily be in many locations in a single day.

Joe gave a stunning example in which a Japanese man had been estranged from a brother for decades. Joe first set his intention on discernable landmarks to guide him and the team to the country and city where the man would be found. Helpful landmarks included unique airports, train and bus stations. Then he remote viewed the next landmark, perhaps one street away, and then the next, so he could essentially create a map of landmarks bringing him closer to where the person would be found. In the case of the missing brother, this process allowed his team to identify a large apartment building where many local workers were housed. The search team staked out the building and were standing there when the brother returned home from work. Since they didn't have the brother's present name, only an old photo, it wasn't possible to ask the apartment building owners if he lived there. If they missed him, they had no other way of finding him because the building was much too large to knock on every door. Remarkably, the brother was located in this manner, through the use of remote viewing paired with a very dedicated team of searchers.

Returning to time's imprecision, we find a surprising parallel in recent ideas in physics. One statement of this counterintuitive notion is "Physicists Find That as Clocks Get More Precise, Time Gets More Fuzzy."[239] As the scientists referred to in the article:

Significance: We find that there exist fundamental limitations to the joint measurability of time along neighboring space–time trajectories, arising from the interplay between quantum mechanics and general relativity. Because any quantum clock must be in a superposition of energy eigenstates, the mass–energy equivalence leads to a trade-off between the possibilities for an observer to define time intervals at the location of the clock and in its vicinity. This effect is fundamental, in the sense that it does not depend on the particular constitution of the clock, and is a necessary consequence of the superposition principle and the mass–energy equivalence. We show how the notion of time in general relativity emerges from this situation in the classical limit.

Ingo Swann also found that time was not sharp and distinct when remote viewing – time lost its precision. Ingo encountered time slippage in what he called "analytics," numerous experiments with letters, numbers and symbols. He observed that when trying to predict the next symbol in a series, the targets were "lining up," waiting to be called, as it were; they formed what he termed a "lump." He could access and convey the order of the targets in the "lump" as long as he did not stray into what he called the "SUMP," a hypothesized Analytical Summation Pool.

As he explained in a report to SRI, the SUMP identifies incoming ESP information "to give it mental-image forms" and uses stored images or constructs mixed images out of what is at hand. Further, SUMP energy is stronger than energy from the ESP information flow itself. The SUMP will create images of both if it is a forced choice between two targets known to the percipient – a point that offers hypothetical constructs (lump and SUMP) to explain displacement.

Ingo felt that in this process "fields" are created and in each field "time and space do not seem to exist in the same line up as we experience them in our normal objective time/space continuum."

> At the end of the session, I had become aware of at least two "energy polarities" which I then called Universe 1 and 2, and then decided to call them "fields". Field 1 was seen to surround the body, and field 2 was seen to extend to and incorporate the target's location. It also began to appear that, within these "fields" the normally expected demarcations between present and future either did not exist or were considerably weakened, allowing for sighting future targets more or less in their correct order.[240]

Swann noted that soon "'lumps' of targets began coming in all together," with emphasis on the next three or four targets. He got a "lump" of 10 targets, eight of which were correct. Ingo described the fields as interacting and felt that a "psychic loop…gets going." He observed that this puts "our problem…directly alongside some familiar quantum models."

The archives at the University of West Georgia unfortunately contain no documents about further development of these provocative ideas. Swann was a man who loved exploring new fields and it may be that the reports to SRI were as far as he took these notions of fields, polarities, configurations, lumps and SUMPS.

The views of Physicist Carlo Rovelli are very much in tune with what Joe and Ingo observed:

Rovelli maintains that our experience of time owes to a blurred, macroscopic perspective of the world that we encounter as human beings. "The distinction between past and future is tied to this blurring and would disappear if we were able to see the microscopic molecular activity of the world," he argued during our interview. In other words, the past and the future are equally determinable at the molecular level.[241]

Could this fuzziness, inexactness or slippage be made use of in remote viewing? It appears it can. We explore this possibility in our chapter on the lottery.

Can we step outside time?

Given this now fluid nature of time, can we step outside it? Ingo Swann thought so. He believed the viewer can not only step outside matter, energy, time and space (MEST) but in fact had to do so to acquire knowledge of the future. He imbibed this perspective from his early out-of-body experiences and from his years as a Scientologist.[242]

> The superpowers of the human bio-mind, of which remote viewing is but one, can be defined as those SPECIES-INHERENT faculties which permit human awareness to transcend the conventionally perceived limits of space and time, and of matter and energy as well (p. 29).[243]

The parallel with science is that in the "block universe" everything is happening at once, to use one everyday paraphrase. Julia Mossbridge had a dream which inspired her to "take seriously Minkowski's idea that time holds all physical events in perpetuity — all physical events in the past, present and future co-exist in a 'block-like' universe with no change or real movement."[244] Yes, we live on the third stone from the sun and we survive on it based on a linear-time continuum, but in reality there is no past, present or future.

According to Swann, we can escape from the linear-time continuum by using our "right brain" and he cites research, new at the time, claiming distinct functions for the left and the right hemispheres.[245] However, recent science has rebutted the oversimplification of hemispheric function, which has become a meme: the left being linear, the right being global and nonlinear, along with other purported unique functions of each hemisphere.

Further, the actual picture is even more mixed – with time again being the critical factor. As SRI Director Ed May points out repeatedly, we actually don't know *when* we receive the psi information. Was it when we woke up this morning? When the tasker decided what the task was? Was it born with us? (Joe McMoneagle believes it is.) We are still at the stage in which we don't know when or where psi information is manifested.

Precognition

The nature of time is obviously of fundamental importance to those who say we can get information about the future to make accurate predictions. To explore that, we turn to precognition, the term used in parapsychology for predictions of the future. This is what ARV is used for 99% of the time.

The idea of linear time has, of course, been held by ordinary people and scientists over the ages, and it is overwhelmingly believed by us today as we go about leading our daily lives, checking the time on our smartphone, watch or clock on the wall. But scientists suspect time is not simply linear. To test that, they have done "presentiment" experiments, in which they measure subjects' responses to a stimulus seconds prior to the stimulus. Dean Radin of IONS has conducted many such experiments with positive results.

Psychologist Daryl Bem and coauthors performed a meta-analysis of presentiment experiments and claimed the experiments showed future events were indeed foreshadowed – and a parapsychological donnybrook followed.[246] Ingo Swann was aware of earlier studies like Radin's and was on board with them:

> They don't say much about the unconscious processes but do show that such processes can access the future, at least a few seconds ahead. Another way of putting this is the information about the future somehow is transmitted back to the present to the percipient or to a remote viewer. Or as Swann put it, "present-time Events are somehow being formed by the future, and the future is somehow looping back into present."[247]

Exploring this further in *The Premonition Code,* Julia Mossbridge and coauthor Theresa Cheung opine that every event could exist as a causal loop in which past and future are pushing and pulling simultaneously, although you cannot prove the physical reality of causality[248] (pp. 41, 44). As David Hume argued long ago, there is only the succession of one event after another; there is no "causal glue" connecting the first and second event. But, Mossbridge wonders, are they in fact bound together? Are they acting on each other? Is there "retrocausation?"

Taking the point about the reality of precognition even further, prominent researchers Ed May and his colleague S.B. Marwaha support the thesis that precognition is the only form of psi. They present a number of arguments for their position. We touch here on only one of them.

Assuming the validity of a signal-based approach, we may consider that, while retrocausal signals are emerging from a future point in spacetime, from the person-centric perspective the information signals are present in real time. This may hold the very concept of 'pre'cognition (a personcentric perspective) redundant, and consider retrocausation as the primary factor in the entire process. (p. 20)[249]

Comment: When a person gets a signal from the future, it is expressed in present tense (gets) and happens in present time (now). For the person, this information *pre-exists* what confirms it, which happens later. The information is precognitive in that she knows it will happen before it does happen. There is no redundancy in the "pre." If retrocausation is "primary," that is a different framework of analysis; it doesn't negate her distinct perspective on what is occurring.

The act of a person giving herself a command to move into the future to access her future self may be "the primary factor" in bringing about the "retrocausation." Or it may be, as Julia Mossbridge speculates, there is a push from the present, or really, the past and a pull from the future, both operating to produce the transfer of information to the present.

One also wonders, as seems to be implied, about a perspective parallel to the "person-centric" one of a remote viewer. Is that the perspective of science? Can science even have a perspective? Perspective is something held by an individual or groups of individuals. Ed May's own perspective on this subject is and must be "person-centric." Perhaps he means this as an allusion to the systems of equations and theories that make up established science?

What about the "perspectives" of the conscious and unconscious minds? Julie Mossbridge presents her views on this in *Time and the Unconscious Mind*:[250]

Most of us think we know some basic facts about how time works. The facts we believe we know are based on a few intuitions about time, which are, in turn, based on our conscious waking experiences. As far as I can tell, these intuitions about time are something like this:

1.   There is a physical world in which events occur,
2.   These events are mirrored by our perceptual re-creation of them in essentially the same order in which they occur in the physical world,
3.   This re-creation of events occurs in a linear order based on our conscious memory of them (e.g., event A is said to occur before event B if at some

point we do remember event A but we don't yet remember event B, and at
another point we remember both events),

4.   Assuming we have good memories, what we remember has occurred in the
past and what we don't remember but we can imagine might: a) never occur,
b) occur when we are not conscious, or c) occur in the future.

These intuitions are excellent ones for understanding our conscious conception
of ordered events. However, they do not tell us anything about how the non-
conscious processes in our brains navigate events in time. Currently, neuroscientists
assume that neural processes of which we are unaware, that is, non-conscious
processes, create conscious awareness as a reflection of physical reality….Thus,
if we wish to understand how events unfold in time in the physical world, we
would do well to attempt to get some hints about how these events are navigated
by non-conscious processes.

Julia's belief is that her calling is to "teach and learn about love and time." Going farther
than into the realms of our inner being than any other psi researcher, she ventures to suggest
that if we feel "unconditional love," this might free us from the usual constraints of how we
perceive time and space. For example, with unconditional self-love, we could perhaps send
information from our future self to our present moment self. (Julia Mossbridge, Marcia
Nisam, RN, Adam Crabtree draft – submission to APA 2020).[251] This is a bold mission!

Cycles of time

Cycles proliferate in life and in many sciences. Wikipedia lists planetary cycles of
astronomy, climate, weather and geology; organic cycles of agriculture, biology, medicine
and brain waves; physics and the electromagnetic spectrum and in math as well. Wiki
also cites cycles in economics, music and rhythm, religion and myth, society and culture,
military and war. As noted, the unit of time is measured in the enormous number of
cycles of a cesium atom in what we call a second. Mathematician and theoretical physicist
Roger Penrose has even written a book about cyclical universes, espousing his Conformal
Cyclic Cosmology.[252]

Turning to remote viewers, Ingo Swann discourses on the cycles in astrology. In his
Agony and Ecstasy of the Signs of the Zodiac,[253] Swann detailed his 30-year study of astrology,
a subject he used to complement his remote viewing. He strongly believed in astrology's
power and had charts done of visitors and clients. To get beyond the stoop where he
often sat outside his home, he would cast a chart to see if a visitor made the grade. He
even undertook a study to find what astrological sign most serial killers fall under. In that

unpublished work and elsewhere, Swann stressed the cyclical nature of time, a theme he expounds on in *Your Nostradamus Factor*,[254] which is his most detailed explication of his views of time and the future.

Cycles also feature in discussions of ARV. One of the main ideas is that viewers reach out to themselves in the future and send information back to the present, creating a "time loop." APP's Marty Rosenblatt has been particularly forceful in presenting this idea. Other cycles involve the idea that participants in an experiment somehow influence each other and create informational loops. Concern about these cycles has led researchers and ARV practitioners to limit the knowledge each participant has. Perhaps telepathy is involved in all remote viewing and, if it is, causal loops or time loops should be minimized as much as possible. It is widely believed that displacement in ARV may be traced, at least in part, to such loops. Hence KISS (Keep it simple, stupid!) finds a place in RV practice and experimentation.

Julia Mossbridge has also written about time cycles. In her review of Eric Wargo's book, *Time Loops*, she shares her experiences and theories about time.

> But Wargo's idea about consciousness displaced in time invited me to revisit an old idea that, for me, was relatively unformed—the idea that consciousness is literally transmitted from the future. The idea is that in a physical but not block-world conception, consciousness is like a physical wave that is transmitted from our future selves to create our present mental objects. This wave may be received by what we call our current experience, or what we call our unconscious mind. But it's not "where" the control tower is—it's "when."[255]

As we noted earlier, Julia Mossbridge contrasts the empirical and the personal – two ways of approaching precognition. She has focused on the empirical (lab, science) but she says Wargo forced her to think more about the personal experience of precog. Julia thought about a taxonomy of precognition, but abandoned the idea for now. She opines that causal loops may be responsible for our sense of our selves (or self). It's not just precog loops at play here.

Are remote viewers more accurate when viewing the past, present or future?

If we are more accurate when viewing "the present" (which instantly becomes part of the very large "past"), then perhaps we should stick to that and avoid trying to peak into the future. If the reverse is true, then all the activity around ARV has a point and is fully justified.

We found contrasting views on this issue. First is a report from the Sun Streak project (part of the Star Gate project) from 1987:

During the period of 17 November 1986 to 2 October 1987, 77 sessions were conducted by six remote viewers against Project "P," a utility assessment initiated to determine a remote viewer's ability to function effectively in a purely predictive mode "P"… Remote viewing efforts conducted under this revised protocol ceased on 2 October 1987. The results at Tab E reflect weak correlations of 13% and 18% respectively for both periods under consideration. Remote viewer enthusiasm in the conduct of project "P" has waned considerably. Suggest this project be halted pending completion of an in-depth review of this effort…

Based on the premise that near-time exposure to future events might enhance remote viewer access to significant occurrences (e.g., the President Kennedy assassination), four remote viewers conducted "free-flight" sessions against events of foreign intelligence significance in the Persian Gulf. (The expression "free flight" merely denotes that remote viewers functioned independently and without the assistance of a monitor or an interviewer.) Although "free-flight" methodology did not produce significant results against perceived underlying future events, the methodology will be attempted against real-time targets when probabilities of success should be somewhat enhanced.[256]

On page 9 Project "P" is summarized. Total number of correlations ("hit ") for Project P weak correlation 14 = 16.87%. Strong correlation 1 = 01.2%. No correlation 68 = 81.93%. Project "P" revised. Weak correlation 2 = 04.08%. Strong correlation 5 = 09.43%, and no correlation 46 = 96.79%. *(Editors' Note: This totals more than 100%).* The conclusion was that precognitive work has slim results for such targets.

Charles Tart found that "present-time" targets were far more likely to be correctly identified than ESP forced-choice targets.

Abstract: 53 studies of present-time ESP and 32 studies of precognitive ESP were reviewed and amounts of information acquired in these forced-choice ESP studies were quantified, using a measure of the average number of bits per trial. It is noted that in present-time ESP studies, the percipient attempted to call currently existing targets; in precognitive ESP studies, the percipient attempted to call targets that would only be generated by a random process at some later time. Upon application of the quantification measure, a difference between present-time and cognitive studies was found: Present-time ESP can work up to 10 times as well as precognitive ESP in forced-choice tests. Three theories are proposed to account for these findings, including a psychological theory, a 2-process theory, and a temporal-break theory. The psychological theory states that there is a generally held bias against precognition in Western culture, so that percipients do not try as hard. The 2-process theory claims that present-time ESP and precognition are 2 basically distinct processes, with inherently different characteristics. The temporal-break theory asserts that ESP is a unitary process, and something in the nature of time attenuates ESP performance that extends into the future.[257]

Tart notes that this finding applies to forced-choice tests. Binary ARV is an example of a forced-choice test (you must decide on outcome A or B, photo A or B).

On the other side of this issue, Joe McMoneagle maintained that dozens of sessions at Ft. Meade successfully predicted events 90 to 365 days in the future.[258] Further, there have been thousands of successful predictions of sports and financial events using ARV, at odds very much against chance, as the chapters in Part 3 show.

Debra's tenets of time related to ARV

As associative remote viewers, we are essentially *Time Bandits*. As in the 1981 film starring Sean Connery, we are like renegades from the future, charging through a boy's closet wall to snatch up information – a treasure map. Only in our case, instead of simply intruding into the past, we are stealthily sneaking into the future and – perhaps – sending information into the past. That is, sending it to ourselves as we remote view an event "precognitively."

And why not? The information is there for the taking in the future after the event – knowable to all spectators and gamblers at the game or horse race, visible to millions of traders who play the stock market. So then, what's wrong with a preview? We have a plan, it's logical, everyone knows their part, and the first few times we are pulling it off with mastermind precision. But then, BAM! The time machine sputters, starts to smoke and stops. Something or someone does a switch-a-roo and our very clear ARV session has failed despite a great match with one of the two photos. We have to take a loss and we wonder, "What the heck just happened here?!" Our rock-solid ARV protocol didn't work, so we begin to question our models. Eventually we wonder about our understanding of the nature of reality and time.

In discussing theories of time, we must acknowledge that our mechanistic view of time is skewed. Our Westernized modern education has taught us that time is something linear, external, fixed, occurring in separate units along a line. Modern civilization has made us slaves to alarm clocks, calendars, "timers," and now Fitbits and other tracking devices. Those who don't put aside their own internalized body clocks and conform to the external time impositions suffer severe consequences, such as loss of income and social prestige. Despite the health risks of putting externalized time over one's own rhythms, whenever economic survival depends on "keeping time," you can be sure it is going to loom large in our lives. It will have an emotional grip on us. It is like Captain Hook, forever stalked by the crocodile that bit off his hand and swallowed a clock: the ticking is an ominous sign. We are primed to be hypervigilant to our ticking, even when on vacation or during down time. We have internalized the essential engine that powers the civilization to which we belong.

While this is the stressful negative side of it, there could be positive aspects as well. Perhaps on an intuitive level we are priming ourselves to anticipate future events on an unconscious level.[259]

In order to expand our thinking about time, perhaps we need something more than the models that scientists and philosophers to date provide. Maybe we need additional tools that can help us step out of what Ingo Swann called "reality boxes" and what Louis Bostwick, founder of the Berkeley Psychic Institute, referred to as "core pictures." I heard a group of mediums recently refer to these as "soul filters." These boxes, filters and lenses have been constructed by Westernized education, which essentially inserts visual representations of things and phenomena that may not really be the best, or even adequate, representatives of the things themselves.

One set of tools that helps with breaking through time is ARV. When you do ARV, especially when you are directing your attention purposefully and consciously at different periods of time, you start to notice interesting things. For example, if you focus on a photo hanging on a wall two days from now, you notice not only aspects of the photo on the wall, but whatever else is happening in the room around it or with yourself two days later.

For example, I was managing a project and a viewer (Dave Silverstein) had a group of Nazis in his transcript. I went to the YouTube feedback video to look at the target, and the ad at the start of the video was a bizarre scene in a beer commercial with Hitler and Nazis. The rest of the video had nothing to do with this. However, when I went back to watch the video again, there was a different ad. I watched it several times but the first ad never returned. It was no longer there, but I had seen it.

This was not a case of displacement, as far as I was concerned. Dave had been told the feedback would be in video form. The beer scene had been there at the start of the video I watched but was not available to send as feedback because the videos continually changed. What this told me was Dave had tuned into a specific future point in time. He was somehow oriented to me, as the manager of that trial, and the intended feedback video I would see.

These striking experiences are why many of us stick with ARV, despite its challenges and frustrations.

Time tenets

What I'd like to outline below is not a full theory, but rather tenets of how I see life operating through my years of intuitive explorations. At best, this might be considered the start of a theoretical framework. These tenets can be seen as natural laws or principles underpinning how consciousness, nature, humanity and the world operate. They are supported by much of the parapsychological and theosophical literature, although sometimes stated in different terms. They also match much of what one finds in Ingo Swann's writings, although again, he used different terminology.

These tenets include:

1. Everything within us and outside us that exists in the world is informational.
    In his 1979 book *Messengers of Deception*, Jacques Vallee called for a "physics of information" that would enable scientists to think in a more nuanced way about a wide range of paranormal phenomena.

    I have always been struck also by the fact that energy and information are one and the same thing under two different aspects. Our physics professors teach us this; they never draw the consequences. —Jacques Vallee[260]

2. Both inanimate things and living systems have their own makeup, their own constitutions, but they are all open systems. As informational systems they extend outward but in less solid form as the distance from their "core" grows.

3. Objects and people (or aspects of them) extend through movement, vibrations, oscillations, emanations. These are not static and not stable. (This echoes remote viewer Simeon Hein's theory of "resonance.")

4. Some of these impact, create or contribute to an entire space, whether physical, communal, or even virtual. (This is on par with Rupert Sheldrake's conception of morphic fields.).[261]

5. These "extensions" form entire fields (i.e., the space of a room or a building) and interface with all that is in the field, including all that is permeant and all that moves through it which is temporary. Individuals' thoughts, emotions and activities

contribute to these fields and the fields in turn impact the individuals on every level. (Environmental and Marxist theory support this perspective.) Some of the above can be measured by current technologies, some may not yet have technologies to measure them.

6. The above can be sensed through a variety of psi capabilities, particularly through clairvoyance (ability to see information visually) and clairsentience (feeling information through the body senses or processes, such as emotion, pain, sensations related to temperature, tiredness, arousal, etc.).

7. The effects can be unidirectional, bidirectional or multi directional. One object or system might impact another, or they may mutually impact each other, or one may impact several and several may impact the one, and in different ways.

8. The impact may have to do with mirroring of relational/emotional aspects. Meaning that individuals that have something in common with others or with the issues or subject matters may impact and be impacted more than other who don't share these similarities. This can operate consciously, or unconsciously.

9. The above processes can happen across time; there are no restraints in terms of linear time. Some may be stronger when in proximity in time to each other, some earlier, some later.

10. All the above tenets operate unconsciously, but people can become conscious of them, either spontaneously or through years of practicing awareness.

If all of the above are true, which I'm quite sure they are, time is mainly a measurement tool.

Our consciousness is always "moving around in time." For example, try this: Open your closet and look at an item. Then notice the contents of your mind. What do you notice? As you look at a certain shirt, or suit, or dress, or shoes, you will probably have multiple thoughts at once about different time periods. You'll see the item as it is at the moment, but you may think back to when you bought it, where you wore it and who commented on it. You might also have considerations about it for the future, like plans to wear it or worries it might no longer fit.

Your mind wanders through time – past, present and future. This is the nature of the mind. At least you are not confused about where your body and mind are right now. You are oriented to time and place. However, imagine if you came across this item in a remote viewing session and had all the same thoughts about it – past, present and future. This could disorient you as to the time of what you are remote viewing and could lead you off on many a tangent from the actual target. This is one reason RV is very challenging.

I conclude with one of my all-time favorite books, Robert Wolff's *Original Wisdom*.[262] Wolff recounts his time spent in Malaysia with an ancient tribal people called the Sng'oi. They seemed to have amazing intuitive powers, such as knowing exactly when a stranger was going to arrive at their isolated village or knowing what was in the ocean while standing on the shore. Every morning, they awoke to share their precognitive dreams in a group huddle. They did not dream for themselves, but for the whole group. They had no devices for timekeeping. Their concept of space was quite different in that they didn't have their own designated home, but took turns sleeping in each others' huts. The only thing they seemed to have of their own was a tree in the desert, which each would adopt as a lifetime partner.

I often think of how native peoples communicated with each other when there was no other means of communications available. I do this because sometimes I need to remind myself how natural and easy it is to access information at a distance. Even that term "at a distance" doesn't seem correct because it denotes the body is at one spot while the psi information is at another. Yet when consciousness is extended, it seems more like a satellite spreading out 360 degrees around us, emitting communications and information, but bringing them in as well. Where telegraph and telephone used to be a model of psi, perhaps what is going on is more "a satellite model."

PART TWO

SCORING AND TARGETING

CHAPTER 8

A History of Targets Used in ARV

In this chapter we will first discuss target materials used in parapsychology experiments and then targets used in applied psi projects. Occasionally there is crossover.

In the experimental projects, target materials are often selected either for purposes of testing the materials themselves or purely for using them to test something else. For example, we can run an Associative Remote Viewing experiment where we want to test whether photos of objects produce better results than photos of locations, or whether viewers prefer videos to still photographs, or whether they have an easier time noticing people as opposed to pets. In all these examples, we are comparing two types of target materials to each other; the materials themselves are under examination.

But what if we wished to test something other than target materials? What if we wanted to find out if female viewers could produce better results when remote viewing than male viewers? For this, we simply need something for participants to turn their psychic functioning to and that would be the same kinds of materials for both groups to ensure the variable being tested was gender differences and not the targets themselves.

Therefore, in designing experimental parapsychology projects exploring intuitive perceptual abilities, one of the foremost questions is "What will the psychic task be?" Implicit in this decision is consideration of what type of task or "target" will bring about the best results in terms of psi conductivity, over a period of time, for the greatest number of participants. While considerations about targets may not be the main research topic, a researcher can never be sure their findings aren't particular to the target material being tested; the subject calls out for more research. One could argue that every study is inherently testing its own methods, including its use of particular materials, as much as it is testing a particular topic.

In projects that are applied or operational (meaning for real-life purposes) and involving RV rather than ARV, target materials are often intricately fused with the problems or questions the project is designed to explore. These materials are not merely a means to the end, but rather the end that makes use of the means. They cannot be replaced or swapped out with others because that would completely change the project's entire purpose. Examples of such applied RV (non-ARV) projects are finding Amelia Earhart's airplane, locating a missing soldier or finding a buried treasure. In these cases, the target material is the airplane, the soldier or the buried treasure.

The exception occurs in applied projects that involve Associative Remote Viewing – the end goal is "associated" with something else, such as who will win a game. That something else is the target material, which could be anything. If we want to know the winner of a baseball game, in regular RV, the target materials are the game. But in ARV, we associate a photo with each team, so now it's not the game that is the target, but rather the photo associated with the game. This opens up many more possibilities for materials to be used to get an answer, which is one reason ARV was developed. In this way, then, ARV protocols are very similar to experimental RV projects, which also means ARV lends itself to combining aspects of experiments with applications.

Below we will give a historical account of target materials to try to answer the question: Which materials have led to the most successful remote viewing performance? This will be followed by an overview of types of applied RV projects, their purposes and materials, focusing on the last 20 years.

Background
Target types in forced-choice tasks
Over the years, a variety of targets (objectives) have been used in psi-based experiments. Here's an overview of the earliest mentions of target material:

Honorton[263] pointed out "the rudiments of an experimental methodology were suggested three centuries ago by Francis Bacon…in Sylva Sylvarum…Bacon discussed experiments in which 'the emotions of shuffling cards or casting of dice' could be used to test the binding of thoughts."

Honorton attributed the "first serious effort toward scientific examination of psi claims" to the Society for Psychical Research, founded in 1882, outlining early experiments by SPR members that involved (purported) telepathic reproduction of drawing at a distance. As early as 1884, Richet introduced "experiments involving card-guessing which involved the application of probability theory to the assessment of deviations from theoretically expected chanced outcomes" (p. 104).

In *Extra-Sensory Perception After Sixty Years,* Rhine[264] summarized a collection of experiments referred to as *forced-choice tasks*:

> There have been two principles determining the selection of test materials: the first is the use of that material to which the subjects are accustomed; and, second, the use of objects which permit a ready evaluation of the test results.

He noted that

> earlier experimenters used familiar objects, diagrams or common items of information as the basis of ESP tests, the reason being that these would require less departure by the subject from accustomed performance…Other materials utilized for targets included small squares to be located on a checkerboard containing 48 such squares, and a row of five small boxes, one of which would be opened by the participant (p. 302).

Rhine further explained: "As it became apparent that there was no essential difference in materials used, the second principle mentioned above, namely, facilitation of evaluation, assumed prior importance. The use of playing cards, numbers and letters of the alphabet followed." Zener cards were later adopted to utilize a fixed p-value. These cards consisted of

> a standard deck of twenty-five cards, five each of five simple designs – rectangle (later changed to square), plus, star, six wavy lines (later changed to three) and a circle. With this design, participants didn't know the order of the targets, but they did know, or quickly learned, the nature of the possible five cards (p. 312).

Rhine's wife Louisa studied various aspects of the cards themselves; for example, examining whether cards with several symbols on them produced better results than cards with just one symbol and finding no significant difference. J.B. Rhine writes:

> Pratt and Woodruff compared symbol size width that provided greater contrast against a white background. While the larger symbols stood out more definitely in visual perception, according to Rhine they were "no better than the others except for the period of novelty which was, for all sizes of symbols, outstanding in score average. MacFarland and George introduced badly distorted symbols in comparison with the normal, anticipating that the weak and erratic mode of perception in the ESP tests might be supposed to miss the distorted symbols

more often than the normal. There was also the possibility that the distorted symbols would arouse esthetic avoidance in the subject. No appreciable difference was found, however, except when the experimenter himself, MacFarland, acted as subject. It is indicated that grossly distorting the symbols does not reduce their perceptibility" (p. 312).

Target types for open-response tasks

At the same time Rhine was dominating the parapsychological scene with his forced-choice designs and emphasis on statistical methods, the French parapsychologists were also conducting formal experiments under blind conditions. However, they were largely using open designs that allowed them to work with a vastly larger selection of images.

In providing a historical overview of this movement, Swann & Puthoff[265] noted that psi-based tasks were no longer simply thought to be "guessing" tasks, but ones involving complex psychological processes. The participants had to allow for the telepathic information to flow through their own minds and bodies, where memory and emotions could facilitate the processes involved or hinder them. While the French parapsychologists did not quantify their results, they focused more on examining internalized experiences of those involved in performing what they viewed to be a combination of telepathic and clairvoyant tasks. This introspective focus was most likely facilitated by the dual experimenter-participant role occasionally taken up by researchers, who noted parallels between normal perceptual processes and psi-based processes.

Upton Sinclair[266] and René Warcollier[267,268] used simple sketches of things one might find in their natural environment. They believed participants should be given targets that would replicate their natural daily experiences and used images such as birds or animals, combinations of shapes, and familiar objects. These images were hand drawn, usually in black ink on white paper, sometimes in color.

Warcollier[269] in particular felt certain principles from the Gestalt school of psychology were applicable to emerging data from his own telepathy experiments. He stated,

We can look to the psychology of perception for other principles that reveal themselves and paranormal behavior … the Gestalt theory is the school of psychology that seeks to understand the wholeness quality or structure of each perception or reaction (p. 26).

Warcollier brought in Gestalt concepts such as "the relationship between figure and ground." He concluded,

What we may include from our knowledge of the normal psychology of perception is that there is always a dynamic core in telepathic perception, not unlike it in normal perception, whether the impression is global or detailed. There is a tendency toward organization, toward a wholeness character of the impression, in order that perception be as simple, as symmetrical, as regular and as meaningful as possible. (p. 27)

Warcollier also found participants responded to different targets. Those who were more emotional seemed to be more affected by the emotionality of targets than those who were less emotional.

Warcollier theorized that even simple drawings that conveyed a sense of movement (such as birds or trains) would be more perceivable than those with stationary subject matter. He wrote, "The idea of movement in the telepathic impression involves the Gestalt theoretical principle of Prägnanz ('good figure') or precision" (p. 28).

Krippner & Zeichner[270] found support for Warcollier's theories in their dream experiments, in which participants had more success describing dynamic than static content, a frequent finding in the parapsychological literature (albeit with some results showing no difference).

Honorton, et al.[271] published the first autoganzfeld study, in which selection of targets was done with the aid of a computer program. A ganzfeld study uses a (purported) sender and a receiver who is placed under conditions of sensory deprivation. For the first time in such studies, short video clips (referred to as "dynamic targets") were used in addition to art prints, photographs and magazine advertisements ("static targets"). The dynamic targets consisted of approximately one-minute excerpts from motion pictures, TV shows and cartoons. Results were significant for trials involving dynamic targets but at chance for static targets.

Honorton, et al.[272] conducted a meta-analysis of all the ganzfeld literature and found the success rate for 1,190 dynamic target sessions was "highly significant," while the hit rate for static targets was not significant.

Watt,[273] building on an earlier study performed by Morris,[274] examined the popular literature on psychic training, comparing it with a survey of spontaneous cases. Her findings seemed to echo those of Warcollier in that she found "there was a narrow focus on physical target characteristics without considering inevitable influences of properties of the perceiver and the environment on what aspects of the target stimuli would appear salient to any individual." Her review of the popular literature suggested that emotional impact and human interest content made good targets. This led her to recommend that target material should be both psychologically meaningful in its emotional impact and interest to participants, while acknowledging that participants would differ in their preferences. Like Warcollier, she also felt these "should be physically salient by standing out from their background properties such as in movement, novelty, complexity, incongruity, brightness and contrast" (p. 247).

Delanoy[275] also conducted an extensive literature review of free-response targets, in search of those that seemed to produce the most successful outcomes. She examined target characteristics related to color vs. black and white, complex vs. simple, novel vs. familiar, abstract vs. concrete, dynamic vs. static, form vs. idea and meaning; emotion; and thematic content. She found very few conclusions could be drawn from the data base, although she reported a tentative finding relating dynamic, multi-sensory targets to ESP success. Other suggestive findings were reported for "novel and abstract characteristics." She defined abstract targets as those that "portrayed a potentially realistic scene or object in either an abstract and or unrealistic manner (to varying degrees) or in a not readily recognizable fashion," although noting that Krippner and Zeichner[276] "found a greater percentage of misses with targets which were described as unrealistic" (p. 235).

OBE research and the birth of remote viewing

Ingo Swann (1933-2013) – was a follower of a vast swarth of ideologies at different points in his life. His niece, Elly Flippen, in private correspondence described him as an "Occultist, follower of the Fourth Way, astrologer, student of the Far Eastern traditions, Scientologist and proponent of Scientology offshoot Avatar, Gnostic and Christ. He attended Christian Sunday School as a child…He was found of saying however so men may approach me, even so, whatever path they choose is mine." He is credited with coining the term and being the "Father of remote viewing." However, in a variety of interviews he stated he was merely developing what others had started a century beforehand.

Starting at the age of three and throughout his life, Swann[277] had a number of exceptional spontaneous experiences related to intuition, out of body experiences (OBEs) and psychokinesis. In his mid-30s he began exploring how to develop and gain conscious control of his intuitive abilities. After a three-year period of self-study and practice, in 1971 he approached the American Society for Psychical Research (ASPR), where he hoped to find guidance from experienced researchers.

According to Janet Mitchell,[278] Swann was "by far the most talented psi performer I've ever had the opportunity to work with and observe" (p. 1). She recounted that Swann continued to "sharpen his ability to see at a distance" with continual improvements, achievements and frustrations. Mitchell and Swann together undertook to develop methodologies that would serve as "perceptual tasks" to "ascertain whether a person could localize part of his or her consciousness in space some distance from the body" (p. 4). These carefully controlled experiments involved placing two boxes on top of a platform 10 feet above the floor where Swann was positioned in a chair with electrodes attached to his body testing blood rate, volume, eye movements, muscle tension, etc., at different times. Research assistants would place items in the boxes to keep both Mitchell and Swann blind to the target material.

Operating with the intention of learning what worked best, built into Mitchell's experimental design was permission to make continued adjustments to protocols related to environmental factors and target materials based on success or failure in earlier trials. They initially found that bright spotlights casting shadows or glare focused on the target material seemed to inhibit results, but results improved after diffused lighting was installed above the boxes. She reported that "some colors were easier to recognize than others" and that primary colors rather than pastel colors were easier to see. This led to using colored construction paper "as the primary target material." Three-dimensional objects were not easier to see. "Flat drawings or cut-outs were preferred" with "whole figures working better than outlines" (p. 5). She found substances such as "leathers, fabrics and clay worked better than metal, plastic, glossy photographs and a glass of water... simple shapes and strong familiar forms were easier to perceive than strange conglomerations or unfamiliar forms of groupings... letters and numbers were often seen only as shapes, or not seen at all" (p. 6). She also noted that Swann's visual capacity fluctuated, with good days and bad days. He was often able to tell in advance of seeing feedback whether he was having a good day or bad day.

On occasion, Swann's attention would wander outside the lab. Once while tethered to his equipment in a windowless room, he said a woman wearing an unusual outfit was standing outside the building. Researchers were able to confirm he was correct. Mitchell also noted that Swann correctly described unexpected weather conditions (rain) in Tucson, Arizona, while he was in New York. He tracked the movements of researchers in real time as they made their way through museum exhibits and, on one occasion, correctly described them getting stuck in a corridor when the museum unexpectedly closed early. Swann was so good his name and fame spread widely among those interested in the esoteric in New York City.

In 1972, Swann joined the US government-funded psychoenergetics program at Stanford Research Institute, under the co-directorship of Targ and Puthoff,[279] which was tasked with discovering the potential for psychic warfare by the USSR and other adversaries.

According to information in the Swann archival collection,[280] over the next two decades Swann received continuous contract renewals with regular wage increases and acknowledgement of his status as a researcher-trainer, in addition to that of research participant. He would go on to take part in close to a million psi-based trials as a remote viewer. As outlined in memos and letters between himself and SRI co-director Hal Puthoff, Swann participated in approximately 19,000 trials in just one year. These involved an assortment of forced-choice tasks related to analytics, chemicals and physical materials, including a study on rock and mineral compositions.[281] Throughout these trials, Swann made notes related to when "learning was taking place" for himself and other participants. His notes concerned target construction, naming, ordering and the approaches taken, suggesting that even for forced trials he was not simply using unconscious guessing, but was experimenting with various somatic and mental imagery approaches. Some of these would become incorporated in his emerging system of remote viewing, which eventually became known as Controlled Remote Viewing.[282,283]

Swann also participated in variety of applied projects referred to as "field work." These projects included searching for sunken ships (Deep Quest), locating buried artifacts (the Alexandria Project,[284] and trying to find treasure (Ft. Huachuca Treasure Project and the Robert Jones Buried Treasure Project). Several projects involved oil exploration – Halbouty Oil Exploration Project, 1976 Ghana Exploration, 1976 Coppermine River Exploration, 1981-1985 Washburn Oil Exploration and Ada Oil Company Sites.[285]

Targ and Puthoff[286] published in *Nature* magazine one of the first public papers about their psi work. Two projects were discussed. One was studying the abilities of famed, but controversial, Israeli psychic Uri Geller, who participated in 13 drawing experiments over a period of 13 days. Geller was tasked with accessing line drawings located at a distant location and selected by researchers unknown to him. Most of these experiments were conducted with Geller being alone in a shielded room. The published study provides examples of these targets and drawings, many of which were remarkable matches. Drawings included grapes (a perfect match), a devil with a pitchfork (Geller drew a pitchfork), a camel (Geller drew what looks like a horse), a bridge (Geller drew a bridge but a different kind) and a bird (Geller drew a bird similar to that in the target drawing, but flying at a different angle).

The second experiment described in Targ and Puthoff's article concerned "remote viewing of natural targets" with former police officer Pat Price as a subject. They wrote, "A study by Osis led us to determine whether a subject could describe randomly chosen geographical sites several miles from the subject's position and demarcated by some appropriate means (remote viewing)."[287] The SRI co-directors constructed their target pool based on "the theory that natural geographical places or manmade sites that have existed for a long time are more potent targets for paranormal perception experiments than are artificial targets prepared in the laboratory. This is based on subject's opinions that the use of artificial targets involves a trivialization of the ability compared to natural, pre-existing targets" (p. 605). They found "Pat Price's ability to describe correctly buildings, docks, roads, gardens and so on, including structural materials, color, ambience and activity, sometimes in great detail, indicated the functioning of a remote perceptual ability. But the statements contained inaccuracies as well as correct statements."

Following publication of this paper, psychologists Marks and Kamman unsuccessfully attempted to replicate the above findings and then attempted to debunk the results of the original study.[288] However, independent researcher Dr. Charles Tart was brought in to reassess results, and he held that the research methods of the original experiment were sound.[289]

Targ and Puthoff[290] wrote a paper that received wide notice, *A Perceptual Channel of Information Transfer over Kilometer Distances: Historical Perspective and Recent Research.* The paper summarized the results of 50 experiments with subjects, both experienced and new to this sort of task, viewing remote geographical locations and buildings located up to several thousand kilometers away. At the locations were "buildings, roads, laboratories apparatus and the like." They asserted, "The development at SRI of a successful experimental procedure to elicit this capability has evolved to the point where visiting government scientists and contract monitors, with no previous exposure to such concepts, have learned to perform well; and subjects who have trained over a one-year period have performed excellently under a variety of experimental conditions" (p. 330). Locations included sites such as a museum, a city hall, a miniature golf course, a nature preserve, the BART transit system and a shielded room. They found no decline in psi when increasing the distance between remote viewer and target.

In this same paper, Targ and Puthoff reported another set of 12 experiments, carried out by five different subjects, two of whom were visiting government officials. The target material included real objects that researchers would visit and interact with while the remote viewers were tuning in. Subjects recorded their responses verbally and in writing. Targets included "a drill press, Xerox machine, video terminal, chart recorder, a random number generator and typewriter." Results were significant across both groups – the experienced remote viewers and the inexperienced visiting subjects. Some of the newer subjects' sketches were said to be "exceptional" but their results were less consistent than those of the experienced subjects (p. 345).

Targ and Puthoff referred to these types of experiments as "outbounder" experiments and wrote extensively about them in subsequent books such as *Mind Reach*[291] and *Limitless Mind.*[292]

Coordinate Remote Viewing

According to a published report by Puthoff,[293] "to determine whether it was necessary to have a 'beacon' individual at the target site, Swann suggested carrying out an experiment to remote view the planet Jupiter before the upcoming NASA Pioneer 10 flyby." As Puthoff recalled, results were dismissed as inaccurate by professional astronomers until "the flyby revealed that an unanticipated ring did in fact exist."

Swann then pleaded with doubtful researchers to conduct a series of experiments that would eliminate an outbounder, who was ostensibly providing a telepathic connection to the viewer. This flew in the face of the predominant theoretical model of the previous 100 years that telepathic transmissions were necessary for successful acquisition of nonlocal information. However, Swann personally didn't feel this was necessary, given he had many spontaneous occurrences of nonlocal perception not related to a sender/receiver situation. He also surmised that unless a viewer could zero in on any location, their efforts would not be of much use to the agencies that were funding them.

After sitting by the pool and hearing a disembodied voice say, "Try coordinates," Swann suggested they use latitude and longitude coordinates as their target references. After much opposition by the entire SRI staff, Puthoff agreed to set up a series of trials with Swann and Pat Price. Swann and others were to successfully describe the target landscapes, structures, objects, etc. Puthoff noted[294]

> Needless to say, this proposal seemed even more outrageous than "ordinary" remote viewing… Suffice it to say that investigation of this approach, which we designated Scanate (scanning by coordinate), eventually provided us with sufficient evidence to bring it up to the contract monitors and suggest a test under their control.

An SRI report by Puthoff, Targ, May, Langford and Humphries[295] described other targets that had been tested at SRI in dozens of experiments to determine how best to direct a remote viewer's attention to the information being sought and to determine "spatial accuracy limitations of the RV process as it might impact on operational utility." These included pictures of individuals, envelopes carried by another person which contained coordinates, matrix addresses, arbitrary map grid coordinates, pictures of a target or portions of target; pictures of a target in a sealed envelope and the word "target." Another series of experiences intended to "determine the degree to which various sensitive mechanical devices are susceptible to remote perception and perturbation." These devices included inertial gyros, strain gauges, etc. (p. 7).

Real objects

While physical locations were most often used for remote viewing experiments, a handful of studies conducted at SRI during the 1970s targeted complex objects in a real-world setting as opposed to photographs.[296] One series of studies sought to determine whether tiny or even miniscule objects could be perceived. The objects were placed in canisters. Targ and Puthoff found viewers did equally well regardless of how small the objects were.

> Targets and target details as small as 1 mm can be sensed. Hella Hammid successfully described 1 mm x 1 mm microscopic picture targets in an experimental series at SRI in 1979, and she once correctly identified a silver pin and a spool of thread inside an aluminum film can, as part of a successful ten-trial series with tiny objects.

Another set of studies involving real objects was designed to test whether clairvoyance could be isolated from precognition and telepathy. The targets included semiconductors and other machines and devices found around the SRI lab. Rather than using photographs as the form of feedback (which is the practice in most studies outside a lab today), the remote viewers and the independent rater visited the actual locations in which the objects were to be interacted with. The rater rank-ordered the feedback photo and the four decoy objects. Viewers scored well above chance, except one viewer was said to have extremely good correspondence to other photos in the set. Some of the transcripts revealed striking points of resemblance to the target.

In summing up over two decades of research, Targ, et al.[297] gave the following advice to those choosing targets, particularly in relation to real objects:

> The choice of appropriate targets is also an important part of successful experiments. In order to limit the universe of images, the target object should be bigger than a match box and smaller than a bread box. It should be geometrically interesting, and extended, rather than compact. For example, a Raggedy Ann doll is easier to describe than an ivory Buddha figurine. A pineapple would be easier to describe than a peach. A hairbrush is better than a nail file. Psychic Ingo Swann used to say to us, "Don't trivialize the ability." By this he meant that a remote viewing object should be attractive, aesthetically pleasing, and experienced by the viewer as equal to the effort involved in describing it: no lumps of coal or #2 pencils. The target should possess a variety of sensory aspects, or what we call "psychic handles." Nothing should be used that might be perceived as frightening or distasteful to the viewer. This is an essential point, since you would not want to violate the viewer's

unconditional trust of you or the process. Above all, the viewer should not feel a sense of disappointment when he or she is finally shown the target. The feedback session should arouse the interest and satisfaction of the viewer. One does not want the viewer to be disgusted, or be thinking, as Hella Hammid once facetiously exclaimed, "You asked me to separate my body from my consciousness for this?!" In the end, a good target is largely a subjective preference of each viewer (p. 372).

Operational Remote Viewing

Riding on the heels of these positive results, a series of projects at SRI was carried out "to determine the utility of remote viewing under operational conditions." One such target was a research center at Semipalatinsk, USSR, known to have ongoing operations. The results of this project were considered successful but much of it remains highly classified. Puthoff writes,

> As a result of the material being generated by both SRI and CIA remote viewers, interest in the program in government circles, especially within the intelligence community, intensified considerably... leading to an ever-increasing number of clients, contracts and tasking, and therefore expansion of the program to a multi-client base, and eventually to a joint services program under DIA leadership[298] (p.10).

Since the declassification of the program, many books and documents have been released demonstrating the types of operational targets that were used. These included a Russian Typhoon class submarine[299] and Pan Am Flight 103, which crashed in Lockerbie, Scotland in 1988.[300]

Buchanan[301] revealed that other operational targets included drug-interdiction cases in alliance with the US Navy Air Stations Joint Task Force; collecting intelligence on foreign military leader's plans and tracking their activities. These included Muammar Gaddafi, Saddam Hussein and General Manuel Noriega. Other targets were missing hostages such as Col. Rich Higgins, who was kidnapped by Hezbollah terrorists in Lebanon; William M. Buckley, who was kidnapped by Shiite guerrillas; and General James L. Dozier, taken hostage by Red Brigade terrorists.

Fuzzy Set theory

Watt[302] acknowledged the challenge of assessing targets, noting there are so many variables involved in any psi-based study that it is hard, if not impossible, to attribute the success or failure of a project to a single factor. May, et al.[303] worked to develop a system that would not only address a variety of these variables but take a holistic, integrative approach to automating them for the purpose of achieving consensus in scoring and rating. This system sought to create material that would serve as a photoset for the remote viewers and from which judges would select the correct photo. The system was designed to ensure orthogonality (extreme difference) of the photos, as well as equality in the level of complexity and interest. The design took into consideration that remote viewers will tend to get a mixture of higher-level features (e.g., "there is a bridge connected to land with water underneath") along with lower-level descriptors (e.g., "there is something round, red, that feels bumpy"). The approach incorporated a system of scoring using a *Figure of merit (FoM)* and a carefully designed set of images consisting of scenes portrayed in 300 *National Geographic* photos.

The photos were grouped into 19 "clusters." The clusters included: flat towns, waterfalls, mountain towns, cities with prominent structures, cities on water, desert/water interface, deserts, dry ruins, towns on water, outposts on water, cities with prominent geometries, snowy mountains, valleys with rivers, meandering rivers, alpine scenes, outposts in snowy mountains, islands, verdant ruins and agricultural scenes (p. 209).

While the researchers suggested consensus among judges was largely achieved, they acknowledged this was sometimes only after rigorous debate and that the system was a work in progress. What was needed was to refine the cluster analysis to ascertain what is meant by "visual similarities between targets" and "refinement of the analysis of responses, in an effort to achieve even greater correlations between the fuzzy set figure of merit analysis and various forms of ground truth." They thought visual experiments using similarities between decoy targets could aid in the development of remote viewing sets that had a higher level of orthogonality.

May[304] refined the system, making it even more automated through computer scoring. While a human "encoder" was still necessary to review the remote viewing transcript and fill out a score sheet, the encoder was not exposed to either the target or other photos in the set prior to feedback.

Working with three highly experienced remote viewers, the researchers observed 32 hits in 50 trials and "the largest Figure of Merit of the three computed for each trial was correct 32 times out of 50. Thus, the primary hypothesis of evidence for anomalous cognition was confirmed."

In succeeding years, May developed the target pool further and described it as follows:

The target pool used in this study was the current result of nearly 40-people-years worth of effort. This pool was based exclusively upon the Corel Stock Photo Library of Professional Photographs. This library of copyright-free images was in digital form and was comprised of 100 images on each of 200 CD-ROMs. The details of this photographic library of 20,000 images were culled to produce the current pool of 300 outdoor images that were arranged in 12 groups of five orthogonal categories.[305]

The present authors and colleagues participated in multiple trials using the above system, including a yearlong Associative Remote Viewing study conducted by Dick Bierman, with Ed May as the coder and Debra as the remote viewer. (See Chapters 9 and Chapter 10). In the Bierman-May-Katz study, 48 single viewer trials were conducted, with a majority of trials resulting in high enough Figure of Merit scores to proceed with a prediction and wager. However, in the end, the overall hit/miss ratio was deemed to be 50 percent – for unknown reasons, considering the multiple high FoM scores and strong matches to some of the photos.

Intrinsic target properties

May, Spottiswoode, & James[306,307] presented what they termed an "intrinsic target property" which they defined as "one that is completely independent of psychological factors and can be associated solely with a physical property of the target." They analyzed static and dynamic targets used in already-completed experiments[308] in terms of the change in Shannon entropy levels. Shannon entropy is a concept coined by Claude Shannon, the developer of information theory. Longmore[309] explains this in simple terms: "Measured in bits, Shannon Entropy is a measure of the information content of data, where information content refers more to what the data could contain, as opposed to what it does contain. In this context, information content is really about quantifying predictability, or conversely, randomness."

May, Spottiswoode, & James studied two previously completed experiments, finding "a significant correlation" between entropy levels within the photographs themselves and the "anomalous cognition" performance. Static targets included the above-mentioned *National Geographic* magazine articles, while dynamic targets consisted of 30 clips from popular movies and documentaries. Although the researchers acknowledged it was challenging to tell whether this was in fact related to entropy levels or to "some other measure related to the static dynamic targets," they suggested the results were promising but "the study raised more questions than answers" (p. 384).

Critics of this approach have suggested that it is often very difficult to ascertain whether a remote viewer is tuning in to the actual photograph, which May, Spottiswoode and James were attempting to utilize through pixel counts, or rather accessing the location, event or concept depicted in the photograph, which has its own entropic and psychological characteristics. In other words, is the target the photo itself or is it what the photo represents? Also May, et al.[310] were attempting to control for psychological properties of targets, suggesting their approach allowed them to do this, while earlier researchers such as Warcollier[311] and Delanoy[312] noted that such characteristics can vary tremendously from participant to participant. Even photographs of targets such as cityscapes compared to snowy mountain tops containing the same rate of change in Shannon Entropy could elicit drastically different emotional responses, as evidenced by the fact that some people prefer to visit one of these environments while wishing to avoid the other altogether.

May would go on to design[313] another experiment that analyzed the Shannon gradient in terms of both informational and "thermodynamic properties." He was inspired by his time at SRI, writing:

> During the time of the Star Gate program, I noticed that there was a class of operational anomalous cognition (AC) missions (i.e., involving intelligence gathering or simulations thereof) that appeared never to fail (May & Lantz, 2010). These included underground nuclear tests, electromagnetic pulse devices, static and dynamic rocket motor tests and rocket launchings, to name a few. One possible explanation might be rooted in psychology... Another view is that these target types share a physical attribute; they all involve an enormous expenditure of energy in a short period of time.

May found results indicated that "AC is mediated via some kind of a sensory system in that all the normal sensory systems are more sensitive to changes than they are to inputs that are not changing."

Target pool construction considerations and their relationship to internal mental processes of participants

Based on surprising findings of two earlier studies by Lantz, et al.[314] in which anomalous cognition results for dynamic targets were initially lower than those for static targets, May, Spottiswood & James[315,316] proposed a theory related to target *bandwidth*, which refers to a target's range of complexity and other physical characteristics. Researchers surmised that the unexpected poor performance for trials using dynamic targets might have been due to differences in the bandwidth of targets and the inability of participants (who were told to provide verbal reports of all impressions) to edit out nonessential imaginings that may have arisen from false expectations about the complexity of the targets. Researchers gave the example of how confusing it might be for participants to combine a set of photos with simple Zener card symbols used in forced-choice guessing tasks and the complex location-based images used in many of SRI's remote viewing projects. Participants might be inclined to find in the static targets more than what is there, while leaving out essential impressions for more complex targets, assuming they had already described enough.

To test this theory, Lantz, et al.[317] conducted a conceptual replication, keeping all variables the same as the original experiment except for the target materials, which were altered so that the dynamic pool conformed to "the topic, size and affectivity homogeneity of the original static targets." Selected targets also were all of "intermediate bandwidth" in terms of their complexity. Dynamic targets included an "airplane ride through Bryce Canyon and a panoramic view of Yosemite Falls" (p. 310). The bandwidths were considered to be identical. Results supported the bandwidth hypothesis, with improvements over the earlier studies' results for both static and dynamic targets.

As noted above, consideration of target materials in free-response experiments extends from the individual properties and characteristics of a target to the relationship between photos within an overall target pool and the effect the ordering might have on participants. For example, as a remote viewer it can be confusing to have two identical or very similar targets in a row, since one can't be sure at the time one is perceiving whether the perceptions are related to the current target or if one is instead having trouble letting go of the experience of a past target. This may be in play even when targets are in different projects. Those who participate in "too many trials" face more of these internal dilemmas, which may account for a general decrease in psi functioning noted by Rhine and others (with the exception of a very few individuals).[318] As we note later, both Ingo Swann and Joe McMoneagle hold that remote viewers have a difficult time honing in on a very narrow present; they may receive impressions from a bit in the future or a bit in the past. This feature (or bug) of remote viewing complicates the situation when viewers face a series of targets. The best approach is for viewers to assume nothing about any targets that are coming up.

Another consideration supporting the concept of target bandwidth is the differing approaches participants have when they remote view. This is rarely discussed in the formal literature, other than perhaps to state that participants are using mental imagery or they are "select" or experienced participants. We found only one study in this vein. Targ and Puthoff[319] theorized that occasionally select/experienced subjects did poorly on a task because their method did not translate well to a new or different kind of target. As May, Spottiswoode & James[320,321] point out, these internal struggles by remote viewing participants tend to not be recognized as factors in free-response projects. We will further discuss remote viewing approaches in relation to target types in our chapter on remote viewing instruction (see Chapter 25).

In conclusion

We would like to offer a few additional observations concerning present-day ideas about criteria for selections for photos in a pool. All one has to do is visit one of the growing number of social media forums to read discussions about which types of targets remote viewers believe one should avoid because viewers supposedly miss them – and which ones make good targets because viewers typically do well with them.

In reading these forums, I (Debra) offer a personal word of caution. Like many of the authors mentioned above, I do feel target type ultimately is quite personalized. I say this because I see on the avoidance lists the very types of target that I not only do well on, but personally love. These include close ups of human faces, eyes, body parts and animals. As a remote viewer, I know if there is a prominent human figure. Not only am I likely to pick up on this in a photograph, but I can tell if the person is facing a certain direction or if there is a special quality to the glance of the eyes, the facial expression, or the arms are in a particular position, such as over the head. Also, if a strong sentiment is displayed on the person's face or in their positioning, that will come through. I've seen these very same things being described well and often by my own students. Many of us love these types of photos and are disappointed when they are excluded from a target pool.

Still, as we have discussed the chapters on displacement, in ARV projects that seek to minimize attraction to the wrong photo, it may be precisely the types of targets we love that we would want to omit from a photo set. This wouldn't be the case if two photos were both of the kind we loved, but the challenge in pairing photos is not only that they should be equally compelling, but they have to be different enough for the judge to be able to tell which one the viewer is describing. For that reason, you'd never want to have more than one photo in a set of a person, an explosion or even a building, unless perhaps the buildings were completely different in shape, color, materials and ambience.

So the question is not simply "What are the best targets?" but "How good is the pairing emotionality and orthogonality?" There are even more considerations: "What is the impact of the order of the potential targets and how do the judges respond to the potential targets?"

While little formal research has been done in this area, one thing we've noticed is it's very easy for a judge to get attached to the very first photo and transcript they see, particularly if it seems a good match. There is a "wow factor." But then the judge may find the second photo is a good match, too! The judge may be biased toward the first photo because of that initial sense of excitement. This is one form of "confirmation bias."

Below is a summary of a recent attempt in remote viewing research to explore the topic of target materials. It was specifically motivated by Debra's observations that various ARV project managers and formal researchers were assigning targets to viewers that many of the viewers objected to (e.g., simple objects against plain backgrounds), interspersed with more complex photographs of real locations.

Effects of background context for objects in photographic targets on RV performance (from Debra)

In 2015, our proposal to test the effect on RV performance of background contexts for objects in photographic targets won the IRVA-IRIS Warcollier Award. This was a project conceived of by myself and longtime pal (friend from high school) Michelle Bulgatz. Later, retired professor emeritus Dr. James Lane from Duke University graciously agreed to join the team as our statistician. This exploratory experiment looked into whether the background in which the object is set within may affect the accuracy of remote viewing. What follows is a summary of our write up.[322]

Project purpose

The purpose of this exploratory project was to perform a comparative analysis of remote viewers' performance when tasked with describing objects positioned in three different background conditions. We also wanted to know whether remote viewers operating under double-blind conditions are more likely to focus on the main object or the setting of the object in a photograph. We sought to decrease possible displacement effects[323,324,325,326,327,328] by using two methods of analysis; the second, which the viewers would be blind to, would occur only after all 30 trials had been completed.

Rationale for current project

As projects increasingly move out of a formal laboratory setting, operating on shoestring budgets but aided by the ease of acquisition of photographs in free or inexpensive photo-sharing sites, more of these projects are using photos of real objects set in different types of backgrounds. These background settings are sometimes real locations in which the objects are typically found or they are computer-edited photos in which the background is devoid of all information, whether a white background or a colored one (black background). Sometimes the objects depicted in the photos are real and other times they have been artificially created.

Despite lack of formal testing of such materials, they are being used, sometimes forming the entire target pool and sometimes mixed in with previously tested types of targets, in both applied and experimental projects. Most of these projects have not been reported on in the formal literature, but the present researchers have participated in them as remote viewers and judges. They have been carried out by parapsychological researchers attempting exploratory studies and by applied RV or ARV project managers as well,[329,330,331,332] the latter to use psi in stock market predictions, sporting events and the like.

Objects within normal settings vs. unusual, unexpected settings

Additionally, Debra has noticed another trend related to target material she used with her students, who were located at various distances and meeting via teleseminar conferencing. Many of the students seemed to have an easier time recognizing larger gestalts, even naming the target, when real objects were set in logical or real-world locations vs. illogical ones. A logical background would be a boat in the water or a piano in a living room. Illogical would be a boat or a piano positioned in a desert landscape. This seemed in alignment with results from earlier studies that Delanoy[333] included in her literature review of target characteristics.

The purpose of this study was to test such materials through a process in which objects set in different types of background conditions would be assigned to experienced remote viewers to discover which might produce better results. Objects within photographs were selected based on the characteristics thought to produce better results, with the objects repeated across different backgrounds to ensure the backgrounds and not the objects themselves were being tested.

A theoretical model for a conceptual replication

A search was conducted of recent cognitive attention literature involving types of photographs that are more easily perceived than others. Of greatest relevance was a series of experiments conducted by Barenholtz.[334] These sought to understand factors involved in visual recognition of objects as they relate to environmental settings. While most visual research has focused on the inherent properties of objects, she wanted to understand the relationship of visual context to object recognition. She based her work on reaction times – How long would it take for a person to recognize an object when first seeing it? Are there certain contexts (settings/backgrounds) that speed up or slow down the identification of an object? To measure this, she devised a system in which images were first pixelated to such an extent that nothing could be comprehended about them. Pixilation was decreased over time until participants indicated that they could perceive what the object was. The reaction time was recorded and reaction times for the object in different backgrounds were compared. The results demonstrated that background conditions did indeed play a role in cognitive processes. Barenholtz reported:

> Participants decreased block size (increasing resolution) until identification. Critical resolution was compared across three conditions: (1) when the picture

of the target object was shown in isolation, (2) in the objects' contextual setting where that context was unfamiliar to the participant, and (3) where that context was familiar to the participant. A second experiment assessed the role of object familiarity without context. Results showed a profound effect of context: Participants identified objects in familiar contexts required higher-resolution images, but much less so than those without context. Experiment 2 found a much smaller effect of familiarity without context, suggesting that recognition in familiar contexts is primarily based on object-location memory (p. 1).

While the study by Debra, Michelle and James does not attempt to replicate reaction time with changes in pixilation, and therefore adopts a very different design from Barenholtz's, it does base its hypothesis on her findings regarding object-context familiarity, seeking to determine the extent to which these translate to nonlocal perception.

Methods

Twelve experienced remote viewers participated in 30 individual open-response, triple-blind trials requiring them to use extrasensory perception to describe the photographic image they would receive via email a few days later. Investigators created a photographic target pool of complex objects set within one of three background conditions: 1) White/ Artificial, 2) Regular/Normal and 3) Abnormal/Illogical.

Participants completed 360 in-depth sessions consisting of 8,460 words and 1,472 sketches. The project used two methods of analysis – the traditional sum of ranks matching procedure and an exploratory method in which each item and sketch was scored by both the participant and an independent judge. These two methods revealed significant, but opposite, results against the background conditions. Scores for the White/ Artificial background were higher for the sum of ranks evaluation but lower for the "item and sketch" evaluation. Researchers were also interested in learning whether perceptions and sketches rated as matches by independent judges pertained to the main object, to the background or to a combination. Results indicated that for all the background categories, the main object was described far more frequently than the background.

Differences found in judging methods

We are somewhat perplexed at the different findings obtained through different judging methods (Phase I, hit rates vs. Phase II, Sum of Rank Matching Tasks). Given that most studies only use one method, we cannot help but wonder if other studies would have found different or contradictory results had they used more than one method of scoring.

One possible explanation for this is that when it comes to matching tasks, judges bring different perceptual and cognitive processes to the task. Both forms of analysis involve comparing impressions and sketches to a photo, but a rank-ordering matching task involves repeating these comparisons multiple times per each photo in the set, and then making a number of decisions and choices. Given the complexities involved, it may be that it is simply easier for judges to make sense of data when there is less information in the background, hence greater success with the white background conditions.

Regardless of the reasoning, these results suggest that if only a hit-rate type of approach is being used, it may be advisable to use photos of objects in their normal or abnormal backgrounds and avoid the use of photos in white backgrounds. Conversely, if a project is going to use a matching sum of ranks form of analysis, it may then be best to choose potential targets with objects against white backgrounds.

A word of caution: plenty of designs use matching tasks but do not use rank ordering – meaning there will be only one score given to the best match and no credit given to the second-best match. We therefore cannot say whether our results would likely be transferable to even slight changes in the analysis protocol.

For our Phase I judging protocols (hit rates), nine of 12 viewers scored their own sessions higher than independent judges did. We cannot say if this was due to a desire to rate themselves more highly or because they had a better ability to know what they meant from their own words and sketches. Viewers were required to input their words into spreadsheets and submit these along with their transcripts prior to receiving feedback and self-scoring, and no data was added after they received feedback.

We looked for patterns and did not find anything striking. One of the present researchers did an informal test of rater reliability for several of the viewers' sessions that showed the greatest disparity between the viewer's self-scores and the scores of the independent rater. She found her own scoring would have fallen somewhere in the middle of those scores. This was true for one viewer-judge team in particular, in which the judge was not a remote viewer himself but served in the dual roles of judge and statistician.

Also, three of the four viewers who generated the highest number of sketches and words had the highest differences between rater self-scores and independent judges' scores. While we do not have an explanation for this, it might suggest viewers involved in projects using independent judges might do better to provide just their best impressions in a session instead of all of them.

Since the main independent variable being tested was background conditions, we felt it was important to control for rater consistency across all trials and therefore made the choice to pair the same judge with the same two viewers. Given that judges may have different tendencies – some restrictive, some permissive – this could account for differences between viewers' scores.

As noted above, viewers described the main object, not the background. Because remote viewers are more experienced in describing locations, we originally hypothesized that they would be likely to describe locations rather than the main object. We hoped this would not be the case, but for years have heard viewers express a dislike for photographs of objects rather than locations. Viewers were told the target pool consisted of objects within a variety of locations. Their goal was to describe the main object, but all correct information pertaining to every element (whether object or background) would be scored as correct.

They proved us wrong. Results showed they most often described the main object. However, since they knew all the photos contained a main object, we cannot say whether their own perception naturally and involuntarily went to these objects or whether they directed themselves through voluntary attention[335] to focus on the objects themselves.

It is possible results would have been different for the project if the viewers had not been told the project involved photographs of objects as the main focal point vs. photographs of locations.

This brings us to one of the main aims of our study: Did results correlate with Barenholtz's "normal" visual perception findings? Again, for item/sketch hit rates, it did, while for sum of ranks tasks, it did not. Her design did not involve judging and, hence, no perceptual aspects of judges potentially affected the process.

As noted above, part of our rationale for using two very different forms of analysis had to do with pressures to use a traditional matching task analysis method (we were applying for a research grant), but also wanting to draw the viewer's attention away from awareness of this method. We feel we were successful in keeping the viewers focused on their own feedback photos and away from the matching task method that took place only after all trials were complete. We did a sampling of the Phase II judging responses for those trials that resulted in misses and also polled the judges, looking to see if there were very close matches to the wrong photo in the sets. We did not find any examples of this. More often than not, the data in the remote viewing session was just not clear enough to make a proper choice. This is exactly what is supposed to happen with this process. In this respect, this study might serve as an example of how to decrease displacement effects.

Still, we (and some of our viewers) may have seen examples of temporal displacement – to targets that appeared at a later date. One viewer, in particular, felt this happened quite a bit, and upon reviewing her sessions, we agreed with a few of her examples. One viewer in an early trial drew a picture of an accordion that was almost an identical match to the photo of an accordion against a white background. Oddly, this picture did not appear in the sequence until several weeks later. We cannot say, of course, if this was a displacement effect or simple coincidence, but the sketch for trial 6 and feedback photo for trial 16 are strikingly close.

In conclusion, our literature review showed two theoretical assumptions seemed to motivate investigations into target characteristics. The first is a "natural sciences approach" that seeks to find a universalized set of characteristics for target material that would enhance nonlocal perception in many (or all) forms of open-response designs. The second is what we might associate with a "human-sciences approach," acknowledging that just as in "normal" perception, what is more likely or easier to perceive through nonlocal perception may be very individualized.

Given that only a few of our experienced remote viewers who have reputations for doing well at location-based targets achieved significant results for these object-oriented targets, future researchers should not expect that a remote viewer who performs well in a project with one set of potential targets and protocols will necessarily perform the same way with other materials and protocols.

Finally, even though our central aim was to study materials and backgrounds, ultimately what we studied was human perception. As usual, it remains elusive, whether we are discussing "regular" or psi-based perception.

CHAPTER 9

Using Computers to Enhance ARV: May and Mossbridge

This chapter addresses the increasing use of computers as hosts for Associative Remote Viewing systems, a development that has had many benefits. Through the internet, millions of potential remote viewers have access to technological wonders that facilitate practical ARV trials and formal research, whereas previously access was severely limited. Photo pools of any size can be more easily compiled and utilized in user-friendly formats. Videos and musical selections as well as photos can be deployed in the pools as potential targets. The computer can become, if not iRobot, a quasi-partner and can play an active role in the structuring and carrying out of ARV trials and research. Hard drives can store hundreds of thousands of photos, and software can automatically and randomly select two or more of them from a pool based on pre-determined criteria and later present the actualized target for feedback after the event has taken place. Automated emails play a role in the process, as well. Thanks to computer storage, computer languages and databases – websites, MS Access, scripted Excel docs, email and even Matlab – the possibilities for ARV are wide open.

Perhaps most significantly, computer programs can "evaluate" potential targets such as photos, enabling viewers, judges and project managers to see or NOT see the target and the potential targets ("decoys"). This capability, while not perfect, can reduce the displacement that plagues ARV by reducing "feedback loops," in particular. by limiting who sees the images. Experts in the field (e.g., Joe McMoneagle, Ed May) advise reducing what each participant gets to see or to know about the potential targets and the particular event in question. A division of labor and the wealth of computer tools can reduce "feedback loops" and displacement.

In this chapter[336] we will look at two such computer setups. The first is Computer Assisted Scoring (CAS), for which Ed May was the lead developer. Ed has used this system with his team of viewers, with success, and several groups in APP have undertaken trials with CAS, as we shall see. The second setup is Julia Mossbridge's Positive Precognition website, which uses an approach similar to CAS – limiting what the viewer and the judge see. Viewers can practice their ARV skills and at the same time contribute to the studies Julia is undertaking.

Computer Assisted Scoring (CAS)

In 2013, Dr. Edwin C. May, former head of the Star Gate project at Stanford Research Institute, gave a presentation at the Applied Precognition Project conference in Henderson, Nevada. The subject was his system of making binary Associative Remote Viewing predictions, which consisted of an Access file, a pool of 300 photographs, and use of Matlab.

Ed conducted 50 non-betting trials with the system. The hypothesis was that "auto-judging" would produce a significant number of hits in 50 trials. It did so with 32 hits, with an effect size of 0.647. The hit rate was 64%, while chance would have been 33%. If a Figure of Merit (discussed below) exceeded the threshold, there would have been 10 correct out of 12 bets, a hit rate of 83.3%.

His presentation also included a comparison using two other parapsychologists as raters after Ed had done his. Ed noted that he was very experienced with this type of coding and also knew the "idiosyncratic responses" of the viewers, whereas the other two raters were not and did not. All raters were blind to the targets. The "good news" was that all three coders produced a significant number of first place hits and above-chance statistical measures, but only Ed "produced significant confidence calls" (that is, would-be bets).[337]

Later, using the system, a series of sports events was bet. There were 30 events and 9 of them were bet, with 8 resulting in a win, a hit rate of 89%. The amount bet was about $1K per game. The presentation at the conference impressed the APP members so much that many wanted to try the software right away. Ed generously offered to provide the system to APP and groups soon began to use the system.

Before discussing those efforts and results, we will outline the main components that went into this system, which was based on many years of work by May and other researchers. It will be of interest to the many people trying to build effective photographic pools for use in ARV.

For those wishing a full understanding of the development of these techniques, we recommend Beverly Humphries, et al.,"Fuzzy set applications in remote viewing analysis"[338] and the many papers by Dr. May that address these issues.[339,340]

In the early 1970s, researchers at the Stanford Research Institute had success in outbounder experiments but wanted to move beyond them. In those experiments, a person would be directed to drive to one of several locations in the Bay Area and the percipient back at the office would attempt to describe the location. Suppose one of the targets was the Golden Gate Bridge. The question arose: Is the target just the bridge itself or the surroundings as well? The same question arose when researchers began to use photographs as targets. Is the target only what is in the photograph or is it the photo plus the surrounding environment? SRI decided the target would be whatever was within the frame of the photograph. As Ed May explained, they started with natural scenes but switched to photos to get targets with definite boundaries, for research purposes.[341]

One popular method to explore this form of psi employed a set of five photographs. One of the photos was designated as the target and the rest were "decoys." Researchers wanted to know if a remote viewer could provide a response that would allow a judge to match the session with the target photo. The judge ranked the percipient's response against all five photos, assigning top rank to the photo chosen as the best match, on down to the fifth photo for the least-likely match. This is called rank order analysis. Researchers went from a small set of photographs (five was common) to larger sets.

At one point, the question was raised – What is the purpose of these exercises?

> If the goal is simply to demonstrate the existence of the RV phenomena, then anything that is perceived at the site is important. But if the goal is to gain specific information about the RV process, then possibly specific items at the site are important while others remain insignificant.[342]

To gain insight into the process, larger sets of photographs were developed:

> This library of copyright-free images is provided in digital form and comprises 100 images on each of 200 CD-ROMs. Each image is approximately 18 MB in size, which corresponds to a landscape format picture of 3200 x 1875 pixels in 24-bit color.[343]

Researchers reached a consensus about photos in the pool: they should not be contrary to viewers' expectations (e.g., a simple circle vs. an outdoor scene which is a mismatch of type, not have negative emotional impact and not ambiguously depict one of the descriptors (e.g., it looks like a lake but in fact is a bay).

As a result of many experiments, using trial and error, it was decided that the most effective photo sets would have the following characteristics: no people, no transportation devices, no small manmade artifacts (toys, tools). Photos were edited to remove such elements (although occasionally not all elements were fully removed due to the limitations of editing technology in those days). There were no mismatches in size between photos. All photos were of outdoor scenes with no odd or unusual camera angles, perspectives or lighting. Of the original set of about 20,000 photos, half were removed by visual inspection of the thumbnails, and a pool of 100 images was constructed. This was later expanded:

> Two laboratory personnel examined all 10,000 images on a high-resolution computer display, and approximately 800 candidate photographs met the above acceptance criteria. After some digital editing, we identified from this set of 800 photographs 12 groups of 25 images for a total of 300 targets.

Fuzzy set theory

"The world is not a very crisp place."[344] This observation led to a further innovation. A branch of mathematics called fuzzy set theory was explored and found to be just the tool needed in our untidy world.

As an example of a fuzzy set approach, take the issue of defining a "very large city." Is it 100,000 residents? 150,000? 500,000? If you define it as 500,000 residents, is a city of 500,001 then not a "very large city?" The difficulty is evident. Precise categories like this fail to do justice to a world with imprecise boundaries. Ed May decided to use fuzzy set theory to characterize photos and to allow for gradations (via a slider) in evaluating the data in the responses.

> Six individuals independently encoded each of the 300 photographs against the Universal Set of Elements…and a consensus was formed to create a fuzzy set representation of each image with regard to how each element…was visually impacting in the image.
>
> One added benefit of having the images all coded as fuzzy sets is that it allows us to see if the various photographs fall cleanly into categories that are different from one another, but at the same time the photographs within a category are as similar as possible.

The approach was visual but "actually the fuzzy set method is completely general. It is easily possible to create an AC (anomalous cognition) target pool that tests for aspects other than visual (e.g., functional, allegorical, poetic, etc.)."

A Universal Set of Elements

To evaluate information in the viewers' responses relative to what was in the photos, several approaches were explored. Russell Targ suggested using declarative statements (called "concepts"). For example, a "small red VW car" was considered a single concept. However, this approach was found to be inadequate to describe elements with sufficient specificity.

To move beyond declarative statements, the idea of a Universal Set of Elements (USE) was introduced. Humphries[345] proposed seven primary and three secondary levels of elements, ranging from the abstract (a vertical line or simple geometric shapes) to the complex (a church, ruin, bridge). A list of descriptors was devised for these categories. Here are a few examples of the categories and subcategories.

Concrete Descriptor Levels I:
Single structures (fort, castle); Substructures (boat, pier, fence, monument)

Concrete Descriptor Levels II:
Settlement (ruins), Elevation (mesa), Land/Water Interface (glacier), Water/ Vegetation or not (desert, forest), Ambience (industrial, religious, urban)

Abstract Descriptor Levels I: Color, Other visual (shiny, cloudy, weathered), Implied texture (smooth, striated), Implied Temperature (hot, humid), Implied Movement (flowing), Ambience (serene)

Abstract Descriptor Levels I: 2D & 3D Geometries (rectilinear, mixed forms, repeat motif)
1D Geometry (stepped, V-shape, arc)

After being tested, these descriptors were found to be an improvement over declarative statements but were considered inadequate as well. Through experience based on thousands of trials, researchers had

a real good idea of what is and is not likely to be part of an AC response...Items were taken from an earlier USE on similar targets that were experimentally shown

174

to be helpful in blind judging. Thus a lot of the low-level elements in the original USE have been dropped here.[346]

The Universal Set of Elements was therefore changed to 24 elements (which has continued to be the set to the present day). These are:

Buildings, Villages/Towns/Cities, Ruins, Roads, Pyramids, Windmills, Lighthouses, Bridges, Coliseums, Hills/Cliffs/Valleys, Mountains, Land/Water Interface, Lakes/Ponds, Rivers/Streams Coastlines, Waterfalls, Glaciers/Ice/Snow, Vegetation, Deserts, Natural, Manmade, Prominent/Central, Textured, Repeat Motif

These 24 elements were selected on the basis of extensive experience, as well as on the formal analysis of a single study. The principal criterion used in the selection was that the elements should not be too "low level" such as lines and geometric shapes, nor should they be too "high level" such as an office building. These 24 elements were an attempt to strike a compromise between these two extremes.[347]

To prevent the viewer and the experimenter from seeing both photos in a binary situation, categories were used to establish a profile for each photo. To develop a coding sheet using categories of photos, the following instructions were given to the coders, using fuzzy set values from 0 to 1.0:

The descriptor list does not contain items that are not in THIS target pool. All elements are to be coded by what is visible in the scene and not from what is implied, but not visible. For example, a view of the Grand Canyon that does not explicitly show the river at the bottom of the canyon must be scored as zero for the water elements. Each element is to be coded with regard to the degree to which it is visually impacting in the scene. Often this is related to the relative area that element occupies; however, it is not exclusively so. For example, a relatively small barn might be visually impacting because it is bright red and your eye is drawn to it. Elements that appear to be opposites (e.g., urban and rural) should be coded as independent items, and they do not have to sum to one. The USE appears to have a natural hierarchy to the items (e.g., Land/Water Interface and Rivers). To the best of your ability, code these independently. For example the land/water interface element might be 0.3 visually impacting and you can clearly see that it is a river (as opposed to a coast), and then perhaps the river element would also be coded as 0.5. Part of this exercise is to see how independent coders address some of the built-in ambiguities.

Another facet of the new pool was: "For the development of this new pool, we chose a different approach. Namely, the analysis of decoy target images would be determined after the AC trial was complete."

To repeat, based on the coder's subjective judgment about the visual importance of an element in the photograph, the coder assigned a value from 0 to 1.0. For example, if the photograph contained a bridge that appeared prominently in the photo, the bridge might be a given a value of 0.8 or .9. If there a road was small, it might receive a value of 0.2 or 0.3. In this way, a profile was built up of each photo.

Then, when it came time to evaluate a session, the analyst would determine the visual importance of an element in the session and assign a value on the coding sheet. For example, if the viewer drew two parallel lines but did not indicate they represented a road, the analyst might assign a value of 0.3 or 4. If the viewer used the word "bridge" and drew something like a bridge, the analyst was obligated to assign a value of 1.0 to the element. In this manner, it wasn't necessary for the coder to compare the transcript to the photo. Instead, the transcript was used to fill out the coding sheet and the scores on the sheet were entered into the computer.

Figure of Merit

The next step in this extended process to develop a highly successful procedure to conduct ARV trials was to assign a Figure of Merit (a score) to the session. This was achieved by programming the computer to calculate the accuracy and the reliability of the transcript against each photo. Both measures were needed. For example, if the response consisted of 100 words, it might contain all the elements in the photo, so its accuracy would be high. But if it also contained many incorrect words, its reliability would be low because it could contain hundreds of nonmatching words or drawings.

Accuracy is the number of elements in the responses that match elements in the photo, divided by the number of elements in the photo. Reliability is the number of elements in the response that match elements in the photo, divided by the number of elements in the response. The Figure of Merit is the accuracy times the reliability.

A Figure of Merit of .455 for this target set, these viewers and analysts was strongly indicative of a match. As Ed May stressed, the Figure of Merit depends upon the particular target pool, viewers and analysts.

The combination of the categories and structure of the USE, fuzzy set mathematics and a consensus approach was thought to provide "a reasonable expectation of the subjective scoring of the same data by a large number of individuals" (p. 207).[348]

Computer Assisted Scoring trials in the Applied Precognition Project

As indicated above, many APP members were keen to try the exciting new system developed by May and associates. (The system didn't seem to have a name, so we asked Ed if he was okay with "Computer Assisted Scoring" and he said he was.) The system is built around an Access database, Matlab (a sophisticated software program), and the pool of 300 carefully vetted photographs.

Over the next several months, eight different trials were undertaken. Jon was the group manager for three trials (CAS OAK A, C and E) and he did one solo trial as well (CAS OAK B) to test the system with a single viewer. Igor Grgić, Chris Georges, Nancy Smith and Sumner W were the group managers for Croatorum, IN8, Sublime and Transcendent, respectively.

The groups did not all follow the exact same set of guidelines, which turned out to make a huge difference. The groups that stuck closest to May's guidelines had the best results.

Results of APP trials

Altogether, 116 events were predicted in the trials conducted in the eight groups. Based on 112 sessions, the eight groups had 14 hits, 14 misses and 88 passes, results at chance levels. (The number of sessions done by Transcendent is not available, hence the Events total is higher than the Sessions total.)

ALL CAS SERIES	Dates	Manager	Viewers	Events	Sessions	Hit	Miss	Pass
CAS OAK A	Sep-Nov 2013	JK	6	20	20	4	0	16
CAS OAK B	Nov-2013	JK solo	1	10	10	0	1	9
CAS OAK C	Jun-Jul 2014	JK	5	20	20	2	2	16
CAS OAK E	Sep-Oct 2014	JK	4	16	16	1	1	14
CROATORUM	May-Jun 2014	IG	5	10	21	1	2	7
IN8	Jul-Aug 2014	CG	2	5	5	0	1	4
SUBLIME	Sep-Nov 2014	NS	6	20	20	3	1	16
TRANSCENDENT	Nov-Dec 2014	SW	10	15	NA	3	6	6
Total			39	116	112	14	14	88
Hit-Miss-Pass Rates						12.1%	12.1%	75.9%
Hit-Miss Rates						50.0%	50.0%	

Figure 9.1 – Events Predicted, Hits, Misses and Passes

The four groups that closely adhered to the guidelines (CAS OAK A, C, E and Sublime) had much more promising results – of 14 wagerable predictions, 10 hits and 4 misses, for a hit rate of 71.4%.

4 GROUPS	Events	Sessions	Hits	Misses	Pass
CAS OAK A	20	20	4	0	16
CAS OAK C	20	20	2	2	16
CAS OAK E	16	16	1	1	14
SUBLIME	20	20	3	1	16
Total	76	76	10	4	62
Hit Rate			13.2%	5.3%	81.6%
Percent			71%	29%	

Figure 9.2a Results from Closely Adhering to the Guidelines

The groups that did not use the recommended guidelines achieved the following results – a 30.8% hit rate compared with 71% when the guidelines were followed.

	Dates	Manager	Viewers	Events	Sessions	Hit	Miss	Pass
CROATORUM	May-Jun 2014	IG	5	10	21	1	2	7
IN8	Jul-Aug 2014	CG	2	5	5	0	1	4
TRANSCEN-DENT	Nov-Dec 2014	SW	10	15	NA	3	6	6
Total			17	30	26	4	9	17
Hit-Miss-Pass Rates						13.3%	30.0%	56.7%
Hit-Miss Rates						30.8%	69.2%	

Figure 9.2b – Results from Not Using the Recommended Guidelines

With regards to three other CAS series: A brief IN8 (innate) series resulted in 1 miss and 5 passes. The group manager considered IN8 an experimental pilot study; after technical difficulties, the series was suspended. Jon was viewer and coder in the CAS OAK B solo series, which resulted in 0 hits, 1 miss and 9 passes. The Transcendent series had 3 hits, 6 misses and 6 passes. To explore the possibilities and limits of CAS, the Transcendent group manager modified the CAS guidelines. We will not expand on these three series given the success of those detailed above.

The CAS trials showed the system can get quite good results with viewers other than the ones May used, whom he always cites as among the best viewers in the business.

The CAS trials did not have as many predictions as the Winner Winner Chicken Dinner group, which achieved a hit rate of 70% over 100 events. Nonetheless, an excellent hit rate of 71.4% was obtained over 76 events by those following the guidelines and with considerably less effort from the group managers than WWCD required.

The very high percentage of passes resulted from few scores reaching the Figure of Merit cutoff score of .4519. When Jon told Joe McMoneagle the large number of passes was negatively affecting the viewers' morale, Joe pounded the table and said, "They are supposed to be pros! They should keep at it."

The small size of the active pool also affected viewers' morale. A photo used in a series is not recycled, not eliminated in future trials. The viewers grew tired of repeats of one particular type of photo. Although the photo pool has 300 photos, only 150 of them would ever be seen in a trial, a fact APP group managers did not know at the time.

May said the photo pool worked for his viewers but he was willing to assist us if we wanted to design a different pool to include a larger set. This would have been a massive amount of work and no one in APP stepped up to undertake it, primarily, I (Jon) believe, because of frustration with the large number of passes.

A further look at the guidelines

The guidelines that were followed in the CAS OAK A, C, E and Sublime series were: One viewer per event. Sports events only. One event per day maximum. The threshold for a call (a decision) to be made was a Figure of Merit of .4519, the only scoring method used. The only person filling out coding sheets was the group manager. The viewer and group manager saw just one of the two photos during the trial and that photo was the feedback photo. Neither viewer nor group manager ever saw the other photo in the context of the particular series. No other person saw both photos during the trial. The viewer got the feedback after each event. (Feedback time was not usually the same for viewer and group manager.)

There were a few differences between the CAS OAK A, C, Sublime trials – we will call them the APP trials – and Ed May's trials (we will call them the Laboratory for Fundamental Research – LFR trials). We do not know the length of the LFR series, but the CAS OAK series were 20, 10 and 20 events. Sublime was also 20 events.

LFR trials compared with APP trials

LFR had three viewers, while APP had about 50. Randomization of the photo association was done by LFR, but not by APP. While the APP group managers were also viewers (that is, outside of the CAS trials), that was not true of LFR's general managers (GMs). Ed May is a researcher and he urges people to choose one path or the other – viewer or researcher. A third person bet for LFR whereas for APP the viewer, the GM or those who had access to group predictions could bet.

The viewer(s) saw the prediction before feedback in APP but saw the prediction after feedback in LFR. The viewer could bet privately in APP but not in LFR. Neither group had shared wagering. One person bet for the group in LFR but not for APP. Instances of sparse coding were negligible for LFR but 3 of 40 for APP. "Sparse coding" is a term Jon used for a transcript with little data, which makes accurate scoring difficult.

Modifications of the CAS guidelines in Croatorum

The Croatorum GM purposely changed the CAS guidelines to see how that would affect results. For example, a trade was based on sessions from more than one viewer – two viewers on average rather than one. The events were Forex trades rather than sports events. The general manager saw both photos after the completion of the series of 10 trials but not during the series.

A Figure of Merit of .4469 was used and Igor averaged the two FoMs when both scores leaned to the same side. Further, a trade was made if "one Figure of Merit was greater than threshold of .4519 OR very close to threshold (>0,43)" or "if all FoM's of two or three transcripts lean to the same side, but MUST have high ABS diff (>0,2)." Croatorum had 1 hit, 2 misses, 7 passes.

Also, the group manager tracked what the results would have been if the SRI scale had been used, which required the GM to look at the unactualized photo. The GM reported results would have been much higher in this trial if the SRI scale had been used to place trades – 3 hits, 0 misses and 7 passes.

Hypothetical results in the CAS trials

Since the most accurate cutoff Figure of Merit might differ from .4519 if there were different viewers, Jon looked at hypothetical results from CAS trials for which he was group manager. Those hypothetical results showed the higher FoM of the two FoMs in each trial would have been a hit only 39 times in 76 events – a 51% success rate.

The hit rate when the higher FoM was above .4 (including values above .4519) would have been 60%. Good as the latter would have been, none of the alternative thresholds would have been as successful as .4519 (which led to an actual hit rate of 71.4%).

Hypothetical hit rates from CAS groups which followed the guidelines:

	Higher	Hit Diff		Miss Diff		High > .4		High >.3 <.4		High >.2 <.3	
	FoM a hit	>.2	<.2	>.2	<.2	Hit	Miss	Hit	Miss	Hit	Miss
CAS OAK A	12	7	5	1	7	5	3	2	2	3	2
CAS OAK C	6	5	1	4	7	2	2	1	4	2	3
CAS OAK E	9	3	6	3	3	3	1	2	2	4	1
SUBLIME	12	4	8	5	3						
Total	39	19	20	13	20	10	6	5	8	9	6
Hit Rate		49%	51%	39%	61%	60%	40$	38%	62%	60%	40%

Figure 9.3 – Hypothetical Results from CAS Trials

The results of using the .4519 value with the recommended guidelines, however, did not reach the 90-95%, level reported by Ed May. In the CAS system and using Ed's viewers, the figure of .4519 was expected to lead to that extremely high level of success. We do not have figures to indicate that that level was achieved. Based on the reports Ed May has given, it was not consistently reached.

The rates achieved in APP may be due to having viewers less skilled than May's, managers less familiar with or as skilled at coding, or for other unknown reasons. The CAS method proved successful in APP, with a higher hit rate than most APP groups have achieved, but as noted earlier, the very high number of passes negatively affected viewer and manager interest and morale.

Julia Mossbridge's Positive Precognition training

Prominent parapsychology researcher and remote viewer Julia Mossbridge, Ph.D., was introduced in our chapter about time. Here we turn to the categories she chose for her Positive Precognition approach which evaluates how well a remote viewing transcript matches a photograph in binary ARV. The categories differ from those May used in CAS. The comparison between the two approaches can be seen through the lens of Embodied Mind theory, about which Jon wrote an article for Daz Smith's *Eight Martinis* ezine. We will compare the two sets of categories after presenting the Embodied Mind theory.

Embodied Mind Categories *and our examples*					
"SupSupSup"	**"SupSup"**	**Superordinate**	**Basic**	**Subordinate**	**"SubSub"**
Creature	*Animal*	*Mammal*	*Dog*	*Fido*	
Creature	*Animal*	*Mammal*	*Dog*	*Great Dane*	*Fido*
CAS scoring categories in bold; *our examples in italics*					
Prominent/ Central Textured Repeat Motif	**Manmade**	*structure*	**Building**	*White House*	
		structure	**Village/Town/ City**	*Copenhagen*	
		Ruins	*castles*	*Heidelberg castle*	
		structure	**Roads**	*Route 66*	
		structure	**Pyramids**	*Great Pyramid*	
		structure	**Windmills**	*De Adriaan*	
		structure	**Lighthouses**	*Pt. Reyes*	
		structure	**Bridges**	*Golden Gate*	
		structure	**Coliseums**	*Roman Arena Arles*	
	Natural	*land feature*	**Hill/Cliff/ Valley**	*Great Riff Valley*	
		land feature	**Mountains**	*Kilimanjaro*	
		body of water	**Lakes/ponds**	*Lake Superior*	
		running water	**Rivers/ Streams**	*Ganges*	
		Land/water interface	**Coastlines**	*Big Sur*	
		running water	**Waterfalls**	*Iguazu Falls*	
		frozen water	**Glaciers/ice/ Snow**	*Mendenhall Glacier*	
		Vegetation	**Grass, trees, plants**	*Gen. Sherman Tree*	
		sand feature	**Deserts**	*Sahara*	

Figure 9.4

Given the centrality of the basic level, should attempts to characterize a photo involve the basic level, too? That is the approach CAS takes. Thus, in CAS we have building, pyramids, windmills, mountains, waterfalls – all basic level terms. However, CAS also uses superordinate categories (land/water interface and vegetation) and categories at an even higher level of abstraction (prominent/central, textured, repeat motif). In short, it is a multi-level approach in terms of Embodied Mind theory and categories. (But CAS does not explicitly utilize Embodied Mind categories; we are overlaying them onto the CAS categories.)

Positive Precog Categories in **Bold** (Julia Mossbridge), with our Examples		
Superordinate	**Basic**	**Subordinate**
Sky/space	night sky	Oakland night sky
Water/fluid	lakes/ponds	Lake Tahoe
Structure, humanmade	building	TransAmerica building
Plants/food	trees, flowers	Gen. Sherman tree
Animal/Human	man, woman	Jane Fonda
Ideas/words	love	Platonic love
(Landscape features)	**Mountains/hills**	Mt. Baldy
Energy	**Fire**	Youngstown fire
		Julia's examples
Sky/space		*earth from space*
Water/fluid		*faucet water*
Structure, humanmade		*pencil*
Plants/food		*hamburger*
Animal/Human		*fish, jellyfish*
Ideas/words		*maps, generic trees, signs*
(Landscape features)		*piles of dirt – any raised earth surface*
Energy		*fireplace, fire, lightning, bright light, electricity*

Figure 9.5 – Positive Precog Categories

Although not mentioned by Ed May and other researchers in the papers leading up to CAS, the Embodied Mind theory, which was developed in roughly the same years as CAS (1970-1990s), pioneered the "fuzzy logic" approach that CAS uses. Eleanor Rosch, George Lakoff, Mark Johnson and others overturned 2,000 years of category thinking going back to Aristotle.

Prototyping and basic level effects

Austrian philosopher Ludwig Wittgenstein questioned the classical conception of categories when he examined what games have in common. "We see a complicated network of similarities overlapping and crisscrossing: sometimes overlapping similarities, sometimes similarities of detail."[349] Games may not share a single common characteristic. He suggested, rather, that what games have in common is akin to "family resemblances."

Similar ideas were explored by Roger Brown in the late 1950s and by cognitive psychologist Eleanor Rosch in the 1970s, and these ideas have been developed by many others since. Rosch described what she termed "prototyping effects" with regard to categories. When asked, people say they consider some members of a category to be more typical members than others. For example, a robin is felt to be a more typical member of the category bird than a penguin or an owl. Such an approach contrasts with the millennia-old idea of categories in which all members have equal status. There are other important differences as well.

Of particular note for our purposes, Rosch went on to postulate what she called "basic level effects." The basic level is of singular interest. Research has shown it has the following characteristics:[350]

- highest level in which category members have similarly perceived overall shapes
- highest level at which a single mental image can reflect the entire category
- highest level at which a person uses similar motor actions to interact with category members
- level at which people are fastest at identifying category members
- level with most commonly used labels for category members
- first level named and understood by children
- first level to enter the lexicon of a language
- level at which most of our knowledge is organized
- level with shortest primary lexemes

The "basic level" is said to be of primary importance as we learn about objects in the external world. This applies to our visual impression of an entity, our ability to draw it, our naming of it, our body interaction with it and more. In this approach, there is a confluence of important characteristics at this "basic level" not found at the other levels. (The term used is "basic"; one might also have said "fundamental" or "central.")

There are two levels on either side of the basic level: the superordinate and subordinate levels, which require conceptualization beyond that of the basic level. They require a further grasp of categories by the learner or modification of information learned at the basic level. For example, generalization is required for a superordinate category such as furniture, mammal or fruit, or learning the particulars of a subordinate category member such as an Adirondack rocking chair, Pumpkin Cheesecake ice cream or a 34" Easton metal baseball bat.

Let's explore the possible relevance of this basic level to remote viewing by visiting the subject of ideograms.

Historically, the great majority of remote viewers have used ideograms, those quickly drawn marks on the paper that are said to represent essential information about the objective, conveyed to the hand by the "unconscious," "autonomic nervous system" or some other agent. The information is in graphic form, which the remote viewer then "decodes" during the session. Ideograms are widely used in approaches to remote viewing that employ a detailed, structured method (as opposed to "simple, natural RV").

As noted above, the basic level is the level at which "category members have similarly perceived overall shapes" and at which a "single mental image can reflect the entire category." This suggests that the basic level does capture a "gestalt" – a shape (graphic form) that is representative of the objective.

However, it is also true at the basic level that the image is associated with (more specifically, is named by) a particular word. The image is correlated in the mind with the word "dog," for example. When a child sees a hairy four-legged creature of a certain size with a big nose and a long tongue, it learns the word that signifies the entity before its eyes is a "dog." It also associates the smell, feel and sounds of the creature with "dog."

The child does not at first know (and therefore doesn't think) the creature is an "animal," and it is most unlikely she knows it is a "mammal." Dog identification takes place when the child is babbling, but not yet speaking.

In this theory, the levels are not fixed. It may be that the child first learns the creature is "Spot," a term at the subordinate level, rather than "dog." But the child learns there are other "Spots" and that they are all "dogs," and then "dog" becomes the basic level category member. "Spot" then becomes the subordinate level member. It is only later that the child learns that dogs and cats are both animals (the superordinate level). Still later, the child learns of categories such as mammal or living creature as the child's experience and knowledge climbs the ladder of generality. The three levels (superordinate, basic and subordinate) are not fixed but do reflect when they are learned, and the specificity of knowledge obtained.

A Pause: Toward better categorization of psi data

During the entire period in which remote viewing was developed and in the early days (1970's-1995) was not known to the public but later was (1995 to the present), the types of data that remote viewers get has by and large been categorized in a very simple bipartite way: "high" or "low" level. For example, in TransDimensional Systems, low level was made up of sensory impressions like colors, textures, sounds, smells, hardness-softness, moistness-dryness – these were considered the "bedrock" data that was essential and could be relied on. High level data were things like church, young man, stadium. Other remote viewing instructors have utilized a similar simple division of the data into two parts (e.g., adjectives and nouns). Could it be that Embodied Mind theory allows us to have a better grasp of the types of data we receive in remote viewing? We think so.

A central tenet of the application of Embodied Mind theory to remote viewing is that the **data is primarily at the basic level**. That includes drawings.

On occasion, the viewer will name the target (Golden Gate Bridge (subordinate) rather than just bridge (basic), *but generally remote viewing words and drawings are at the basic level.* It would be terrific if a viewer could get the specific name of an individual, building or other target, but this is rare. Beginning and accomplished viewers too appear most comfortable at the basic level. The viewer may resist being explicit (that is, naming) a word or drawing at the basic level because in CRV and offshoots, you are told "Describe, don't name!" This is good training, since most such "namings" are incorrect.

Given the centrality of the basic level, should attempts to characterize a photo involve the basic level, too? That is the approach CAS takes. Thus, in CAS we have building, pyramids, windmills, mountains, waterfalls —all basic level terms. However, CAS also uses superordinate categories (land/water interface and vegetation) and categories at an even higher level of abstraction (prominent/central, textured, repeat motif). In short, it is a multi-level approach in terms of Embodied Mind theory and categories. (But CAS does not explicitly utilize Embodied Mind categories; we are overlaying them onto the CAS categories.)

By contrast, Julia Mossbridge's Positive Precognition approach uses primarily *superordinate* categories – structure, plant/food, animal/human, energy, etc., (again without explicit reference to Embodied Mind theory). It employs two basic-level categories: mountains/hills and fire. One could consider sky/space to be at the basic level, as well, but combining the two elevates them to the superordinate level, the same as animal/human. Recall that the basic level is the highest for which you can draw an image representing the category (e.g., dog, cat, human). When you combine human and animal, as Positive Precog does, then you no longer have the basic category: you can't draw an image which represents animals and humans.

The Positive Precog approach would seem to have substantial advantages. For one thing, it has fewer levels of categories. Also, the categories match commonly used ideograms such as water/fluid, energy, structure, lifeform and mountain. Pru Calabrese even maintained that these five and one more were all the basic gestalts one could have! The importance of making ideograms central to your categorization is that if you were able to match the photo by drawing a single ideogram, ARV would be a simple and straightforward matter indeed.

On the other hand, if the Embodied Mind framework has relevance, CAS has an advantage since so many of the CAS categories are at the basic level – the main level at which remote viewing functions. So in practice, which is better? The Positive Precog results are not in yet but we know CAS has achieved very good results with top-notch viewers and quite good results in APP when the original guidelines were followed.[351] As we shall see in the next chapter, though, CAS also has its pitfalls.

We hope this long chapter explaining the background of the CAS system, including Embodied Mind theory, has been helpful in offering insight into a more differentiated and useful set of categories with which to examine and understand remote viewing data.

CHAPTER 10

Debra's Experiences with a Year-long CAS Project

Introduction

From August 28, 2013 through June 13, 2014, I participated in an ARV experiment using Ed May's CAS System. I was the sole viewer. The project was spearheaded by Dick Bierman with Ed May as the coder. Dick Bierman is a longtime professor with the department of Psychology, University of Groningen in the Netherlands. According to the university website, he started his academic career with research in Atomic and Molecular Physics, and then became engaged in consciousness research. As noted previously, Ed May was one of the original creators of the CAS system. He stated however, for this particular project, he was purely acting as coder, with no authority over the management of the project.

Procedure

All communications for the ensuing trials took place digitally. Bierman initiated the process using an online database program that generated an email stating a task was ready for me to complete. I logged onto a webpage that had the target number and did the session from my home. Then I'd scan my paper and upload the transcript. Given this project was testing whether a computer could replace the need for a human to eyeball any of the photos involved, Ed May, as coder, would not ever compare my transcript to either of the photo options. Instead he'd look at my transcripts, compare them to his coding sheet and assigned a value from 0 to 1.0 to categories of data (i.e., waterfalls, land/water interface, rivers, bridges, ruins, etc.) based on the degree to which he felt the transcript manifested those categories. I should emphasize at the time I had no idea what these categories were comprised of or what types of data I needed to supply that would facilitate the process. There were times I asked about this, since I knew I could adjust what types of descriptors I'd supply. However, all I was told was to keep my responses short. Sometime into the project however, May did respond to my ongoing inquires that providing higher level information rather than broken up bits of data (such as colors and shapes) would be more useful.

Once coding was completed, CAS generated a notice to Bierman, who then played an online casino game resulting in a binary outcome. I am not sure what kind of game that was. Once the game's outcome was known and entered, the computer system generated a photo associated with the winning outcome. This served as my feedback photo. Within a week or so, I'd receive an email that my feedback was ready. I'd log back into the system, look at the photo and compare it with my transcript. I'd then take some notes and organize these in a file on my desktop for future reference. Initially my feedback came somewhat inconsistently but then we got on a schedule that remained consistent, until the project's last couple months when researchers indicated they were traveling and feedback/communications became less so.

Results

In all I completed 37 trials. Nineteen were bet on – that is, they were deemed to have a high enough Figure of Merit score for one photo and a low enough one for the other to place a bet. Of the 19 wagerable predictions (about half of the trials), the results were 9 hits, 10 misses and 17 passes.

This project was not written up. However, upon my request, he was kind enough to share the data with me.

A few years later when we met at a conference again, I asked him if he had personally ever looked at any of my transcripts or compared these to the photos. He advised he never evaluated the data beyond compiling it into a spreadsheet and never compared the remote viewing sessions to the actual photos once the project was complete to try to understand what might have aided in the successful predictions and what might have led to misses. He stated as a statistician he equates lack of statistical significance within an overall project with lack of evidence for psi. I will present some examples of my session work and transcripts below for the reader to decide whether or not this is an appropriate assumption.

Jon later examined the score sheet to see if the guidelines shared by May and McMoneagle at the APP conference of not issuing a prediction unless the FoM scores were greater than .4519 was followed. Jon found three misses could have been avoided if they had passed instead of bet (since .4519 had not been obtained) but following that guideline would also have eliminated three hits.

My personal tracking of viewer awareness of trial outcome

One of the most interesting things that came out of this study for me personally was that I seemed to have been able to predict outcomes (that is, success or failure) during my sessions, not in terms of the ARV trials (beyond the 9 hits) but in terms of an intuitive sense of how the session and trial was going to go. Early on I became aware that I was having emotional reactions that seemed to be indicators of whether a trial would be successful overall, or if the session would be a good match to the correct target, or if something was going to go wrong. Bierman told me to keep track of these notes on my own, which I started to do more consistently about a month or so into the project. In all I had notes from 19 trials. From my notes I discovered:

At least three misses would have been avoided if I had been asked how I felt the trial would go. Also, I was confident about at least seven of my hits. I had marked that I was not confident on six trials that were marked as passes. On one pass I gave it a score of 5 out of 10 on an informal confidence scale. One trial was actually eliminated due to my predicting there would be a problem. They listened to me and did not go through with the trial. There was only one trial I was incorrect about when it came to these notes. In all I was correct 18 out of 19 times.

I'd like to emphasize that this was not at all about me knowing the outcome of the casino-style game on the online wagering site. In terms of the game itself, I didn't really know what it entailed except I thought it was probably simulating something like roulette with a black, red, or green possible outcome. I just could tell if things were going to go well based on whether I felt super relaxed during a session or highly anxious or stressed. I wasn't trying to do anything to get this sense – just noticing a correlation as the session went on. All of this may have been intensified by the fact I really was striving to do my best job possible, whereas I tend to be much more relaxed with other types of RV projects, where I haven't had this awareness, except perhaps when it was going to go extraordinarily well. Sometimes at the end of a session that resulted in a hit I was inclined to put a smiley face in my email when I submitted the session.

I shared this all with Bierman and he also agreed this was very interesting. However, he said since it was not part of the original design, he would not be able to include mention of any of it in a formal write up.

Discussion

New challenges lead to new awareness and perceptions of nature
When Ed May mentioned at an APP conference that his (the CAS) location-based target pool was devoid of people, animals and vehicles, I made a rather undiplomatic comment that the targets sounded quite boring – I tend to miss natural scenes without

much happening at them. Ironically, when I ended up working with the targets, I gained quite an appreciation for them.

With targets I typically would focus on stripped away, I found myself tuning into other aspects of targets I didn't usually pay attention to. This opened up an awareness of aspects of the natural world that has only continued to blossom. Being exposed to certain repeating aspects of landscapes allowed me to experience variations between them I had never noticed with regular perception.

For example, typically I focus on the top of a waterfall. Therefore, I'd expect that to also be my focus in a remote viewing session. Instead, it was the opposite – I'd notice the water hitting below. I learned Victoria Falls was more tumultuous and intense at its base than other waterfalls, even compared to Niagara Falls.

Before this project, if I had been tasked with a birds-eye view of a mountain range, I would have missed it completely. I would have expected something more compelling. The project acquainted me with the smokestacks of European cities. Noticing the smoke from these alerted me that I was likely being shown the tops of buildings rather than ground level. I learned to notice details – a 3D or a bird's-eye view perspective or one object positioned to the left of another – and became aware that at times I tuned in to angles and positionings shown in the photos rather than the locations themselves. At other times, I had sensory experiences of locations not shown in the photos. I also began noticing subtleties of shadow and light and color of various nature targets, verified by the feedback sessions. I then paid much more attention to these in everyday life. These targets helped me to begin to develop more of an artist's eye, which I never thought would happen in this lifetime.

Unfortunately, I suspect few, if any, of these details I perceived made the slightest difference in the final scoring. After the project was completed, I discovered the coding sheets don't ask about any of these finer details.

Some issues with the target pool construction and photo composition

I had assumed when Ed said the photographs in the pool didn't contain people, animals and transportation devices, it meant those things had never been at the locations. However, I often had impressions of these in my sessions. The first time it happened, I studied the feedback photo closely and noticed a person had been edited out so only a body part remained. I think it was part of an arm and a head but I can't recall exactly. My concern was not so much about imperfect editing techniques – the technology was limited in the 1990s – as the idea that researchers would think editing lifeforms and objects from a photo would prevent viewers from perceiving them.

This suggest the original creators of this system assumed viewers would only tune into the physical aspect of the photo itself and not what is really there at the location. It also suggests they were ignoring ways in which objects at a location help viewers to recognize and make sense of a location, in the same way this is true for those using regular visual perception.

For example, if a viewer encounters a straight strip of asphalt, it's going to be easier to know it's a runway instead of a city street if an airplane is hovering above or taking off from the runway. (This is less true if the airplane is parked on the ground of the runway, where it might be mistaken as another kind of vehicle that tends to be parked – after all, the first thing we think of with airplanes is that they fly, even though most do tend to spend more time grounded.)

As another example, for one target in this CAS project, I couldn't figure out if the area I perceived was inside or outside. However, I had a flash of a person sitting at a table, which reminded me of someone sitting outside a restaurant. I knew the target probably wasn't a restaurant – and the project prohibited a person in the photo – but I was certain the location contained a courtyard or patio where people could sit under an awning, just beyond some doors and windows. This helped me determine that the location was outside, but within an enclosed space. Without perceiving the people, I doubt I would have gotten a sense of this scene.

As much as a viewer might want to describe only what is in the photo itself or the confines of the frame at least (two different things), it might be as impossible for a viewer to avoid aspects of a location that do lie outside the parameters of a photo, particularly if the photo is a close up shot of a location. For example, let's say there is a closeup photo of a bridge. The bridge is over an ocean but we can't see the ocean in the photo. Well, in regular perception, even a blind person who walked over a bridge would likely still hear the water, feel the ocean breeze, hear sea gulls, feel the mist rising up. They would certainly smell the salt and fish. Therefore, there would not be a way to avoid noticing an ocean. This speaks to the idea that a viewer isn't just visual, but uses all their senses, along with an additional knowing ("clair-cognizance") that something is present. Likewise, if rows of colorful buildings bordering a Venetian canal have been cut out of the photo so that only the canal is displayed, how would one who was strolling past these still not glimpse these structures? In the same way, if a field typically has animals grazing in it but the photo isn't showing them presently, how would one avoid the smell, or sounds, or just an overall sense of animals being around?

In the same vein, if a photo is of a ruin where tour buses of people arrive 365 days a year, the location may be experienced very differently than ruins located in a very remote area that receive very few visitors. The tourists and tour buses may very well be

192

perceivable to the viewer because they are present and leave imprints, whether or not they are visible to the project manager, judges, or eye when looking at the photo. This is why if I'm going to give ruins to viewers and I don't want them to perceive a lot of tourists there, I'd only choose those that remain at very isolated settings. This didn't always seem to be the case with the target pool utilized in CAS. All of this is quite clear to viewers who have been doing session work for a while, but would not be understood by non-viewing researchers, particularly those who design a project thinking a viewer is only going to describe a photograph.

This brings us to what is a photograph? A digital photograph truly only has shapes and color in it. It is a representation of something three-dimensional that has so much more in it. Now it would be one thing if this system required viewers to only describe shapes and colors and not provide meaning. However, to the contrary, CAS is designed for viewers to recognize elements such as bridges and waterfalls. This is requiring the viewer to use higher level processing and get many more details, so that in order to excel they really do need to be free to allow their natural perception to work for them. They shouldn't have to restrict this to meet the expectations of those who programmed a system in the past – which is always the case when a computer is involved unless that computer system can actually learn with the viewers and improve upon itself, something present day software used for remote viewing projects does not have the capacity to do.

Homogeneity within target pool

Another challenge I had was that often target photos contained structures made of natural materials similar to those found in the photos of natural landscapes. Many were of older European cities or structures made from crumbling stone or older bricks that were tan or a lighter color. During a session, I might come across something hard, crumbly, tannish or light brownish, rock-like, structural, with edges, but this could describe a building, a mountain, a cliff, a ruin or even a bridge.

I didn't see this as a huge problem, but it meant I had to work much harder to get enough details to work all of this out. The only way I can know if I'm describing a building or a mountain made of similar materials is if I see a door or windows with glass or metal. Otherwise, what seems like a door could be a natural archway or opening to a cave. For this reason, I would have preferred if the buildings in the photos had been made of metal or more modern construction materials and in more artificial colors.

Loneliness and isolation long terms as a viewer

As noted earlier, one of the main reasons CAS was designed was to eliminate the need for researchers and raters to see the various photos involved. This means that when they look at the viewers' sessions, they have no way of telling whether the viewer's transcript truly does match the photo options. All they can do is see if the viewer has described anything that is listed on their score sheet. They will leave it up to the computer to decide on whether there was correspondence.

Whether a trial ends up receiving a high enough FoM score by the computer to place a wager, and whether this wager pans out to be a hit or a miss is all they can say. Bierman sometimes wrote to me after a trial and said "good job". He was also always responsive to my emails discussing scheduling and other practical issues. However, because of the predetermined protocols, this was the extent of communication about my remote viewing work. I was not getting paid for this work other than the knowledge that I was "contributing to science." However, I was putting in quite a bit of time and effort into every session.

Prior to the session I would clear my schedule and I would take time to meditate. Then once I started a session, I would usually spend between 30 minutes to an hour or sometimes even longer to see what relevant information reemerged (something that one can't do with a very short session). Then I would take time to recreate a summary, choose the sketches to submit (or redraw if messy), then I'd upload these to the system, and then take time to do a feedback session. I'd then make sure I combined the feedback with my transcript and keep this all organized on my desktop since I knew I'd likely not be able to get this all back in the end from the researchers themselves. Additionally, I needed to make sure that I was not doing too many other remote viewing sessions during the week, as I didn't want the feedback photos to get confused with each other, so I turned down participating in several other projects during these several months.

On top of time and effort spent, it mattered to me greatly what these prominent researchers thought. I expected this to be written up in the end; I knew wagering was going on and therefore my performance might impact whether money was lost or won (although the computer wasn't supposed to allow for a prediction to be issued if scores weren't high enough towards one photo and low enough for the others). I knew too that a certain number of predictions needed to be made before we could conclude the project. I also feared if I did poorly on this, word would get out about this to others. Of course, these concerns are not unique to CAS, but they all factored in to creating a level of stress. Further, as noted above, there were times I knew in advance that no matter what I was doing, something was not going to work out with the trial. Any time one has a precognitive sense something bad is going to happen, and at the same time is doing everything they can to avoid it, and also gets a sense they can't control the outcome – one experiences a high level of stress. This was the case with these sessions as well.

There were many times I'd receive my feedback and be quite pleased that my transcripts were close matches to the feedback photos. During these times I would feel excited and yet a bit let down, since unlike other remote viewing projects (especially those done with my Sublime group or done within APP projects) there was just no one to share these successes with. Part of the fun of working with a group is discussing what has just worked and where one got derailed – and laughing at silly drawings or words that emerged on the paper that were great matches or didn't match at all. This is what relieves the stress inherent in any project. None of that was possible with this computer-assisted scoring project.

Recommendations

After this study, I participated in some CAS trials with our RV Sublime group. Our acting project manager had continued difficulties retrieving the correct feedback photos. Then I served as a judge at an ARV conference that was using the CAS system. I finally was able to look at the coding sheet, which viewers weren't supposed to see, and that alerted me to the various categories. From that point on, when I attempted to do a few more trials I felt too frontloaded, so it was more like a forced-choice task than free response. That ended my practice with CAS.

In light of these experiences, I'd have some recommendations.

I think any remote viewing project involving computer assistance needs to put the human connection back in to ensure the viewer stays motivated and engaged instead of alienated. Viewers for long-term projects need to feel celebrated and appreciated. If the protocols mandate that no one in the project sees their sessions, at least allow for it later. Or permit something productive to be done with the sessions when the project is wrapped up.

One solution for projects that seek to replace human judging with computers would be to do alternative trials, with half using computer judging and half using human raters, who could then give comments to the viewers after judging. Not only would this provide more emotional fulfillment, but data could be compared between the two groups to help demonstrate which form of judging is more effective.

I also think any system, computer-based or otherwise, has to be designed to work with a viewer's individual strengths, to allow for learning curves and to actually grow with the viewer. Otherwise, even if it is useful at first, it won't be for very long.

Also, I believe a viewer must be more advanced to do well on these kinds of trials and project managers using CAS should be aware of this from the onset. While ARV trials with human raters allow for brand new viewers who just need a simple matching sketch and color to help them determine which photo is being described, this system requires viewers to name, or at least tune in to a limited number of specific elements (such as windmills, waterfalls, lighthouses). At the same time, the viewer needs to be able to focus their attention without perceiving anything outside the photo. This is virtually impossible, both in RV and in regular visual perception, but some viewers are more adept at this than others. Also, those who have learned controlled remote viewing methodologies really need to adjust their protocols, so they are only turning in summaries and compilation sketches, as opposed to pages and pages of raw data of broken down perceptions (no matter how correct these are).

In closing

The underlying research question in utilizing a system such as CAS is: Can artificial intelligence replace, or assist with, human-based analysis when the protocol requires the use of human consciousness for the remote viewing phase? We remain with questions like: What are the consequences of such a replacement? What are the benefits vs. costs...and to whom?

CAS doesn't replace human judging; it forces the human to choose predefined categories, which means all perceptions, words, images that lie outside these categories get disregarded, or still need to be interpreted. For example, if a viewer draws a curve shape and two lines on either side, does this get credited as a bridge? This is something the coder still has to decide – and a permissive coder would say yes, while a more conservative judge would say no. Since a coder is still needed to translate the data into something the computer can understand, I feel that it essentially turns a coder into a blind rater of sorts. Instead of having more tools at their disposal – the tools being all the words a viewer produces and all the data in a photo – now they have far less than before to work with.

Apart from this particular study, I often hear RV enthusiasts and parapsychological researchers voicing enthusiastic over the possibility of automating as many processes as possible within RV, and specifically Associative Remote Viewing project designs, in hopes it may lead to better results and higher efficiency. Unfortunately, they appear to give little consideration to how this might affect the remote viewer, especially with long term projects.

This automation trend is a far cry from the days in which former SRI director, Russell Targ, sat side by side, knee to knee on a sofa with Hella Hammid – her eyes closed, deep in trance. He visualized along with Hella, moment to moment, softly murmuring prompts and encouragement as she moved her attention to another researcher (the outbounder) who had traveled to a distant location so she might perceive their surroundings and activities. This was followed by a fun outing. Researchers would drive Hella to the site so she could see for herself what was there. The success of many such RV projects using these approaches has been well documented[352,353,354,355,356,357]

Yet, instead of a call for a return to these times in which researcher/tasker and their relationships with the viewer were seen as vital components in a telepathic process involving sender/receiver-viewer, now such human relations within the RV protocol are often seen as unnecessary. They are perceived to require too much time, effort and resources, or to lead to displacement effects, and lack of control over the many variables.

I don't wish to dampen the enthusiasm for automated RV systems in this golden age of emails, automated databases, drop boxes, online drives and endless spreadsheets, or to suggest I'm any less addicted to these technological wonders than anyone else. I do, however, think some are moving in the wrong direction when they want to cut out the human connection, and this feeling is very much backed by my own personal experiences as a viewer during this project, and others I participated in where researchers purposely did not engage with viewers because traditional (materialistic) scientific principles call for separation of roles and controlled and limited communications. This isn't to say there aren't creative ways to maintain these while bringing in the human element – I try to do that in the experiments I run (regardless of the technology involved) – but it does take extra resources in terms of planning, time, personnel and overall funding.

Samples of sessions that were hits:

Viewer Impressions: Natural, Large peaks (two at least), valley, Mountainous. Water, Ice, snow. Clumps of snow over water and earth, water flowing downward and in various directions, water bubbling coming up, sheaths, trees, leaves, some flat land drifting. Arch., movement, blowing, spraying maybe a fence type structure

Figure 10.1 – Trial: #310 – Hit

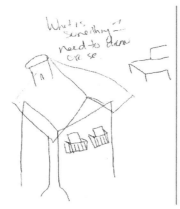

Viewer: "Manmade Structures. Tall, Populated area, Rows, Strong Columns feature, Angular, diagonal lines. Roof tops. Chiseled decor part of the buildings like either statues or part of building design. Smokestacks. Windows. Shutter/Structures around windows, maybe balconies or bars or something. Stairs. Connecting, possibly multiple structures, light colored natural materials. Inside/outside elements as if people could be under a top but still outside balconies. Stacked, wedding cake like structure. Seems like there may be a river type/water way area, separates foreground from background maybe"

Figure 10.2 – Trial #315 – Hit

Nature, Natural, elevated, intense movement of water flowing downward, splashing up, hills, mountains, living, tree like objects, different shades of green as if in a natural scene. My ideograms were indicating more biologicals than I've ever had shown up which of course is disturbing since there aren't supposed to be people or animals in the photos. But they kept coming up so thought I'd mention.

Figure 10.3 – Trial #321 – Hit

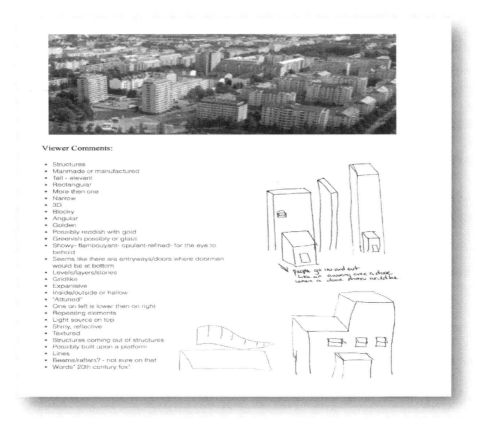

Figure 10.4 – Photo of City and Viewer Sketch with Comments

Figure 10.5 – Trial #323 – Hit

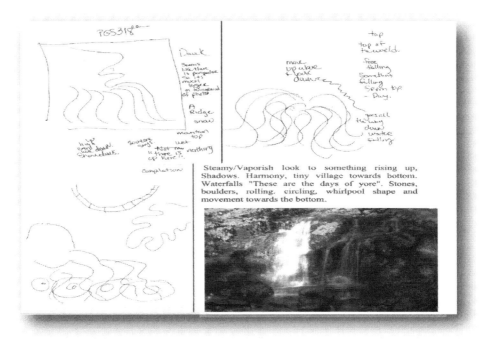

Figure 10.6 – Trial #318 – Pass

Debra's Summary submitted with sketches below, a few days prior to seeing feedback photo.

"Structure is: " knight" "armory." Reminds me of a Bavarian Design castle-like, medieval, it has been here for a long time. Massive, angular, tall, has an archway that can be passed through, brick like, reddish, blocks, spire/tower part, green foliage around. In natural Setting. Stone/Rocks around too. Part of it seems to be wall like, reminds me of a tall wall. High up, may have black iron fittings, bars or something like a gate. Structure and nature interface/integrated, nature seems to be part of this, so materials are very natural and it may be built into the landscape or the landscape kind of encourages it. Landscape may have high and low elements, not sure if it's the landscape that does or the structure. Boxy, but also may have a top that is like a spire or narrow triangular (this part may not be correct at all, couldn't tell if it was flat at top or if there was a bell/gong type ringing, like a really big one, this is what made me think there may be a bell tower nearby, but might not be in photo, didn't see it. At the top of this it seemed slippery, wet, mossy, treacherous, trickling, pouring. A sense that people may go up to the top."

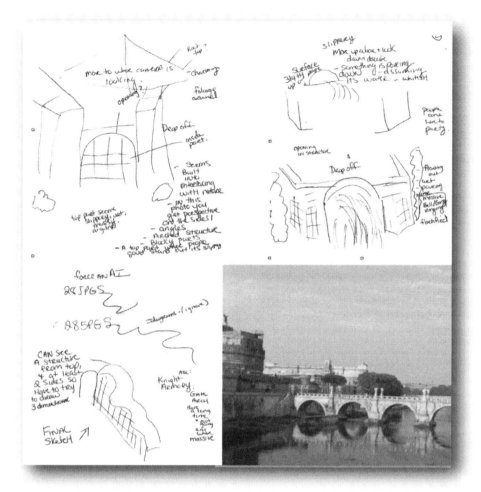

Figure 10.7 – Viewer Sketch and Feedback Photo of Castel-SantAngelo
Feedback: https://www.britannica.com/topic/Castel-SantAngelo[358]

One month later Debra was surprised to learn she had been assigned this same photo again, not realizing that was possible as it was assigned by a computer that had hundreds of options to choose from. She had less time and patience, so this session took about 1/4 amount of time as the first.

Debra's Exact Description Submitted to system for researchers:

Heavy Stones, Rocks embedded in cement type of inset, this is repeating

Water, Ocean crashing against. Heavy Structure or part of one, 3rd rectangular base with possible.
Tapered top

"Fort or Fortified"
Tower like
Prison Like
Armory
Gunnary
Embarkment
Has "a basement". a basement?
A wall around a structure
Very tall structure, one that seems rectangular with other
Rectangular attachment
Progressively getting taller
Blocky

Has like a rim around it, maybe where people could have stood but something in Photo that seems like could be more of a circular tower shape, tapered so narrower at the top.

Not getting any sense of foliage around, lacking in green or foliage I think, mostly a gray stone like feel. Harsh environment with violent water, history of violence and death and guns and protection. Windows or areas for viewing. People can be up on it, walking around or focusing on guns or canons. May be an issue with not all of the structure being in the photo, like part of it.

word, "planetary," "Divide," "tattooed," "guns," "ladder"

Water fountains, some statues, some stone steps, the side of a buildings with a triangular shape."

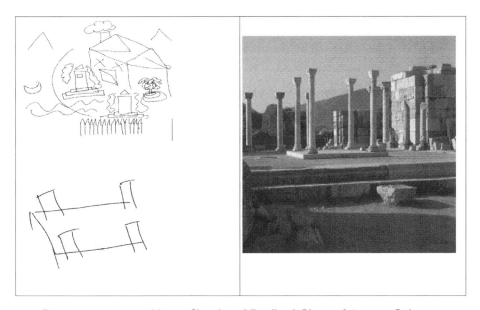

Figures 10.8 & 10.9 – Viewer Sketch and Feedback Photo of Ancient Columns

CHAPTER 11

The Dung Beetle Scoring System

Alexis Poquiz is a former US Marine, a very talented remote viewer and founder of the largest and longest-running RV-related Facebook group. He is creator of the Dung Beetle system, which over the years he allowed us to refer to as "the Poquiz Methodology" in formal arenas. Although he has always preferred the term "Dung Beetle," he understood journal editors and peer reviewers would take issue with this moniker – which is probably why he prefers it. He's a rebel at heart, but a reasonable one, when all is said and done.

The Poquiz Methodology was introduced in social media and at various remote viewing conferences and was later adapted for use in formal projects such as Katz, Beem & Fendley[359] in which 39 remote viewers described a microscopic organism, with ratings of the sessions by qualified scientists. A revised version of the Poquiz system was used in a remote viewing project designed to predict the outcome of a US presidential election.[360]

We believe this scoring method stands as the most sensitive available, although it requires a lot more time, effort and concentration than others and therefore is not always going to be the method of choice. While versions have appeared on the internet, we wanted to make sure the complete version was in print and believe readers will benefit from having access to it.

We invited Alexis to present the full version in his own words, along with an extensive history of revisions and the rationale behind them. He briefly speaks about other methods; for our description of them, please visit the respective chapters.

The following is by Alexis Poquiz, System Creator

To begin talking about the Dung Beetle System, we first need to begin with the SRI Confidence Ranking System. Dung Beetle did not attempt to rebuild the wheel. It attempted to evolve what was already there. The Dung Beetle system is simply the evolution of the SRI Confidence Ranking System.

SRI Confidence Ranking Score – 7 Point scale

Score	Description
7	Excellent correspondence, including good analytical detail (e.g., naming the target), and with essentially no incorrect information.
6	Good correspondence with good analytical information (e.g., naming the function of the target) and relatively little incorrect information.
5	Good correspondence with unambiguous unique, matchable elements, but some incorrect information.
4	Good Correspondence with several matchable elements intermixed with some incorrect information.
3	Mixture of correct and incorrect elements, but enough of the former to indicate that the viewer has made contact with the target.
2	Some correct elements, but not sufficient to suggest results beyond chance expectations.
1	Little correspondence
0	No correspondence
	From paper by Wenden Weigand, Dean Brown, Jane Kantra, Russell Targ

Figure 11.1 – SRI Confidence Ranking Score

The genius of the SRI ranking system was that it allows for the concept of subjectivity. However, the problem with the SRI ranking system is that it allows for the concept of subjectivity! Yes! An absolute paradox! Make no mistake about it, the SRI ranking system was revolutionary in that it scientifically attempted to make that which is subjective, non-subjective. This is exactly what science needed to attempt to measure the immeasurable.

The most glaring problem with the SRI ranking system is how imprecise the language is for each confidence ranking level. Some ranking levels are more subjective than others. There are seven ranks in the SRI ranking system; eight if you count "rank zero," which is CR= 0. Except rank zero, which represents no correspondence, every other ranking level is subjective. Even the gradations between each level are subjective. This is a huge problem.

Let me give you instructions on how to bake a cake using the language used in the SRI ranking system. First gather all your ingredients. A little bit of butter, some salt, a mixture of eggs, a good amount of sugar, and a lot of flour. You don't need to be a chef to know that the cake is going to be disastrous! Can you imagine trying to bake a cake using those kinds of instructions? That is exactly what you are doing when you use the SRI Confidence Ranking System.

This was the problem with the SRI ranking system, it was too subjective. If subjectivity is the problem, then reducing the level of subjectivity was the natural solution. This leads us to the military method of evaluating remote viewing sessions (See Chapter 4). The military methodology for evaluating RV sessions is a heavily quantitative approach.

However, the weakness of the military methodology is that it loses contextual data. Furthermore, the process of tabulation was too time consuming for what it was worth. It was just too impractical.

Enter Bruce Miller's CRV Score Spreadsheet. Miller took the military method scoring sheet and converted it into a functional Excel spreadsheet. The digital spreadsheet made processing a session less laborious by automating the tabulations of Yes/No perceptions. However, you still had to manually input the tabulations into the scoring sheet. So, despite alleviating the problem of being too impractical it was still a time-consuming process for very little benefit.

Now, I must say that the military methodology of evaluation made absolute sense, in that the methodology was designed with an operative mindset. If remote viewing sessions were to be trusted, then understanding a remote viewer's capability in perceiving certain categories is of vital importance. The contribution of the military methodology of evaluation is in the strength of its quantitative approach.

With the emergence of new technology, an entirely computerized approach for judging remote viewing sessions entered into the picture. The Laboratories for Fundamental Research (LFR) developed Fuzzy Set Analysis, which was given the name CAS – Computer Assisted Scoring.

You have to understand that CAS was designed to systematically judge a remote viewing session in a laboratory setting. Now, don't make the mistake of thinking that CAS will work in any kind of lab setting. No. CAS only works within a specifically defined isolated pool of remote viewing targets. Additionally, the remote viewers are restricted to using a narrowly defined list of categories to describe. Furthermore, judges are given a set of isolated responses to score a given session. The entire methodology is contained in a completely closed system. What this means is that the methodology is completely useless beyond the constraint of the closed system.

The significance of CAS is that it produced a variable concept known as the Figure of Merit. The Figure of Merit is composed of two valuations – accuracy and reliability. Accuracy is a quantitative valuation and reliability is a qualitative valuation. Furthermore, CAS leveraged the power of technology to automate calculations. Perhaps the most unique aspect of CAS is that it built into the system the concept of orthogonality. However, despite all these advancements, the fact that CAS is dependent on such a niche set of constraints, it is ultimately impractical for evaluating a typical remote viewing session.

It was at this point that innovation ended. The SRI Confidence Ranking System became the de facto standard for evaluating remote viewing sessions in the remote viewing community. This result was easily arrived at, due to the impracticality of CAS and the tedious nature of the military methodology. Despite the massive problem of subjectivity with the SRI ranking system, the RV community continued to use it. There simply was no better alternative. This is where I decided that if I wanted to make the SRI ranking system better, I would first need to somehow reduce the subjective element of the ranking system. Little did I know, I was creating what I would later define as The Dung Beetle system.

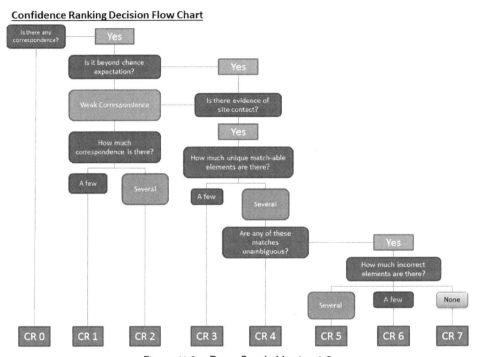

Figure 11.2 – Dung Beetle Version 1.0

It was not originally called Dung Beetle. I originally called it the Confidence Ranking Decision Flow Chart. It started off as an attempt to take the SRI 7-point ranking system and process it into a decision flow chart. My intention was to reduce the subjective nature of the ranking system by formalizing the process in which a judge comes to the conclusion of a confidence ranking score. As you can see in this diagram, step-by-step it systematically walks you through the SRI ranking system.

CR = Confidence Ranking

Start the decision flowchart at the upper-left corner with the question: "Is there any correspondence?"

- If no, then the score is CR = 0.
- If yes, then continue to, "Is it beyond chance expectation?"

- If no, then proceed to "Weak Correspondence" and then proceed to "How much correspondence is there?"
  - If yes, then proceed to "Is there evidence of site contact?"
  - …and so forth and so forth

So, in this manner you continue throughout the decision flowchart, eventually coming to a definitive CR score.

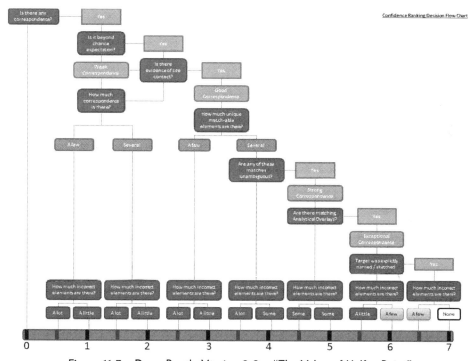

Figure 11.3 – Dung Beetle Version 2.0 – "The Value of Half-a-Point"

Dung Beetle Version 2.0 attempted to formally introduce the concept of half-a-point.

A problem with the SRI Confidence Ranking System is that there were no in-between values. There was no such thing as a CR 1.5 or a CR 4.5 or any decimal value of a CR score. The SRI ranking system used whole numbers only. However, in practice people naturally gave a score that was between two likely CR scores. For example, a judge using the SRI ranking system might evaluate that a session was worth a score of "either" a CR = 3 or a CR = 4. Rather than constricting themselves to either a 3 or a 4, users would just score the session the average of their estimations, which in this example would be a CR = 3.5

To formally create a systematic evaluation if something is worth half a point more or half a point less, I inserted the question: "How much incorrect elements are there?

- If there were "a lot" of incorrect elements then the CR Score would be reduced by half a point.
- If there were "a little" bit of incorrect elements then the CR Score would not be penalized.

In this manner, Dung Beetle systematically incorporates into the confidence ranking decision flowchart the confidence ranking valuations for half-point differences.

CR = 0.5, CR = 1.5, CR = 2.5, CR = 3.5, CR = 4.5, CR = 5.5, CR = 6.5

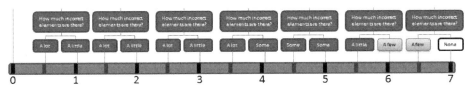

Figure 11.4 – Ranking Valuations for Half-Point Differences

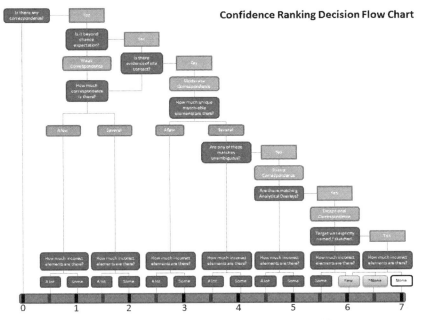

Figure 11.5 – Dung Beetle Version 3.0 – "Too Many CR 7.0s"

One of the more challenging aspects of judging happens when you encounter a really excellent remote viewing session worth CR 6.0 to CR 7.0. As a judge, a part of you wants to give the viewer the BEST score possible, but there is also a part of you that wants to remain objective, firm, fair, consistent and accurate.

As a judge, it is your responsibility to give a score that properly reflects what a session deserves. You don't want to hand out a CR 7.0 when a session only deserves a CR 6.5. The problem was not that there were too many people giving out too many CR 7.0s as opposed to CR 6.5. The problem was that it was very difficult to consistently identify where that line of separation was that separates a CR 6.5 and a CR 7.0.

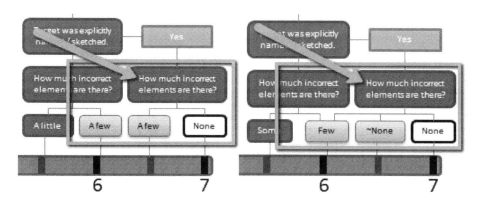

Figures 11.6 (left) Version 2.0 and 11.7 (right) Version 3.0

In order to make the flowchart more intuitively reflect the true value of a CR 7.0, I made a subtle update to the decision flowchart. I made an effort to emphasize the concept of no incorrect elements by allowing a judge to filter a session through "None," "~None" and "Few."

"None," represented no incorrect elements and led to a perfect CR 7.0.

"~None," which represents "approximately" no incorrect elements, reduces a session to a CR 6.5. What does "approximately" no incorrect elements even mean? It recognizes that some elements may/may not exactly fit into a right/wrong criteria. It also allows a judge to flexibly state that the session did, in fact, have incorrect elements BUT the incorrect elements were not significant enough to downgrade and penalize a session to CR 6.0 but were enough to definitively state it was not a perfect session.

"Few" represents that although the session explicitly states what the target is, the session contains an indisputable mistake and is deserving of a downgrade to a CR 6.0.

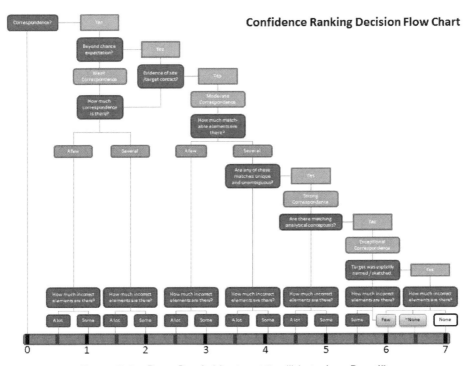

Figure 11.8 – Dung Beetle Version 4.0 – "Meticulous Detail"

Version 4.0 is really not so different from Version 3.0. In fact, the changes made were simply cosmetic in nature. For example, I changed the wording of "Is there any correspondence?" to simply "Correspondence?" That may not seem like a significant change, but to me, it was. No detail was insignificant.

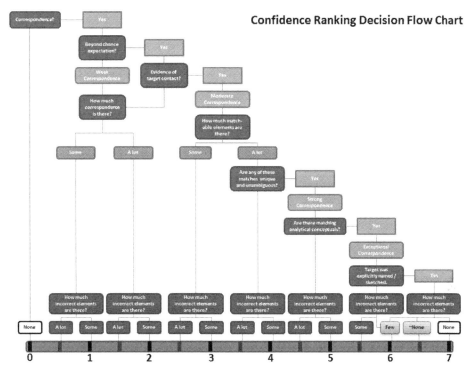

Figure 11.9 – Dung Beetle Version 5.0 – "The Difference Between a Few, Some, Several and a Lot."

What is the difference between "a few vs. several" and "some vs. a lot?"

After much contemplation, I realized that both phrases "conceptually" expressed the same meaning, that one is lesser than the other. The expression "a few" generally represents something that is less than three, similarly the verbiage of "some" generally represents something that is around three. It is slightly different but similar enough that "a few" is vernacularly interchangeable to "some."

Meanwhile, the expression "several" generally represents that you have more than the quantity of "some," likewise the expression of "a lot" represents that you have more than the quantity of "some." Again, it is slightly different but similar enough that "a lot" is vernacularly interchangeable with "several."

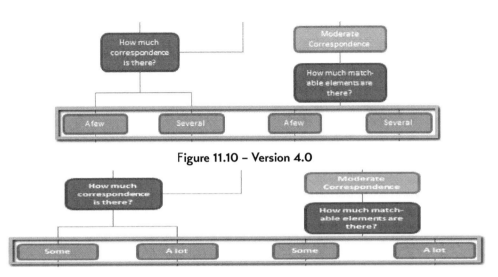

Figure 11.10 – Version 4.0

Figure 11.11 – Version 5.0

This was an opportunity to simplify the flowchart by replacing "a few" with its equivalent of "some" and replacing "several" with its equivalent "a lot." Unbeknownst to me at the time, this simplification ends up playing a significant role in the final iterations of Dung Beetle.

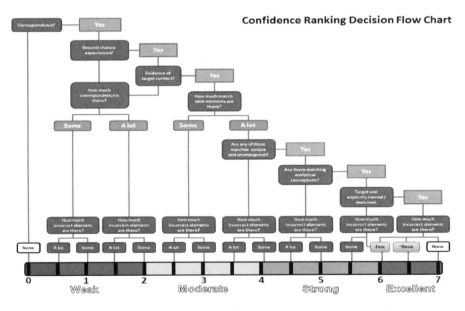

Figure 11.12 – Dung Beetle Version 6.0 – "Weak, Moderate, Strong and Excellent."

As I continued to use the decision flowchart, I began to be aware that in practice, it was unnecessary to identify correspondence and label them "weak, moderate, strong or exceptional." A weak score will have weak correspondence, a moderate score will have moderate correspondence, a strong score will have strong correspondence and an exceptional score will have exceptional correspondence. It became clear that it was unnecessary to identify and label the level of correspondence in the flowchart. This resulted in the following score categorizations.

Weak Scores:	CR 0.0 - CR 2.0
Moderate Scores:	CR 2.0 - CR 4.0
Strong Scores:	CR 4.0 - CR 6.0
Excellent Scores:	CR 6.0 - CR 7.0

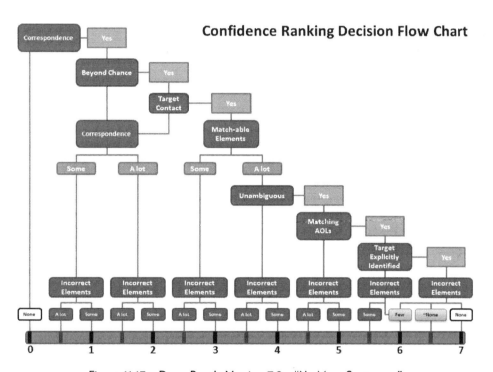

Figure 11.13 – Dung Beetle Version 7.0 - "No More Sentences"

Dung Beetle was developed with the judge in mind. I wanted the flowchart to be as simple as I could possibly make it. So, I worked to simplify the flowchart by getting rid of all the sentences and reducing them into one recognizable word.

Figure 11.14 – Dung Beetle Version 8.0 – "What Score Represents a Hit"

One of the most important concepts in judging is knowing when a score represents a hit. This is when things start to get interesting, and this is also when I began to question what it really means to get a "hit."

What score represents a hit?

Now, let's be very clear, first of all, that a "hit" in the context of a typical remote viewing session is when the viewer has "hit" the target. This means when a remote viewer has expressed enough perceptions to prove they are describing something beyond the level of chance, then we can say the remote viewer had a "hit." Anything at or below the level of chance means the viewer may have made "contact," but not enough to show they only made contact simply due to chance.

If we use the definition from the SRI Confidence Ranking System, a CR score of 2.0 is considered the level of chance. The system does not have 0.5 increments, so the next possible SRI ranking score is a CR of 3.0. This means that under the SRI ranking system, a CR score of 3.0 is considered above chance.

Therefore, one can state that a CR score of 3.0 means a viewer has expressed enough perceptions to prove they are describing something beyond the level of chance. We can then state that a remote viewer who earns a CR score of 3.0 or greater represents a "hit." This discovery was a big deal in that it established the numerical representation of a hit. That is why this information is included in the legend. To better understand this insight, we need to look at the exact definition as stated in the SRI ranking system.

The exact verbiage for a CR 3.0 in the SRI ranking system is: "Mixture of correct and incorrect elements, but enough of the former to indicate that the viewer has made contact with the target." If we roughly translate the verbiage to its numerical equivalent, we can interpret the verbiage of "mixture of correct and incorrect," to mean 50/50. When the verbiage states "enough of the former..." it is implying there is more correct than is incorrect.

This reveals another insight! A session with more than 50% incorrect elements will receive no higher than a CR 2.5. This is true because at the 50th percentile, any more incorrect would not result in a CR score of 3.0 because it would violate the principle that there is more correct than there is incorrect.

A hit is not a hit

A word of caution about "hit" in the context of Associative Remote Viewing (ARV). In ARV, the term "hit" has a different meaning than "hit" in the context of a typical remote viewing session. In ARV, the term "hit" usually means your prediction resulted in the "correct" prediction. An ARV "hit" is not the same as a remote viewing "hit."

This confused me a lot! It was hard to wrap my head around it. How could an ARV "hit" be different from a regular remote viewing "hit?!" Why isn't a "hit," a "hit?"

All this time the community has been treating an ARV hit as the same thing as a remote viewing hit and it's wrong! They are not the same! During an ARV session, your remote viewing could be absolutely fine, but because the judge failed to correctly match your remote viewing session to the correct "outcome," your remote viewing session is considered a fail!? This is wrong. It is entirely a different independent variable.

What we are looking at when we talk about an ARV hit is actually the result of three different variables. The first variable is the remote viewing session. The second variable is the association. The third variable is the judgment. At any point of the process of completing an ARV session, if any of those variables fail, then it results in a miss. The problem is that people are attributing a failed association or a failed judgment as the result of a failed remote viewing session.

Dung Beetle makes it clear what exactly a hit means. A hit is when the remote viewing session achieves a CR score of 3.0 or greater. This means that in a given ARV session, it is entirely possible you can have a session where you see both targets. Seeing both targets means your remote viewing session achieves a CR score of 3.0 or greater for BOTH possible-outcome images.

A hit is not a hit. I believe this is still a common misconception and a huge problem in the ARV community. Even those who have been practicing ARV for the longest time may still not truly understand the difference between an ARV hit versus a remote viewing hit.

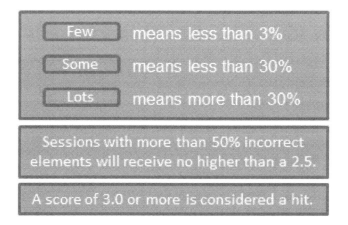

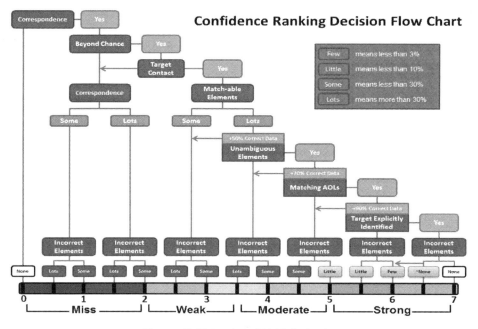

Figures 11.15 (top) and 11.16 (below)
Dung Beetle Version 9.0 – "The Significance of Unambiguous Elements"

"Unambiguous" is one word that received a lot of attention and resulted in a lot of community discussion. This word alone is what determines whether or not your session was considered strong. If you had a very high percentage of correct elements but did not have an unambiguous perception, your session was considered weak or moderate at best. In fact, if you did not have an unambiguous perception, the highest score you could earn was a CR 3.5.

A CR of 2.0 is considered the level of chance. What this means intuitively is that from CR 2.1 to CR 3.5, it is very close to the level of chance. However, it is still considered above chance. It is not until you reach a CR score of 3.5 or greater that one can confidently state that you performed moderately well in your remote viewing session. This insight can be practically applied when making ARV wagers.

When wagering on an ARV session, you will want to take action only when you are confident that you performed well-enough during your remote viewing session. We have just defined that a CR score of 3.5 or greater represents that you performed moderately well in your RV session. Therefore, we can make the principle that you should only wager on an ARV session, when at least one of the possible outcomes results in a score greater than CR 3.5. If you remote view and fail to achieve a CR score of 3.5 or greater, it means your remote viewing performance was not good enough to justify wagering. We have to remember that a CR score of 3.0 or higher is considered a hit. A CR score of 3.5 or greater means you can be confident that your remote viewing session, at the very least, made an unambiguous contact with the target.

The problem of scale and the usage of a quantifier

The meaning of a word or an adjective can be interpreted wildly differently. For a simple adjective like the color "red," the interpretations of what constitutes red can wildly differ between judges. A judge will say red means passion. Another judge will say red means anger. No, red means life! Even with seasoned judges, the interpretation of basic adjectives is easily in disagreement, so you can imagine that complex adjectives can cause significant disturbances in analytical agreement.

A quantifier is a complex adjective. The SRI ranking system utilizes nothing but quantifiers. This results in a system that is highly susceptible to potentially creating massive disagreements in analytical agreement.

Let me provide an example highlighting the difference between using a percentage and using a quantifier. The meaning of "a little" in the context of a quantity of 100 perceptions is different from the meaning of "a little" in the context of a quantity of 25 perceptions. A "little wrong" could mean about 12 errors in the context of a 100 perceptions. But if you had 12 errors in the context of 25 perceptions... you literally had almost half wrong. Getting only 12 perceptions wrong in a remote viewing session where you described 100 perceptions is pretty damn good. Getting 12 perceptions wrong in a session where you only described 25 perceptions is basically 50/50 chance. This is why we need a concept of measurement that retains its meaning regardless of scale.

The value of using percentages is that percentages retain meaning despite varying amounts of quantity. The meaning of 50% when there is a total of 100 perceptions, carries the same meaning of 50% when there is a total of 25 perceptions.

One of the most complex concepts about Dung Beetle is how I converted the adjectives, "few," "little," "some" and "lots" into their respective percentages. How did I arrive at the value of these percentages?

Why do I represent "Few" as less than 3%?

Why do I represent "Little" as less than 10%?

Why do I represent "Some" as less than 30%?

Why do I represent "Lots" as more than 30%?

The following is the sequence of thoughts that I mentally encountered as I arrived at my values.

- I began with 100% to represent CR 7.0
- And then, 0% to represent CR 0.0
- The definition of CR 2.0 represents chance, so I conveniently set that at 50%.
- The verbiage "some" and "lots" typically represents three or more elements in the real world. So, I counted "some" of my fingers up; one, two and three. I then intuitively thought, three out of ten is 30%. So, just like that, "some," was defined as anything less than 30% and "lots" was defined as anything more than 30%.
- The definition of CR 3.0 represented something that was a mixture of correct and incorrect, but enough of the "former" [correct]... to indicate that the viewer has made contact with the target. I simplified that in my mind to mean that a CR 3.0 represents something that has more correct than incorrect. That intuitively felt like a passing grade. A passing grade is 60%.
- "Good," for most of society, is anything that is normal behavior. Something that is above average is good. Additionally, the jump from CR 2.0 to CR 3.0 represented a difference of +10%. So, it intuitively made sense to progress from CR 3.0 to CR 4.0, by 10% and onwards to be increments of 10%.
- CR 5.0 became 80%, and CR 6.0 became 90%.
- CR 1.0 was defined as "Little correspondence." Each increment is worth 10%. So, from CR 0.0 to CR 1.0 we can represent it as an improvement of 10%, or in other words, a "little" can be defined as 10%.
- Now, the concept of a "few" equating to 3% has to do with interpreting a "few" to have a lower value than a "little." Naturally, I gravitated toward half of 10%, which

was 5%. However, I thought 5% was too much. So I took half of 5% to offset the over-calibration. This set the value at 2.5%. I then rounded up to simplify it to 3%.

I was apprehensive about the intuitive logic I was following. I was also cognizant that someone could potentially look at all of this and criticize how childish it sounded. However, despite how ridiculous it seemed to me, there was something about it that intuitively made sense. I knew Dung Beetle wasn't perfect yet, but I was confident I was making progress because the significance of stepping away from quantifiers toward percentages meant we were reducing the variability of interpretation.

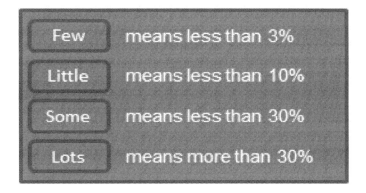

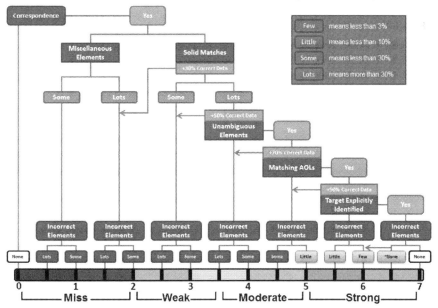

Figures 11.17 (top) and 11.18 (below)
Dung Beetle Version 10.0 – "Quality Not Quantity. Cheating the System."

At around the tenth iteration of Dung Beetle, I was beginning to notice a pattern from judging and analyzing so many different remote viewing sessions. Specifically, I began to notice that not all perceptions deserved equal weighting.

What do I mean by this? And how do I represent this insight in such a way that you could represent it graphically? Is it possible to represent the concept that "quality" matters? Is it right to penalize a session that is too basic? Is there a way to game the ranking system in such a way that a session could be worth more than it is actually worth?

While judging some remote viewing sessions, I came across a particularly interesting remote viewing session where it became blatantly obvious to me that the remote viewer was off. It happens, it is very common and there is nothing devious about a remote viewer being off target. Remote viewing is a skill and sometimes you have your off-days and sometimes you do really well! I took this concept to the extreme.

Suppose I was not a trained remote viewer? How can I fake being a remote viewer AND trick people into thinking I was a remote viewer? I know that statistically I can hyperinflate the number of correct matches in a given remote viewing session simply by listing a whole bunch of generic-all-encompassing perceptions. Perceptions that I know are probably going to be more correct than incorrect. For example, I perceive a perception of the concept of "big." I then begin to list out words similar to "big" – wide, gigantic, expansive, universal, not-small, etc. etc. etc. In this way, I can create a buffer for my remote viewing session because I can now allow myself to absorb enough incorrect elements and my score will not be penalized by too much. Furthermore, I understand that if I state enough correct elements, more than incorrect elements, then I can achieve a score seemingly greater than chance.

The problem is that people were unconsciously doing this. Your mind naturally works by association. So, it is entirely natural for your mind to go from "big" to wide, gigantic, expansive, universal, etc., etc., etc. So, how do we combat this potential loophole? My solution was to introduce the idea that some perceptions are really just miscellaneous, fluff data.

Now, if you have ever worked with a remote viewer who utilizes Coordinate Remote Viewing (CRV) or a derivative of CRV, you will understand what I mean when I state that CRVers will write pages and pages and pages of fluff. This shotgun approach of writing down every perception that comes to your mind, indiscriminately, is not what I would call higher perception. It is called "guessing."

The problem with a purely quantitative ranking system is that you could unknowingly create an environment where people who took the time to "guess" more would eventually get a higher average score than people who actually perceived impressions and weren't just "guessing." This problem is very difficult to identify because it is very difficult to know for sure if you are actually "perceiving" or…if you are actually just "guessing." Purely quantitative ranking systems are susceptible to being exploited by beginners and "experts."

A beginner can exploit the system by simply making more guesses. The problem is that beginners are encouraged to make guesses. The CRV process encourages remote viewers to write down any impressions you perceive without discriminating whether or not those perceptions are correct or incorrect. The progression from a novice remote viewer to an intermediate remote viewer is that the intermediate remote viewer begins to develop an actual, perceivable, discernable sensation when describing perceptions. A novice remote viewer will shotgun their perceptions while an experienced remote viewer is more capable of controlling and directing their attention to intentionally perceive. A novice perceives by accident. An expert perceives by choice. It is not that beginners are intentionally cheating. It is just that methodologies that are purely quantitative in nature penalize experienced remote viewers who rely on quality over quantity.

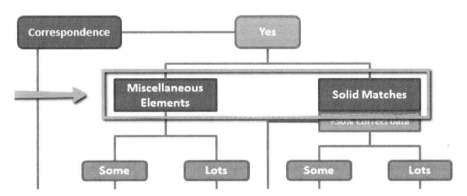

Figure 11.19 – Version 10.0 – Miscellaneous Elements & Solid Matches

The most significant update for Dung Beetle version 10.0 is the introduction of "Miscellaneous Elements" and "Solid Matches." It was updated specifically to address the possibility of exploitation. It is significant because it establishes there are perceptions that, although you may get them correct, they aren't necessarily relevant or significant. This update caps the benefit of having a really high number of miscellaneous perceptions. In order to progress to higher scores in the Dung Beetle system, you have to express solid, meaningful perceptions. Putting down pages and pages of superfluous perceptions is now going to indirectly penalize your score.

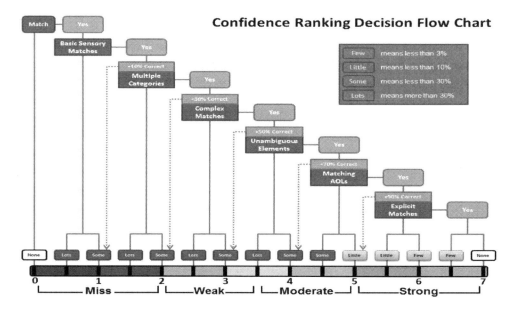

Figure 11.20 – Dung Beetle Version 11.0 – "Variances Between Judges"

You learn a lot from engaging with other judges. The more judges you engage with, the more variability in interpretation you will encounter. One of the lessons I learned while interacting with multiple judges is that some judges have a proclivity to be "too lenient" and some have the proclivity to be "too literal."

A remote viewer once put down on their session "something feels furry, it has a tail, it reminds me of a cat." The target was actually a "beaver." The question is, how will judges score the session?

A collaboration with multiple judges revealed that some judges scored the session as low as CR 2.0 and others as high as CR 6.0. The range of interpretation was immense. How could judges differ by so much?

There is no perfect solution to this. People see the world with different lenses based on their life experiences. So it is inevitable that there is going to be variance between judges. However, there are known practices in judging that cause these extreme judgments from being more common. There is a practice of judging sessions where the judges primarily look at the number of correct perceptions and downplay the importance of incorrect perceptions. This type of judging mindset will result in positively inflated CR scores. Nice judges are very susceptible to having this kind of mindset.

However, there are judges who are just too stone-cold literal. A cat is not a beaver. Never mind that a cat and a beaver share common traits, and never mind that a cat and a beaver have similar visual silhouettes. A cat is not a beaver. This is too literal of a mindset. These types of judges do not have an understanding of basic literature regarding extrasensory perception. It is well understood that the unconscious mind tends to express itself through symbolism. A symbolic representation of a concept carries so much more information that a literal translation cannot fully express. Yes, it is true that a cat is not a beaver, but it is also probably true that this kind of mindset negatively inflates CR scores.

It is important for a judge to positively increment the value of a session based on correct perceptions. It is also just as important to penalize the value of a session based on incorrect perceptions. If a judge has the mindset that they will only increment a CR score, but not penalize scores, it will lead to inflated CR scores. Likewise, if a judge is too literal and is too inflexible, then it will lead to deflated CR scores.

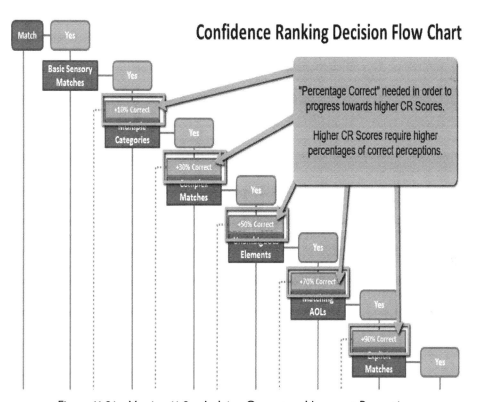

Figure 11.21 – Version 11.0 – Judging Correct and Incorrect Perceptions

Dung Beetle version 11.0 further emphasizes this dynamic understanding of correct and incorrect elements. Progression through the Dung Beetle flowchart follows a methodical approach in which every level progressively requires a higher percentage of correct elements in order to proceed into higher CR scores.

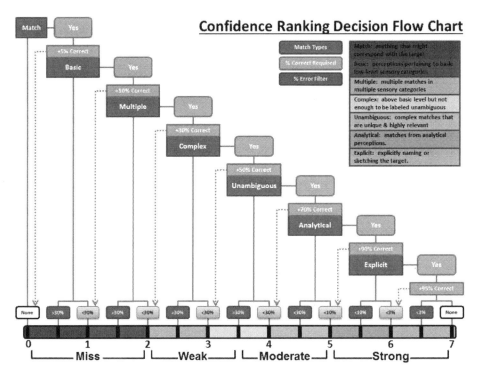

Figures 11.22 and 11.23
Dung Beetle Version 12.0 – "Driven by Quality, Tempered by Quantity"

The modifications in Version 12.0 were driven by data.

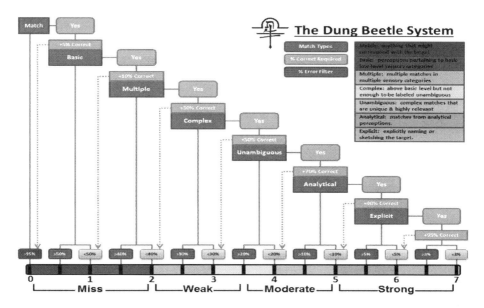

Figure 11.24 – Dung Beetle Version 13.0 – "Deviation from the SRI Ranking System"

Ahead of its time, Dung Beetle was often misunderstood.

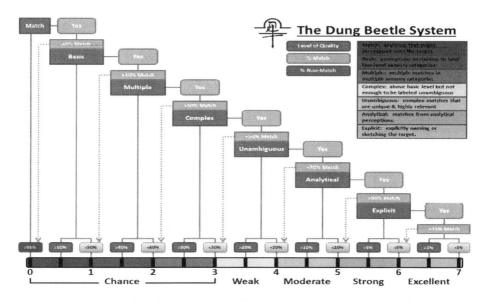

Figure 11.25 – Dung Beetle Version 14 – "Dung Beetle Demands a Higher Standard."

Dung Beetle Related Topics for Consideration

- The First Confidence Ranking Calculator
- The Khepera Ken Trials
- The Khepera Tate Trials
- Project Shamumu: Attempting to Harness the Power of Global Judging
- Anomalous Cognition Ratings: A True Skill Performance Rating for Precoggers
- Algorithm 1: The correlation between low confidence ranking scores and hit rates
- Algorithm 2: Correlation between disparity, base CR scores and Hit rates
- Algorithm 3: Using Machine Learning to develop a threshold value for Hit rates
- Implementing Solar Wind Data
- Implementing Moon Data
- Implementing Geomagnetic Activity
- Advance Warning of possible Bleedthrough
- Advance Warning of possible Displacement
- Advance Warning due to personal biases
- Advance Warning due to Historical Performance
- Dung Beetle and the Dream to enter Cloud Computation
- Understanding Uncertainty and How Dung Beetle Produces CR Ranges
- Empowering Dung Beetle to use AI Image Recognition to Automate Judging
- The first attempt at utilizing a Computational Artificially Intelligent Judge
- of West Georgia.

PART THREE

APPLICATIONS

ASSOCIATIVE REMOTE VIEWING TARGETS THE WORLD

CHAPTER 12

Entangling with the Future: The Applied Precognition Project

In 1998, three years after the DIA-CIA Star Gate program closed, the focus in the remote viewing field was not on Associative Remote Viewing (ARV). It was at that time that Marty Rosenblatt, a retired physicist who had worked in the aerospace industry, became interested in ARV and launched what has grown to become the largest active remote viewing group with close to 1,200 members.

Marty created a website called P-I-A (Physics-Intuition-Applications) and began a newsletter called *Connections Through Time*, both of which explored the relationships between remote viewing, physics and applications of intuition. Marty and early subscribers began working together practicing Associative Remote Viewing. Marty also gave workshops at IRVA and elsewhere to spread the word about ARV, the results the group was obtaining and presenting material on physics and metaphysics to try to understand what made viewing of the future possible.

Predictions by some of Marty's groups and viewers were made by P-I-A using an online interface Marty developed called Precog, later Precog 10, and then WE – Winning Entanglements. Viewers log in, receive a tag or identifier for the target, do their session, self-judge or use a third-party judge and the resulting score is used to pick a side associated with the outcome of a game or financial trade. A revised version still exists and is used by APP.

In 2000, Marty shared the results of a P-I-A experiment called "the AVM project" that predicted stock market closing points. As reported on the P-I-A website and confirmed in subsequent interviews with participants, seven viewers were paid to do a staggering 500 sessions each, for a total of 3,500 predictions funneled into 700 investment targets. "Up,"

"Down" and "Near-Neutral" stock changes were randomly associated with the "Animal," "Vegetable" or "Mineral" nature of five AVM photo targets. According to Rosenblatt's report, overall performance was just about what you would expect based on chance. In two instances, the group produced a very high "prediction cluster," at the 99.4 percentile compared with chance.

In 2013, along with Tom Atwater and Chris Georges, Rosenblatt founded the Applied Precognition Project (APP) as a successor to P-I-A. The group expanded on the work that Marty had been doing and recruited more viewers, a total of 83 members that year. APP held its first conference the same year featuring top remote viewer Joe McMoneagle and former Star Gate project head Edwin C. May.

APP is organized into groups of viewers. Some viewers and groups used the WE software while others did not. Membership increased to about 1,000 in 2016 and more groups formed. By 2020, the discussion group had grown to 1,200 members, with paying members numbering about 460. The main activity of the groups was and has remained viewing and making predictions, while a few (like Sublime) developed strong friendships and built a subgroup culture.

Besides building the group, doing outreach, organizing conferences and daily betting, Marty has used his programming skills (Perl, R) to maintain statistics on the results of the intense amount of viewing that was being done. These were not, of course, formal scientific studies being tabulated but a detailed record of hits, misses and passes by the groups and by individuals in the groups. It is the largest and longest-running such effort.

Marty is that rare individual who gets up at 6:00 every morning, goes to his computer, looks over the data that has come in from the viewers, evaluates it, then makes a financial trade or bets on a sports event, or passes. He often tweaks the pick based on his wide experience and intuition. Marty has been doing this for years on end with phenomenal dedication – and, consequently, APP has a well-deserved reputation as the leading group doing ARV.

Marty has produced numerous videos about ARV, many of which are free online, and others are reserved for paying members ($55 a year). In many of the "Talk With" videos, experts present their views on a wide variety of topics related to remote viewing, intuition, physics and lately esoteric subjects as well.

In addition to the managers of APP groups, a few individuals have over the years assisted Marty with administration and organizational tasks – managing membership, doing outreach, undertaking the logistics of the biannual conferences and other tasks. APP started with one conference a year but moved to two per year since Marty felt that frequency was needed to maintain cohesion and spirit in the group. APP hosts a major conference in June and a smaller workshop in the fall called APP Fest. Both of the present authors worked extensively in APP as viewers and group managers and have presented at APP conferences. They also designed and helped maintain the APP website (Debra with friend Michelle Bulgatz) and assisted with membership administration and programming (Jon).

Accounts of a few of the APP conferences have been published in remote viewing magazines. (For links, please visit http://www.arvbook.com)

APP has had many component groups over the years: Financials, Sublime, Sage, First Groove, Omega, Pegasus, 1ARV, CAS-Oak A-E, Croatorum, In8, Direct Psi, Khepera, Lively!, P7B, Transcendent, ARV Team, RusShining, Sweet Dreams, Winners Circle, KARV, Alpha, Beta, EarthSuit, Qmagic1, Magi, Tzunamy and Crazy Wind. In addition, the APP Institute (APPI) – a nonprofit organization – has several groups (APPI 1 through 5 as of 2020).

The APP groups have individual group managers who conduct trials using Marty's online WE software. A few used Ed May's Computer Assisted Scoring software for a time.

The APP group with the most successful long-term record was one with the quirky name, Winner Winner Chicken Dinner (WWCD). We will next present details of how this group functioned, which is representative of how other groups in APP were conducted, as well as detailing the methodology of the most successful APP group to date.

In sports betting and other fields of gaming, the established firms that have a wealth of information can, by report, achieve a maximum success rate of about 53% to 55%. The APP statistics show many group and solo efforts over the decades have achieved a hit rate of 60% or higher, clearly showing a decided advantage to using psi in making predictions. In fact, counting all the predictions APP has made, the odds against chance have been about 1.6 billion to 1.

In line with the above, a hit rate of 60% to 65% occurs very frequently in the APP data. This range has often been encountered in other psi research, too. It is almost as if there is a narrow range above which sustained success is quite rare. Only one group in APP has achieved 70% over a significant period of time. Individuals have done so, as well, with Marty reporting in 2020 that 72% was the highest rate the Precog Pros of APPI have achieved for a stretch of 25 events. For 10 successive events, a success rate of 80% to 90% has been achieved. We were unable to learn the exact figure, but approximately 20 people have earned modest profits ($22K as of 2016) from these, including one of the present authors (Debra; Jon has not taken part).

Meanwhile, an assessment by a few project managers independent of APP's management who had access to the data (e.g., Igor Grgić and Mark Samuelson) found some WE groups that had focused only on predictions of financial instruments and used the software program did not fare so well, with less than a 50% hit rate. (This is discussed further in Appendix 1, Project Firefly.) That project involved contributions of hundreds of ARV trials from multiple viewers and groups for close to a year. Peer analysis suggested more submissions were made by underperforming WE groups as opposed to non-WE groups, which may have contributed to the extensive misses and subsequent financial losses suffered by Firefly; however, this has not been confirmed. It was also suspected that some participants in the WE groups were newer viewers at the time, which may have been a contributing factor.

Winner Winner Chicken Dinner – The 70% solution

WWCD is the only APP group that sustained a greater than 70% hit rate over 100 events. The hit rate after 80 events was 76% with about 40% passes and in 2015, WWCD achieved a 90% hit rate for 25 consecutive predictions. After 100 events, the hit rate was 72%, including events prior to Sumner (a pseudonym) becoming the group manager in May 2015. Sumner was interviewed that year and described the details of his approach, including his opinions about why the group had such success. In an interview with another group manager and viewer T.W. (Teresa) Fendley, Sumner cited two main factors: how games were selected and the weighting of viewers' sessions.

With her permission, we have drawn on Teresa's article in this account. We also include the list of factors Sumner cited in an interview with Jon.

1.   Previous experience as a group manager. Sumner had been group manager for three years before becoming manager of WWCD. In one previous group, Transcendence, at first Sumner felt good about the group, but it didn't end up well, with three hits, six misses and six passes. That effort used the Computer Assisted Scoring software, which was discussed in Chapter 9.

2.   Creating a supportive group atmosphere. Sumner was personally involved and felt good about the WWCD group from the outset. With a teacher's mindset, he strove to write things in language people could understand. He wrote out the taskings so they were clear, complete and accurate. He emailed each viewer about their session and stayed in touch. He knew some of the viewers in person. When recruiting at gatherings, he would mention how it was fun. You may make money. You will become a more complete person. You get to hang out with cool people. It's cool to be able to see the future and measure it. He offered to help and train people.

3.   Getting enough data from the viewers. Sumner preferred to receive data from six to eight viewers per event. He felt it was harder to judge if there were fewer sessions. Greg Kolodziejzyk thought having just a cross for a church target was fine for his ARV, but Sumner preferred more data. It's better to have eight or nine data points

that match than just one type of data. Sumner noted that he saw some symbolic data from the viewers, but it was mainly literal.

4.   Automated interface. The automated Winning Entanglement (WE) software Marty developed was very attractive to Sumner and was one reason he joined APP. He felt the software helped prevent group manager burnout. When the viewer logs in, the software presents a reference tag and photos for each event. These can be accessed during the trial if viewers decide to judge their own transcripts. The software enables them to examine their past record of hits, misses and passes, along with the notes they made at the time. Three hit rates are plotted – 10-day, 25-day and cumulative – and are available online after the viewer logs in. The group manager has access to the data, as well, to inform viewers of their evaluations and predictions.

5.   Type of event. To Sumner, the type of event was important. The group did better with sporting events than with financial events. Sumner used the MLB.com schedule to choose a baseball game midweek. He felt baseball is more predictable than football. Baseball has fewer changes of players on teams, fewer injuries and has detailed stats of how each batter has done against a given pitcher, which helps predict future performance against the same pitcher.

      "Before, I was just choosing the games that were interesting, or the latest one on Sunday, to give viewers the weekend to work on their transcripts," he said. "But in games with good pitchers, the score would be so low, it would be hard to meet an 'over' prediction." (Over/Under predictions are based on a game's total score.)

6.   Use existing statistical information about the event. "It does help to know something about a sport. You wouldn't want to pick a game where the best team is up against the worst."[361]

      As examples of how Sumner used statistics and betting lines to help select the game, for each team he would find out the runs or points allowed and scored per game in their recent games. He compared the total with the point spread for the Over/Under for all candidate games that day. He chose the game for which the estimated Over/Under was closest to the line at the sportsbooks Bovada, 5 Dimes and Betfair. He preferred the line to be at 0.5 rather than say 40 or 41. Sumner found that too often he had chosen games for which the Over was not likely to occur; for example, if both pitchers had been outstanding in recent games, the Over might not occur. He used the above method to get a handle on that.

7.   Supplementary technique – the pendulum. Sumner used the pendulum as a complement in selecting a game. For example, he might ask, "On Monday when the result is in, if I pick this game, will it result in a hit?' Sumner made his pick about an hour before a game. He felt it is better to view closer to the game than way ahead

of time. He used a pendulum in daily life, draping it over one finger and resting his elbow on a fixed point. Clockwise rotation indicated Yes and counterclockwise No. His first question to the pendulum was, "Am I in a good state to do this?"

8. The target pool. The photo pool is important. WWCD used the APP pool, which consisted of about 800 photos. The pool was convenient to use and it was being slowly improved. Sumner kept track of animal and food targets, which he felt many did not do well with. He was open to viewers describing elements just outside the target image, which is consistent with APP usage. The target is a "photosite" – it's the photo but also what is at the site itself.

9. The time when the target photosite is chosen. It's important that the computer chooses the target after the transcripts are uploaded. This forces the viewer to look to the future for the target. Sumner believed it was better to have a separate target for each viewer to help avoid "bleed-through." In this case, that referred to one viewer's session possibly affecting another viewer's transcript, although good results were obtained at APP workshops when everyone had the same target.

10. Lengthy and meticulous preparation and implementation. Here's how Sumner approached a game. He put considerable time into selecting the game. He generally did the Sunday game around 1:00 p.m. and spent three to four hours that morning preparing and judging. The transcripts were due Sunday morning. Sumner accessed his information about the viewer (see point 12 below) and made his pick. He would place the bet about an hour before the game. (Sportsbooks close down the betting about a half hour before the game.) After the game, he would email each viewer to discuss the event.

11. Scoring method. Sumner used a variant of Joe McMoneagle's method. He gave greater weight to viewers with better results. He feels it's better if the viewers do not know details of the method. Sometimes when he was unsure, he would use a pendulum to make a final decision. He allowed both self-judging and independent (himself) judging. Some viewers wanted only one target given to them as feedback and he abided by that preference.

The big innovation was how Sumner weighted the viewers' transcripts. Referring to the 7-point SRI scale used by many remote viewers to judge their sessions, he said he didn't always use the sessions with the highest confidence rankings (CRs): "I don't count newbies' CR scores the same as those with high hit rates." Sumner assigned a separate score to each viewer's transcript based on his judgment of their strength, separate from the confidence ranking. For instance, a strong transcript would get an 80% rating, whereas a weak one would be rated 55%. He added this figure to the viewer's hit rate, then divided by two to derive the weighted score. "It wasn't all stone-cold statistics, though. For instance, if the 10-day trend showed a viewer was 'hot,' he might have given that session more weight."

12. Recordkeeping. "On a single Excel spreadsheet, he displayed all the game data and information from all the viewers – the CRs which they gave their transcripts, their cumulative and 10-day hit rate, and their weighted scores. He drew a line across the page to separate the higher-weighted sessions from the lower-weighted ones and relied on those above the line to determine the group prediction."

13. Tiebreaking method. Sumner occasionally did a session himself as a tiebreaker. He also used the pendulum to confirm the prediction would be correct. He asked the pendulum: "Go with this answer or pass?" He would never change the side to bet but he might pass, based on the result of the pendulum assay. "I never let the pendulum switch me from one side to the other," he said, "but twice as often, it will keep me from choosing the wrong side."

The pendulum helped correct for such things as when Sumner forgot the "confusion factor" that might have plagued tasking, uploading transcripts, etc.

14. Passing. "Even though I don't like passing," 35 of the 80 predictions were passes. "What led to passing? If the viewers were split – with half choosing one side and half the other – it could be a pass, depending on the strength of the transcripts. Other factors could include confusion in the process, trouble with scoring, comments from the viewers, difficulty connecting to the target or photosites that previously caused misses or displacement." Sumner admitted it was hard to tell why a certain strategy worked.

"It may be just spending more time and energy in the process helps the group. I have a feeling the more time I spend on any part of the process contributes. Somehow, I'm telling the universe I care about this and I want it to be right, so show me the future I'm asking for."[362]

15. Smooth flow and avoiding confusion. If there was confusion in the process for a given event, Sumner was more likely to pass. If there was an error in tasking, an automatic software glitch, viewer confusion, photo issues or changes in the betting line, those could lead to a pass.

16. Breathing exercise. Sumner used a Rosicrucian-derived breathing exercise. (Inhale 5 seconds, exhale 10 seconds, repeat). This exercise and the pendulum improved his viewing, he felt. "Tells me when I'm in the right state of mind to start viewing." He experienced a slight gasp or sigh after 10 seconds, which told him he was ready. Sumner has done craniosacral therapy and found it initiated a different breathing cycle, resetting the body and mind.

17. The viewer should do a feedback session. Sumner's advice:

> The feedback session should be done in the context of Remote Viewing. So you could do a "cool down" meditation with music before you start. Besides creating something for your past self to view, you are also Remote Influencing yourself to notice and "feel" all the great connections between what you drew and the winning target image.
>
> It is important to feel excited about what you accomplished and how you remotely saw the winning target in the future before the game event actually happened. You could even redraw or trace the image in a way that communicates the basic elements to your past self. If you do that, draw it on a different paper. This will give you valuable improvement for your artistic sketching skills, and develop an artistic eye toward everything you see in the world.
>
> When you open the page, you will see the Winning Target Image. You can compare this to the paper copy of your 1st Transcript.
>
> Below the winning image is a text box where you can "Describe and save your thoughts, feelings, ideas, etc. for this Winning-Side FB Session." Here you can write how you feel about the session or even list all the ways the Winning Target matched the 1st Transcript.
>
> If you chose to look at the "Other" (losing) image that was mailed to you, you should also compare your 2nd Transcript to that image, and notice how what you wrote and drew matched that target image. You can list those matching elements in the text box also.
>
> Remember, this feedback step is where you communicate the target image to your past self. Everyone agrees that this will improve the accuracy of your present and future sessions.[363]

18. Keep an even keel. When he first became a group manager, Sumner told viewers of his excitement when the group had strong sessions, but he later decided that can be counterproductive.

Sumner understood his role to be:

> My job as a group manager is to communicate strength and responsible predicting …if some of the top viewers in the world get a miss 30 percent of the time, then our group can, too…Getting excited about a prediction BEFORE we know the answer works against you when it goes wrong – and viewers get discouraged. Now

I just say we've got good congruence, and we only celebrate AFTER we know the prediction was correct.

Other noteworthy APP groups and projects

In 2011-2014, APP undertook a major effort using a protocol Marty devised called 1ARV. Another major effort used software Ed May and his team developed, which APP called Computer Assisted Scoring (CAS). This is explored in Chapter 9.

As noted above, the largest undertaking by APP members was the yearlong Project Firefly (2014-2015). This massive undertaking involved all of the APP groups and a few solo viewers making Forex predictions. Firefly failed financially but many lessons were learned. (See Appendix 1.)

Another successful group in APP has been the Sublime group, which has done a variety of types of viewing, all involving dreaming. (Please see our Chapter 17.)

Precog Pros and APPI

We asked Marty about the APPI Precog Pro program, in which viewers can get paid as professional remote viewers. (For links, please visit http://www.arvbook.com)

APPI, a nonprofit affiliate of APP, has had a Precog Pro program since September of 2015. APPI's main goals are to educate people about consciousness and psi with a focus on precognition and fostering/teaching interested people to apply ARV (Associative Remote Viewing) and become Precog Pros.

Debra was one of the first viewers to join APPI for a short run in which she and her RV tasker, Chris Georges, generated a 66% increase in profits ($1,300 across 10 trials).

Report for Debra Katz and Chris Georges as Partners with APPI								
Date	Trade	Profit/ Loss	Joint Account Profit	Joint Account Value	$Invest Nominal	Nom. Factor Invest	HMP	Cum Hit
20150920		-		$2000				
20150924	NZD. USD	$197	$197	$2197	$200	0.100	Hit	
20151001	NZD. USD	-	$197	$2197	$253	0.115	Pass	
20151008	NZD. USD	$(253.00)	$(56)	$1944	$253	0.115	Miss	
20151015	NZD. USD	-	$(56)	$1944	$190	0.098	Pass	
20151022	AUD. USD	$190	$134	$2134	$190	0.098	Hit	
20151029	NZD. USD	$240	$374	$2374	$240	0.112	Hit	
20151105	AUD USD	$307	$681	$2681	$307	0.129	Hit	
20151112	AUS. USD	$147	$828	$2828	$399	0.149	Hit	
20151119	Pass	-	$828	$2828	$483	0.171	Pass	
20151203	NZD. USD	$483	$1311	$3311	$483	0.171	Hit	0.8571
20151210	NA	-	$1311	$3311	$651	0.197		
Year End		-	$1311	$3311	$651	0.197		

Figure 12.1 – Report for D. Katz & C. Georges APPI 2015

Terrific work :-) Our joint account is showing a profit of $1311, a 66% increase in less than 3 months.

Figure 12.2 – Debra's "Silvery Fish" Session

Professional Precoger Partnership Summary				
Hit Rate (%)	Proft/ Loss	ROI (%)		Invested
75	$2775	139	APPI	$2000
82	$2561	128	APPI	$2000
80	$1567	78	APPI	$2000
86	$1311	66	APPI	$2000
65	$4900	49	Private	$10,000
62	$155	8	APPI	$2000
<50	$34	2	APPI	$2000
<50	$(560)	-28%	APPI	$2000
Total	$12,743			

Figure 12.3 – Professional Precoger Partnership Summary

Since September of 2019, **we have new qualification requirements and a new compensation package.** These are first summarized below and then an overview of results.

Below is the summary of Qualifications for becoming a Precog Pro for newbies (on the left) and those already having 25 or more predictions (on the right). The basic qualification is a Hit Rate of around 70%. We are happy to bring in viewers with as low as a 64% Hit Rate if they can maintain that for 5 predictions in a row. What is important is not only a good Hit Rate, but a consistent Hit Rate. The predictions are entered online using the Winning Entanglements (WE) online software.

Precog Pro Qualification using the WE (Winning Entanglements) ARV Online Program	
ARVers with 10 to 25 Predictions*	**ARVers with 25 or More Predictions***
• 7 Hits out of your first 10 Predictions* (HR = 70%) • 11 Hits out of your first 15 Predictions* (HR - 73%) • 14 Hits out of your first 20 Predictions* (HR = 70%) • 17 Hits out of your first 25 Predictions* (HR = 68%)	• Maintain a 64% or more 25-Day Hit Rate for 5 Predictions in a row
*Non-Pass Predictions, Passes are ignored for Hit Rate (HR) = #Hits/(#Hits + #Misses)	

Figure 12.4 – Precog Pro Qualification

Next is the Compensation chart showing seven Levels of increasing compensation as viewers successfully complete 25 ARV (non-pass) predictions. The details are on the chart below, but the key is for viewers to have a 72% or higher Hit Rate after 25 predictions to move up one level. The 25 predictions encourage a viewer to integrate their ARVing with their life. Some viewers are doing five predictions per week while others are doing one a week. When viewers do their predictions is entirely up to them.

Some do self-analysis and some do independent analysis. The idea is for viewers to just focus on ARVing as professionals. They will get paid based on performance.

[Name] APPI Precog Pro

Compensation Schedule as of September 4, 2019

For more information contact
marty@p-i-a.com

The APPI Precog Pro
Compensation Schedule for 25 Non-Pass Predictions

Precog Pros start at Level 1 and are paid based on their Personal Hit Rate after 25 predictions in Financial Groups.

After completing 25 personal predictions, the viewer moves Up or Down one level based on being in the Green or Yellow zone, respectively. (The viewer stays at the same level if in the Gray zone.)

Viewer's number of hits	Hit Rate (%)	Level1	Level 2	Level 3	Level 4	Level 5	Level 6	Level 7
10	40%	$ 54	$ 109	$ 163	$ 218	$ 272	$ 327	$ 381
11	44%	$ 82	$ 163	$ 245	$ 327	$ 408	$ 490	$ 572
12	48%	$ 123	$ 245	$ 368	$ 490	$ 613	$ 735	$ 858
13	52%	$ 184	$ 368	$ 551	$ 735	$ 919	$ 1,103	$ 1,287
14	56%	$ 276	$ 551	$ 827	$ 1,103	$ 1,379	$ 1,654	$ 1,930
15	60%	$ 414	$ 827	$ 1,241	$ 1,654	$ 2,068	$ 2,481	$ 2,895
16	64%	$ 620	$ 1,241	$ 1,861	$ 2,481	$ 3,102	$ 3,722	$ 4,343
17	68%	$ 931	$ 1,861	$ 2,792	$ 3,722	$ 4,653	$ 5,583	$ 6,514
18	72%	$ 1,396	$ 2,792	$ 4,187	$ 5,583	$ 6,979	$ 8,375	$ 9,771
19	76%	$ 2,094	$ 4,187	$ 6,281	$ 8,375	$ 10,469	$ 12,562	$ 14,656
20	80%	$ 3,141	$ 6,281	$ 9,422	$12,562	$ 15,703	$ 18,844	$ 21,984
21	84%	$ 4,711	$ 9,422	$14,133	$18,844	$ 23,555	$ 28,266	$ 32,976
22	88%	$ 7,066	$14,133	$21,199	$28,266	$ 35,332	$ 42,398	$ 49,465
23	92%	$10,600	$21,199	$31,799	$42,398	$ 52,998	$ 63,597	$ 74,197
24	96%	$15,899	$31,799	$47,698	$63,597	$ 79,497	$ 95,396	$111,296
25	100%	$23,849	$47,698	$71,547	$95,396	$119,245	$143,094	$166,943

Example 1: Viewer completes 25 predictions in **Level 1** with a Hit Rate of **76%**. APPI pays him **$2,094**. This viewer is now in Level 2 (up 1 level) for the next series of 25 predictions.

Example 2: Viewer completes 25 predictions in **Level 6** with a Hit Rate of 68%. SRF pays him a fee of **$5,583**. This viewer is now in Level 5 (down 1 level) for the next series of 25 predictions.

Figure 12.5 – APPI Precog Pro

Overview

- A total of 62 viewers are, or have been, in the Precog Pro program.
- They have earned over $22,000.
- We currently have active Pre Pros in Levels 1 and 2.
- We are looking for as many Precog Pros as are interested in this program. If you qualify, let us know since we cannot follow all viewers' stats.

Summing up

Overall, APP has been an enormous and very positive presence on the remote viewing scene. APP has helped bring ARV to the fore. Ten and 15 years ago, ARV was a minor factor but due to the efforts of Marty and other APPers, ARV is now the most common form of remote viewing and is very frequently discussed online. One reason is it offers a chance to get a clear, immediate outcome of your remote viewing effort. Your choice, bet or pick either won or it did not. Another reason is that people want to make money and ARV offers the most direct, although still risky, way to do that. One reason Marty said he got into ARV is that "money talks" and he wanted ARV and RV to become widespread and valued in society. Another factor is it is much simpler to do ARV than regular remote viewing sessions, which generally require a lot of time. For ARV, you need only a few correct impressions to decide between alternatives, although some prefer more data than that. Finally, viewers can participate in many ARV-related projects and social groups for virtually no cost or join APP at $55 a year for a complete group experience.

CHAPTER 13

ARV Programs & Applications: Gattis, Grgić, Hilleard & Ferrier

In this chapter we will cover four developers and their desktop- and Internet-based programs designed specifically for Associate Remote Viewing applications. These include Bill Gattis of *New Intelligence*, who developed the first ARV desktop program. Next is Igor Grgić, who created *ARV Studio*. This was followed by Sandra Hilleard of *Project X* and finally Michael Ferrier of *RV Tournament*.

Figure 13.1 – New Intelligence Software

Bill Gattis, president of New Intelligence Inc., developed the first ARV desktop program. During the late 1960 s and early 1970s, Bill was employed in the aerospace industry working on classified projects for the US Navy and the US Air Force. He has authored numerous articles on technology, computer graphics, instructional systems and speech synthesis. Bill is chairman and CEO of Intersect Systems, which specializes in the development of records management and retention schedule management systems for school districts, local governments, state agencies and commercial businesses. He has also been chairman of the International Conference on Technology and Education, a nonprofit organization he founded with the University of Texas in 1982. ICTE conducted annual International Conferences on Technology and Education in the US, Europe and Africa for 22 years. Prior to that, Bill was vice president of the Education Division of Tandy / Radio Shack, leaving Tandy in 1989 to lead a research effort that established cross-platform methods for software interoperability on personal computer systems, developing early concepts of "virtual CPUs," "virtual engines" and "software CPUs" – precursors to platform-independent programming languages such as JAVA.

From the New Intelligence website:

The Remote Viewing Interactive Analysis System is not designed to teach the process or techniques of Associative Remote Viewing; a number of experienced and capable individuals formerly associated with one of the government's remote viewing programs offer training classes to teach remote viewing. This software is designed to facilitate the process of remote viewing using the Associative Remote Viewing approach, with the additional objective of helping the remote viewer address three of the challenges often encountered in remote viewing: identifying specific locations, identifying numeric values and identifying points in time. The methods implemented in the system take advantage of Windows-based software, including easy-to-use, point-and-click user interface screens and state-of-the-art database methods. The system provides a means of keeping records of the outcomes, the persons participating in a remote viewing session and related notes about each session. A unique reporting and analysis function using "auto-query" double-click queries is included.

Bill said, "During the early experiments with the software, I considered the ARV software primarily benefitting those experimenting for the first time with remote viewing – with the goal of perhaps moving on later to controlled RV."

The program has received positive comments and support from Russell Targ and former IRVA President Pam Coronado, as well as instructors Paul Smith and Lyn

Buchanan, the latter of whom told Bill, "That software is fantastic!" Bill has implemented their suggestions to improve the program, such as adding additional textures, a dowsing component, a "demo" testing option and illustrating both target pre-selection and target post-selection.

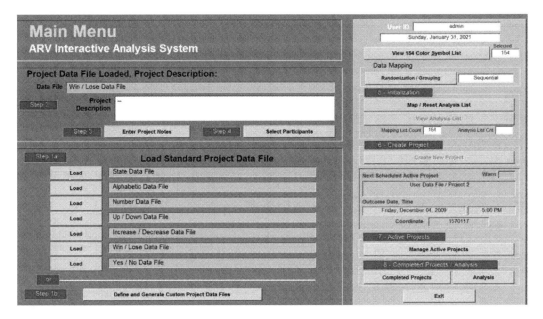

Figure 13.2 – Main Menu

The New Intelligence ARV program allows the user full control to undertake solo or group viewing. It allows management of participants on an email list with user ID and password administration.

The program has a wealth of features. For example, you can choose among 11 color fields, 14 black- and- white symbols, 154 color symbol, 10 common tastes, 34 sound selections, 14 classical music selections and 116 color images. You can explore directional movements by use of a compass and you can use coordinates.

Another feature is a button that allows the user to learn the relevant LST (Local Sidereal Time), which researcher James Spottiswoode in 1997 initially found to be a factor that might influence remote viewing, but later was not able to replicate in a follow-up experiment with a larger dataset.[364,365] However, Spottiswoode reported that review of a much larger dataset indicated a "clear max" in Effect Size at 13:00 LST.[366]

New Intelligence also has a "GMF" button for keeping track of geomagnetic influences. Several studies and empirical results suggest that solar weather and the moon may affect psi results. (The link to the GMF site is out of date but the user can enter links to other geomagnetic sites.)

The program does have some limitations. For example, it does not exit utilities unless you click on the NI logo (an eye). Each time you go from the start page to utilities you have to log in. While there are sounds and music selections, you cannot add more of either, nor photos. Bill noted that the random number generator is not truly random and sometimes a target will repeat. However, these are minor limitations in a very powerful and varied program for conducting ARV trials.

ARV Studio – Igor Grgić

Figure 13.3 – ARV Studio

Another stand-alone ARV desktop software program is Igor Grgić's ARV Studio. Although we have referred to Igor previously, we have not mentioned his wide range of undertakings and accomplishments. Igor Grgić is a SAP Basis senior consultant (IT) based in Croatia and an experienced Forex trader. He is an award-winning remote viewing researcher (PARE 2017 Award), a published remote viewing author,[367] an RV project manager and the developer of the ARV Studio software. Igor is also a coauthor of the article on the Project Firefly (see Appendix).

In his free time, Igor is occupied with his "true life passions." These include remote viewing, dream interpretation, exploring the original teachings of the Kabbalah and Forex trading. He recently launched the Precognitive Trading Group (PTG), which is presented in a later chapter. He writes:

My involvement in remote viewing started in early 2014 when I took a home-based remote viewing class at one company trying to train new remote viewers. Soon after, I joined online RV community and organization called "Applied Precognition Project." Also in 2014, I formed my Associative Remote Viewing Group named P7B involving up to nine remote viewers…In my RV work, I conducted and was involved in numerous ARV trials and ARV projects as a Manager, Tasker, Analyst and Trader. I analyzed around 700+ ARV transcripts/sessions. I conducted test series using ARV software known as CAS (Computer Assisted Scoring), which was developed by Mr. Ed May, Ph.D., former director of US government-sponsored remote viewing project Star Gate…As a result of all my work with ARV, I designed and developed a computer program to help me conduct and manage ARV trials.[368]

The P7B Group in the Applied Precognition Project

After Igor joined Marty Rosenblatt and his Precog group in 2014, he proposed a trial using colors and asked if anyone was interested in being a viewer. Jon replied that he was and shortly thereafter viewer Mark Samuelson indicated his interest, as well. Igor called his method P7B (in English: protocol of seven colors) because seven colors were involved. He suggested the colors white, yellow, red, green, blue, brown and black be used. No blended colors or colors other than those seven were to be used. Still, this was standard binary ARV using photos as targets. But the potential targets were photographs in which only one color dominated. For example, a photo of a green garden with green grass and green trees; a blue whale in deep blue waters; brown leaves on dark brown soil in autumn (brown is dominant). Igor suggested the viewer write down the color he would be shown as feedback – "Don't analyze, simply write down one of the seven colors that you feel is involved."

In July 2014, a pilot study began with Igor as coordinator and Mark and Jon as viewers. Two more viewers soon joined, Teresa S (T.W. Fendley) and Carlos M. As an example, the tasking for July 21 was "2136-8241: Name one highly dominant color in the photograph which will be sent to you as feedback after this event takes place."

Every two days starting July 22, the team provided a color to Igor. By Aug. 7, the record for the team was five hits, two misses and two passes. This concluded the pilot series. During the last five predictions, the SRI 7- point scale was used in addition to color, and this led to four hits in a row. Apparently, an SRI score was an important factor.

A second series began on Sept. 8 – the "The Fellowship of the Diamond Ring" (FODR). The events were spaced out more, with about a week between them. In addition to the regular transcript, the viewers reported and ranked colors by their importance in the session. Igor had concluded that colors by themselves were unreliable for making a prediction, hence use of SRI scores. By November 18, 2014, the tenth event had taken place and the record for the first two series was now 11 hits, 4 misses, 4 passes: a total of 19 events, with an excellent hit rate of 73.3%.

Further trials of the Fellowship of the Diamond Ring continued in 2015. By May 4, the results were 17 hits, 10 misses, 11 passes, for a somewhat lower but still good hit rate of 63%. Igor proudly noted that July 22 marked the first birthday of P7B and there was a hit that day. By this time, a total of 10 viewers had been involved.

In April 2016, the P7B group finished a short pilot study of 10 standard binary ARV events. Igor called it P7B 2016 (Odysseus).

For this study I was advised by parapsychologist Mr. Patrizio Tressoldi Ph.D., (Dipartimento di Psicologia Generale, Università di Padova – Italy) where he proposed:

– to select and use photo targets with a high level of informational entropy for both photos in the binary pair. In absence of computer program for calculation of the informational entropy in the photo targets, a pragmatic approach was used to select the photo targets, referring to the notion of degree of change included in the photo. For example (from E. May's paper) a faint satellite as it moves across the night sky contains a higher level of informational entropy with respect to a star of the same intensity. So, target selection was based on best visual assessment of the Tasker/Judge.

– that the remote viewers rate their mood and self-efficacy using scale 1 (Very negative/low) to 10 (Very positive/high). This is the Likert scale. (Note: self-efficacy is a personal judgment about how well one performed in the RV session.)

Five remote viewers participated in this pilot study with 3.5 participants (on average) per event. After 10 events, the results were 2 hits, 1 miss and 7 passes.

Final Results						
Final Group Results	**P7B**	**HIT**	**MISS**	**PASS**	**Number of events**	**HIT Rate**
		2	1	7	10	66.7%
Final Viewers Results: (Individual results)	**Name**	**HIT**	**MISS**	**PASS**	**Number of Events Participated**	**HIT Rate**
	RV1	0	0	1	1	N/A
	RV2	3	0	4	7	100.00%
	RV3	2	2	5	9	50.00%
	RV4	3	2	4	9	60.00%
	RV5	1	1	7	9	50.00%
	TOTAL	9	5	21	35	64.29%

Figure 13.4 – Final Group Results

Igor noted there were hits both with low (<6) and high scores (>6) of Mood and Self-Efficacy. The role of these variables was not very clear, so another study was proposed for further exploration.

By May 3, 2016, the Odysseus series was complete and on July 14 P7B finished a second series called Poseidon. Patrizio Tressoldi chose the pictures for the second series and the hit rate after 10 events was 3 hits, 6 passes and 1 miss. The combined hit rate of the two series was 5 hits, 13 passes and 2 misses. Tressoldi said there were no statistically significant results due to an insufficient number of trials to draw conclusions.

Igor said P7B had "some of the best ARVers around," such as Loraine Connon, Teresa S and Tom Cunningham. After two series, Teresa's hit rate was 100% with 6 hits, 0 misses and 9 passes. Tom was at 66% and Loraine at 71%, while in the previous series (Poseidon) they were both 100% (2 of 2).

Here is a 2014 transcript from Teresa in which there is no doubt she hit the target!

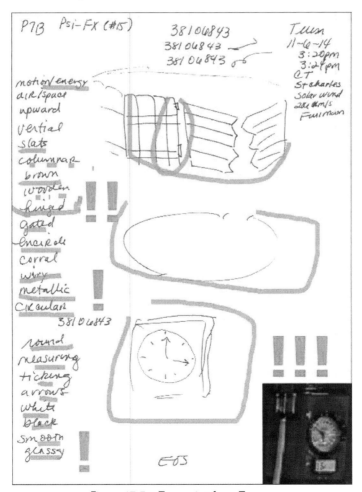

Figure 13.5 – Transcript from Teresa

Key: Motion/energy, air/space, upward, vertical, slats, columns, brown, wooden, hinged, gated, encircle, corral, wiry, metallic, circular, round, measuring, ticking, arrows, white, black, smooth, glassy.

Summary of the two series

The group hit rate for Poseidon was 3 hits, 1 miss and 6 passes for a 75% hit rate, while the statistics for Poseidon plus Odysseus totaled 5 hits, 2 misses and 13 passes, for a 71% hit rate, both excellent rates.

The hit rate for individuals in Poseidon was 81% (9 hits and 2 misses), while the individual hit rate for the two series combined was 72% hit rate (18 hits and 7 misses).

Series 1 and Series 2 Combined						
Final Group Results	**P7B**	**HIT**	**MISS**	**PASS**	**Number of Events**	**HIT Rate**
	5	2	13	20	71.43%	
Final Viewers Results: (Individual results)	**Name**	**HIT**	**MISS**	**PASS**	**Number of Events Participated**	**HIT Rate**
	RV1	0	0	2	2	N/A
	RV2	6	0	9	15	100.00%
	RV3	4	2	10	16	66.67%
	RV4	5	2	12	19	71.43%
	RV5	0	0	1	1	N/A
	RV6	3	3	12	18	50.00%
	TOTAL	18	7	46	71	72.00%

Figure 13.6 – Series 1 and 2 Combined

These series explored variables such as the viewer's mood and self-efficacy and the effect of creating target pairs with images of high physical entropy (activity/movement/energetics). The result was an improvement in accuracy in the last series with this kind of photo pairing.

Based on his two-year experience with P7B, Igor offered the following tips about how to get a high ARV hit rate:

1.   Work with remote viewers who have greater than 60% ARV accuracy.

2.   If you work with a group of remote viewers, then don't make the group too big. In group mode, it is best to have three to four participants (all greater than 60% accuracy) for each prediction.

3.   Solo mode is a better way to do ARV.

4.   Make your ARV setup as simple as possible. Whatever you can make simpler – do it.

5.   Standardize the steps you perform in an ARV trial. This way you just do it, without thinking about it, so you can focus on what matters most: your feedback photo and judging phase. But this shouldn't lead you to boredom! Both as group manager or solo, you have to care about every trial. Don't worry about anything, be positive, be happy with the transcript, trust the transcript, expect a winning transcript!

6.   Create and define a purpose or goal for your ARV project. Have a desire, motivation and interest to deal with every trial with intention and expectation of results you wish for. Define the length of the series (three months maximum with one trial per week). Then have a break and rest.

7.   Make sure you are selecting good and dissimilar photo targets. Or better, have an independent target selector (human-experienced or computer/software)

8.   Respect every transcript, approach it with objectivity and don't get misled by some perceptions, because 30% of all perceptions can fit the target due to chance only. Pay attention to gestalts rather than AOLs. Make sure you are comfortable with the side you picked.

9.   Count all perceptions and sketches in the transcript. Then count all the matches. If the number of matches is around 30%, be very careful. Don't let yourself get deceived! Take time to assess the quality and significance of those 30% matching perceptions!

10. Never hurry while judging. Don't let anything interrupt you when in your judging flow. While judging, be in the zone, so to speak. Don't lose your focus and concentration.

11. Don't risk it if you have an inconclusive judging session. Be patient and pass. Having more passes is better than having more misses.

12. If the confidence ranking scores are close and both are high, then pass. If you notice a mixed signal, meaning that the transcript is a good match for both photo targets and especially if this leads to very tight scoring (e.g.: Side A 4.0 and Side B 4.5), then pass.

P7B continued throughout 2017 with an event every week. By Aug. 14, 2018 – using ARV and a hybrid trading system – the results were 14 wins out of 29 events. However, the hybrid ARV trading system showed 22 wins out of 30 events or a rate of 73.3%. Igor found a way to extend the trade to sometimes recoup a loss or convert an ARV failure into a monetary gain. In other words, if ARV alone was used to execute the trades, such an approach would result in a small overall loss (e.g., 14 wins out of 29 trades where the risk-to-reward ratio is 1:1, meaning with each $100 trade you either lose or win $100). On the other hand, with a hybrid ARV system that involves intellect as well as psi, the result was 22 winning trades out of 30 trades, yielding a significant profit. This and his experiments with direct remote viewing of financial charts led Igor to the eventual launch of his Precognitive Trading Group.

ARV Studio

Igor explained how he came to develop ARV Studio (ARVS) as a result of managing P7B. In a typical ARV trial prior to developing ARV Studio, he had to do a lot of manual work to carry out the steps needed for each trial. At one point, he had nine remote viewers and managing the group was getting very time consuming. He decided to code a computer program to automate most of the steps in a typical ARV trial. The purpose was not just to save a great amount of time, but also to reduce the possibility of human error to zero and keep a record of all the data. After working on the software for six months, Igor released ARV Studio on April 29, 2016. From his introduction:

> ARVS allows the viewer to conduct blind remote viewing from a pool of 1,000 targets. It also offers different modes – a solo ARV trial with either self-judging or a third-person judge. ARVS also had a "group mode" to task and manage a team of viewers. Included is an option to do a solo prediction based on sounds. Other features are automatic TRN generation, random target pairing assigned after viewing and, based on an algorithm, automated emailing of the tasking, prediction, feedback and outcome. The program also stores data in a csv file of the event, keeping track of ratings of the sessions, predictions and outcomes.

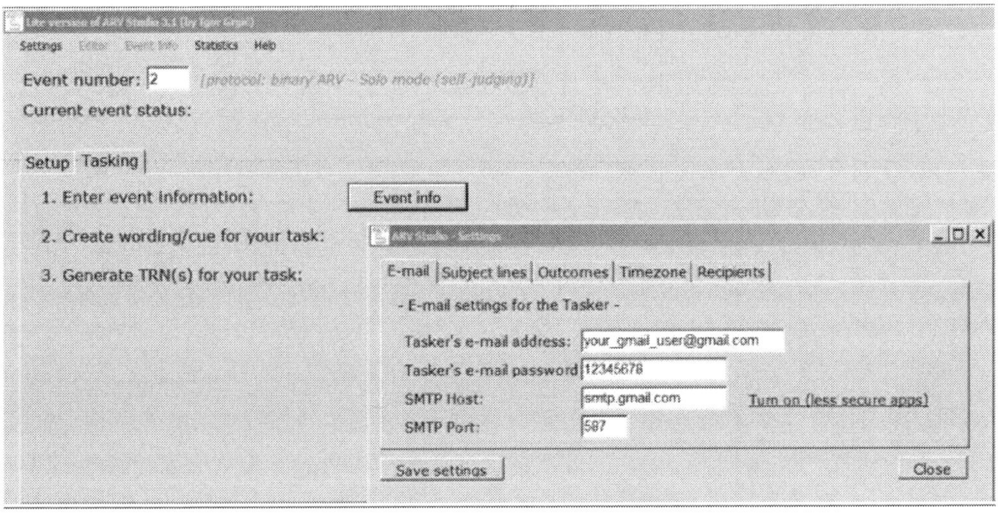

Figure 13.7 – Event Number 2

Since the initial release, Igor has released updated versions. Version 2.3 added the capability of using colors as targets. After three and one-half years collecting and analyzing more than 600 ARV trials, Version 2.4 was released. The target pool was expanded to 1,200 photos. Igor claimed Version 2.4 made sessions easier to analyze and, with improved target pairing, reduced ARV misses by up to 10%.

Jon was one of the beta testers of the original version of ARV Studio and found it very user-friendly and without bugs. In revisiting the software in 2020, Jon found it even better with its additional features. Igor explained improvements he had made:

Along with other factors, a poor-quality binary ARV pair is considered to happen when:

a) both photo targets include identical gestalt or elements like water, natural element, human/animal, building and other. This kind of pairing is not such a problem for the viewing part of the process, but it causes difficulties for the judge, who is trying to determine which of two targets is a better match for the remote viewing session.

b) one target's complexity and visual/physical entropy is not equal to the other target's complexity and entropy.

c) targets are of poor resolution or have blurred or cut off parts thereby not allowing unambiguous identification of what the target actually is.

After observing that numerous poor ARV target pairings cause session judging/ analyzing to be more difficult and thus put judges into a tough decision-making position, the author has created a powerful ARV pairing algorithm called ARVOPTIMAL to create conditions to be able perform at optimal ARV accuracy level ("hit-rate") reflecting the remote viewer's current skill level and experience. Poor target pairing is a significant factor creating unwanted judging/analyzing difficulties and frustration and producing ARV misses. ARVOPTIMAL algorithm ensures dissimilarity of the selected photo targets, meaning that they are as different from each other as possible.

Igor also implemented ARVGUARD, a built-in safeguard algorithm, which ensures nonrepetition of a selected photo target in the next 300 remote viewing practice sessions and the next 150 ARV trials.

In addition to ARV Studio, Igor released a free Windows program "Lottery ARV":

I'd like to continue contributing to RV community with new version 1.1 of my Lottery ARV program, which is free. I achieved so much in past few years and many thanks to all wonderful RV people I met and worked with, and my desire is to give something back for free…The program is now able to assist you in any lottery type or game. You can create custom lottery of any number of balls and any number range.

More specifically:

Lottery ARV is a simple, free and standalone MS Windows program designed to assist during the "remote viewing" (nonlocal perception) process for predicting future lottery numbers (0 to 9) in games like: Pick 3 (Cash 3) or Pick 4 (Cash 4). Before mastering this skill to accurately pick all three numbers of Pick 3 or all four numbers of Pick 4 it is highly advisable (by top remote viewers in the field) to begin your learning and practicing process by first accurately predicting one number only (e.g., the first number).

The program is based on sensories like tastes, smells and sounds. Igor commented:

I do hope someone (Ok, all!) wins the lottery using it! In meantime, a lot of other ARV ideas are popping up in my mind, like creating automated judging system, so the need of human judging or self-judging is completely removed from the ARV process. This can be done but not with photosite targets (image-recognition systems are not good enough yet) with using targets that combine several aspects like color, basic textures, basic emotions. If the RVer reports to the system a standardized "3-word input," let's say: "red, soft, happy" then the system can do computer judging and also deliver feedback and outcome without other human involved. I don't know if this kind of study or experiment was done before. Eliminating human judge may reveal us new insights in the field of ARV or at least keep judge away with his/her intentions and subconscious interactions away from the ARV trial.

In April 2021, Igor announced a new version of ARV Studio Lite, version 3.2 (Windows). It includes additional features such a larger and improved target set with 1300 Premium Targets and a stronger algorithm to sharpen the dissimilarity of targets. The program is now able to generate 358K binary photo pairs.

In August 2016, Igor was proud to announce his ARV Studio program had been accepted for an IRVA presentation.

ProjectX – Sandra Hilleard

ProjectX Showcase Display Public Sessions

The 10 most recent examples of Remote Viewing Sessions

Target Image	User Name	TRN	Gestalt	Summary
	Potter62	3848 - 7963	Land	Glassy,green,bland,wide,open,straight,red,orange,clear,fine,c
	Potter62	8683 - 1331	Water	Vast,open,large, grey,blue, clear,tangy,green,small,upwards,c

Figure 13.8 – ProjectX

Sandra Hilleard is a longtime practitioner of remote viewing based in Perth, Australia. She co-founded the Remote Viewing Unit in 2008. Sandra is the author of *Anomalies*, which details her childhood experiences and eventual psychic detective work. She has trained in a variety of disciplines and speaks four languages. In 2019, Hilleard launched ProjectX, Remote Viewing Target Practice & Research Database. She explained the project this way:

The field of remote viewing (the controlled use of extrasensory perception) has always faced the challenge of proving its value to the community and business. Remote viewing has long proven its value in investigations and military intelligence. We only need to look at the declassified US "Project Stargate" files to conclude that this tool has been applied since 1971 and was used up until 1995! If it was not deemed useful for intelligence purposes, the program would not have lasted for 24 years.

Consistency, Accuracy and Value – There has been a lot of speculation about the "scoring" of remote viewers and their remote viewing sessions. The question is: "What is being measured and how?" During the early SRI (Stanford Research Institute) experiments and many others, the remote viewing session was conducted under a double-blind protocol and the feedback was given after the remote viewing session was completed. Sometimes multiple independent "judges" were used to determine if the session matched the target. Former US military remote viewer and database expert Lyn Buchanan designed a system of measurement within different categories. His system measures the number of perceptions correct against the total amount of measurable perceptions. This system is great for measuring individual performance. However, all these approaches have one major problem; it requires human interpretation after the feedback is available.

A critical question is: "Are the judges, the researchers or the remote viewers making the data fit the target feedback?" No human being is completely unbiased, no matter how hard we try. A better way to measure would be a binary computer system! It is either "yes" or "no," "correct" or "incorrect," there is no "maybe, possibly, could be, might be." However, the computer cannot interpret natural human language and there are many ways to describe one and the same thing. The challenge is to design the best possible computer-aided scoring system. A system that will, beyond a shadow of a doubt, determine the remote viewer's accuracy and consistency.

The Importance of Double-Blind & Consistent Scoring – It is quite difficult for remote viewers to gauge how they are performing or to demonstrate their performance is really consistent and accurate. Some have stood up to the challenge of remote viewing on live television and performed really well, only to hear that: "Their result must be a coincidence or is not quite confirmed!" It is also difficult to gauge how your performance measures compared to others. Unbiased scoring of double-blind remote viewing sessions over a longer period of time is the solution. Of course, human feedback will also be available within the system, but the most important part is consistent and accurate (as accurate as possible) computer-based scoring/measurement.

Remote Viewing Track Record – Are you one of those talented top remote viewers? Do you want to see how far you can go? This online tool will not only give you the opportunity to test yourself under double-blind conditions, but also to demonstrate (if you wish) your consistency and accuracy over a long period of time.

The most famous people in remote viewing history (e.g., Ingo Swann, Pat Price, Joe McMoneagle) have all been hired as remote viewers or "consultants" because they have been rigorously tested under a double-blind scientific protocol. Their remote viewing data was/is proven to be above average in accuracy and consistency. Not everyone gets the opportunity to be tested in a laboratory environment and not everyone wants to! You can now be part of a research project from the comfort of your own home and build a proven track record! Do it for yourself or show the world what you can do... It's up to you!

We asked Sandra about the targets and sessions for ProjectX and she responded as follows:

My system is based on CRV, but the principles could apply to ARV. As you know, ARV is not my main focus. ProjectX is designed for both testing and learning. The system measures two things:

1. The "Gestalt," which is the intended focus when the target was created. This is either land, water, mountain, life form, structure or energy. Most targets in this world will consist of multiple land/water, land/structure/lifeform etc. The viewer is to select one of 6 options in regard to the main focus of the target.[369]

2. Each target has a lengthy description, and this description is stripped from words like "the, and, or, etc." The remaining words are captured separately in a separate table. These words are then put through Princeton University Wordnet Database to find all possible synonyms. A car could be a vehicle, motor vehicle or a BMW Z-series. The synonyms are manually examined for anything that is not associated with the target. Water, for instance, is also used as pee or piddle in the SynSet database, but that is not associated with a body of water. Humans have rich language and it's trying to account for as many

possible ways people could describe the target (including spelling variations of US-English, British English, Canadian English and Australian English).

The remote viewer can enter their summaries as text and a text comparison algorithm checks their descriptions against what is known in the database about the target's description words (much like a plagiarism checker). It will highlight exact hits as green when the feedback is given and orange if a synonym is used. It will count the number of words matching in the remote viewer's summary.

As you can imagine…setting up targets is a lengthy process. Each target has a feedback photograph that is checked for usage rights, suitability as a target, if it includes a date, time and location the photograph was taken, if the photo hasn't been edited or altered and the location on Google Maps or Google Space has been included and, of course, the information available about the target as text. Plus manually examining and importing the Wordnet SynSet.

It is still not the "be all and end all" because some of the words remote viewers use that are indeed correct are still not captured by the system. At least it captures a great deal more than, for instance, the PSI test by IONS.

It is my goal to eventually create or link up an AI system that can capture the nuances. I do not have this skill set at the moment but I'm hoping to get some help developing this in the near future. (Or add it to my skill set in the future.)

An important development for ProjectX

In her Jan 5, 2021, *Newsletter* #5, Sandra wrote:

Up until recently, the system was used for learning and practice only. However, in November 2020, I received a request from a Dutch team of researchers to provide them with research data from ProjectX.

Professor Dr. Dick Bierman from the University of Amsterdam and his colleague Dr. Fred Melssen from the Radboud University in Nijmegen asked if they could use the ProjectX database for the analysis of LST, Local Sidereal Time. We discussed the types of data ProjectX collects and if this would be suitable for their research. Indeed, the data was suitable!

Previous research on LST was conducted by Dr. James Spottiswoode and was published under the title: "Apparent association between anomalous cognition effect size and sidereal time" in the *Journal of Scientific Exploration*, Vol, 11, No. 2, 1997. Interestingly, there appeared to be a significant increase in correct descriptions (a.k.a. hits) around 13.5 h LST.

This was a very promising indicator of a possible way of improving remote viewing sessions for practical use. The 13.5 h LST seemed the optimum time to do a session, giving remote viewers a better chance of accurate data! However, additional studies indicated that 13.5 h LST did not make much difference at all.[370]

Personally, I wanted to test the 13.5 h LST theory for myself, when I first read about it, however, there is one problem; all the research had been done in the Northern Hemisphere and I am based in the Southern Hemisphere. Would that make a difference?

Well, even our water spins clockwise in the Southern Hemisphere and counter-clockwise in the Northern Hemisphere on a larger scale, this can be seen in storms or cyclones. The Coriolis force accounts for why cyclones are counterclockwise-rotating storms in the Northern Hemisphere but rotate clockwise in the Southern Hemisphere. We have our seasons reversed (winter is like summer and summer is like winter) and we also look at a different night sky! Since everything seems opposite "down under," would there also be an opposite, optimum-LST for Remote Viewers in the Southern Hemisphere? We don't know the answer to this question until experts have analysed the data.

ProjectX has collected remote viewing session data from the remote viewers in both the Northern and the Southern Hemisphere. This provides a unique opportunity to potentially discover something about the functioning of remote viewers all over the world.

Since there was no previous Southern Hemisphere data available, it means that this needs to be tested multiple times by independent researchers to ensure that the results are a real effect and not a proverbial "glitch in the system." In short, we need more researchers and more participants willing to practice and contribute to remote viewing research in the Southern Hemisphere.

I feel incredibly excited and honoured that ProjectX is now assisting world-renowned researchers in their efforts to find answers. I am proud of the CRV students who diligently practised by using the ProjectX system, learning valuable lessons on a personal-experiential level and contributing to science on a different level.

I want to thank everyone who participated in ProjectX by submitting their sessions for the practice of their skills! Without your participation and support, the researchers would not have the data that could potentially provide us with answers. You are supporting the entire RV-community by contributing to research in the field. We all learn from each other and together we can uncover what works and what doesn't, both on a personal, individual level and perhaps also on a more scientific level.

Remote Viewing Tournament – Michael Ferrier

Figure 13.9 – Remote Viewing Tournament

In February 2019, Michael Ferrier introduced a mobile app called Remote Viewing Tournament (RVT). It quickly gained an enthusiastic group of users, with hundreds of ARV predictions being submitted each day and lively ongoing discussion in the app's Facebook group.

Ferrier has a master's degree in cognitive science and a background in education and game development. He was a developer on the early MMORPG Asheron's call, and has long had an interest in psychic phenomena. There are two excellent articles about Ferrier and RVT, which we recommend for background about him and the development of RVT. (For links, please visit http://www.arvbook.com)

In this chapter we focus on the features of RVT, the results and statistics, and aims.

When Ferrier introduced RVT, he described it as offering a step-by-step tutorial to develop skills, as well as the opportunity to practice targets daily and to earn monthly prizes. As the app caught on and he got feedback from users, he added and modified features to improve the experience of using RVT.

One of the first problems Ferrier encountered was *displacement* – again and again RVT's users would find they were receiving impressions from both image choices (and sometimes only the incorrect one) rather just than the target image. Of course, this makes sense; immediately after RVing, the user is shown both image choices and only the next day are they shown the target image. It's understandable that the user's subconscious would treat the immediate display of the two image choices as "feedback" more so than the actual feedback of the single target image that is received the next day.

Ferrier updated the app's tutorials to include various methods for minimizing displacement, such as focusing on the coordinates, spending more time examining the feedback image when it's received and being careful not to react emotionally to seeing the image choices as validation of the impressions received. Displacement continued to be a problem, however, so over time he added several new methods of judging that would prevent the user from having to see both image choices while judging which one best matches their RV session.

The first of these was the *Sequential* option, introduced in March 2019. This method still allowed viewers to self-judge, but instead of showing both images at once, only one image was shown. The viewer could then make a decision with a self-selected confidence level based on that image or could choose to see the second image. However, this option was later withdrawn due to poor results.

A second new self-judging option called *Descriptions* was added in April 2019. Text descriptions of the two images are presented rather than the images themselves. This way the viewer does not have to see either image choice. This did cut out the possibility of direct visual comparison of a drawing and an image, but many users have preferred this option and it has produced slightly better results than the default self-judging option.

In June 2019, *Independent Jury* judging was introduced as an option. Rather than self-judging, viewers could now opt to have other RVT players compare the viewer's transcript with the two image choices. Using this method, the viewer would never see an incorrect image choice. Judging other viewers' entries proved to be an enjoyable activity in itself for RVT users, and so typically a participant's transcript is viewed by several dozen judges whose opinions are averaged to produce a final judgment of which image is the best match. While some RVT users avoid the Independent Jury judging option because it removes the choice of image from their control, others prefer it for the decreased likelihood of displacement. It has so far produced better results than both the original self-judging option and the *Descriptions* method.

Finally, in January 2020, an option called *Rate Tags* was introduced. Instead of being shown the two images, the participant is shown a list of words and phrases called tags that describe an element of one of the two images. The viewer rates each tag as being more or less likely to match the impressions they have of the image. RVT then selects the image choice that best matches the tag ratings. This self-judging method has also performed well, producing results similar to the *Independent Jury* method.

Being the default option, the original self-judging method of selecting between both image choices remains the most popular and has been used in more than 95% of submissions. The newer judging methods tend to appeal to longer-term users who are interested in maximizing their success rate, and each accounts for about 1% to 1.5% of submissions.

Typically all users who submit an entry using the default self-judging method on a particular day are shown the same two image choices. Half the viewers are assigned one of the images in the pair as their correct target while the other half receive the other image as their target. This was established to cancel out any effect from one image being inherently more attractive as a choice. The two groups are assigned different sets of coordinates.

An optional change to this system was made in July 2019. Players can now join a "group," and all players sharing the same group name always receive the same correct target image on a given day. This was introduced to avoid the possibility of two people who are close being assigned different target images, and one perhaps having their remote viewing session contaminated by receiving impressions from the others.

The Remote Viewing Tournament photo pool was selected manually by Ferrier, and images were paired together, attempting to minimize common objects, colors or other characteristics between the two images in each pair. A pair is randomly selected by the computer each day and there is only one target each day.

Ferrier collects detailed statistics: Of 355,956 entries through March 2021, 50.06% predicted the correct image. While that overall hit rate isn't very impressive, more experienced users tend to do better, with the top scorers averaging more than 52% correct. The highest standard deviation for an individual is 3.66 with 58.1% correct more than 515 rounds at odds of 1 and 7,700 by chance.

Ferrier uses the predictions by viewers to make stock picks. He began with $12,500 as an initial investment and it grew to $19,235. He has 53.7% accuracy with 201 correct in 374 investments. The predictions are whether the Standard and Poor 500 average will move up or down. Ferrier's goal is to help bring RV and the psychic spiritual side of human nature further into mainstream consciousness. He would like to set up a system so viewers could be paid for their efforts according to their skill.

The RVT app has gained an enthusiastic following. Examples from those playing are posted daily on the RVT Facebook page and frequently on the Reddit forums and Discord channels devoted to remote viewing.

As noted above, RVT results are very close to 50%. Part of the reason may be the design by which half the viewers get one target and the other get the other target as the correct result. This setup creates many feedback loops for the hundreds of players taking part each day. It does prevent one photo from being dominant on a given day (if it is intrinsically more interesting), but this may be offset by the "mixed message" the setup creates. Unless a viewer joins a "group," each photo is correct for some viewers and incorrect for the others. We submit that this is a recipe that will keep the results very close to chance, which so far they have been.

Several players commented that they get the next day's target instead of the current day's. As we have noted, the phenomenon of displacement in time has been observed from the very beginning of attempts to view targets. This phenomenon will likely affect Ferrier's plan to use the crowd-sourced viewing in RVT to make predictions of the S&P 500 (or any financial indicator).

Another observation is that many players do very quick drawings with fingers or fingernails, often with little more than a rough outline or two (this includes ourselves). A minimum of data is sought and offered. This is in agreement with getting just one or a few gestalts about one of the images to make a decision between the two photos, but it does not help develop the viewer's full RV abilities. For that, self-training or a course in "regular" RV is highly recommended.

Are there ways to filter RVT's thousands of users to get better results? Ferrier notes that Descriptions, Independent Jury and Rate Tags judging have always done (slightly) better than 50 percent, while the Scrambled and Sequential options have done worse (the latter so much worse that he dropped it as an option). Results might improve if more viewers used the zero-confidence option when they weren't sure about the pick. This makes the entry essentially a pass (a fact we were not aware of until Ferrier pointed it out to us).

One obvious idea to try to improve results predicting the up or down of the S&P 500 would be to use predictions only by viewers with the best hit records. Marty Rosenblatt has explored this in APP. He tracks hit rates for the last 10 events, last 25 and overall. Doing the same for RVT might be of benefit. Marty has stopped tracking individual group rates and focuses on individuals' success rates. However, the massive number of viewers in RVT may make for a different overall dynamic and might be successful.

Ferrier looked into using AI for this purpose and wrote:

I agree that it's very appealing to use machine learning on group predictions to find patterns and combine them into a single (hopefully) "best" prediction. I tried this myself with the RV Tournament data. I'm not an expert in deep learning but I picked up the very readable book *Deep Learning with R* (by Chollet and Allaire). It goes through how to use the R programming language, along with the Keras and TensorFlow libraries (all free tools), to approach just this kind of problem (along with a lot of other interesting applications). By feeding it each entry's prediction, along with a lot of other data about each entry such as the prior record of that user, I was able to get some positive results out of it; depending on the period of results it was trained on and the period it was used to predict, it would correctly predict 50%-60% of trials. However, I was never able to get it to make consistently better predictions than I could myself using the patterns I had identified by hand. So I haven't been using it regularly, but I agree it's worth pursuing further.

Despite the overall low group stats, it should be noted that some viewers have done remarkably well. One of these is Grin Spickett, one of the moderators of the RV Reddit group, who is also a writer for various RV-related publications. In a recent article on remote viewing for Medium.com he wrote,

Are the all-time RV Tournament high scorers also those who perform the best against chance? Right now two of the top 20 high scorers are also in the top 20 for z score (the measure of how many standard deviations their performance is away from what's expected by chance). This is because a high score depends not just on percentage correct but also on the number of rounds played, and it can take playing at least 200 rounds to crack the top 20. That's done so that players have to show good performance over the long haul to be a top scorer. For example, right now the second-highest z score, 3.32, belongs to a player who played 11 rounds, got all of them correct, and then quit. That's impressive, but there's still a 1-in-2,048 chance of someone getting 11 rounds in a row correct. With the many thousands of players RV Tournament has had (3,557 of whom have played at least 11 rounds), it's likely to happen at least once, just by chance. The current highest z score, 3.65, is more impressive. Though their percentage correct (58.4%) is lower, they've kept this up over nearly 500 rounds. The chance of getting that high a percentage correct over that many rounds is only about 1-in-7,600. Even this could still possibly be attributed to chance – what's needed to determine what is and isn't meaningful, again, is more and more data. That same percentage over 1,000 or 2, 000 rounds would be much harder to write off to chance.

CHAPTER 14

How About One Target? Unitary ARV

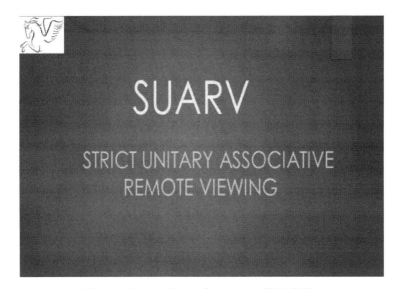

Figure 14.1 How About One Target (SUARV)

We have discussed displacement and the many attempts to overcome it in traditional binary ARV. These include a variety of ways to reduce the number of potential targets from two to one. Such methods can be considered forms of unitary ARV – there is only one target; no other potential target(s). We define unitary ARV as any Associative Remote Viewing protocol that has only one potential target.[371] There are several forms of unitary ARV, including what Jon calls Strict Unitary ARV (SUARV).

In this chapter we will discuss a format called 1ARV, which was employed in a long series of APP trials. 1ARV was sometimes referred to as UARV. In this format there were two groups of viewers (let's call them Group A and Group B). Each group had a photo to remote view as their target, and each photo related to the same event. For example, one photo was associated with the Houston Astros winning a baseball game (Group A) and the other photo with the Washington Nationals winning the same game (Group B). If Houston wins, Group A is shown their photo and Group B gets no feedback. If Washington wins, Group B is shown their photo and Group A gets no feedback. There were many trials of the 1ARV protocol, as we shall see.[372]

Strict Unitary ARV

SUARV also predicates a single target that is associated with the outcome of an event. Jon started using the term to distinguish it from 1ARV, which was not truly unitary. As an example of SUARV, the target could be emotions of people who have a keen interest in who wins or loses a game, or it could be a photo associated with the outcome of the game. Examples of the emotions form of SUARV (emotions) include the feelings of a devout fan, a pitcher or a coach at the end of the game or 15 minutes later in the locker room. The viewer notes the emotions (e.g., happy if a win) and the viewer, if self-judging, or the tasker uses the emotional response to decide which outcome to bet on or whether to bet at all.

Another form of SUARV uses a single photo and you score the session against the target photo. If the score exceeds a cutoff score you have established, you place a bet. If not, you don't. Alternatively, you could use any one of the senses (smell, taste, touch/feeling, sound; perhaps others – there are more than five senses) in Strict Unitary ARV.

Dr. Don Walker, a skilled remote viewer and a member of the TransDimensional Systems public demonstration team (c. 2001-2003), came up with the idea of trying unitary ARV. He expounded his idea in *Viewing outcomes for fun and profit or How to be a Zen monk while in Las Vegas* published in *Eight Martinis*.[373]

For about a year before the coming together of the Aurora group,[374] some friends of mine, Ken, Roma Zanders and myself, found ourselves with some viewing time on our hands and turned towards viewing outcomes. We all had history together from the old days, working and viewing at TDS. Besides the obvious excitement and anticipation of hitting it big on the lotto, there was also a desire to test some boundaries we were told existed in the remote viewing realm.

We had been taught that you couldn't view numbers much, if at all. We were told that too much ARV or "outcome" viewing was unreliable over prolonged periods of time and potentially bad for your viewing mind. There were other limitations about degrees of blindness and frontloading that we thought we needed to check on as well.

We proceeded to test a few of these boundaries, to see if they were in fact laws of the viewing terrain we were on, or just beliefs, limiting us as if they were laws.

Don goes on to describe his efforts to predict the lottery (he had some success on the Pick 3) and his use of hypnosis, which he felt improved accuracy. On Dec. 29, 2005, Don wrote to Jon:

I'm wondering about tasking my reaction or association, to reactions of fans / not the athletes, as they are pros and who knows how they react. But how about "viewer will focus on the overall primary emotional reaction that the 'Lakers' (could be team) fans would have open learning that the Lakers did lose, if they lose on Sunday's game. Viewer will reveal the viewers response or association, to the fans response of the Lakers losing, if they do."

I'm wondering if primary emotional reactions could be easier for the Subcon to pick up on, that a winning or losing scenario…and then determining how I (a viewer) would react to that reaction MAY/could be easier to report, than an accurate report of what someone else's reactions were to an event. Just thinking about tasking "outside the box." This sort of thinking…is exactly what I was talking about a few months ago when I was talking about getting out of the paradigm we've ALL been in about what RV is, how it's done, and what we can and CANT do with it.

Don's viewpoint was based on extreme sensitivity to emotions while remote viewing – he could pick them up easily. As a chiropractor, he was in close touch with his body, where emotions are felt. In fact, he suggested the term "Knosomatics" (body knowledge), which Pru Calabrese incorporated since the TDS method is extremely body-oriented and involves sensing, probing, standing up, moving around, etc.

Shortly after Don's letter, he and Jon sought out two other viewers. Roma Zanders became the third member of the team. Roma had been at Farsight with Pru and Courtney Brown and she was an excellent viewer. Along with Don and Jon, she was one of three interns Pru had selected from the class of a dozen viewers who had gone through the intensive TDS "Bananaslam" training program. She readily agreed to take part in trials using emotions to predict outcomes of games. A fourth viewer and friend, Liz Ruse, who was a natural psychic, also agreed to participate. Liz's method focused on team kits (standard equipment), colors and symbols (she did not include emotions in her approach).

Trials were undertaken in 2006 with Don, Athena (Roma) and Liz viewing and Jon tasking. The results were published years later in another article in *Eight Martinis* titled "Trailermakers in the Forest: Results from Two Team ARV Trials."[375] Extensive excerpts from the article follow.

Members of what became the Aurora Remote Viewing Group experimented with ARV during the past decade, particularly from 2005 on, as individuals and in teams. This article presents the results of two of the team trials, in 2006 and 2007. The prevailing method was ARV, but two of the participants were natural psychics who used their own psi method within the team framework…

The 2006 trial had very marked successes, in both accuracy and wagers won. The 2007 trial did not achieve these results, but it was useful in testing other variables than those explored in 2006. The contrasts between the two trials suggest some trail markers in these largely unmapped woods…

There were many factors to assess in undertaking these team ARV trials: what kind of events to choose; what kinds of targets to associate with them; characteristics of the viewing and coordinating team; tasking method; viewing method; feedback for the viewers; judging the sessions; methods and strategies of betting; coordinating across far-flung time zones and other variables. Within each of these categories there are sub-categories. For example, targets may be landscapes, scenes, people, objects, smells, tastes, symbols, shapes, colors, patterns, ideograms, emotions, music, etc. Elements of viewing methodology include the type of viewing (e.g., CRV, TDS), partial or full sessions, affirmations and intention, and cooldown. Taskings may be worded sparsely or in complete sentences. Also, there is the issue of the degree of standardization to be applied within the team.

In deciding which variables to test, we were aware of two major trends widely encountered in ARV. The first is initial success followed by a decline in accuracy. The degree and duration of success vary, but by both measures results can be very encouraging. After a break of days or weeks and a restart, accuracy may again be initially high, but then will be followed by another decline. A second notable tendency is displacement. That occurs when the viewer's data exhibits elements of both targets, leading to erroneous matching of the session by the judge. Both issues have proven persistent, not to say intractable, although at least two people claim to have overcome them.

Regarding the first issue, there is no easy solution. By testing different variables in the two trials, we thought we might be able to eliminate some causes of declining results in later tests. Regarding the second problem – one way to tackle the issue of displacement would be to modify the ARV setup. According to one viewpoint, having two targets, as traditional ARV does, sets up a binary situation – the two targets are bound together; the setup is asking for spillover to occur. Put another way, standard ARV can be construed as creating a binary "thoughtform." From another standpoint, traditional ARV reinforces a particular kind of dialectic. In Marxist praxis, for example, everything that exists is considered as a "unity and struggle of opposites." It may be that each "side" has more than one "opposite" but specifying two and only two alternatives may be "playing into the dialectic."

However it is framed, if having two objectives is a problem, the obvious alternative is to posit just one objective – call it unitary ARV. For example, one could task the main emotions felt by players or fans of the winning or losing team at the end of a game. This would still be associative RV, since one would not be viewing the game itself but something closely associated with the game. But there would be only one target/objective. This is the approach we took in the two trials.

After considering the many variables and a tryout period from April through June 2006, we conducted a run from July through October. We conducted a second trial, with different variables, a year later.

Setup of the July-October 2006 trial

Participants: Three very experienced viewers/practitioners – Don Walker, Liz Ruse and Roma Zanders (DLR) – took part; the author was the coordinator. All of us were members of the same group and had known each other via the internet for several years, along with some in-person acquaintance, as well.

Type of event: We decided to focus on one type of event only – a binary sporting event. That is, the rules of the games we selected allowed for only one outcome (win or lose, no ties/draws).

Length of trial: We did a run of 61 games over four months.

Targets/Indicators: For two viewers, the emotions of a group or a (usually) anonymous individual vitally concerned about the outcome of the game were used as the targets, e.g., a group of fans or a bettor at a particular casino. For the third viewer, the target was usually one team, focusing on the team logo or team colors.

Number of sessions per game: Don and Roma received three different tags and did three viewing sessions for each game. (Later in the trial, Don did two sessions per game.) Liz viewed the target as many times as she felt necessary to obtain a clear result.

Taskings: The taskings were customized for Don and Roma. We experimented with different wordings during the course of the trial. As noted, Liz viewed only one objective, which often was one of the teams in the game. She was given only the name of the team along with the tag.

Example of a tag and a tasking for Don and Roma: Tag: Tammo2.

This was all the viewer had to go on in doing the sessions. After the sessions were done, the viewer would look at what the target was:

Tammo2: Focus on a non-professional gambler at the Sahara Hotel and Casino in Las Vegas who places a straight bet on the Seattle Mariners to beat the LA Dodgers Tuesday, June 20, 2006. Focus on this gambler at breakfast time the day after the game and how he/she feels emotionally about his/her straight bet on the Mariners to win this game.

Betting information provided by online sports books was included in some of the emails that conveyed the tags and taskings emails.

Judging the sessions: Each viewer received the tags and did their sessions. They then looked at what the objective corresponding to the tag was. Then they made a judgment about what their sessions indicated about the outcome of the game and reported their judgment as to the winner to the coordinator.

Concurrences: We tracked concurrences among viewers – when all three viewers (DLR) were in agreement, when two viewers agreed and the third disagreed or had no pick, and when two viewers disagreed and the third had no pick.

Feedback to viewers: Viewers found out what the target was by looking at the bottom of the email containing the tag and tasking. Later they found out the outcome of the game via email from the coordinator or by checking the game result themselves.

Betting: Two viewers and the coordinator placed bets. One viewer placed bets for the third viewer. There was some consultation among team members prior to the game, but betting itself was left up to each individual. Bets ranged from small amounts up to thousands of dollars.

Observations

Several concurrences among the viewers generated higher accuracy rates than the rate of each viewer considered separately. Individual accuracy per viewer was 56%, 66% and 59% over the course of the four-month trial. All these individual rates were above chance and the overall average of 60% matches or exceeds that of many long-term sports prognosticators. But most group accuracy rates were even higher, including 90% for DLR agreement, 67% (two concurrences), 75% (one concurrence) and 100% (one concurrence).

The total number of these concurrences was not large; the sample was quite small. But these correlations were very suggestive and often led to winning bets.

Going into the trial, we hoped that the highest rate of accuracy would occur when all three viewers agreed. In the event, "DLR agree" produced the second-highest accuracy rate – 90% in 10 games. In a way, the ARV trial built up over time to this tenth DLR concurrence. We had experimented, viewed and tracked accuracy for months, including the warm-up period, and here was our best indicator. (RL v D, which was 100%, had not yet been tracked). So, come the day, some team members bet quite a bit of their accumulated winnings on this 10th concurrence.

Unfortunately, this turned out to be the first failure of the three-way agreement. Considerable money was lost and team morale took a large hit. Later, looking over the particulars of the 10th instance of the concurrence, I noted that some steps had been done differently than earlier. For example, this was the first time each target number for each viewer referred to only one team. Ordinarily, two target numbers would refer to one team and one tag to the other team. Also, the tasking procedures were run a little differently than previously.

Results of July–October 2006 Trial

Viewer	Don	Liz	Roma	DLR Agree		%	LR v D		%	Roma 3-0		%
Number of Games				Right	Wrong		Right	Wrong		Right	Wrong	
July	11	11	17	3			1			12	14	46%
Aug	9	11	14	1			1	1		Roma 2-1 or 1-2		
Sep	7	3	8	2						24	9	73%
Oct	23	10	22	3	1		1					
Total	50	35	61	9	1	90	3	1	75			
Number Right				DL v R			DR agree & no L pick			D v R & no L pick		
July	7	8	10	3			1	2			1	
Aug	4	5	8		2			2		1		
Sep	3	2	6							1		
Oct	14	8	12	1			5	1		3	3	
Total	28	23	36	4	2	67	6	5	55	5	4	55
Percent Right				DR v L			LR agree & no D pick			L v R & no D pick		
July	64	73	59	1			Right	Wrong			1	
Aug	44	45	57				2			1	1	
Sep	43	67	75				2			1		
Oct	61	80	55	1	1		1			1		
Average	56	66	59	2	1	67	5		100	3	2	60

Figure 14.2 – Results of July-October 2000 Trial

KEY: *D, L, and R stand for Don, Liz and Roma. DL v R indicates Don and Liz were correct 67% of the time (4 right, 2 wrong) when their choice disagreed with Roma's. DR agree and no L pick - means Don and Roma were right 55% of the time when they agreed and Liz had no pick. D v R and no L pick – means Don was right 55% of the time when he and Roma disagreed and Liz had no pick. There were no instances of 1) D v L and no third pick or 2) DL agreed and no third pick.*

The lowest rates of concurrence occurred when two viewers agreed, with no third opinion being offered. In other words, a 'mixed opinion' (2 versus 1) produced better results than when two viewers were in agreement with no pick from the third viewer (2 versus 0). These lowest correlation rates were 55, 55 and 60 percent. An exception was the 100% when Roma and Liz agreed and Don had no pick (only

5 instances however). This last may have been a statistical quirk or it may reflect subconscious interaction within the viewing team.

Comparing the 2006 and 2007 trials

We ran a second trial in 2007, in which we altered the setup in order to explore other variables. For reasons of space, I will contrast the two trials briefly here, without going into many details of the 2007 trial.

Participants: In 2007, the viewing team was Daz, Glyn, Roma, M and briefly, Don. In 2006, viewers were Don, Roma, and Liz. Members of both teams used a variety of psi methods (CRV, TDS, ERV or the participant's own method). The coordinator was the same for both trials. All participants were very experienced with psi, but some were more experienced with ARV than others.

Type of event: In 2006, there was only one type of target: physical sports games. In 2007, we chose three types of targets: sports games, financial targets and other. Other consisted mainly of political targets (to retain some focus within the category). The change to three types of targets was made in part because potential clients wanted RV projections about more than sporting events. It is clear that the viewers did much better, individually and as a team, having one type of target rather than three. We can't say for sure that this was a contributing factor to the better results in 2006, but it may well have been.

Length of trial: 61 games over four months in 2006. 32 events over two months in 2007. The differences in time and scope are not likely to have contributed to the different results.

Targets/Indicators: In both 2006 and 2007, nearly all of the targets were emotions related to the game. To our knowledge, this is the first time emotions have been the main indicator in a fairly extensive practical ARV/psi trial.

Number of sessions per event: In 2006, there were often three taskings per event, sometimes two, and for one viewer as many as nine. For 2007, only one tasking per event. This variable, multiple taskings, certainly warrants further testing.

Customized taskings: In 2006, we used Don's method (emotions as the target), tweaked by Don and Roma to accord with their preferences as viewers. Liz utilized her own unique method. In 2007, we also used emotions as the target focus but with little or no customization of the taskings. One participant, M, used his own psi method, which has been extremely successful in his private and public work worldwide. Likelihood: customized taskings contributed to the success of the 2006 trial.

Judging the sessions: In both 2006 and 2007, viewers decided what their sessions indicated about the outcome of the game and conveyed that to the coordinator via email. There were no independent judges, as is often the case in group ARV.

Accuracy rates and concurrences: For 2007, the overall ratings for the three types of taskings were: Games: 48% (16 right, 17 wrong), Financial 39% (15 right, 23 wrong) and Other 45% (11 right, 13 wrong). While viewer accuracy overall was less than 50%, in three instances over 50% accuracy was obtained: Glyn had 60% for Games (6 right, 4 wrong). Roma had 70% for Financial taskings (7 right, 3 wrong) and M had 67% for Other taskings (4 right, 2 wrong). Again, the total sample was small, and made even smaller by being broken into three types of objectives, so these percentages are at most suggestive. In the context of generally sub-50% overall accuracy, there appeared to be no useful concurrences (either positive psi or psi-missing) among viewers in 2007, which was in marked contrast with 2006.

Feedback, motivation and stress: In 2006, the viewing team viewed nearly daily and many bets were placed over the four months. This put pressure on the viewers to produce since significant cash was being won or lost. The stress may have contributed to burnout by the viewers and could have contributed to the one major failure at the end of the trial. For the 2007 trial, we wanted to see if delaying the disclosures till the end of the trial and not risking money during the trial would produce as good or better results. As it turned out, the results in 2007 were inferior to those in 2006. We can't be sure if the pressure on the 2006 viewers was a positive factor in some way, but for further such testing it would make sense to include some wagering or other strong motivating factor.

Location of the subject(s) of focus: When the tasking involved a casino, results were tabulated regarding the specific casino the bettor(s) were in. The 2006 results indicated no notable differences among the casinos, so this variable was dropped from the 2007 trial.

Time interval from session to game: This variable was introduced in 2007. The results indicate there may be a small positive effect (less than 20 days), but the sample is quite small.

The article concludes with lessons from the two trials and suggestions for future unitary ARV efforts. By the time of the 2007 viewing, Jon thought that what was being done was in fact Associative Remote Viewing because we were viewing the event solely to determine the outcome of a sporting or financial event, not for the variety of reasons one would view an event in regular remote viewing.

1ARV in the Precog group and in the Applied Precognition Project (APP)

In 2010 Marty Rosenblatt, who had been exploring ARV for several years, read the two articles on unitary ARV in *Eight Martinis* and addressed the subject in his P-I-A magazine:

The idea of Unitary ARV has been explored by others. I do not have a complete list, but there are two articles in EightMartinis.com that discuss ARV wagering applications, including Unitary ARV, that are worth reading:

1. Remote Viewing outcomes for fun and profit, or How to be a Zen monk while in Las Vegas, by Don Walker in Eight Martinis Issue 2.
2. TRAILMARKERS IN THE FOREST: Results From Two Team ARV Trials, by Jon Knowles in Eight Martinis Issue 4.

As the name implies, Unitary ARV has only one *potential* Target that is rigidly associated with one future *potential* Outcome. Examples include: The outcome of a sports game wager in one direction (Team A wins the game; or Team A wins the game by three points or more; or total game score is Over 43 points). Financial Instrument ABC will triple in value in X months.

In UARV, an Indicator PhotoSite (IPS) is the *potential* Target. If the Outcome occurs, i.e. the Outcome is actualized, then the IPS is actualized as the Target and sent/shown to the RVer as FeedBack (FB). If the Unitary Outcome does not occur, then there is no Target to show…there is nothing to declare as a Target.

In Unitary ARV, there is no entangled 2nd PhotoSite to displace to! There is only one IPS which you may or may not see as FB. Your task during your RV

Session for coordinate 123456 is to view and describe your future FB Session for coord 123456.

"Indicator Photosite" was Marty's term. In the above piece, he conveys what he considers a form of unitary ARV. Single photos were not used in the 2006 and 2007 unitary ARV trials; however, that is what Marty decided to use for the Precog group:

> The coding for Unitary ARV is in place as an option in the PRECOG program. The binary approach is still available for those who are still working to stop personal displacement in a binary mode – go for it. However, we strongly urge you to do sessions with the UARV approach, and especially with the 1ARV approach.

Marty's idea was to explore a side-by-side unitary approach: there would be two groups, each with a unitary target. Jon and Don were on board with the idea of exploring this modification of the unitary ARV and Jon wrote to Marty:

> As I see it, our two groups/sets of people have come together to work on Unitary ARV, and this is a very good thing. The two groups/sets are, on the one hand, former members of the Aurora group and on the other, you, the people you have been working with in Precog, and some contacts you are bringing in. (I realize that not all the ex-Aurora people on the list have been taking part in 1ARV.) With our joint efforts, we are moving toward having a substantial cohort for some substantial 1ARV work. And from my point of view (and Don's, as we spoke this morning), it is great to be working with someone who is devoted to and very good at the organizational side of this!

Thanks to modern internet technology, we have access to the many emails that discussed implementing 1ARV in APP, and we offer highlights here (generally in chronological order). We present this extensive account of the 1ARV series in detail since it is a rare close-up look at how an ARV protocol evolved during trials and because the protocol was used in the largest ARV group with trials lasting over a year. In other words, this was a major trial in the explorations of ARV. We begin with Marty Rosenblatt's description of the 1ARV protocol itself.

1ARV (1Associative Remote Viewing)

1ARV involves multiple independent Unitary ARV predictions on every Side of a sports or financial event/wager. Each prediction is based on an association, i.e. an entanglement, between the Remote Viewing Session and the associated FeedBack Session.

For now, we are focusing on events that have two Sides or potential outcomes, e.g., a sports game has a total of Over/Under X points, or a financial trade may be Up/Down during a specific time period. The sports or financial events are secondary to the more important objective of developing Personal Precognition skills which is Remote Influencing yourSelf (RIS) skills.

1ARV is all about reliably predicting the future using intuition. This is called precognition. RIS is all about influencing your present from your future. You learn to influence your Remote Viewing Session from your FeedBack Session. To enhance reliability, the 1ARV protocol includes RV and FB Sessions for both Sides of a binary event.

The 1ARV mindset and basic philosophy is shown below and follows directly from the psi Vibration Model where a Universe of Collective Consciousness (UCC) is developed. This UCC is based on the assumption that Consciousness is The Fundamental.

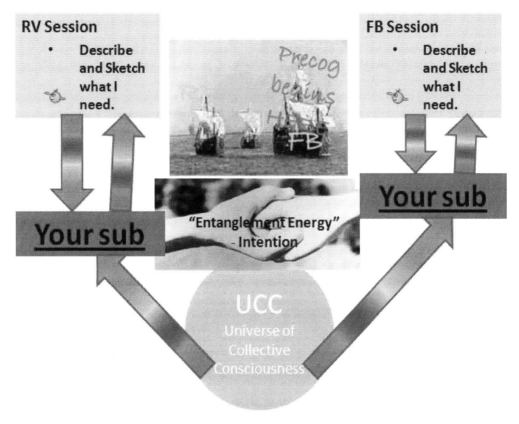

Figure 14.3 – 1ARV in the Precog Group

Describe what I need ![bird icon] is a simple Tasking Cue showing importance and yet light-ness. This cue is similar to what Joe McMoneagle often uses, *Describe what my Tasker needs*, except now the viewer is in charge at all levels, including being the tasker!

In Group 1ARV, a Group of RVers take on predicting the Outcome of a Future Event – for example, Up vs. Down in one day in a financial instrument. Redundancy is built into the Group Prediction to increase the reliability. A single Group 1ARV precognitive prediction consists of two RV and two FB sessions from each viewer.

The RV Session has as its primary intention to gather information from the FB Session. The FB Session has as its primary intention to share, or entangle with the RV Session. The truth is known at the FB Session and it is thus the perfect opportunity to Remotely Influence your earlier Self (RIS). Thus, there is real meaning to the phrase, 1ARV Begins with the FeedBack Session.

In 1ARV, FB is considered critical. We recognize that FB is not necessary, but we designed the 1ARV protocol to encourage meaningful and fun FB sessions to make the precognition easier, more reliable, and encourage long-term growth/transformation in the RIS area of personal consciousness!

In 2010-2013 the Precog group conducted a series of 1ARV trials; we named it as an alphabet series, Abel through Juliett. During this period, Marty consulted with team members and modified the procedures as he felt necessary, such as after a few missed predictions. We will trace most of the modifications. It was entirely appropriate to modify the protocol since the trials were not intended to be a formal scientific investigation but instead a detailed effort that kept statistics and aimed for a successful practical outcome.

On Nov. 15, 2010, Marty invited several people to join a 1ARV group, with Marty, Don Walker and a third person as organizers. By December, the team consisted of 17 viewers, including well-respected Ft. Meade viewer Mel Riley. There were two groups, each doing unitary ARV. At the time, one viewer (Loraine Connon) pointed out this was like having two dart teams, but she preferred single-team ARV.

In the initial months, and later, there was confusion as to just what the protocol was, and Marty and Don took pains to make the method clear. As Marty wrote (in an email to Jon on Dec. 14, 2010):

So, the unitary team that happens to be the team associated with the winning wager gets to do a full FB Session with the FB PS (Feedback Photosite). And yes, the other team does not do a FB Session and will never get their FB PS. However, no issue with anyone or everyone seeing the FB PS for the actualized, true Outcome…the more the merrier :-) Perhaps this makes the viewing easier… we all light up the correct FB PS. The 1ARV protocol is robust.

Your Under vs Over group is chosen randomly for each task. The protocol is for all to go to the link showing the transcripts of the group doing FB Sessions based on the winning wager. This further entangles the correct FB PS with the transcripts:

The direction for today's trade was up. This direction was randomly chosen this morning. At the same time, a unitary indicator photosite was randomly chosen, in this case the man/red/white/beard photosite. The randomly chosen direction is rigidly associated with the photosite, thus, if the EUR/USD actually moves up, the photosite becomes actualized as the Target. Today the market moved up and here is your man/red/white/beard, but not the Santa Claus, FB Target PS.

Marty offered a further explanation on Dec. 16:

> The Outcome is a Hit :-) Both groups did a wonderful set of RV Sessions. Congratulations to all of us. 1ARVers rock!
>
> The Under Unitary Group (Glyn, Martin and Jon, who were randomly chosen for the group) will find their Target FeedBack PhotoSite (FB PS) as well as all the group transcripts at (link).
>
> The Target FB PS is labeled 'Indicator PS for Under Wager.jpg' since that is what it was in linear time until the actual Outcome occurred…until the Predicted Outcome was Actualized.
>
> Anyone who wishes to view the Target and the transcripts can do so at the above link. However, remember that this is the Target FB PS for the Under Group members only.
>
> The unitary Over Group also did just what they were supposed to do; they provided clear contrast. We will not show or discuss any of their details, except to say, bravo. Each of their subs took them someplace quite different than their unitary Indicator PhotoSite, making the Analyst/Judges (AJ) job easy – again, well done.

While trials were proceeding apace, Loraine Connon continued to raise a question about the setup:

> My disagreement with Marty is only about the logical soundness of 1ARVx2 protocol (I'm good with single 1ARV). Marty is suggesting that you can just lash two 1ARVings together via the same football match event, then assume the groups are independent and conclude the protocol is robust. I don't agree. I think they are actually only separate, which certainly does not equate to being independent. His idea goes against the grain of what has been general(ly) established about the importance of remaining blind and seems a strange choice of project to leap to, to me, especially without first exploring ways to apply two groups in a genuinely independent way.

On Dec. 28, Marty modified the protocol:

You three were in the "Over" Group and your Analyst/Judges leaned in your way for the wager, which was the successful wager. Congratulations to you and here is the link for your transcripts and Target PhotoSite. We are moving even further toward a pure Unitary ARV approach for each group in that the Under group will not be given the following Over link...it is just for you, enjoy :-)

A few days later Marty wrote:

I believe it is valuable after each wager, especially when we have a miss, to evaluate whether it is time/appropriate to improve the protocol. The 1ARV protocol is still in its infancy and we as a group, are nurturing it along as we learn about our new baby. This morning, after reviewing the data to date, the Analyst/Judges realized that we could improve the protocol by doing the following: Provide more time for AJing.

Each wagering group will be AJed by a different person(s). The AJers will not have access to the "other" PhotoSite or transcripts.

On Jan. 9, Marty initiated another change. Scores were now to be shared with viewers prior to the event. And again, on Jan. 27, Marty wrote: "We recently went to fixed-viewer groups because of the nature of targets that you Aurorans seem to like for ARVing...the 'simpler' ones as opposed to more 'complex' targets with lots of material to potentially describe."

On Feb. 8, Marty expanded on this as follows:

The groups have been divided into more or less definite, fixed groups. One group gets Don's relatively simple targets (simple by design). The other group gets more complex, ordinary RV photosite targets. Some viewers prefer to have complex targets; others, the simple targets...Of course, it isn't known ahead of time which group is going to be the actualizing-target group and which the complementary group. One group, say, is said to be the Over group, when that is the bet, and the other group is the Under group.

So the above changes were implemented and trials proceeded. On Feb. 10, 2011, Marty proposed a 1ARV workshop:

The first Independent 1ARV team was started by bringing together Don Walker, Jon Knowles and Marty Rosenblatt, in November 2010. Don was the first to

publish (July 2009) a report, as far as we know, on Unitary ARV predictions/
experiments. Jon was the second to publish (August 2010) Unitary ARV results.
Marty's focus had shifted (after more than six years in applying standard Binary
ARV and then in applying ARV for up to 20 choices in a horse race) to developing
what he calls 1ARV, which is an application using Unitary ARV. Don and Jon
will be active participants at the workshop including discussing what they learned
from that early work and from our current 1ARV Inde Team.

A polarity factor

On Feb. 19, 2011, Marty proposed an additional scoring idea, a "Polarity Factor."

The "Polarity Factor" uses sum of predictions with CR >= 3.5 plus passes on
"other" side." The largest of the resulting sums provides the wager direction
and the factor. For example, in our last Independent (better word I think than
"Private") 1ARV prediction, we had the following, from the database.

Polarity Factor									
Polarity Factor uses sum of predictions with CR >= 3.5 plus passes on "other" side. The largest of the resulting sums provides the wager direction and the factor. For example, in our last independent (better word I think than "Private") 1ARV prediction, we had the following, from the database:									
GlynF	Don	Over			2	1	1.5		Pass
JonK	Don	Over			3.75	2.5	3.125		Pass
GaryG	Don	Over			5	3.5	4.25		Hit
Alpha2f	Don	Over			2		2		Pass
Essie	Catherine	Under	2.25	2			2.125		Pass-S
Tmaya	Catherine	Under	2.5	1			1.75		Pass-S
Tomath	Catherine	Under	2	1			1.5		Pass-S
InsightRV	Catherine	Under	2.75	2			2.375		Pass-S
This shows 5 points for the Over side (1 with CR>3.5 plus 4 Passes on the Under side) versus 3 points for the Under side () with CR > 3.5 plus 3 Passes on the Over side). Thus the factor =5/8=0.625, the cutoff factor of <0.6 is what I recommend we use since this leads to the 11 Hits and 2 Misses for our data. It is also consistent with, but does not help much, the public data.									

Figure 14.4 – Polarity Factor

*This shows 5 points for the Over side (one with CR >3.5 plus four Passes on the Under side)
versus 3 points for the Under side (0 with CR > 3.5 plus three Passes on the Over side). Thus
the factor = 5/8 = 0.625. The cutoff factor of <0.6 is what I recommend we use since this leads
to the 11 Hits and 2 Misses for our data. It is also consistent with, but does not help much, the
public data.*[376]

Russell Targ suggests simplification of 1ARV

In preparing for the workshop Marty had proposed, Russell Targ weighed in. From an email by Marty Rosenblatt, Feb. 22, 2011:

Russell Targ, who is a participant in the public 1ARV experiment, has simplified the Background and detailed 1ARV tasking as follows:

Definitions and Background

In 1ARV, a Group of RVers take on predicting the Outcome of a Future Event – for example, Up vs. Down. This is done by breaking the event up into two Unitary ARV predictions based on the possible outcomes. Each Outcome has its own Indicator PhotoSite (IPS). What's new here is that some viewers will be randomly assigned to the Up Outcome and others to the Down Outcome.

If the Up Outcome occurs, then everyone on the Up Side will see the Up IPS. If the Down Outcome occurs, then viewers on the Up Side will only get a thank you, but no IPS; in this case, the people on the Down Side will be shown the Down IPS. A person on the Up Side who submits a blank transcript, when the Down Side occurs, has submitted a correct response. Thus, this viewer is being 100% supportive of the actualized Down Outcome.

Detailed 1ARV Tasking Statement:

Your task is to describe and sketch what your subconscious provides to your conscious mind during your RV Session. Your sub shares the information from the future time when it knows, for sure, whether you are going to see an IPS. If your side is actualized, then the associated IPS will be shown to you for your sub to describe at an earlier time. If your side is not actualized, then there is no IPS for your FeedBack, and your sub is free to provide whatever info it chooses… including nothing.

What was new was the idea of randomly assigning viewers to the two groups and simplification of the protocol.

Throughout the 1ARV trials, Marty kept statistics from all the 1ARV groups, private and public. The accuracy rates as of March 6, 2011, were 65% hits for the public viewers and 56.5% hits by the private viewers. For the first two alphabet series – Abel and Baker

– the results showed 12 hits, 5 misses and 6 passes (70% accuracy 12/17) with three games to go. The 70% results were definitely an achievement since it is quite rare for a group to reach this level of success.

The "pass-supportive" feature in 1ARV

The protocol was complex, which Marty fully understood, but in the interests of exploring the ins and outs of this new approach (1ARV), he did not hesitate to try out new tools and techniques, as we've noted. In March 2011, Marty wrote:

> I have been asked to better define the meaning of Pass-Supportive (Pass-S) and how it is used in 1ARV. Let's first summarize the Hit, Miss, Pass and Pass-S data from all the viewers for the 12 games we have predicted so far:

> Viewer Stats: Hit: 40 Miss: 17 Pass: 18 Pass-Supportive 29

> The 29 Pass-S come from being on the Complementary Side of a 1ARV prediction. In the 1ARV public approach, there are two Sides as the possible Outcome for the sports-related predictions. For example, Over or Under a certain number of total game points. One side will actually occur as the Outcome and this side is called the Actualized Side.

> Thus only after the game is finished, do we know whether Over or Under is the Actualized Side, and then of course, we also know that the other side is the Complementary Side. Both sides are equally important and are part of the Polarity Factor analysis in reaching a prediction. Depending on what Side you are on AND the Targ confidence ranking (CR) your judge provided for your RV transcript we have the following Hit, Miss, Pass and Pass-S possibilities:

> Actualized Side: Hit (CR greater or equal to 3.5)
> Actualized Side: Pass (CR less than 3.5)
> Complementary Side: Miss (CR greater or equal to 3.5)
> Complementary Side: Pass-S (CR less than 3.5)

> Thus, if you are on the Complementary Side, you are supporting the Actualized Side when you get a CR less than 3.5 because we do not send you any Target as FeedBack.

The Viewer Stats above are showing statistical significance for the predictions being truly precognitive – foretelling the future outcomes! This can be looked at in several ways when compared to the random expected result of odds vs. chance.

- 40 Hits vs. 17 Misses has odds vs. chance of 624 to 1 (this is usual definition of Success Rate)
- 40 Hits vs. 18 Passes has odds vs. chance of 372 to 1 (this is just looking at Actualized Side)

Interestingly, the Complementary Side stats are not quite statistically significant:

29 Pass-S vs. 17 Misses has odds vs. chance of 18 to 1 (this is just looking at the Complementary Side)

This is clearly related to the displacement issue. Increasing the Pass-Supportive percentages for viewers who happen to be on the Complementary Side will further improve our current excellent Game Success Rate of 75% (9 Hits, 3 Misses). Go Viewers :-)

After noting six predictions were a miss for the game on 3/17/2011, Marty wrote:

I believe that the next stage in getting reliable predictions is to fully apply the power of Unitary ARVing and 1ARVing by aggressively improving the way we handle the Complementary Side. I would like to propose the following protocol when getting "nothing"… a simple thank you, as FB when on the Complementary Side.

Later that month (March 31, 2011), Marty wrote:

I just had an interesting conversation with Russ Targ concerning his recent Public 1ARV submission. I had AJed it as a CR = 3 and he points out that CR = 3 does include psychic data…so why count it on the other polarity side. This is an interesting observation about the logic concerning 3.0 or so CRs.

For example, we could treat Passes as follows:

CR >= 3 Pass CR < 4 Thus 3.0 to 3.99999 could be a Pass. Count in opposite direction – toward other pole – only if CR < 3. This would then mean, no contact with the IPS. (4/1/2011)

As the Precog groups were growing, along with enthusiasm for 1ARV, on April 6, 2011, Marty suggested the ***Applied Precognition Project*** be formed. It was launched 18 months later. APP would grow over the ensuing years to become by far the largest ARV membership group.

In May, Marty wrote that he had a "personal insight," a feeling that "1ARV is really simple and I am making it appear too complex with all the quantitative stuff."

In June 2011, the first 1ARV Workshop was held. Using 1ARV, there were two hits, one miss and three passes at the workshop.

Through the summer of 2011, the alphabet series continued. Delta achieved four hits, five misses and one pass. Marty consulted with Mel Riley and Mel suggested a "joyous celebration of hits" could provide a boost.

For Echo 3, Marty put everyone on the Over side and thus used one indicator PhotoSite. For Echo 7, Marty asked webinar participants to do two 1ARV predictions, an Over and an Under. They would get to experience both the Photosite and the Wildcard as feedback. In October, Marty made further changes to Echo 10A and 10B.

In April 2012, Marty posted the 1ARV statistics to date and they were impressive.

Our stats since we started the Group 1ARV Protocol with the Echo 8 prediction are summarized below as part of all the stats. We had 10 Hits, 5 Misses/SideKicks and 3 Passes for a very respectable 67% Hit Rate.

Group Stats				
Group ID	**Hits**		**Passes**	**Summary Info**
1ARV	10	5	3	
Oct2004	9	4	4	
LARV2011Dec	6	3	5	
R7/pers.	13	5	5	
ARV2012FEB	3	3	2	
TOTALS	41	20	19	0.004927 Ptail
Hit Rate	67%			201.9746 odds
				0.672121 Hit Rate

Figure 14.5 Group Stats

He explained further:

Prior to Echo 8 (Oct. 6, 2011), we were still actively experimenting with protocol changes. Some of these were invisible to you as viewers. For example, I experimented with using the same IPS (Indicator PhotoSite) for both Unitary ARV predictions. After Echo 8, the protocol was frozen until displacement began seriously rearing its head.

By July 30, 2012, the statistics for the seven existing 1ARV groups had risen to 69%, again an impressive result.

1ARV	*15*	*8*	*8*	*65%*	
Oct2004	*11*	*8*	*11*	*58%*	
1ARV2011Dec	*11*	*6*	*11*	*65%*	
R7M	*18*	*6*	*4*	*75%*	
1ARV2012Feb	*3*	*0*	*4*	*100%*	
Pegasus	*6*	*1*	*1*	*86%*	
Sublime	*3*	*1*	*1*	*75%*	
Totals	*67*	*30*	*40*	*69%*	
Hit Rate	*69%*			*9135*	*Odds*

Figure 14.6 – 1ARV Group stats 2012

Having presented the development of the 1ARV protocol and results in some detail, from this point on, we present a few highlights and summaries.

In the fall of the previous year (2011), the charts showed a large fall off. After some rebound, the 25-day moving average for all groups from October 11, 2011, through August 14, 2012, was 68% – still extremely good.

A comparison of groups doing ARV

As it happened, Jon was active – probably too active – in three different ARV groups at the time. One was in APP and two were outside APP and all were using binary ARV, not unitary ARV or 1ARV. On Oct. 12, 2012, Jon wrote Marty with the statistics for the three groups to that point.

By Jon's count, all the 1ARV groups in APP had made 81 predictions and had a combined 64% hit rate.

The second group was Mark White's ARV Club, which had a 63.5% hit rate with 63 predictions. (Further details about the ARV Club are presented in Chapter 24.) It was remarkable that the two groups had a nearly identical hit rate. It could be expected that well-run groups might be in the 60-69% range because success rates in that range have been noted in many groups over the years (almost as if it is a natural limit), but for two groups to be within half a point was striking.

The third group Jon took part in was called the Wager Team. This large group had made 63 "primary predictions" with a success rate by Jon's count of 47.6%. That figure does not include three "secondary predictions" that were misses, and it does include five successful predictions that were not given to the whole group. However, full statistics were not made available by the group manager, so these figures may not be accurate.

	Hit rate:			
1ARV grp	59.3%	37 outcomes (Foxtrot onward)		
Wager Team	47.6%	69 outcomes		
ARV Club	63.9%	72 outcomes		
1ARV all groups	68.0%	118 outcomes		

Figure 14.7 – Hit Rate Table

On Oct. 1, 2012, Jon wrote to Marty and 1ARV participants as follows:

1. One possible way to move from c. 60-67 percent accuracy toward 80% would be
    a. Have just one unitary target for an outcome
    b. Vary the target pool from time to time

The 1ARV method has accomplished a lot already – it's the longest group effort at this high a percentage that I am aware of: currently 64% with c. 118 outcomes. Way to go Marty and all of us! I'm not suggesting we move away from the present 1ARV method we are all part of. Maybe a new group could attempt this at some time – but if we want to avoid displacement, this seems the logical way to go. (We did use such a method in Aurora and other work and had some pretty good success with it. E.g. in 2006 we had a 60% success rate in one trial with three viewers doing between c. 35 and 60 outcomes.)

By Marty's description, as I understand it, 1ARV consists of two separate unitary ARV assays/trials/games. I see the two possible outcomes (Over/Under) and the two feedback photosites as semi-independent rather than separate (not sure if Marty would agree). As we saw in Tom's recent example, we can get unfavorable displacement from one photosite to the other in the 1ARV methodology. (We can also get useful bleed-through – and we have so far gotten a lot more of this than unfavorable displacement.)

As mentioned above, to eliminate the possibility of displacement, one could have just one target. How would that work?

One Target number, one outcome (Cowboys win game Oct.1), one feedback photosite. If the session is scored high enough to be wagerable, you bet. If not, you don't bet. That's it. (You don't bet the "other side" – we tried that; it does not work.) If the Cowboys win, the viewer is shown the feedback photosite. If the Cowboys lose, the viewer gets a wildcard. This could be set up for one judge and one viewer. Or it could be set up for multiple judges and/or multiple viewers, but scoring would best be more convergent/standardized by the judges before doing this. Obviously, scoring is key here. (It's possible the Targ levels could be made less open to interpretation by glosses on it or by revising it.)

Target pools: Each target pool I've ever taken part in or constructed has certain characteristics – those consciously and subconsciously utilized by the person who builds the target pool. The viewer gets a feel for that as you do more and more targets. I suggest that varying the target pool from time to time, perhaps by having different people build them, along with discussion about what may or may not go into them, could help keep the viewers' subs fresh and ready to go. E.g. more items with emotional content in one pool, less so in another.

The 1ARV experiment had enough stability and momentum that groups entered their second year on Oct. 11, 2012.

On Nov. 12, 2012, Marty wrote that he was broadening the pre-selected group of potential Targets from just PhotoSites.

PhotoSites will be mixed in with any other target that I can think which I believe you can view with your RV Senses and which will have FB. This is based partly on our quite successful example of "Abraham Lincoln" as a Target at Friday's webinar… just the words were enough to solicit good wagerable RV sessions. I also want you to go into your RV Sessions with no frontloading at all – zero, zip frontloading. The only frontloading is that you will receive your Target and you will do a FB Session. Your tasking and intention: Describe/Sketch My {coord/ TRN/tag#} FeedBack Target.

Applied Precognition Project is founded

On Nov. 21, 2012, Marty announced the formation of the APP:

APP is a brand new Nevada LLC founded by Tom Atwater, Chris Georges and me. As the three 1ARV group admin/facilitators, we feel it is the right time to more aggressively increase public awareness concerning the reality of: "precognition can be successfully applied" and "consciousness is the fundamental." We are just now formulating our plans for our first workshop/webinar event.

APP will be a data-collecting umbrella organization for as many public precognition groups we can form or attract. We believe a grass-roots approach is best for our part in the consciousness paradigm shift. Working publicly, maintaining a database, and exchanging ideas will assist all of us. This is a long-term project, which will automatically include the data from all the existing 1ARV groups.

Here is APP's mission statement: "The Applied Precognition Project's mission is to publicly explore and apply logical and intuitive ways to predict future event outcomes, enabling participants to evolve personally while contributing to the elevation of global consciousness."

Keep Precoging and Enjoying the Journey,
    Marty, Tom and Chris

Around Dec. 10, 2012, Nancy Smith wrote about her group's (Sublime) development of a 1ARV modified protocol, which achieved seven hits in a row.

It's very simple. And I think because there are so few people looking at the same photo, IT somehow keeps it cleaner, only a guess on that.

Here's how we did it today:
I did all the AJing last night, like normal. Then made the prediction decision – but didn't send it out. Then I invited Chris[377] to take a look at his WC photo. He pulled his WC and matched it to the coordinate that he thought it looked most like. For fun he then had his daughter take a look and she confirmed that it was indeed a match to the 58425 coordinate. The WC preview is not assigned a CR, it's quick, simple and fun. Chris tells me about this match, then I tell him 58425 is "Under." So the WC is "Under" by his preview, and we are very happy. That supports and confirms my group CR prediction of "Over." This creates redundancy. I send that "Over" Prediction out to the group. I don't really need to see Chris's WC photo at all, however we both agree that more eyes and excitement gives it more importance to our subs.

At the end of 2012, Marty provided a summary of the data through Dec. 31, 2012. APP had grown to 50 members in nine groups.

Summary of Data							
	Hits	Sidekicks (misses)	Passes (ignored)	Hit rate	Prediction	Protocol	Group Manager
1ARV	18	14	15	58%	Sports	Basic 1ARV	Marty
Oct2004	16	13	15	55%	Financials	Basic 1ARV	Marty
1ARV2011Dec	18	9	16	67%	Sports	Basic 1ARV	Marty

Web/misc.	19	8	8	70%	Misc	Basic 1ARV	Marty
MultAjers	3	2	2	60%	Sports	1ARV	Tom Alexis
1ARV2012Feb	4	4	14	50%	Financial, QQQ	Push Forward 1ARV	Marty, Russ
Pegasus	13	6	3	68%	Sports	Basic 1ARV	Marty
Sublime	10	6	9	63%	Sports	Wildcard Preview 1ARV	Marty
Sep2012	3	2	3	60%	Sports	Basic 1ARV	Marty
Totals	**104**	**64**	**85**	**62%**			
Ptail	0.0013		odds vs chance 793 to 1				

Figure 14.8 – Summary of Data

There were three significant protocol modifications to the basic 1ARV approach:

- Multiple judges, including a "Dung Beetle" Analysis/Judging approach
- "WildCard Preview" with four hits in a row at the end of 2012
- "WildCard Push Forward" tested in 2012 and formally starting Jan. 4, 2013, with the QQQs.

APP kept doing 1ARV through early 2014. On Nov. 7, 2013, Marty switched to a new format, *Winning Wildcard (WWC).*

There is only one Target, (a Winning WildCard PhotoSite) provided by the tasker (currently me). The WWC Target is Associated with the Winning Side of a binary Outcome Event.

There is a second Target provided by the Viewer <u>after</u> the Outcome is known.

There are two coordinates/Transcripts generated by the Viewer (that is inherited from 1ARV); however, it is up to the Viewer to Analyze/Judge the 1 WildCard PhotoSite versus the 2 RV Transcripts <u>before</u> the Outcome is known (that is inherited from WCP).

On March 17, 2014, Marty announced that *viewers could now access their own APP stats online.*

Winning Entanglements (WE) is a protocol designed to permit you to improve your long-term precognitive Hit Rate above the 65% level.

That is a pretty bold and controversial statement that can be tested, with your assistance. We now have the ability to show you your Personal Hit Rate history in a chart and table. Also, your comments from previous WE predictions are available to you. These basic data represent a type of ongoing biofeedback which you can use both intellectually and intuitively to improve your precog Hit Rate. You are the one who does the RVing, Judging, and FeedBack; a perfect situation for improving communication with your submerged consciousness using intentions and actions.

SUARV 2016–2020

To close out this chapter we return to SUARV trials that were conducted a decade after the initial ones in 2006 and 2007. We quote from Jon's article in *Eight Martinis*, "Putting SUARV to the Test: Remote Viewing Future Emotions":[378]

Box. Contained space. Wall, ugly yellow paper. Feels like there is something on the walls & I want to touch the walls. I feel like I am in a casino. Single life form. Male. Casino. Feels like he is jumping off the chair and pumping his fist, saying "Yeah, yeah, yeah." Papers fall out of his hands. Celebrating. Chubby. Pot belly. Middle age. Bar. Red hat. Bird on hat, Cardinals or something. Amateur bettor or maybe just drunk. Dominant emotion: Yeah! Yeah! Yeah!

This is the majority of the data produced by a viewer in an Associative Remote Viewing task using a method I call Strict Unitary ARV (SUARV). The objective was:

Target (8-digit number)
Cue: Focus on the person in Las Vegas who is the ideal bettor for Elisa Lagana to view to know how this bettor feels about his or her heartfelt bet on the Oakland Athletics to win their Major League Baseball game on Saturday, July 23, 2016. Focus on this bettor after the game at the peak moment of their feeling about how their bet turned out.

The viewer is asked to peer into the future and sense the emotions of an "ideal bettor" who has a strong interest in the outcome of a game, in this case the A's Major League Baseball game on July 23. You might wonder – Is this really doable?

Yes, it is – the viewer was in fact able to identify not only the emotions which indicated the A's would win – they did – but also the kind of person who had them – an anonymous bettor. It takes a very good viewer, but it is definitely doable.

In this form of ARV, there is only one objective – the emotions of a person or persons in the future – and the viewer knows that going in; she is frontloaded to that extent. But she does not know if it is a player, fan, announcer, coach, vendor, bettor, security person, bat boy, trainer, owner, etc., nor whether the person is at the game, watching on TV somewhere, in a casino or elsewhere. In operational remote viewing, as opposed to the lab, some frontloading may be used (and in some cases is necessary).

For each event in these series, the viewer was tasked with not one but three different individuals who had an interest in the result of the game. For this game, the other two people were an (anonymous) avid, emotional female fan of the A's and the final pitcher for the Tampa Bay Rays. The viewer sensed that one of these two was in fact a happy female fan in the crowd (totally correct) and that the other person was a very disappointed male close to the Tampa Bay players (the viewer's impression was consistent with this). In other words, the viewer's data about the emotional state of three separate individuals, two wishing the A's to win and one wishing Tampa Bay to win, was correct. The impressions lined up and clearly pointed to an A's win. The bet was made and the A's did win.

Binary ARV

Ordinarily, in the great majority of experiments and informal efforts, binary ARV is utilized – there are two targets, most often two photos.[379] One photo is associated by the tasker with one outcome (Dodgers win) and the other with the other possible outcome (Nationals win). A judge (or the viewer herself) decides which photo the session best matches and assigns a score to it. If the score is high enough, then a bet is placed on that outcome. In the Applied Precognition Project (which the SUARV participants take part in), the minimum score for a bet is 3.5 on the SRI/Targ scale of 0 to 7. This binary ARV method is the basic one used in almost all of the games and financial trades in the APP.

This method of using two targets (or even more targets) can produce satisfactory results. In fact the most significant financial gains with ARV have used such a multi-target method. Further, APP's statistics show that overall, using a great variety of viewers and group managers, the success rate is at 57% since 2003, at odds of many billions to one. Further, the APP Institute was recently formed for viewers who have compiled exceptional track records. Marty Rosenblatt has released figures indicating that four of these viewers have hit rates at 75% or above and 7 of 8 such viewers are making money. These are exceptional results.

Strict Unitary ARV

With *some* viewers achieving such excellent results, why SUARV? Well, for one thing, very few viewers have achieved a high level of success over a substantial period of time. Is 57% really the best we can do for large numbers of viewers? That is 2-4% above what professional sportsbooks and touts can achieve over time, but is that an upper limit? Many have initial success with standard binary ARV, but, apparently without exception, a decline effect sets in. Repeated displacement is probably the main factor leading to reduced participation or simply dropping out.

Why not try ARV with just one target and see how we do? I have long had the feeling that unitary ARV will eventually be found to produce better rates than binary ARV. The main reason is that in binary ARV you posit two objectives (usually photos) and the viewer may get data from both photos or produce a session which better describes the wrong photo (the one that corresponds to the result which does not occur).

If you have only one objective, as in SUARV, there is no second target to displace to. Of course, it is possible, if using three targets for each event as in the above SUARV series, that the viewer could displace to another person, rather than the one designated by the cue. In the limited series that have been done so far, though, this does not appear to have happened.

Further, in a 2006 SUARV series, using a single target and focusing on future emotions, we did achieve satisfactory results and in fact when all three viewers agreed, we had 9 hits in a row – which is pretty rare in ARV. A good deal of money was also made.

The current SUARV series

Now, 10 years later, I wanted to try another single-target series. I wanted to implement lessons from the 2006 series and also follow some of the recommendations of leading researcher Ed May and #1 viewer Joe McMoneagle. They have achieved remarkable success, albeit using custom software – with two photos but with both the viewer and the tasker seeing only one photo. Their main lesson could be summarized as "Eliminate all feedback loops!" That is, cut down or eliminate all arrangements in which knowledge of the target, the game, session data, betting, etc. can circulate among the participants.

Hence, the guidelines for this SUARV series included having a small team with well-defined and limited roles, no cross-talk during the series, a designated bettor, and other measures and safeguards. Three series were conducted. In the first two, Elisa Lagana was the viewer, I was the tasker, and Alexis Poquiz was the bettor. For the third, the viewer and tasker remained the same, while the bettor was (a third person).

I decided to go with one viewer, rather than three, which we had done in the 2006 series, since there were indications Elisa was exceptionally good and Ed and Joe's efforts used only one viewer.

I also decided to have three targets (anonymous individuals) per game as in 2006. The one time we had just one target in 2006, we had a significant miss – one that ended the streak of 9 hits in a row. Another change was that in 2006 Don and Roma did self-judging. For these SUARV series, the viewer did not judge; the tasker (me) did the evaluation and made the pick. On a couple of occasions I did solicit the opinion of the viewer since I was not sure.

Results of Three SUARV Series						
Team	Series	Hit	Miss	Pass	%Hit	%Pass
EJA	Trial	1		1		
EJA	Series 1	5	1	1	83.3	14.3
EJA	Series 2	1	2	3	33.3	50.0
EJR	Series 3	4	2	1	66.7	14.3
Total		10	5	5	66.7	25.0

Figure 14.9 – Results of Three SUARV Series

The hit rate was 66.7% for predictions made – which in the world of ARV is quite good. Elisa's hit rate exceeded that of the two viewers using emotions in the 2006 series (those rates were 28/50= 56% and 36/61 = 59%). Her hit rate was the same as the third participant in 2006, a talented psychic from Australia who used team colors, logos, etc. (Good viewers all!)

To sum up very briefly, unitary ARV aka SUARV has shown that it can get results equivalent to those of standard binary ARV. When using emotions there is no evidence, so far, of displacement. Displacement would consist of "displacing" to the "wrong" emotions; that is positive emotions attributed to those on the losing side and negative emotions to those on the winning side. It can't be said that this doesn't occur, but the range of success (around 65%) overall suggests it is not doing so. Unitary ARV using one photo is also possible and early results by the CryptoViewing team of remote viewers are quite positive. Very few trials of unitary ARV have been conducted compared with the vast number for binary ARV and we hope the results shown here encourage more exploration of this alternative form of doing ARV. *(End quote from Jon's article.)*

CHAPTER 15

Using ARV for Financial Trading

In this chapter, we take a look at the financial predictions of Greg Kolodziejzyk, a two-person German team (the Müllers) and Italian remote viewer Nicola Laurino. We also explore the use of ARV in financial ventures in other chapters, such as the decades-long work by the Applied Precognition Project, Project Firefly and two chapters on cryptocurrency.

Greg Kolodziejzyk is a phenomenon, and not just because of his last name. You are probably wondering, how do you pronounce his name? As he said when presenting at an APP Conference: It's "Grayyyyg." His other name sounds like "kolochessik." Greg was a software millionaire at 30 and had an innovative and successful career prior to venturing into ARV. One of his early enterprises was Image Club Graphics during the digital desktop era when CD-ROM discs were state of the art. The company sold stock photography, typefaces and digital art to publishers. Greg sold this company in 1994 to another company that was being acquired by Adobe Systems, and this allowed him to "retire" that same year.

We put "retire" in quotes because Greg has been abundantly active since his "retirement." To improve his health and lose weight, in 2000, Greg began competing in running, swimming and biking events and set several long-distance endurance records. In 2008, he launched Human Power and worked as a motivational speaker. In 2014, he developed and sold an automated trading program called AlgoLab, which had been quite successful. In 2019, he became a Commodity Trading Advisor and proprietor of AlgoLab Capital Management, which manages trading accounts using 24-7 automated software for clients with a minimum entry amount of $100K. In his famous ARV experiment,

Greg drew on both the stock photo industry that he helped create and his knowledge as a trader.

Greg is a pioneer in Associative Remote Viewing. He created a stir in the remote viewing world when he reported that from 1998 to 2011 he conducted 5,677 ARV trials with a profit of $146,587. This is the longest and most successful solo ARV effort to date. Greg documented the results in the *Journal of Parapsychology.*[380] .

Let's turn to the ARV process Greg used in the long, arduous run that resulted in profits of $146K. He describes his process as follows:

Step 1: He sits back in a chair, relaxes, closes his eyes and imagines himself in the future looking at his feedback picture at the exact time specified in the tasking. Greg does many trials for each event and times the feedback, e.g., every two minutes he will get feedback on one of his trials. He lets his mind fill with random thoughts. The more surprising they are, the better.

Step 2: The AlgoLab software chooses two random photos from hundreds of thousands of photos on his computer. The software also randomly selects a trade out of a basket of a "dozen different futures and commodities such as Corn, Wheat, S&P 500, Bonds, Oil, Gold, and a few currencies." No stop is used to exit a trade. Nor are there any moving averages, target goals, charts or any other indicators.

In advance, Greg randomly selects a period during which to hold the trade. For example, a trade is automatically entered on Monday at the opening of trading and exited on Friday at the close of trading. The software performs this function and randomly associates one photo with Buy and the other with Sell. Greg does not know which market has been selected nor the trade within the market. He does not read the financial news – he doesn't want any conscious knowledge of the field to enter his mind.

Step 3: Greg self-judges and does not know the associations of the two photographs while judging. He does very brief trials so there is only a little data in each session. He pays particular attention to any surprising or unexpected ideas he got in Step 1. He chooses one of the two photos as the best match for his session.

Step 4: The trade takes place based on the Buy or Sell. The automated software is connected to Greg's broker so it can access the results. The software records the result and selects one of the two photos. When Greg logs in, the "winning" photo will be shown to him. The other photo is not shown to him. That photo pair will never be used again.

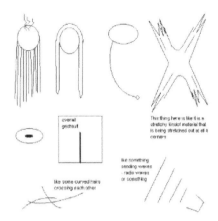

Figure 15.1 (left)– Octopus in the Water and Figure 15.2 (right) – Sketches of Octopus

Greg has explained on his website and in interviews how he thinks this works.

Your life consists of an infinite number of random number generators – from the quantum particles bouncing around in your mind, to decisions and choices that you make. That universe of randomness is available for influence by your thoughts and intentions at every moment…I hope I have demonstrated (this) to you using my Associative Remote Viewing experiment, which creates a signal in the past which allows us to take action to create a future that we desire.

YOU create your universe. Earl Nightingale said – You are and become what you think about. But I think it's more like – Your world, your universe, you and everything in your life IS and will BECOME what you think about.[381]

This standard, very short ARV trial was repeated a number of times for the same future event. Greg was both the viewer and the judge. The average outcome was calculated for all the predicted outcomes and a prediction was made if there was a strong consensus using a modification of the majority vote method Targ and Harary used.

After all trials had been remote viewed and analyzed, the ARV application calculated the sum of all the confidence scores for outcomes A and B. The outcome with the higher sum was the final prediction for the project question. If actual capital was to be risked on a trade, then the difference between these two sums was a consideration, with a larger differential resulting in a larger effect size.

Greg K's statistics

Let's review the statistics from Greg's 13-year experiment. The capital in his trading account averaged around $50,000. He conducted 5,677 trials in all. He undertook so many trials because he wanted a sufficient number for each prediction to convince himself that one outcome was more probable than the other. Of those trials, 52.65% correctly predicted the outcome of their questions, with a Z score of 4 (very high). That equates to a probability of only 1 in 31,547 that the result was due to chance. In other words, he would have had to conduct 31,547 trials to be sure to get by chance one result as good as the results he actually got.

In these 5,677 trials, there were 285 project questions, most of which were to predict the outcome of a futures market. Profits of $146,587 resulted from 60.3% of the trades being profitable, for a Z score of 3.5. He would have had to repeat the entire 13-year experiment 4,299 times in order to find one result of 60% successful trades by chance.

Greg also used a filter – the more trials, the more confidence he had in a prediction. A consensus of six trials would lead to 70% expectation of a hit. If it was a consensus from 20 trials, it had a 75% expectation of success. If 50 trials, 90% expectation – that might take a couple of weeks to do.

> By increasing the number of trials in a project question and giving more weight to higher subjective confidence scores reflecting the quality of the match between the remote viewing and one of the two target images, the success rate increased to above 70%. One hundred eighty-one project questions resulted in actual futures trades where capital was risked. Of these, 60% of the trades were profitable, amounting to approximately $146,587.30.[382]

Greg hoped ARV could have other uses to benefit society, such as predicting hurricane strike zones, finding alternative energy sources and making medical breakthroughs. Although Greg succeeded in showing ARV could be used to make a great deal of money over 13 years, he no longer uses ARV in his automated trading business. He says it takes too much time to do the preparatory meditation and the many sessions. His algorithmic software does a sufficient job using stops. He sums up his current approach this way:

> We use a simple trailing stop. We found the best way to exit trades. There are many different ways of doing it. That's the best one. We get out of losing trades very

very quickly. We try to hold on to winning trades as long as the trend continues. Generally, we have about 42% winning trades. So most of the trades are actually losing trades, but the average win to loss ratio is 2 to 1. So you can see why we are profitable.[383]

Figure 15.3 (left) – Smokestacks and Figure 15.4 (right) – Greg's Sketches

A few thoughts on Greg's project

Greg attributes his success to combining his knowledge of the stock market with the use of logic paired with the intuitive practice of ARV.[384] This differs from the majority of ARV projects that have relied exclusively on remote viewing to make predictions. Greg had access to specific platforms and tools, paired with his expert knowledge in these areas, so we can't say that another remote viewer who did not have the same external and internal resources would be equally successful.

While Greg has fluctuated in how much weight he gave to the ARV sessions, his strategy included making wagers when both his own ARV sessions had stronger scores towards one option AND when logic pointed to a particular outcome. This, along with being a single operator, made this project unique at the time. Also, because he was not an experienced remote viewer and he was conducting so many trials, he has explained his own remote viewing sessions were often conducted very quickly and were not that detailed. In conference presentations, Greg reported he went for quantity rather than quality in the sessions. While he did not do any kind of formal analysis, he felt photos with objects against white backgrounds devoid of information were easier to view than others.

Further, his photo options were chosen with the help of a computer, which meant some of their pairings lacked optimal orthogonality (strong difference between photos). Still, these limitations did not seem to harm his overall success.

It is hard to say what his results would have been if he had not used RV but only an informed logical method or if he had gone for quality over quantity in his RV sessions and the overall project. What can be said is he put in a tremendous amount of work on this project for a very long time and it netted a significant amount of money. In our opinion, he is representative of the caliber of self-motivated individuals who have done a fantastic job integrating and combining remote viewing activities with the rest of their life. If anyone thought using psi was just some kind of woo-woo, Greg's work should dispel that impression.

(On the website for this book, we have provided Greg's links for those who want to try to replicate his approach: http://www.arvbook.com.)

German Stock Index Project (DAX): Predicting the stock market: An Associative Remote Viewing study

In 2017, two German researchers, Maximilian and Laura Müller, won the IRVA Warcollier Prize for an ARV-related proposal. The $3,000 prize helped finance their stock market wagering. The main research objectives were to determine the hit rate for predictions of the German stock index DAX (*Deutscher Aktienindex*) using Associative Remote Viewing, to test whether feedback is necessary for ARV predictions and to explore factors that might influence the quality of the viewers' perceptions in ARV sessions. In addition, they wanted to "identify a design for subsequent studies in the sense of a proof of principle study" (p. 2). The team provided the following table from their literature review:

"Overview of relevant ARV studies which tried to predict a financial market. Some studies were excluded (e. g., Smith C. Exp. B (2009) and Kolodziejzyk, 2015) because they used a computer for target selection and/or the association process."

Study	Correct Predictions	Hit Rate	p-value (binomial distribution
Puthoff (1984)	21 out of 30	70%	$p = 0.01$**
Harary & Targ (1984)	9 out of 9	100%	$p = 0.0019$**
Targ, et al. (1995)	6 out of 7	86%	$p = 0.054$
C. Smith (2009) Exp A	14 out of 17	82%	$p = 0.005$*
C. Smith (2009) Exp C	8 out of 11	72%	$p = 0.08$
Smith et al (2014)	7 out of 7	100%	$p = 0.0078$**
Total	65 out of 81	80%	$p = 1.39 \times 10^{-8}$ ***

Figure 15.5 – ARV Studies' Correct Predictions

In their initial comparison of the studies, the team noted that:

…because of the different experimental setups and uneven number of trials in the reported studies, it is unclear which factors might influence the hit rate and which hit rate is possible with a specific experimental setup. At first sight, it seems that the studies with less than ten trials[385,386,387] were more successful than the studies with more trials. Statistically, there is a negative correlation between the number of trials and the hit rate (*Spearman's Rho* = -.81, p = 0.13), which means that the more ARV trials are conducted in a study, the lower the hit rate. However, this correlation is not significant because of the small number (n = 6) of studies. There are hardly any studies which conducted a reasonable amount of qualitative trials to determine a baseline hit rate for predictions in the long term (p. 328).

One of the variables they wanted to test was the importance of feedback. They postulated that viewers' intention was important, rather than feedback. To test this, they split their trials in half whereby each viewer would receive feedback for only half the trials. They used a single remote viewer per trial for their predetermined 50 trials instead of groups of remote viewers per trial. Their project was unique in that the entire trials were completed within an hour rather than over the course of 24 hours or more as has been the case in other projects.

Another seldom-found aspect of their design was that rather than viewers working solo, as has typically been the case in ARV, the study used monitors in a way similar to their use at SRI. Another uncommon feature was that the monitors were aware of both photo options (each paired with a possible outcome) while the viewers were working. Although the viewers were mostly new to these tasks, they had been trained in the Controlled Remote Viewing methodology as taught by former military remote viewer Paul Smith, using Stages 1 – 4. Scoring was not done using a scale but merely with instructions to go with the first impression of the best match for a photo.

The hit-miss ratio was quite impressive; in fact, it is probably the highest ever achieved for group predictions with this many trials. The researchers indicated that "the ARV method used in our study predicted the near future of a stock index above chance level" with 38 correct of 48 predictions amounting to

a highly significant result (p = 2.3 × 10-5, binomial distribution, B48 (1/2); z = 3.897), reflecting the hit ratio of 79.16%. The z-score divided through the square root of n = 48 trials corresponds to an effect size (ES) of 0.56. In contrast, a true random number generator (RNG; random.org) was not able to predict the stock index significantly (24 of 48, binomial distribution, B48 (1/2), is p = 0.11; z = 0) (p. 335).

However, their wagering did not fare as well, earning them only (237€), which was not significantly higher than the profit the RNG would have produced. The average profit per trial for the ARV predictions was 4.93€ and for the RNG predictions 1.60€ ($t = 0.722$, $p = 0.472$).

The authors of the study explain:

We discovered that the average DAX point difference for the hits ($n = 38$) is 13.89 points and for the misses ($n = 10$) 29.1 points. This difference is significant ($t = 2.603$, $p = 0.023$) which means that we lost more money for the 10 wrong predictions than we gained for 38 correct predictions. Financially spoken, we lost on average 29.10€ for a wrong prediction and gained on average 13.89€ for a correct prediction, which is a highly significant difference ($t = -7.361$, $p < 0.001$) (p. 336).

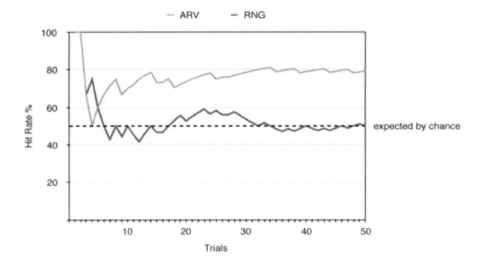

Fig. 4: Hit rate with ARV over 50 trials in contrast to predictions with a random number generator (expected by chance). The two invalid trials (trial 16 and trial 27) are shown in this illustration, but have no influence on the overall hit rate.

Figure 15.6 ARV RNG Graph

Graphic is from M. Müller, L. Müller, and M. Wittmann.[388]

They concluded it was not necessary for a viewer to receive feedback in the near future in order to do well, at least for this particular protocol, which had a short turnaround. They wrote:

...24 out of 48 trials were sessions with a feedback for the viewers, the other half was without feedback. Both conditions were independently significant: In the feedback condition the viewers succeeded 20 times and failed only 4 times ($\chi2 = 10.667, p = 0.001$). In the non-feedback condition, the viewers succeeded 18 times and failed only 6 times ($\chi2 = 6.000, p = 0.014$). A Chi-Square test for the frequency of hits and misses shows that there is no significant difference between both conditions ($\chi2 = 0.505, p = 0.477$).

One thing we wanted to know was if the viewers were aware in advance whether they would receive feedback. Max Müller advised that, to the best of his recollection, they were not told in advance.

Another interesting finding has to do with clarity:

The viewer's perception is clearer and less ambivalent when the stock index also has a clearer outcome at the end of the prediction period. Therefore, the quality of Anomalous Cognition as the underlying construct depends on the prediction object (DAX) irrespective of whether it is a hit or not (p. 337).

They theorized this could be because:

It can happen that a future event predicted at present changes over the course of the delay due to unforeseen influences. This could be an indicator for a phenomenon called "retro-causality" because the effect (alteration of viewer perception) precedes its cause (volatility of the future DAX course) in time. If the viewer's perception was altered by the stock index in the future, it would support the assumption that the actual future event is to some extent variable and not completely clear at the time of the session...A possible explanation for this finding is the consideration of a probabilistic future which could be the most determining factor for future predictions.

Following this thought, during an ARV session the viewer would not describe the actual outcome in the future (through the associated target stimuli), but rather the most probable outcome from his position in time at the time of the session. Consequently, the actual outcome in the future can change over time and a prediction which indicates the most probable outcome at only one point in time, can become a wrong prediction when probabilities change after the session.

For instance, at the time of the session the viewer describes the picture which is associated with a rising stock market. After the session an event happens (e.g., an influential person impulsively releases economic information) which was not clear at the time of the session but influences the volatile stock market to such an extent that the stock market has a falling course in the prediction period. The prediction would become a miss and it would not be possible to determine whether the cause for the miss was the viewer's performance or some probabilistic event that changed the course of the market after the session.

If these assumptions were true and the future is indeed probabilistic and only partially predictable, this should be taken into consideration regarding achievable hit rates with ARV. An opportunity to test this hypothesis is a comparison experiment in which the hit rate of ARV for targets existing at the present moment is identified. All other variables in the ARV process (target selection, data collection method, judging, etc.) should be kept constant to ensure that the observed error variance (misses) can definitely not be explained by the probabilistic future. The new hypothesis would be that the hit rate of ARV with binary outcomes with targets existing at the present moment is significantly higher than the hit rate of ARV with binary outcomes in the future. If the results were positive according to this hypothesis, the probabilistic future would be an additional factor for predictions with ARV leading to more misses. As we can show here, a relatively high hit rate is nevertheless achievable (p. 343).

Because of this they suggested their project be replicated:

The new hypothesis would be that the hit rate of ARV with binary outcomes with targets existing at the present moment is significantly higher than the hit rate of ARV with binary outcomes in the future. If the results were positive according to this hypothesis, the probabilistic future would be an additional factor for predictions with ARV leading to more misses. As we can show here, a relatively high hit rate is nevertheless achievable (p. 343).

They also suggested future studies should continue to attempt to predict the stock market on an hourly basis, noting that while

this is even more difficult by conventional means because of the high volatility of the market across a given day. Generally, if the ARV method is properly conducted, it has the potential to become a probed and tested paradigm for the research field and can convincingly prove that Psi effects are robust and replicable.

The following considerations would guide future studies:

The overall ARV hit rate for future predictions is primarily influenced by target selection, data collection and judging. These factors are mainly controllable, and it would be simple to conduct a replicable ARV experiment, if the necessary experience and human resources were available...The most fundamental stage of the ARV process is the target stimuli selection. A good selection ensures a simplified judging process whereas a poor selection complicates the judging especially when the viewer performance is poor. If the target stimuli are not selected on the basis of maximal distinguishability, it increases the probability that the judge makes a wrong prediction decision because of the overall ambivalence of his associations...that "target selection, data collection method, judging, etc., should be kept constant to ensure that the observed error variance (misses) can definitely not be explained by the probabilistic future" (pp. 339-340).

In 2020 we contacted the authors after their study was published in a German journal to ask a few follow-up questions. First, we wanted to know how individual viewers performed and whether they noticed any learning curves. They replied that some participants who did repeated trials showed a substantial learning curve during the project although this progress was not defined in their write up. Maximilian Müller explained, "We did not formally evaluate any individual stats. But I can share your observation that people improve in RV in general through practice and training – so it was in our study."

We also asked them, "Given how informed and well thought out this was, had you and your wife already been involved in remote viewing or ARV for a while?" Max responded:

Yes, my wife and I practiced a lot for ourselves with RV practice targets and ARV sessions on the stock market/sporting events. During the conceptualization of the study, we thought about the best possible conditions for an ARV study (achieving a preferably high hit rate + testing the feedback hypothesis). Of course, relying on the state of knowledge at that time. Moreover, I often think that the information about how to design a study that will achieve a specific outcome just come(s) to me intuitively.

Financial experiments by Nicola Laurino

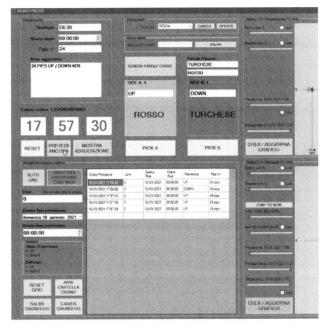

Figure 15.7 – Financial Experiments by Nicola Laurino

Nicola Laurino provided the following information about himself and text and graphics about his experiments:

Nicola Laurino was born in Salerno, Italy, in 1983 and is now an architect who has developed a strong interest in paranormal phenomena and esoteric physics. He has been working since 2017 on several precognition projects which aimed at finding an effective predictive model for financial markets and shedding light on the inner nature of space and time.

In the history of Associative Remote Viewing, we can find several experiments where it has been shown that it is possible to successfully forecast financial market trends based on one's precognition skills. However, the majority of these trials were limited to a short time span or a small number of predictions, while it's very difficult to find experiments which were consistently successful in the long term.

The most interesting research paper describing long term ARV attempts at precognition applied to financial markets is probably Greg Kolodziejzyk's 13-year associative remote viewing experiment results. My research in this field aims at finding new approaches to get a higher reliability for precognition when applied to financial markets.

1 – Classic Financial ARV

This precognition experiment took place for approximately 3 years, from July 2017 to June 2020, making use of classic Associative Remote Viewing to predict financial market trends. Predictions were made about several currency pairs in the Forex market and stock market indices. Normally, one or two sessions was carried out daily. A typical event description would be this: "Will the EURUSD currency pair, from 9:00 CET to 11:00 CET, go UP or DOWN?"

Methods: All the steps in the ARV sessions were managed by me with the help of ARV software. The software used for the majority of the predictions was ARV Studio, developed by Igor Grgić; however, for the last few sessions different software developed by me (called Precog) was employed.

In both cases a typical ARV session was composed of the following phases:

Tasking: A new task was created with the help of the software making an association between the two sides (possible outcomes of the event) and two raster images (grid of tiny pixels). The association was stored without the user being aware of it.

Viewing: This is the remote viewing proper phase, where I tried to describe and sketch the correct image associated with the event outcome that I would see during the feedback phase after the conclusion of the event. The remote viewing part was preceded by a brief focusing / relaxation practice which could last a few minutes, using several breathing techniques to clear the mind and get it ready for ESP perception. The remote viewing phase lasted typically from 10 to 15 minutes and was very similar to a classic RV session as described by David Morehouse in his book *Remote Viewing*,[389] but going only as far as Stage 4 (instead of Stage 6). I started by taking the coordinate of eight digits given by the software and then drawing an ideogram. Four stages were usually sufficient to gather enough information to identify the most probable outcome. During this phase my perceptual focus was always on the feedback image. For most of the sessions the viewing phase was concluded approximately 10 or 15 minutes before the beginning time of the event, leaving just the strictly necessary time for evaluation.

Evaluation: The two images were shown to me by the software and compared with the remote viewing perceptions written down on paper in text and sketches. All similarities with one or the other target image were highlighted with different colors. Particular similarities (the so-called "unique matchable elements") were marked with a circle to indicate higher relevance. After counting all the matchable elements, a rating was given to both images according to Russell Targ's Confidence Rating System (from 0 to 7 points). If the difference in rating was equal to or greater than one point, the higher-rated side was chosen as the predicted outcome; otherwise, it was a pass.

When working in "solo" mode (self-evaluation), the judging phase is a critical one because the viewer is forced to see both target images instead of seeing only the one associated with the actual event outcome (at feedback time). Both the evaluation and feedback phases are located in the future with respect to the remote viewing time, so there could be some interference, meaning that the viewer could receive perceptions from the judging stage and from both pictures, resulting in possible displacement effects. While this certainly happened in my first ARV trials (prior to this experiment), I tend to believe that in most cases I was able to focus on the feedback image.

There are upsides as well working in solo mode. When the evaluation is done by a third person, it is based solely on what was written on paper by the viewer, without any knowledge of the internal processes occurring in his or her mind. But when I was judging my own sessions I could remember details, impressions, and shades that I hadn't been able to write down. Also, the viewer knows themselves and their specific way of having remote perceptions better than anyone else. For instance, I knew that when I wrote down "land/water interface" in a session it could refer to snow, based on my previous experiences.

Feedback: In this phase the software showed me the image associated with the winning side and I did a couple of things in order to make this stage unique in the spacetime continuum and differentiate it from the evaluation phase. I would smell the scent of two spices which I would normally never encounter in everyday life, and make a quick sketch of the feedback image. I always use the same two spices. They are meant to create a unique experience which marks the feedback as different from the judging phase or any other moment of the day. They are similar to what Marty Rosenblatt calls a "beacon."

Results: The experiment concluded with a 55.07% hit rate. The deviation of the hit number from the expected value was 2.88 standard deviations, which can be considered statistically significant. The probability of such a deviation, calculated with binomial distribution, is 1 in every 2,290 attempts. There was also a theoretical gain of 1859 pips (a pip is a unit of measure for prices of financial assets) obtained after 808 predictions.

Protocol Total Attempts										
Protocol	Total attempts	Passes	Actual attempts	Hits	Hit rate	Standard deviation		Critical factor	Probability (binomial distribution)	This happens randomly every ... times:
Financial	902	94	808	445	0,5507425	14,2126704	41	2,8847499	0,043670%	2289,92139
Deterministic	36	8	28	21	0,7500000	2,64575131	7	2,6457513	0,441089%	226,711476
Instant Precog	460	64	396	231	0,5833333	9,94987437	33	3,3166247	0,016185%	6178,42430
Free will (simple choice)	26	2	24	20	0,8333333	2,44948974	8	3,2659863	0,063336%	1578,88349
TOTAL	1424	168	1256	717	0,5708598	17,720045	89	5,0225605	0,000007%	13.788.099
FILTERED	1424	168	1256	717	0,5708598	17,720045	89	5,0225605	0,000007%	13.788.099

Figure 15.8 – Protocol Total Attempts Table

Comments: During the course of the experiment, I found a few areas with room for improvement.

One of the main issues is image pairing. Working with ARV Studio the target images selected from the available library by the pairing algorithm were occasionally quite unbalanced. One image could be much more interesting than the other because of a higher density of information (colors, materials, elements, energy), or just because it was taken by a more highly skilled photographer. Personal preferences could influence perceptions as well. For instance, I tended to do better with elements like cars and other artificial items, while it was difficult for me to perceive animals. All these cases could result in displacement in favor of the most interesting picture. In other cases, there were two images which were just too similar to each other.

I finally solved all those issues by developing the Precog software with the specific purpose of having an optimal image pairing as well as improving some other details. I didn't test the Precog software long enough to see a difference in the hit rate, though, because after a few weeks I discovered a quicker and more effective way to make the same predictions. I called it Instant Precog.

2 – Using the B component of CRV

This method was discovered while analyzing classic ARV sessions. In standard remote viewing there is something called the "B component," belonging to Stage 1, according to David Morehouse's manual. This is a simple perception of the prevailing aspect in the target between natural and manmade. After a number of ARV sessions I had become quite good in spotting this component, at least when the target images were very different from each other. I just felt it. So I thought, If I can get a substantial aspect of the target in a minute, why would I spend 15 minutes to complete a 4-stage session?

I had noticed many times during classic financial ARV experiments that any prediction, even the ones I was very confident about, could be overthrown by a change in the course of the events. Sometimes there was a noticeable and sudden movement in market price opposite to my forecast. At first, I tried to reduce the event duration in order to lower the probability of a "change in the future," but obtained only limited improvement in the hit rate. I was looking for a prediction method that could be performed quickly and right before the event, instead of taking half an hour to get a forecast. Therefore, taking inspiration from the B component experience, I devised a very simple associative method for precognition.

Instant Precog: Instant Precog is based on a simple association between two random words and the two event outcomes. I developed a new algorithm, integrated in the Precog app, which makes an association between two random color names (taken from a pool of 20) and the event sides. For approximately 5 minutes before the event, I repeatedly try to feel which color is associated with the winning side of the event. When I feel confident about my perception, I hit the button for the chosen side so that the software keeps count of the choices, and I go on with a new couple of randomly selected color names. The total number of these choices per session has been on average 21. If just before the event (around 30 seconds before) one side has been chosen at least 62% of the time, that side is recorded as the prediction side; otherwise, it is considered a pass. Initially the consistency threshold was set to 66%, but then I adjusted it according to the hit-rate statistics.

Since this precognition method requires only 5 minutes for each prediction, it was possible to carry out several predictions per day (in some cases more than ten). Therefore the event duration was lowered to approximately 25 minutes, while my use of classic financial ARV averaged 293 minutes.

Results: This is an ongoing experiment. As of today (02/01/2021), after 412 attempts the hit rate is 58.01%. The deviation of the hit number from the expected value is 3.25 standard deviations. The probability of such a deviation, calculated with binomial distribution, is once every 5,080 attempts. The theoretical financial gain is 1,278 pips.

Total Attempts									
Total attempts	Actual attempts			% actual attempts	Critical factor	Gain in pips per attempt	Total gain in pips	Total loss in pips	Net gain in pips
540	396			73,333%	**3,3166247**	3,056389	4429,791	-3219,46	1210,330
Hits	Standard deviation	Deviation	Happens randomly every ... times:	Hit rate	Probability (binomial coefficient)	Gain/Loss ratio	Average gain in pips	Average loss in pips	Pips +/- ratio
231	9,94987437	33	6178,4243	58,333%	0,01619%	1,375941	19,2599	-19,6308	0,981106

Figure 15.9 – Total Attempts – Actual Attempts Graphic

Comments: Instant Precog presents several advantages compared with standard ARV. The time necessary to obtain a prediction is approximately five minutes, instead of half an hour. The mental effort required is lower; consequently, more than 10 predictions can be done in a day, instead of two. The association is made with simple words which are more neutral than images, so there can be no displacement effect due to personal preferences towards one target image.[390]

When we asked Nicola whether there might be a decline effect over time, because only one method is used, he replied:

I don't think that using the same method is a factor of decline. The basic rule is that one should always be motivated. As Ingo Swann said, PSI activity should always have a practical application in real life. For instance when I worked on the deterministic project I described to you in an email, I lost interest after it was clear to me that my hit rate was much higher than the Financial protocol. I had the answer I was looking for, so I started to have sessions with poor viewing that resulted in a pass. For this reason, shortly after, I decided to end the Deterministic protocol.

In conclusion

The work of the three individuals and teams displayed in this chapter shows the inventiveness, scope and depth in applying Associative Remote Viewing to financial trading. In the next chapter, we take a look at the originality and simplicity of direct drawing of graphs of financial movements.

CHAPTER 16

Direct Drawing of Financial Charts – Chartograms or "Wowsing"

The following projects all have one thing in common: the designs call for the viewers to create a graph that mirrors a chart representing either movement of an individual stock or movement of markets of which a trader is seeking foreknowledge. The viewer's sketch should match the chart on the trading platform after the stock transaction has taken place. This approach is quite different from ARV projects, which pair unrelated photos with potential outcomes. However, in direct viewing of graphs/charts, as in ARV, the viewer can be blind to the trader's activities while trusting the trader has chosen a specific stock and period of time as the target.

To start, the trader or tasker gives the viewer a trial number or other indicator of the task, and the viewer then draws or visualizes a line or lines with the intention of matching the line to the movement of the stock or other instrument. The idea is that this allows the trader to predict the movement (such as up or down) at a pre-designated closing time or alternatively predict fluctuations over a period of time.

The practice of direct graph viewing, while not new, is just coming into popularity among ARV enthusiasts. In fact, when inquiring of a few RV professionals, we learned much of their experience began in the past year (2020). We might be seeing the "100th monkey effect" since it did not appear that people we reached out to had talked to each other about direct drawing.

In the following discussions, we'd like readers to keep two things in mind – that in any psychic task there is the overall set up of the project and there is the way psi is accessed and reported. In this case, some knowledge of both the trading protocol and the ARV protocol is needed by one or more of those involved. We will share the information we have about both of these aspects in reporting on the following projects; however, much information is missing since we don't have access to each person and their role (in particular the traders). We wanted to know the different methods and ways in which the viewers received or expressed the information. For instance, Debra's and Pam Coronado's visual approaches are quite different from those of Daz Smith, Julia Mossbridge and John Vivanco.

The three psi approaches we have identified could be called the clairvoyant approach, the transit line approach (writing, somatic) and dowsing.

Debra's foray into graph reading – A phenomenological account (by Debra)

Recently, I was approached by a new client who was interested in seeing the movement of stock trades over the course of a trading day. He said he was living off an inheritance and had some money to play with. The client was very new to remote viewing and just wanted to see what I would come up with. I was curious about this myself. He offered to pay me in advance, but I didn't know how well I would do so I told him he could make a donation later if we were successful.

I had recently read an article about a project in China in which people attempted to read text through the use of a "visualized" reading screen. The article noted that one subject looked at his screen for 45 minutes on average before the words began to appear. While I tend to visualize a screen (like a TV or movie screen) as my general approach in psi work, I use other approaches as well.

I don't sit there waiting very long for something to happen. So I was interested to see what might happen if I hunkered down and did nothing else except continually refocusing on a blank screen for 45 minutes. (I say refocus, but all attention is about refocusing; the mind simply can't stay static for more than a second or two without having to be redirected.) Usually, people get discouraged by the fifth time they realize they have lost focus again.

In 45 minutes I might have to refocus 1,000 times, just as I'm going to need to inhale and exhale every few seconds. Imagine if you had the silly idea that one breath was supposed to last for an hour and then when it didn't, you beat yourself up about it like people usually do. Anyway, I decided this would be my approach for looking at whatever graph he wanted me to see.

To begin, I went in my bedroom, set a timer and laid down. I chose to do the session in more of an ERV mode – the deeper brain wave pattern would border between being asleep and awake. One final thing I did before focusing on the screen was to ask my subconscious to give me something that would stand out above all the mental noise (in the form of doubts, imagined objects, whatever) that would surely arise. That way, a lot of mental noise didn't matter. I'd somehow know when something important emerged because it would stand out in a different or noticeable way. This seemed like a pretty fail-safe way to proceed – total permission for mind wandering, for mental noise to remain, to take as much time as needed and to be as comfortable as possible.

To my surprise, I only had to wait about three minutes before what appeared like the letter W wrote itself on my screen! I then decided since that was a little too easy, I'd focus on the last part of the day to see what the last leg of the movement would look like. This direction seemed to correspond to what I had already seen, with the last leg of the W shape going upward. I sketched the letter W. Below is a picture of my transcript and then the feedback the client sent at the close of that day. He wrote "Wow" in red letters and superimposed my drawing (see red line) on top of his graph.

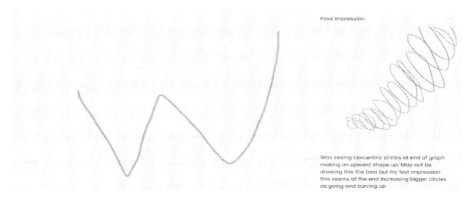

Figures 16.1 – Drawings of a W by Debra

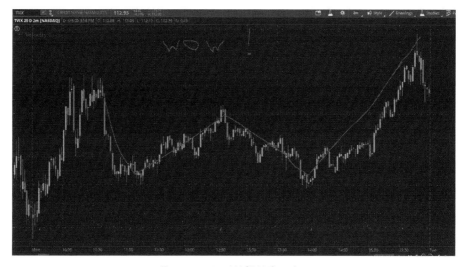

Figure 16.2 – WOW Graphic

319

We decided to repeat the exercise. He wanted to make an adjustment on his end to see how he might make better use of what I was producing for future trades – he was still figuring out what this predictive information could be useful for.

For this next trial, the letter "T" appeared to me on my visual screen. I felt this indicated a large drop. I didn't draw the T, I just typed the letter T. He responded, "You predicted a T for Task N2 and clearly it went straight down, as straight as it could possibly go in the stock world. So you clearly got it right again."

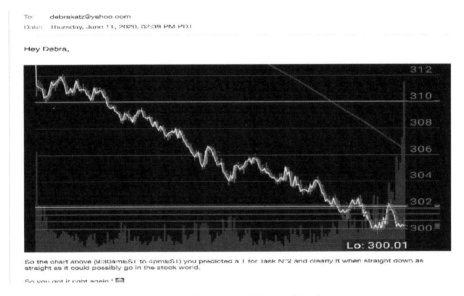

Figre 16.3 – Downward Zigzag Graphic

For the third trial, all performed in the same week, I repeated the exercise. This time something seemed complex. I felt mental strain as I repeatedly tuned in to get more so I could be confident about what I was reporting. When doing a number of trials like this, one can relax after some initial success. Yet it's easy to get into a mind game with oneself – maybe the initial success was just a fluke – so it's a continuous game of "prove it" to oneself and others. I thought I had gotten over this a long time ago, but it was back in full force with this new approach. This time I had an image of fingers on a hand, and then for the closing of the graph, a pointy shape. He seemed happy with this, as well, although it wasn't an exact replication.

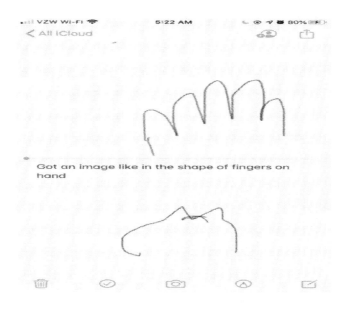

Figures 16.4 (top) and 16.5 (bottom) – Fingers Hands

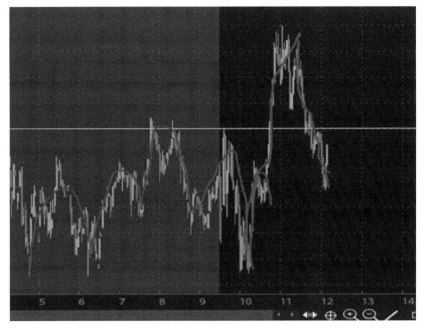

Fgure 16.6 – Dark Graphic Peak at right

At this point, however, I think he got overly excited. He started to add questions about opening and closing movements the day directly before and after, along with what was happening on the current day. I started getting multiple images and could feel my mind become tangled and stressed as I attempted to work out what I was sensing. I also started to get information about him personally, including messages for him. I fell asleep not feeling confident. I began falling in and out of sleep, dreaming of him, of the charts, of odd things. This was really disruptive, and I woke up with a headache and feeling more stress. I sent him several pages of graphs and he said he really couldn't decipher them. I think he was a little unnerved I was reading him, too (that's what I do for my day job). Of course, I didn't have to work this way in the evening before bed. I might have had a very different experience if I had done it in the middle of the day at a desk.

The client then told me he was convinced I was able to see these graphs, but he was a bit perplexed as to how to set up the project in a way he could profit from it. He had never represented himself as an expert in this area, but I was already feeling frustrated from having lost sleep for a week when I had a lot of other things to focus on (like completing my dissertation, this book and three research papers). At this point, I realized the client was being forced to act as project manager even though remote viewing project management is something that takes quite a bit of time to master. By now, he was promising to share a percentage of his inheritance with me, but I told him I didn't think it was going to work out, at least until this present book comes out. He sent me

322

$100 and promised to send more to cover my time when he got his next check, but that didn't happen, and I did not reach out to him again.

And that is the extent of my work with graphs so far.

What I learned from this experience is that the trader has to be very clear about what movement he wants the graph to represent, such as a particular rise or fall at closing time. He should know ahead of time the specific financial system, the date and when he'll send a photo of the chart to the viewer as feedback.

In my example, the client took it upon himself to superimpose the line I drew on top of the stock graph. I had mixed feelings about him doing this, but it did allow for easy judging of how closely the predictive drawing matched the computer-generated graph. It might have even strengthened the session from a retrocausal perspective. It ensured that the client/trader reconnected with the feedback photo and the viewer's work. So overall, I think it was great, but I had one concern. If I was to receive an image of my feedback and he included my sketch in the feedback, what about a situation in which my sketch was not a perfect match? Would this produce a "false image" for feedback? I didn't do enough trials to find out. I liked the idea of him showing me both at the same time and wonder if superimposing a viewer's sketch on the feedback photo could be used in other remote viewing projects in which a viewer is tasked to describe an object or structure – something I've never thought of before.

This experience is an example of why I frequently agree to do short-term projects – there is always something new and sometimes it's because the client was not trained or "indoctrinated" into a method or mindset, so creative ways of doing things emerge. Sometimes, however, a client ends up trying to reinvent the wheel and wastes everyone's time. This project sparked the idea of asking other remote viewers if they have used a similar approach to visualizing graphs. I was pleasantly surprised at what I learned.

Pam Coronado

Pam is known for her psychic detective work on the popular TV show *Sensing Murder*[391] and is a past president of IRVA. She responded:

> I use graphs all the time in personal readings when a client asks about their financial state. I look at the whole thing on a graph, where they are coming from, where they are now and where they are going. I can do this with anything, like the virus. I have done it with stocks, but then dates and timing becomes an issue.

When asked if her approach is more visual or somatic, she said visual.

Daz Smith.

The CryptoViewing company is now the largest employer of remote viewers and support personnel in the United States. The team has about a dozen full and part-time employees and has grossed considerable money from their "infotainment" using the Patreon subscription platform. Three of the viewers – Daz Smith, Dick Allgire and Edward Riordan – have built up substantial online followings, which has no doubt contributed to their success with Patreon.

Daz Smith, a top-notch viewer and publisher of *Eight Martinis* magazine, has tried sketching the movements of currency. He does this by "just free hand movement like an ideogram…It's all a first attempt and very experimental – but does show promise – each sketch only took seconds. You can see the sketches mirror the monthly progression of each crypto."

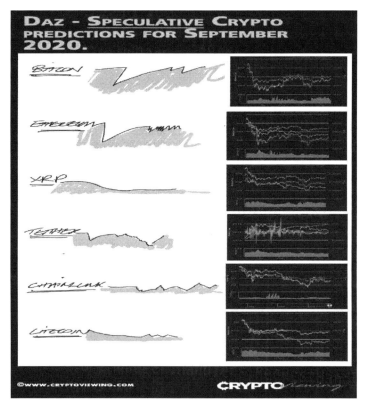

Figure 16.7 – Daz Speculative Crypto

Daz continued his development of the method and in March 2021, he drew a very unusual ideogram.

Figure 16.8 – Daz Page 1 S1 D on Lined Paper

Daz:

It's the first time in three years I have had an ideogram flow go backwards (usually a combination of up and/or down, but always forwards) – it startled me as in the session and I felt uneasy/not confident in the target because of this detail, and its backtracking.[392]

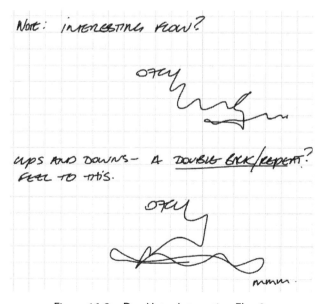

Figure 16.9 – Daz Note: Interesting Flow?

Text: Ups and Downs – a Double Back / Repeat? Feel to this

Remarkably, this same coin was involved in an extremely rare "double deposit" within four days of Daz's ideogram!

> According to developers at Filfox and FileStar, Binance processed a "double deposit" of FIL on Wednesday worth millions of dollars. This is not a true, on-chain double spend, but Binance credited the account for filecoin miner 6Block (the parent of Filfox and Filestar) twice for one deposit due to a "serious bug" in Filecoin's remote procedure call (RPC) code.[393]

Daz wrote: "The fact I wrote double back and in the same week we have this DOUBLE news even has shocked me with its accuracy." After doing these ideograms for coins regularly each month, Daz thinks he may have reached the point where they reflect the movement of the coins quite well. In his session for Filecoin, he also drew a suggestive representation of the Filecoin logo, and his adjectives describe the founder. (Admittedly many of the crypto tech geeks would have generated similar descriptors.)

Julia Mossbridge and John Vivanco

Julia Mossbridge, Ph.D., is both a highly trained researcher and an active remote viewer. In addition to being the Co-founder and Executive Director of TILT: The Institute for Love and Time, Julia was a Visiting Scholar in the Psychology Department at Northwestern University (2013-2020) and is now an affiliate professor in the Department of Physics and Biophysics at the University of San Diego. She is a Fellow at the Institute of Noetic Sciences and an Associated Professor in Integral and Transpersonal Psychology at the California Institute of Integral Studies. Her Ph.D. is from Northwestern University in Communication Sciences and Disorders; her M.A. in neuroscience is from the University of California at San Francisco and she received her B.A. with highest honors in neuroscience from Oberlin College. While we could go on for several more pages, what is most applicable here is that she has become an avid remote viewer and psi dreamer herself, is co-author of *The Premonition Code*,[394] and she now co-leads a team of remote viewers who do projects for clients and researchers, some of whom are our friends and colleagues appearing in this book.

Julia and the team call their approach to viewing graphs "Wowsing." When she and others drew graphs of stock movements, the investor involved said "Wow" after seeing them. Her husband Brooks (a team member) coined the term "Wowsing" ("Wow" plus dowsing). It wasn't until drafting this chapter that we realized how appropriate the term "Wowsing" is, given the "Wow" written on Debra's graph by her client!

Julia describes her personal psi method as follows:

> As to my psi process for wowsing, I put myself into RV mode (cool down, binaural beats), then I recorded my ES (emotional state), PS (physical state) and my intention, then did the wowsing almost as a transit line to the target (which is probably why it went opposite). Then I exited.

The transit line is a feature of the TransDimensional Systems method in which you let your hand draw freely on an entire page, rapidly curving and moving until you arrive at a point, which signals you are ready to proceed with the session proper. The transit line carries you into the tasking.

Another team member is John Vivanco, author of *The Time Before the Secret Words: On the Path of Remote Viewing, High Strangeness and Zen.*[395] When asked about his approach, John replied,

> The process I used was to first stand up and feel in my body any abstract sensations. In a sense, it is a tuning of the body. I don't treat that as "data" though. Then I would dowse with my hand the movement of the line on the graph.

Julia reports that the team correctly predicted the direction of the graph (Up, Down or Flat) four out of five times over five days. During pre-testing Julia found she predicted the opposite of what would occur while her co-viewers predicted the correct direction of the curves. John was particularly consistent with correct orientation, while Julia was particularly consistent with the opposite curve or direction. In actual trials, they always flipped the direction described by Julia and used a majority rules vote to predict the movement. If there was a conflict in the graphs, they tended to go with Julia's and John's dowsing data only, as these usually matched once Julia's was flipped.

However, for one stock she got two humps and the investor went against their advice, which was to sell at the second peak. He would have made money if he had followed their advice. Here are their graphs for Stock D and the chart of the stock. Recall that Julia gets results opposite to the actual movement.

Market Summary > **Bed Bath & Beyond Inc.**
NASDAQ: BBBY

11.62 USD +0.15 (1.31%) ↑
Closed: Jun 28, 5:15 PM EDT · Disclaimer
After hours 11.62 0.00 (0.00%)

| 1 day | **5 days** | 1 month | 1 year | 5 years | Max |

Figure 16.10 – Market Summary Bed Bath

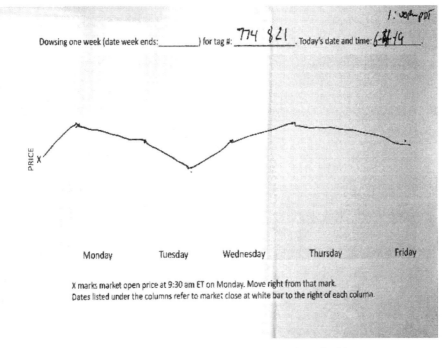

Figure 16.11 – Dowsing One Week – Transcript by John Vivanco

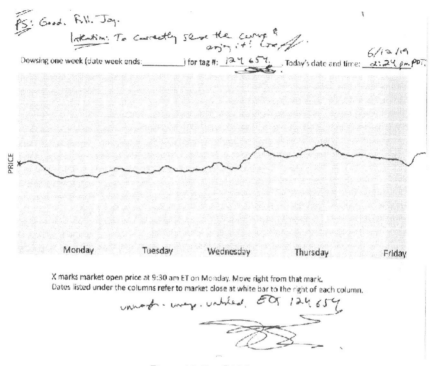

Figure 16.12 – R.V Joy

On the graph with the Up then Down then Up – Stock D – we got clear prediction of the two humps (we had a total of 9-12 sessions on each stock) from multiple sources (wowsing and RV sources both) so we told the investor to sell at the second peak, which would be higher than the first and would occur later on (Friday). He sold at the trough – frustrating![396]

As a result, they stopped working with the investor. Her suggestion for future use is:

Another way to approach this, and a better way I think, is to create online software where people can trace their impressions, then it could determine whether averaging the impressions came out with something better than our approach to data analysis.

Igor Grgić

An extensive attempt at direct drawing, visualization and similar methods was undertaken by Igor Grgić a few years ago:[397]

Back in 2019, I and group of remote viewers made an effort to directly remote view future movements for Forex charts and currency pairs. I wanted to test this new approach since after many years of doing by ARV method, I wanted to explore other possibilities.

A team of 11 remote viewers was assembled, some I knew and worked with years before, some completely new people that were interested to join my public call on this experiment. My role was as tasker, analyst and feedback/outcome provider for the viewers.

There were a total of six RV taskings, once per week. The task for the viewers was to *locate the start of the most significant WW move on QQ 1-hour chart for the next week YYYY-MM-DD1 Monday to YYYY-MM-DD2 Friday.*

WW and QQ were parameters which were randomly chosen. WW was either UP or DOWN. While QQ was a randomly chosen Forex currency pair. Therefore, here is one example of a tasking:

Locate the start of the most significant DOWN move on the Forex EUR/USD 1-hour chart for the week YYYY-MM-DD1 Monday to YYYY-MM-DD2 Friday.

Viewers were given a template with a blank rectangular area. They were asked to view/sense/dowse to identify the most significant WW move. On purpose, I put QQ here – viewers were blind to the WW and QQ parameters, as was I (the tasker/analyst). Both WW and QQ were determined by random selection & stored in a file, both being blind to all participants during the viewing process. After viewers submitted their sessions, I analyzed the data and marked the predicted area(s) for the next week's Forex chart.

Analysis of the results – with several images of the actual charts/predictions:

Week 1: spot-on correct prediction of the week's most significant 70 pips move

Week 2: spot-on correct prediction of the week's most significant down move of 90 pips. Any trade entry at predicted & marked chart areas would produce significant profit in desired direction. See chart and prediction below. Note: an actual Forex chart image was merged with group's predicted areas which are marked in PURPLE. Some image parts are intentionally hidden.

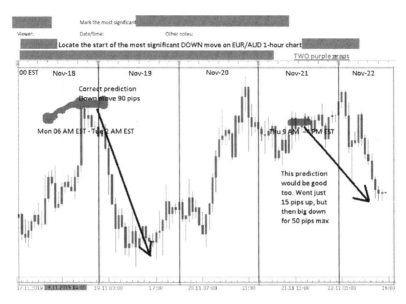

Figure 16.13 – Locate the Start of Down Move – Week 3

Week 3: correct prediction of the week's most significant up move 60 pips

Week 4: most significant 63 pip move predicted with two additional moves (45 + 50 pips) in correct direction. Interesting, all these three moves were marked precisely at areas for the most profitable trade entry (from the time perspective of the X axis (horizontal).

Figure 16.14 – Locate the Start of UP Move

Week 5: mixed result with 3 predicted areas, but 3rd area was an exact trade entry for the most significant 70 pips down move.

Week 6: three predicted areas – one wrong, one ok and one/third was an exact and correct entry predicted for 115 pip move.

Figure 16.15 – Locate the Start of DOWN Move EUR/CAD

These predictions were not traded; they were analyzed as to what could happen if traded. My estimate is that after six weeks the profit would have been around 500 pips of profit. I tried to be conservative; besides trades that were winners, I also calculated failed trade entries, which were exited with a small loss.

Precognitive Trading Group - Igor Grgić

The following is by Igor Grgić, who reported on his development of his Precognitive Trading Group:[398]

I was able to develop a methodology for predicting future price movements for Crypto and Forex charts. It includes a direct remote viewing of the charts (price action) to come up with a binary style decision/prediction; as opposed to the ARV (associative remote viewing) which is usually used for the binary type of predictions. So, instead of that, the two possible outcomes (such as Up or Down price outcomes) are associated with two photo targets (ARV) – the Precognitive Trading Group does a direct chart viewing, which later helps me as a judge/analyst to decide on the final prediction in the form of a simple Up or Down prediction.

Over the years I have done 1,000+ binary ARV trials as a judge – judging the viewers' transcripts. Overall, the ARV hit rate was somewhere around 60% - 65% long term. Some of these ARV projects produced 55% hit rate, others were 60%, and some even close to 70% hit rate for a simple Up vs Down (Long vs Short) financial & Forex types of predictions. The one and only problem is – of course, the displacement problem in ARV – sometimes even with a spot-on correct viewer's descriptions of the wrong ARV target. By wrong, I mean that the viewer sketched and described the photo target that was not actualized target/outcome, meaning it produced a miss and a loss in the Forex trade.

Displacement is the one and only issue which is hard to overcome in ARV, not impossible, as I had witnessed many hits in a row (9 by one viewer or 10 by other remote viewer). I still love ARV, but was time to do something else :)

Now, in Precognitive Trading Group/PTG we have several remote viewers producing very high accuracy with the method of direct chart viewing, something I very rarely witnessed in a classic ARV approach.

PTG's current Crypto and Forex predictions are at 86% correct. We also use Google Groups, which serves as a time stamped proof-of-predictions.

And several of individual viewers' accuracy is on about the same level of around 82% - 85% – that's for Coral and Karsten. They both have own approach for direct chart viewing. Some insights and session examples:

For Karsten it can be a simple as sketching instinctively the price action, without doing a CRV session. It also includes sketching through an unconscious muscle reaction. The general idea is to let the body react to the RV task without thinking about it. It is like a freehand sketch in Stage 3 of CRV. It's fast.

This example shows the resulting sketch:

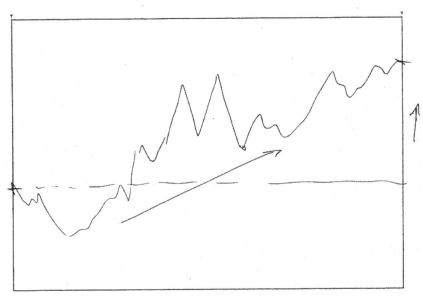

Figure 16.16 – Example of a Freehand Sketch (Karsten)

Coral has a different approach to the PTG RV task. Easiest here is that I quote her from the email:

I trained in CRV with Lyn Buchanan, this means I start from an ideogram. I feel out the ideogram to get a first idea and then in the first stage I start describing the ideogram. This gives me an idea of the general feel of the market. I feel when the market is calmer or more nervous or when it feels heavy or negative. I keep touching the ideogram and receiving and writing down perceptions, until I get to stage 3 when I see a flash of a drawing, I try and reproduce this on the paper.

When the feedback comes in, I go over it very carefully and I often retrace it. Sometimes I open my drawing over the feedback in Photoshop so as to correct the line in my subconscious.

I think going over the feedback is almost as important as the viewing.

Sometimes I will go over my drawing with a dowsing rod to check on the place of entry and the place of exit, but the dowsing rod can also be influenced by analytic overlay. The important thing is to write down perceptions without ever naming anything consciously.

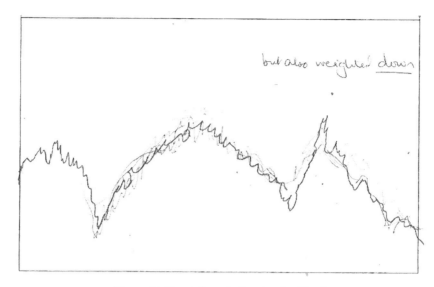

Figure 16.17 – A Sample Session by Coral

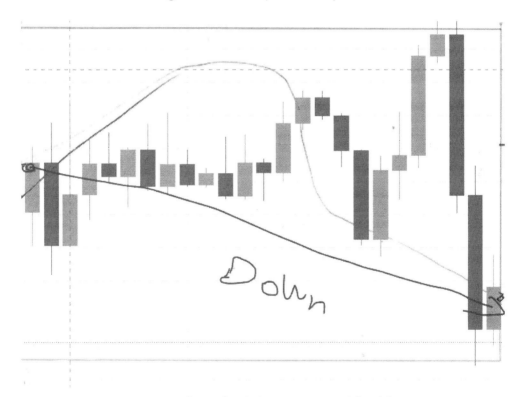

Figure 16.18 – Down Graph. Image courtesy of Coral Carte.

Is direct graph viewing an ARV task, just an RV task or something else?

Should this type of psi approach to accessing information about financials be considered an associative task or a direct remote viewing task?

We could argue it is associative because you don't see the movement of the trade or market itself, you see a graph, which is a representation of it. Many people enter trades – buying or selling – and these collectively are reported in the form of a graph. The graph represents collective acts that are reported as having moved upward, downward or stayed the same. In this respect, the graph on the trading platform is representative of the movement of a particular stock or an entire market.

When the psi practitioner creates a representation of the graph by sketching it, they are attempting a direct representation – to get as close as possible to what's on the trading platform screen. Therefore, one could argue direct drawing of graphs is not Associative Remote Viewing at all – it's direct remote viewing.

Instead, one could reason the viewer's sketch replicates both things. Or it may be one or the other, depending on the remote viewer's intention. If I set my intention to see the trading platform screen, that is direct remote viewing. If I set my intention to see a graph as a representation of what the actual stock trade or market is doing, that is an associative representation of the activity. Both intentions may lead to the exact same result. This is uncharted territory and we lack sufficient data to come to conclusions about it.

Some observations on the psi methods

The approaches described above could be categorized as:

- Clairvoyant (visual) – Debra, Pam
- Somatic (wowsing, transit line, dowsing) – Julia, John
- Somatic ideogram-like – Daz, Coral
- Various, using a grid (11 viewers) – Igor

These are all simple to do. The clairvoyant method just requires more mental focus. The somatic (transit line, hand dowsing, ideogram) methods are extremely easy for those used to starting every session with an ideogram. However, those new to the idea that the subconscious can essentially communicate something vital through the hand, bypassing the analytical and conscious mind, may need to practice to "get out of their own way" by trying to control their hand and instead let the "intuitive self" take over.

A concluding story from Debra: I once tried to teach my significant other how to do an ideogram. He had no interest in doing this but was humoring me since I had just complained I'm always willing to go along on his death-defying activities (like surfing, ocean kayaking, snowboarding, motorbike-quad riding and let's not forget shark cage diving, etc., etc., etc.). Meanwhile he seemed to always have an excuse to not sit down and do remote viewing.

I had chosen an object from my son's room as a target. The object was a medieval-looking box with smiling animal skulls all over it. I did not let on what it was. As we were sitting there, I told Danny to close his eyes and make a reflexive mark. He said he couldn't do it. I was like "What do you mean? You just close your eyes or look away and say you want your hand to make a mark that is a symbol of the object – then we'll have you trace or touch it to get more impressions. The mark isn't supposed to look like anything. Just scribble."

He was confused and getting frustrated. Finally he blurted out, "Do you realize what you are saying here…I'm a mechanic, I need to know what I'm doing and be in control of it. If my hand starts to just do whatever it wants to do, I'm going to get into trouble, who knows what it might do…!"

"Oh my God," I yelled at him. "Just make a mark on the freakin' paper!" Now mad, he was like "Okay fine, here you go, anything to get out of here!"

He proceeded not only to sketch a simple line, which was all that is required for an ideogram, but kept drawing, which I thought was going to be like a transit line, but he continued to furiously sketch. I could tell he thought he was being funny and derailing the whole exercise. "There! That's it. Can I go now?!"

I couldn't believe my eyes. "What is that?" I asked.

"I don't know. A skull" he said. "I guess it's an animal skull."

This obviously is not just an example of somatic sketching – it speaks to the idea that we can all do these types of activities if we can get out of our own way. Also, it's possible we may be expressing knowledge of the world around us through our bodies all the time and not be aware of it. This entire chapter, in also sharing examples from experienced remote viewers, demonstrates there is always something new to learn and how this work requires repeated attempts, experimentation and adjustments.

CHAPTER 17

ARV Is a Dream
A Quantitative and Qualitative Analysis
of a Novel Precognitive ARV Dreaming
Experiment

This piece was written by Debra Katz and Dale Graff, with extensive contributions from our fearless and tireless project manager Nancy Smith, as well as Michelle Bulgatz, James Lane and remote viewers-turned-dreamers Dave Silverstein, Sam Smith, Chris Georges and Marty Rosenblatt.

In this chapter, we present what happened when a group of remote viewers replaced their waking remote viewing protocols with dream ESP in what could be considered the first-ever Associative Remote Dreaming (ARD) experiment.

Introduction

In spring 2019, the quantitative assessment for our Associative Remote Dreaming (ARD) Study was published in the *Journal for the Society for Psychical Research (JSPR)*. As in most formal journal articles in parapsychology, there was not space to discuss our personal phenomenological experiences and findings related to photographs we would see in future dreams. This was the first time a group of remote viewers had attempted over a period of time to systematically replace an Associative Remote Viewing protocol with an ESP dream protocol. Most of the experienced remote viewer participants had never attempted to have an intentional dream that would allow them to produce a transcript that would, if successful, match a photograph he/she would be shown at a near future date. The exception was Dale Graff, former DIA Director of the Star Gate project and author of *Tracks in the Wilderness*[399] and *River Dreams*,[400] who joined the Sublime remote viewing group specifically for this project.

This is a brief overview of the project's methodology and results, which hopefully will inspire readers to read the full *JSPR* paper. We will then focus on our qualitative-oriented findings, which include compilations of individual and group transcripts illustrating themes that emerged during the 56 weekly trials.

The idea for the project was born in a taxicab between Bourbon Street and the airport in New Orleans. Debra and Nancy were still reeling from the presentation Dale Graff and his research associate, Patricia Cyrus, had made during the weekend at an International Remote Viewing Association conference. Graff and Cyrus described a decade-long project[401,402] in which the two used a precognitive dreaming protocol they had developed for the purpose of intentionally describing a specific newspaper or magazine page they would be shown at a near future date. They showed a number of sketches produced through dreaming and some quite extraordinary matches to the photo and sometimes to the headlines in the articles.

Nancy and Debra were members of a small group of viewers that had been working together, mostly on ARV projects, for a number of years. The Sublime remote viewing group consisted of seven members: Dave Silverstein, Nancy Smith, Sam Smith, Debra Katz, Michelle Bulgatz, Chris Georges and Marty Rosenblatt.

The Sublime group had originally formed under the umbrella of the Applied Precognition Project, but operated independently from APP as well. The group was used to trying out new protocols. That day in the cab Debra and Nancy speculated whether other members would be up for replacing the usual remote viewing approaches with the Graff-Cyrus dreaming methodology, but keeping within an ARV protocol.

Debra mentioned how great it would be to get some extra training from Dale. Nancy, being her typical proactive unabashed self, whipped out her smartphone and began dialing. She didn't know Dale personally, but she did have the phone number of one of the former military guys. She called him and it turned out he and Dale were together at that moment touring a World WII Museum. Dale graciously offered to discuss our project then and there, and even offered to be a dreamer for it. Thus, the very first ARV Dream project was born.

The Sublime group members, with Graff on board, designed and participated in a yearlong study of 56 trials in which they attempted to have precognitive dreams that would enable them to produce descriptions and sketches matching a photograph they would be shown at a future time.

Summary of the project design

In order to ensure our design was solid, we submitted it to John Kruth's team at the *Rhine Research Center* for a proposal review they offer to researchers interested in initiating parapsychology-based projects. Over the years we've found it imperative to have outside professionals look over our plans as there are always adjustments to be made that if not caught early could easily invalidate an entire project after extensive work has been done.

Once the project was initiated, Debra (who for the rest of the chapter will be referred to as "Katz") and her pal, Michelle Bulgatz, moved from their positions of researchers to join the other Sublime members as dreamers, leaving the Sublime Group leader, Nancy Smith, to run all other aspects of the project. Smith recruited an experienced team of photo selectors to choose the photos and create photo sets. The order of these sets was randomized, assigned random target numbers using an online random number generator, and hidden in a folder on Smith's computer so she would remain blind to the photo sets till after the dreamers completed their sessions.

Smith gave the viewers their assignments each week, sent them reminders with due dates, collected the dream transcripts via email and gave ratings to the photos using the 7-point SRI confidence ranking scale.[403] The transcripts with a confidence ranking score of 3.5 or higher for one photo and with at least a two-point spread between each photo were often considered wagerable. If a transcript was weak or strong in equal measure for each photo, a pass was called and no prediction was issued.

If the trial resulted in a prediction, Smith then chose a sports event via an online wagering site using a pre-determined selection protocol and placed a wager of $110, which could potentially result in a payout of $100. If a pass was called instead of a prediction, no wager was made. Once the sports event was over and the outcome known, Smith then emailed the photo associated with the winning outcome to the remote viewers, completing the "precognitive feedback loop." She then encouraged the viewers to do a "feedback session" in which they would compare their transcripts to the feedback photo and write up their comments and feelings about what worked and what did not.

Dreaming protocol

The following excerpt is taken from the original *JSPR* article.[404] We share it here as this is the exact protocol we recommend for anyone seeking to have a dream about a photo they will see at a future date – this could even be a newspaper article for those wishing to dream about national or global events:

> The dreamers purchased a dream journal of their choice. On the Saturday evening, they would write out an intention statement on a page in their journal with the

freedom to add to it as they saw fit. Some dreamers found it useful to add in a congratulatory message about what a great job they were about to do while conjuring up enthusiastic emotions over this imagined success.

Dreamers were advised to write down the target number in their journals or papers, as they have all been trained to do at the start of the remote viewing session, except instead of proceeding with a remote viewing session, they would go directly to sleep. This intention/tasking included telling their subconscious to have the needed information appear during the final dream of the evening, prior to waking, so that it could easily be recalled and distinguished from the earlier dreams of the night. It included the intention to have visual information come into the dream that could easily be converted to a sketch upon awakening. It was also recommended that, when possible, they give their dream a title.

Dreamers were instructed to record all dream impressions of the evening without delay upon awakening into their dream journal or onto a piece of paper if they didn't have a journal nearby. If they awakened prior to having a dream that could be recalled, they needed to either try to go back to sleep if time allowed, or to simply send an email to NS with the words "no dream" … All dreamers were required to turn in their transcripts or report "no dream" by 9:00 a.m. CST time. They did this by taking pictures of the page(s) of their journals with their camera phones, uploading these to their computer and emailing them to the manager (p. 75).

Results

Five out of seven remote viewers/dreamers were able to consistently produce dreams at will. Their 278 dream transcripts were utilized to make predictions and wagers on the outcomes of sports events. They produced an overall rate of 17 hits out of 28 predictions, which a binomial test showed to be at chance levels (one-tailed). Nevertheless, the overall monetary gain was just under 400 percent of the initial stake. Further, one individual dreamer had a 76 percent correct hit rate with 13 hits, 4 misses (and 20 passes), while another dreamer had 16 hits and 9 misses (hit rate of 64%).

As was noted in our *JSPR* write up:

While $400 is a small amount as far as earnings go, one has to take into account the conservative amounts of $110 that were wagered. Prior ARV projects may have made more profit but this relied on individual wagers of as much as $10,000. Given this, and the reported success rate reported here, one could postulate that had a more ambitious wager been made of $1,000 per bet, a profit of $4,000 could have been made. This would translate into a potential profit of $40,000 if wagers of $10,000 had been placed per bet. Hence, future research needs to give consideration to the amount of money that should be wagered. Future projects therefore might consider wagering higher amounts.

Also, given that a few individual dreamers' stats were higher than the group's stats as a whole, with one dreamer's hit/miss ratios as high as 76%; future projects might incorporate using selected individual dreamers versus a group aggregate approach. For example, JL, the project statistician, performed simulated calculations on how much money could have been made for the one individual remote viewer who had a 76% hit rate. He calculated that if $110 was risked on each bet to make a $100 win, or a loss of $110, individual performances would have been quite profitable, growing the investment to $860. One could then postulate that if $1,000 had been wagered each time $8,600 would have been earned.

The overall average hit rate at 61% was only marginally above chance at 50%. Given that this was only marginal and one-tailed it needs to be interpreted with caution. Nevertheless, this is an encouraging trend. As mentioned earlier, this marginal effect may have been due to a lack of statistical power given the low number of trials. Therefore, it is suggested that one major adjustment to the protocol for future research is to have a set number of predictions rather than a set number of trials (p.81).

Several other recommendations were made for future projects. These concerned issues about scoring and decision-making in issuing a prediction. Targ, et al.[405] suggested that a CR score of 4 was sufficient to proceed with a prediction, since in their study close to half of the transcripts earned scores ranging from 3.5 to 6.5. Yet many of these higher scores resulted in predictions made for the incorrect photo option, resulting in 10 misses of 29 predictions and led to many passes when the higher scores were split between the two options. While we can't be sure, we suspect the reason we had so many high scores for the incorrect photo may have been overly permissive judging. Judging biases about a particular dreamer may also have played a role in scoring and making predictions. To mitigate this, it was recommended that future projects use a team judging and management approach.

Further, it was speculated that in addition to correct data, the dreams may have produced a substantial amount of extraneous and irrelevant content compared to what is typically produced during waking remote viewing efforts by experienced participants, especially since multiple dreams were often reported. This data may have matched the wrong photo (or the correct photo at the wrong time) simply due to chance. Occasionally the researchers found displacement to the wrong photo occurred. Therefore, it was recommended that for ARV trials involving dreaming the prediction threshold on the SRI 7-point scale should be raised from 3.5 or 4 to 6. If the threshold was set higher, however, it would be imperative to adjust the experimental design so a specific number of predictions are made.

Another suggestion was to disallow assignment of half points or other partial percentages to the SRI 7-point scale (i.e., scores of 4.5 or 5.75). This change would force the judge to make a decision in one direction or another. Half scores were not included by the originators of the SRI 7-point CR scale and no precedent suggests they add anything to the judging procedure, except indicating where the judge may be conflicted.

The authors further noted that Targ, et al.[406] in their redundancy experiment suggested that predictions and wagers should not be made and instead passes called in situations where one or more remote viewers have the same (or close) scores for both photos. Nor when one team member has a high score for one side and another team member a high score for the other side. In the present project, the judge did not always follow this rule. The analysis showed some of these trials did result in misses; however, had the rule been followed in all cases, some of the existing hits would not have occurred. The decision on whether to follow this rule should therefore be based on the financial stakes and risks involved, with the bottom line being how important it is to avoid such misses.[407]

Qualitative findings

The following discussion will focus primarily on our experience of what we encountered after changing our psi protocol from remote viewing to dreaming. We will highlight themes related to dreaming on demand, dreaming and sketching, and dreaming within a collaborative group environment. We will also contrast RV with dreaming as a protocol. Where appropriate, we will provide samples of the dream transcripts.

Ability to dream

As reported in the original *JSPR* article,[408] out of 357 possible transcripts,

278 total dream transcripts were submitted by the seven viewers. The bulk of missing transcripts came from just two dreamers – one dropped out after the first 10 trials due to being in a place in his life where dreaming was not practical and the other participated sporadically. Missed transcripts by other participants were attributed to changes in sleep schedules related to house guests, traveling and special circumstances out of their control (p. 80).

Dreamers were told to not try to dream if they were too busy or stressed, so lack of turning in a dream was not necessarily indicative of an inability to dream on demand but – according to their post-experiment survey and interview responses – was often due to disrupted sleep cycles from the above-noted factors. Most did not feel they had a problem going to sleep on schedule yet failing to have a dream. Three of the participants new to this type of dreaming task expressed being pleasantly surprised they were able to dream on demand. Katz reported at least three times when she had woken up within an hour of the deadline without having had a dream, and then told herself (and her family) she had to go back to sleep to retrieve a dream and was able to do so in time to turn in a transcript. On one occasion this was accomplished within 30 minutes of the deadline, producing a transcript resulting in a hit. This was despite her often having difficulty falling asleep outside of the project.

Distinguishing content related to photos vs. other dreams.

Some of the viewers noticed they could occasionally distinguish dream content relevant to the target from other dreams, giving them a high degree of certainty a particular element would be present in the photo. Some dreams had a different quality than other dreams. Sometimes this awareness was related to the dream's visual content, particularly unique shapes, patterns, positioning and colors that were not typically present in regular dreams. At times, parts of conversations (words or thoughts) stood out, drawing the dreamer's attention to aspects highly relevant to the future feedback photo. Both Graff and Katz shared that their dream experiences would often mirror the degree of complexity and detail within the feedback photo. Complex dreams might result in several dreams, each providing input into a different aspect of the target, while a very brief dream might indicate a target with only a few elements (See Figures 1, 2 & 3).

Figure 17.1

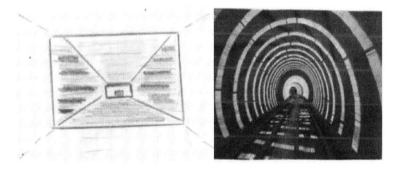

I am in an underground corridor or tunnel that leads to a large underground complex. The sides of the tunnel have many rows of shelves with varying thicknesses of brown putty-like paint. I go to the end of the tunnel. Dale Graff

Figure 17.2

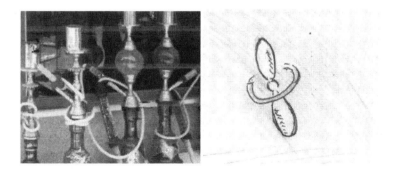

Figure 17.3

Several viewers noticed that their precognitive dreams contained details that emphasized physical aspects of a place they normally wouldn't pay attention to in a dream. Sometimes these aspects were present in the photo but reversed. Again, these details alerted the dreamer to their importance. During one dream, Katz wrote, "OK there has to be a building with windows in the photo!" She also wrote, "my dream was from the inside of a building. I'm with people and a guy that is working on the outside (construction) with the window." The photo turned out to be a man falling off a ladder inside a room with windows that surrounded him. Other dreamers had correspondence with this same photo (see Figure 4) by sketching a man falling and words like falling and drifting:

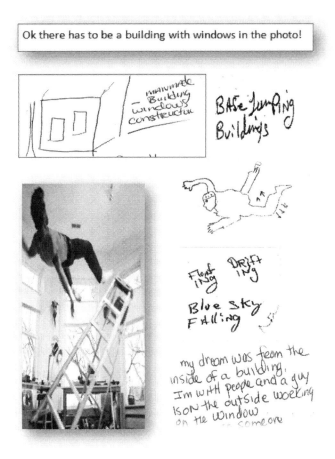

Figure 17.4

Some dreamers occasionally reported they became the figures in the future photos they would see, as if from a first-person perspective. Figure 5 is an example Graff shared. He wrote,

In the dream I became the fisherman (as sketched) on the left side who is holding a net (the tan streaks). *I pull out a soft flexible plastic or rubber-like sheet…drag it to the center in shade or shadows…I adjust its volume by tightening or loosening a*

346

rope or chord by twisting an attachment to the chord that keeps the ends together. The
adjustment creates a 3-D configuration that resembles a large shallow tub…

Later he explained, "These two dreams were exceptionally clear. Attention to tiny details
in the dreams also gave me confidence that the dreams were spot on."

Figure 17.5

Occasionally dreamers reported having lucid dreams in which they were aware of
being involved in a precognitive experiment. In Figure 6, Katz submitted the following
description with very simple sketches that match her typical ideogram for mountains.
In her transcript she wrote: "I had two short dreams, really just scenes about a mountain
top…in the dream I said, 'that was easy the photo is of the top a mountain,' and it was."
Upon awakening, she reported this dreaming process was much easier and less time-
consuming than her remote viewing sessions tended to be.

Figure 17.6

Some of the participants reported interesting dream narratives that gave clues about an important aspect of the feedback photo. After awakening from one dream, Bulgatz wrote: "I was working or helping a team remote view a treasure. Debra was there. I was monitoring. This is what her *sheet* looked like…" She then sketched a building and wrote the word "railings" twice with the word "spaces" in between. Prominent in the actual photo is a courtyard filled with railings with spaces. Furthermore, hanging on the railings are "sheets." (See Figure 7.) For this same trial and photo, Katz produced a transcript highlighting another example of shapes and positioning. Although the phrase "circle within a circle within a circle" was not correct – it should have been squares within squares – she sketched circles and then filled them in with multiple, repeating squares and also stated that in the dream many people were coming together into small, sectioned off spaces to sleep together, having brought their own bedding.

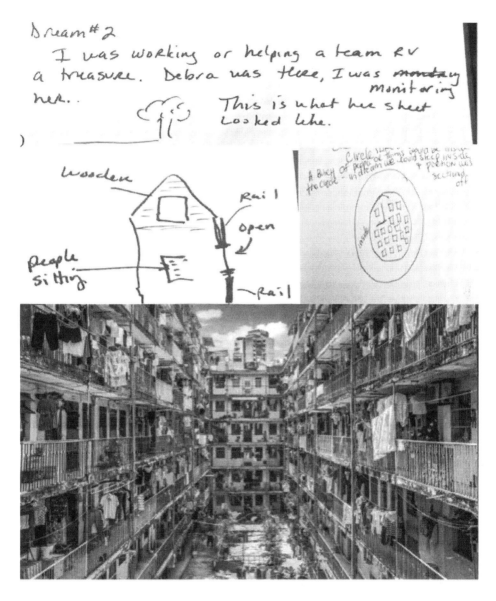

Figure 17.7

Timing of dreaming

Since the viewers didn't have access to dream lab equipment and were operating independently, there was no way to formally monitor their sleep cycles or be sure they were waking up in the middle or at the end of the dream. Still, some of the viewers had a sense of whether the significant dream content came earlier in the evening or just before waking. Some who decided to dream one night earlier found one aspect of a dream came Friday evening, while another important aspect of the dream arrived on

Saturday night. Still others reported that occasionally a single important aspect of the photo would appear on both nights, but in two radically different dreams or occasionally in two similar dreams. Some of the dreamers would suspect they knew which content was important but worried if they only reported the final dream, they might miss something, so they tended to report all dreams on the night preceding feedback.

Graff reported that as the project progressed, he started to withhold reporting information from earlier dreams of the night. While occasionally he regretted this, it seemed to help avoid reporting irrelevant content, which he saw as more important to avoid misses. Graff explained:

> Dreams are great story tellers that will likely create interesting but non-relevant deviations to the psi material in the dreams. I found that keeping a pre-session intent to "present only the facts" helped filter out extraneous and personal dream content, resulting in brief to-the-point dreams. I also learned that very often additional dreams occurred that were clearly of personal situations that should not be included in formal submissions. Even if some relevant data is tossed out, over time more noise will be discarded than relevant data.

Other dreamers also reported observing themselves moving through a learning curve. One of these tended to have vivid dramatic dreams throughout the entire designated dream evening with extremely acute details. As the project progressed, she reported having an easier time distinguishing relevant content from extraneous content relevant to her own life.

Many of the dreamers and the project manager found sketching helped pare down the essential dream content. Some dreamers reported that upon awakening and recalling the dream, the idea of summarizing would initially feel confusing and overwhelming. But then whatever made its way onto paper as a sketch was oftentimes the strongest match, containing a mixture of aspects both consciously decided upon and what just happened to emerge with less intent or conscious awareness. Graff also found tasking himself to name or label the dream as if one were giving a film a title was a very useful technique in getting to the heart of what the dream was about (see Figure 8). He also used colored pens or pencils in his sketches. The colors often had close correspondence to one or more elements in the photo, as illustrated in Figure 9.

Figure 17.8

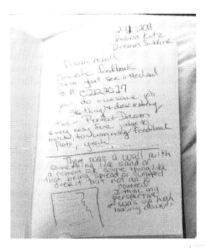

- Dream sketch to the top right is by Dale Graff showing Color match to dream, Sketch to top left is a page from one dreamers Journal of same target showing tasking night before with words <u>under stating</u> *"There is a wall with something like sand being spread over it"*.

Figure 17.9

351

Displacement

The manager and all the viewers occasionally observed what could be called instances of displacement – defined as describing something that is going to happen using one's precognition, but not that which was intended. For this project, three categories of displacement seem likely to have occurred by describing:

1. the wrong photo in the set (the unactualized one, associated with the losing outcome)
2. an aspect from the dreamer's life that would occur the next day
3. something from the manager's life that would occur the next day

Dale Graff shared a synchronistic incident illustrated in Figure 10. On his transcript he wrote:

> In the dream I am suddenly looking at a vivid scene that resembles a Southwest, USA landscape. Higher cliffs surround a wide-open desert terrain that has several mesa-like formations or similar structures. Some of the structures are thin and tall. A tan-colored mountain range is in the distance.

Later in a post-experiment survey, he recounted:

> That afternoon, after I had submitted my dream input, I went hiking on the Appalachian Trail in the mountains near my home in eastern Pennsylvania. On an impulse, I took a side trail that I had not previously been on. After a long hike, I came to a plateau that had a vista of the distant mountain range. As I stood at the vantage point, I had an "AHA" moment! The scene resembled the vista in my dream! I felt certain that the setting of my dream sketch would be accurate, even spot on. A synchronistic sense, a déjà vu feeling, surged through me. I felt as if I had been drawn to this vista in order to complete the dream. I returned home feeling confident that I had already seen the target picture. That evening I received the feedback target picture. The dream scene, the target picture's scene and the view of the mountain ranges that afternoon were similar. My subconscious connection with the target picture had continued into my afternoon hike. Even though the dream did not clearly present a boy standing at a ledge, it seems that I became that boy later that day when, like him, I was standing on a ledge overlooking a similar mountain scene.

Figure 17.10

Another seeming case of displacement from his life occurred when Graff dreamed about "an incredible acrobatic performance by a dark-skinned bare-chested man." Graff recounted, "On Sunday morning, after I had submitted my input, I saw the front page of the local newspaper. It had a full-page photo of a bare-chested Muhammad Ali, who had died the day before."

Figure 17.11

Project Manager Nancy Smith felt dreamers quite often described elements from her own life that took place on the day of feedback. While one can always assert these are just coincidental, some of the elements were quite specific and out of the ordinary. For example, Katz stated in one dream there was a large bookcase with shelves empty except for a single box of Cracker Jacks. This was very strange given she hadn't thought of Cracker Jacks in years and wasn't sure if they were still sold in stores. She wrote a note suggesting that Nancy google "Cracker Jacks" to see if the figure in her dream might pertain to the future feedback photo. A sailor is on the front of the box. In her dream, a single person was in the room while everyone else had already left. Smith would later explain that she felt the sailor matched one of the photos, which was of people being rescued from the water by the Coast Guard, with one man in the air and one person still in the water. What was more interesting to her (and amusing) was that in the few hours between the submission deadline and sending the feedback, Smith was invited to run in an uphill marathon, where they provided snacks for the runners at their "rescue stand." Oddly enough, boxes of Cracker Jacks were the only thing they had on the tables for the runners!

Another example of possible telepathic displacement was when Graff dreamed about a guy "telling me something about a duck and I see a duck and wonder if it is real or a carving." The man also mentioned something about "ducks mate for life." Smith advised soon afterward in her feedback to Graff that she had been discussing ducks with her husband Sam a few hours prior to the submission due date, and Sam actually said, "Ducks mate for life." Graff felt this was a case of telepathic submission into his dream.

Dreaming vs. remote viewing

Several dreamers reported they initially had difficulty avoiding accessing visual information about the targets outside of dreaming (since this was a dream experiment and not a remote viewing one). They noticed even a momentary thought of "I wonder what the photo will be?" could initiate a spontaneous picture or flow of information. They had to exert effort to not even think about the target outside of setting their intention and then forgetting about it. Occasionally receiving information while awake could not be avoided and in these cases, they reported that and did not submit it to the project manager unless it truly was a case in which they were between the dream and waking stages and had not consciously attempted to tune in.

Some of the Sublime members stated they really enjoyed dreaming, finding it was more efficient than remote viewing, which requires more effort and time. Others felt they had less control over the information gathering and reporting process than they typically do when remote viewing. While most of the dreamers felt the task of intentionally dreaming future photos disrupted their sleep cycles, only one reported this to be problematic enough that he might be weary of repeating another long-term project. Most of the participants felt intentional precognitive dreaming was disruptive in that they slept less soundly, awoke more frequently, had to take time to write down their dreams in the middle of the night, causing disruption to others in their household, etc., and were more tired the following day. Still, they felt it was worth it because of all they learned from the process. A few felt something was missing after they stopped the trials and were therefore looking forward to starting a new project.

Final thoughts

This project was initiated with a few questions in mind: Can dreaming replace remote viewing in an Associative Remote Viewing protocol? Can experienced remote viewers new to intentional precognitive dreaming consistently produce transcripts through dreaming that match a future feedback photo and on a par with transcripts created during a conscious remote viewing session? And would it be possible to earn a profit and achieve statistical significance over the course of 56 dream trials?

The answer to the first question is "mostly yes" considering the number of highly descriptive transcripts that were consistently turned in from the start of the project by five of the seven remote viewers and with little prompting or incentives (beyond wanting to contribute to the goals of the group over a one-year period). This applied to most of the dreamers but two of the participants felt their lives were not conducive to dreaming at the time. It is not known whether they were simply unable to dream or whether they just choose not to participate beyond some initial attempts in the first several weeks.

In response to the second question, the answer is a somewhat cautious "Yes" in terms of profit. In terms of hit ratios and statistical significance, the overall hit group hit rate of 17 out of 28 predictions was still considered to be statistically at chance levels, despite having more hits than misses. This was due to having had more passes than was typical for the group. (We would have had to have a few more trials and hits to show significance statistically.) What this suggested was that using a dreaming protocol, at least for this group, can result in more passes and therefore fewer predictions than when using a remote viewing protocol.

It should be noted, as well, that some individuals performed at a higher rate than others while some displaced to the wrong photo more frequently than others. That being said, one does have to consider that judging and making predictions are huge factors in projects such as this. It is possible the judge may not have been consistent at times in rating viewers' sessions. For example, given Dale Graff's reputation as an excellent dreamer, the judge sometimes was more inclined to trust his transcript as correctly pointing to a particular outcome when, in fact, his transcript didn't quite warrant this. If any session is given a higher score than it warrants, leading to a wagerable prediction, this puts it in danger of being a miss. In retrospect, it was felt this happened on at least a couple of occasions.

Several of the viewers experienced a *learning* curve over the course of the project, feeling they were beginning to understand which parts of their dreams would be relevant to the photo as opposed to unrelated content. While they were not able to evaluate whether this curve showed up statistically, several felt "learning" did take place. This particular group of dreamers and others like them who have a chance to practice what could be called "the art of precognitive dreaming" are prime candidates for future dream-related experiments.

Further, we suggest repeating the study with individual dreamers working one-on-one with a project manager to compare how well they do as individuals compared with an aggregate of dreamers. Finally, in a post experiment survey, Graff attributed some of his own misses to photo composition and artificial coloring of photos. He wrote:

> Photo composition has a significant effect on how our psi dreaming process perceives them. Pictures with elements that have indistinct boundaries or poor contrasts will very likely result in psi dreams that have very little resemblance to the target picture. The subconscious mind strives to perceive the target picture's imagery and will present the closest match to similar imagery in memory. This "match" may result in dream imagery quite different from what is seen in the target picture. At a conscious level, our visual system integrates diverse imagery into a coherent recognizable picture. In psi perception, a process resembling a bit-by-bit access of the picture occurs and if the element boundaries are indistinct, a variety of dream images are possible that differ from what is consciously seen in the picture and will likely be a miss.

While Graff's theory suggests his mind may have produced an image that was wrong, Nancy Smith, who – unlike all the dreamers – had access to the unactualized photo in the set, theorizes Graff and other viewers unconsciously described the photo that was easiest to comprehend, which unfortunately in the above-mentioned cases wasn't the one associated with the winning outcome.

Upon further examination of the photos involved in some of the misses, it is clear some photos contained a single hue. Everything in one photo was blue; in another everything was pink – most likely due to photo-editing software. Therefore, we recommend that photos be chosen that have not been color-corrected.[409]

CHAPTER 18

Off to the Races – ARV and Horseracing

The first use of Associative Remote Viewing was an attempt by Stephan Schwartz to predict the winner of a horse race at Bay Meadows in San Mateo, Calif., in September 1977. Examples well prior to that describe attempts to use psi to pick winners in "the sport of kings," such as the claim by a woman who was put to the test by the Society for Psychical Research. Although he didn't call it ARV, Ingo Swann used a phonetic method to try to pick winners at the Meadowlands and Aqueduct in 1973 and 1984. In 1980, Elisabeth Targ was successful in predicting the winning horse in the sixth race at Bay Meadows using multiple-choice ARV. From 2006 to the present time, horse-racing enthusiasts Tom Atwater and T.W. "Teresa" Fendley have used psi and logic in hundreds of trials and with considerable success. Recently Tunde Atunrase applied his Primary Pool method to pick the horses. In this chapter, we examine the methods and results of all of the above practitioners.

1938 – "Miss X"

"Miss X" approached the Society for Psychical Research in England stating she possessed the ability to predict the winners of horse races by symbolic means. As reported at the Meeting of the Council of the *SPR* in 1938:

> Before the race she would take careful note of any unusual images which suggested themselves to her mind and would then examine the lists of runners in the race, where she was generally able to find certain names which corresponded with the symbols. She suggested that (1) these selected horses proved to be winners

more often than if they had been chosen by chance, and (2) that the fact that she was usually able to find names corresponding with her symbols (obtained before consulting the lists) was in itself significant of paranormal faculty. She wished these suggestions to be scientifically tested, and freely placed herself at the disposal of the Society.

The society carried out this experiment. First, to test for chance outcomes, six days of racing were chosen at random. Two horses were randomly selected from probable starters by drawing the number of the horse from a bag. Hypothetical bets were placed. Wins and losses of all horses that ran were calculated. Of 54 horses that were selected and ran, the result was a net loss of 8.5%. Miss X submitted her predictions on special forms, predicting 154 runners, three times more than in the chance experiment. The result was a net loss of 114 units. At first sight, it appeared Miss X possessed a "negative pre-cognitive faculty." But evaluation showed she preferred "hot favourites and rank outsiders" as opposed to middling choices.

Next the Society examined if she got symbols of actual names of horses. They selected 15 days' racing at random and compared symbols with the names of horses that ran a week later. They concluded names of runners were so numerous and varied that any fairly general symbol could apply to any day's racing.

The conclusion was: "There seems therefore to be no reason to suppose that Miss X possesses either a faculty for selecting winners or a faculty for producing symbols which apply to the names of runners more aptly than is explainable by chance."

1973-1984 – Ingo Swann

Ingo Swann tried his hand at the horses in 1973 and 1984 and left many pages of notes in the Swann archives at the University of West Georgia that detail his attempts. From these materials, it does not appear that any of his choices were winners. If they were, he did not write about it. Ingo's approach was to try to get the name of the horse that would win. This is related to an advanced stage of CRV (phonetics), although he had not developed CRV by 1973 or 1984.

An example will make his method clear (quoting directly from his notes in the archives):

Words got for win, place, show horses for Oct 27, 1973, at Aqueduct
First race: 1 elo olo orbit 2 hell 3 Fr…(French?)
Second race: 1 ovbil 2 factor 3 elaias
Third race: 1 welfare 2 dravidian's chance 3 elegant

Fourth race: 1 defeat (defiat?) 2 Rich 4 Fletcher
Fifth race: 1 elsewhere 2 Texas 3 Defiant
Sixth race: 1 terrorize 2 love's lay 3 alamagordo
Seventh race: 1 Tvexit 2 annahilate 3 lovers lane
Eighth race: 1 alignment – along the line 2 wax 3 truth
Ninth race: 1 alone along 2 drums, drops 3 alexis day

Perhaps the most interesting aspect of Ingo's attempt to pick horses was his use of the terms and methods of Scientology. That Ingo was a Scientologist is now well known, but that he was inspired by many of their techniques in his development of CRV is less well known. Joe McMoneagle knew about the connection:

> What the argument has been about is specifically the validity of the 'Ingo Swann Methodology' of training as effective. I do not believe it is. It is born right out of many of the techniques used by Scientology and I find these to be psychologically confining.[410]

While we do not agree that CRV is ineffective as a training method, we mention McMoneagle's comment as confirmation that Swann built on techniques from Scientology.

Ingo's connection with that group began in 1966.[411] He took a Saint Hill Special Briefing Course, then the Clearing Course and from there went up the OT (Operating Thetan) levels. He felt the training benefited him greatly:

> My greatest win on the OT Levels I think was the ability to separate myself with great conviction from matter, energy, space and time. To have gained the ability to hold my viewpoint without quivering. To have gained the ability to be appreciative of the need for an ethical, philosophic philosophy and to want to help produce towards its fullest use.[412]

It was this training in separating his mind from his body, he said, that enabled him to move his attention to the highly shielded "impenetrable" magnetometer buried beneath the floor of a Stanford Research Institute lab. He believed that shifting his attention to the device is what modified its output, an action that shocked the scientists and was instrumental in leading to CIA contracts with SRI and eventually to Ingo's development of CRV.

The Scientology "Grand Tour" in 1952 gave Ingo practice in mentally visiting the planets, a precursor to his successfully remote viewing Jupiter in 1973. He reported it had rings – a fact unknown to science until 1979 when a satellite made a bypass – a feat that brought him attention and fame.[413]

Other former Scientology members at SRI included Director Hal Puthoff, Pat Price and Director Ed May. Based on correspondence between himself and Swann, the scientists at SRI did not function as a Scientology cohort. In fact, when they first got together, Puthoff and Swann were concerned that their past membership in the organization would not go over well with the intelligence agencies funding the programs, so they stayed mum about their past membership.[414]

In the notes illustrating his attempts to predict winners at Aqueduct and the Meadowlands, Ingo used Scientology terminology as he tried to understand what was happening in his mind.

As explained to us by Russell Pickering, a remote viewer and former Scientology member:

Ingo: "8. Prediction then means being ____or to game – no game - LF Bd"

This is an important concept in Scio.[415] It essentially means being free to have a game, or not have a game, by choice (to be, or not to be, literally!). Many beings trapped in the games matrix MUST have a game. Many other beings "hovering in bliss" WON'T have a game – games frighten them. The optimum state is to be able to have, or not have, any activity at will with no compulsion either way. Again, you can see this was followed by the long fall (LF), blowdown (BD) of charge.

The gist is this. Let's say that an earlier iteration of Ingo decided to use his "powers" and had a bad experience. As an example, let's say that in a previous biography, when his powers were less diminished, he said, "I wish that guy would die!" And then the guy died. Then Ingo says to himself, "I will NEVER use my powers again!"

Now he wants to use his "powers" to predict horse races. That old incident will kick-in (i.e., NEVER use your powers!). This will create a conflict between wanting to use them and having earlier decided to "never use them."

It is a matter of goals in conflict. And thus inhibitive counter-intention to being successful now by using them.

Ingo: "The horses – having to deal with sets such as letters, numbers, running order – this will cause predictive AT/Us (?) to synthesize. right."

ATTENTION UNIT, 1. a theta energy quantity of awareness existing in the mind in varying quantity from person to person. (HCOB 11 May 65) 2. actually energy flows of small wavelengths and definite frequency. These are measurable on specifically designed oscilloscopes and meters. No special particle is involved. (Scn 8-80, p. 45)"

He is asking the question on the meter. It appears that he is wondering if predictive attention units will blur together (synthesize) regarding the multiplicity of numerical values involved in horse racing, presumably causing a bad outcome.[416]

1977 – Stephan A. Schwartz

In Chapter 2, we mentioned how Stephan Schwartz came to invent Associative Remote Viewing. Here we focus on its first use, which was in a horse race. Schwartz settled on a two-team approach, with one experienced viewer and one new viewer. He chose Hella Hammid for the former and Neddie Pena for the latter. He invited SRI researcher Ed May to take part and Ed agreed. They chose the sixth race at Hollywood Park in Los Angeles on Sept. 9, 1977, and Ed May flew down for the event.

> We created two target sets of Los Angeles locations and ran each session independently. The assigned task was "At 4:30 PM tomorrow, we will be standing somewhere. Please describe, using all sense impressions, where you are." The session data were judged in a blind, rank-ordered assessment of the data against the target images. Both women unequivocally selected the target associated with the 6th horse in the 6th race at Hollywood Park. That night, we went to the racetrack and placed a two-dollar bet. We won $14 and jumped around, clapping each other on the back, as if it were a hundred times that amount. ARV had made money. And people have been using it for that purpose ever since.[417]

1980 – Elisabeth Targ

Elisabeth Targ, the daughter of SRI researcher Russell Targ, conducted two ARV experiments in 1980. The first was a successful prediction of the winner of the 1980 election. In the ARV trial, she identified a small object in one of four boxes, the object having been assigned to be associated with a win by Reagan. Similarly, for a horse racing experiment, six objects were selected as potential targets. Each horse was assigned a number and Elisabeth was told that after the race, she would be given the objects corresponding to the winning horse's assigned number.

As Elisabeth began her remote-viewing description, she saw something "hard and spherical." It reminded her of an apple. She continued, "If I hold it up to the light, I can see right through it." Since one of the targets was a spherical apple juice bottle with a raised apple leaf design around its edges, everyone thought we had a great success. Of course, at that time, the race had not yet been run, so the most anyone knew was that we had a good description of one of the objects in the target pool."

Since Elisabeth's description so accurately described one of the target objects, students from all over her college dormitory contributed money to a betting pool for horse number six, whose name was Shango, in the sixth race at Bay Meadows. The next day, Shango won and Elisabeth got to see her apple juice bottle as a reward for excellent psychic functioning. Shango paid six to one.

The important point here is that in these experiments the medium is not the message. Analytic information can be obtained, but it requires that the medium (protocol) be non-analytic.[418]

Targ and Harary make an important point about ARV: it provides "intermediate information" indicating whether or not psychic functioning is present. If Elisabeth had "a clear mental picture" of a moose, yet there was no moose among the targets, that would have been evidence psi was not present. Whereas, if Elisabeth had said "I see a six," the experimenters would not know if psychic functioning was present or not.

Off to the Races: ARV and Horse Racing – 2006 to 2020

(The following was written by Tom Atwater and T.W. Fendley)

Background

Tom Atwater has been doing Associative Remote Viewing since 2006, for horse racing, sports and Forex trading. He has received RV training from Pru Calabrese, Marty Rosenblatt, Skip Atwater, Julia Mossbridge and John Vivanco, but mostly learned on his own how to access his intuition in this way.

Tom's intent in any financial ARV/intuitive venture is always to maximize the return-on-investment (ROI) as opposed to the usual focus on predicting winners (hit rate). ROI takes into account both win percentage and returned offered (e.g., the odds), and is the proper measure of financial success. Tom prepares for new experiences in applying intuitive methods for financial gain by taking time to clearly state his intents, convince himself that he is good at what he does again, and relaxing and having fun with it all.

In addition to horse racing, Tom had a winning track record in Forex and sports predictions for the ARV-based Applied Precognition Project (which he co-founded but has since departed) and in some other Forex groups.

T.W. (Teresa) Fendley is a horse racing enthusiast, long-time remote viewer and speculative fiction author. She learned about ARV in 2009 at a Monroe Institute class on intuitive investing taught by Marty Rosenblatt and Paul Elder. Since then, she's been an active member of Rosenblatt's Applied Precognition Project, including managing two groups and serving as APP webmaster. She has enjoyed many successes using ARV for horse racing, sports predictions and Forex trading. Her website http://arv4fun.com features examples of ARV hits.

Teresa studied Controlled Remote Viewing with Lori Williams (Intuitive Specialists) and honed her RV skills with the help of many other luminaries in the field such as Angela Thompson Smith, Tom McNear, Lyn Buchanan, Sean McNamara and Gail Husick. She is also a co-researcher and co-author of various remote viewing formal research projects published in RV magazines and peer-reviewed journals.

Tom and Teresa used both ARV and non-ARV methods over the years to predict horse race outcomes in several different ways. The primary intent in most of the experiments discussed here was for individual viewers to make money, betting on their own. Not all viewers bet on all races, however, with some participating to practice their intuitive skills.

Using ARV to predict horse race outcomes

Two different ARV protocols were used: one that associated a different photo for each of the four to twenty horses in a race, and one that divided the horse race field into two parts, so that the usual binary ARV methods could be used.

Comparing ARV to logical horse race handicapping

Unlike the usual binary ARV used for sports events and stock market trading, horse racing ARV requires a set of four to twenty photos to be associated with each horse in a race, if one wants to predict outcomes for individual horses. This makes judging much more difficult than for binary ARV, because the judge must determine the best single photo from up to twenty, not just two. In the experiments given in this section, the horse racing outcome associations were done with photographs, and viewers and judges were blind to which horses were associated with which photo until judging was complete.

During 2006-07, Tom performed an ARV experiment on 109 North American horse races, consisting of between four and 14 horses per race, in an attempt to show a profit using only ARV predictions.

Protocol. Using photosets of up to 14 photos generated using rules to randomly search Google images (http://images.google.com), spreadsheet formulas were generated to associate each runner in a given race with a photo, in such a way that the viewer (Tom) was blind to the association. The viewer then performed a session, drawing a sketch with his impressions. He self-judged the sketch against the four to 14 photos, using the Targ CR 0-7 scale.[419] The highest CR rating above 3 was then the prediction; if none were above this value, the race was Passed.

Logical predictions. Tom also used publicly available horse race handicapping data together with his twenty-five years' experience handicapping to predict the winner by the usual logical means. This was done prior to the self-judging step in the ARV protocol. This logical prediction could then be compared to the ARV prediction. Tom was always blind to the race details until after self-judging of his ARV transcript was complete.

Expected results. Since all the horses in each race were treated equally, the random expected hit rate was 1/f, where f was the field size, or number of horses in the race; for this experiment, the average f = 8.7, so the random expected hit rate was 1/8.7 = 11%. In North America, the track takes an average percentage of about 18% off the top of the total money wagered to win, so the random expected ROI was about 18 cents lost per dollar bet.[420] Figure 18-1 shows how the data compared with these values.

			Hit Rate			ROI per dollar bet	
Type	Races	Hits	Actual	Expected	Avg Odds	Actual	Expected
ARV	109	25	23%	11%	9.0	+$1.30	-$0.18
Logical	111	42	38%	11%	3.3	+$0.29	-$0.18
ARV+Logical	14	9	64%	11%	3.3	+$1.78	-$0.18

Figure 18.1 – Results for associating a different photo with each horse in a race, 2006-07

Results. For ARV predictions, the actual hit rate was more than double the expected hit rate; the actual ROI was even more impressive, the ROI being more than double the amount bet per dollar, far above the 18% expected loss per bet.

The logical performance shown was about as good as any reported in the horse race handicapping literature; the logical horses had much lower odds than the ARV horses (3.3 to 1 vs. 9 to 1), accounting for the higher hit rate than for ARV.

ROI is the real measure of financial success, and that was much higher for ARV predictions – 130% profit vs. 29% for logical. This was the result of more longshots (very high-odds horses) selected by the ARV method.

Also shown in the above table are results for the 13% of races where the ARV prediction and logical prediction were the same horse ("ARV+Logical"). Although 14 races are a small sample, the data indicate even greater success than either ARV or logical alone – 64% winners and 178% profit. Not shown in the table is the fact that the ARV+Logical prediction "placed" (finished first or second) in 13 of the 14 races; one would expect this to happen only about three times at random.

Conclusions. Tom concluded ARV that associates each individual horse in a race with a photo may be a viable means to make money betting in horse races. Where the ARV prediction agreed with the standard handicapping logic prediction, the result was a very high win rate and ROI. That finding needs to be verified with more data, but these results indicate logic and intuition working together may be the best method of all.

ARVing horse races for the Applied Precognition Project (APP)

In 2009 Tom formed a horse racing prediction group in conjunction with Marty Rosenblatt for Physics-Intuition-Applications (forerunner to APP). Rosenblatt expanded his original binary-choice PRECOG software for managing ARV photosets to 10 and later 20 possible photos to associate with each outcome for an event, to accommodate horse racing predictions for fields of up to 20 horses. T.W. Fendley, Lincoln Lounsbury and Loraine Connon were among the early participants in the group, which had fun doing the predictions with modest success and a few winning wagers by individual participants.

Work with the group continued after Tom left APP in 2013; Teresa took over as APP's horse racing group manager. From 2016 onward, Teresa led an APP group ("First Groove") in predicting horse race outcomes using ARV.

Protocol. Unlike horse racing predictions discussed thus far, this experiment forced a binary choice to use standard binary ARV protocols and judge only two photos against the session transcripts. Side 1 of the binary choice was postulated to be the Morning Line Favorite (MLF) in the race; Side 2 was the remaining three to 19 horses in fields of four to 20.

The intent was to incorporate logical handicapping with ARV. The MLF is a track employee's estimate of which horse is expected to be the favorite at post time; it is used as an indication of the horse in the race with the highest win probability. (The actual race favorite, or horse with the most money bet on it, cannot be determined precisely until the race has already started.)

The probability of either side winning was irrelevant to the viewers, who were only making a binary choice; therefore, the expected random probability was the same as for any binary ARV experiment – 50% for each side.

Teresa used APP software to obtain the ARV photosets. Each viewer's coordinate and binary photoset were different from those received by the other viewers. Each viewer also received unique feedback – the photo associated with the winning side from their own binary photoset. The number of viewers participating over the four years was 42; all sessions were self-judged by the viewers using the Targ scale of 0-7.

Results. Of the 354 trials between 2017 and 2020 that resulted in a Win prediction for either Side 1 or 2, 176 were hits, or 49.7%. Comparing this to the expected hit rate of 50% for a binary choice indicates nothing anomalous happened for all 42 viewers as a whole. No pattern was found among the viewers to indicate some performed better than others beyond chance. Hit rates did not vary significantly from year to year.

Issue with protocol. Due to a misunderstanding about race setup parameters in the APP system, Side 1 (MLF Wins) of the binary choice was always associated with the first photo the viewers/judges saw. This issue wasn't known until the data was analyzed in 2020. Two independent statisticians who reviewed the data expressed concern that the lack of randomization may have skewed results. They recommended ensuring randomization for any future projects.

The actual impact on this project is unknown. A survey in 2021 showed few of the judges (who were also the viewers, since they self-judged) realized the association was not random. Twelve of the 16 respondents, including the top viewer/judge with 60 sessions, chose the answer that Photosite A was either "Randomly associated with either Side 1 (yes) or Side 2 (no)" or "I don't know."

Beyond ROI. The survey also revealed that the main intent of *all* First Groove viewers who responded was to "practice ARV and improve [their] hit rate" or simply to "have fun." And fun it was! While in Las Vegas for the IRVA (International Remote Viewing Association) conference in 2018, Teresa joined a rowdy crowd in the casino's sports betting area to watch the Belmont Stakes, the third jewel in American thoroughbred racing's Triple Crown. Armed with First Groove's prediction, she rooted for Justify to become the thirteenth winner of the Triple Crown since it began in 1919. She cashed in a small winning bet, but for Teresa, the most fun was bragging to the other remote viewers at IRVA about the APP group's winning prediction. In fact, one of the viewers – Lori Golden (with her husband David as judge) – had correctly predicted all three races.

ARVing for horse race handicapping contests

In 2010, Tom and several group members began participating in horse race handicapping contests, using ARV occasionally supplemented with logic to make contest selections. Much of this is documented by Teresa's blog entries at ARV4Fun.

Tom qualified on his own in 2010 for the National Thoroughbred Racing Association's National Handicapping Championship (NHC), finishing third of 200 entrants in a qualifying contest. To successfully pick longshot horses, he used the same ARV technique of associating one photo with each horse and self-judging all horses in the field, together with intuitive knowing through listening to guidance, in combination with his logical horse race handicapping skills and experience. For this contest, he felt this was a direct result of his strong intent to combine logic and intuition to produce profits, both for his own benefit and to teach others by example. Tom also feels his work in shifting his vibration about winning and allowing the universe to inspire his winning predictions had a positive impact.

For the two-day NHC championship contest in January 2011 in Las Vegas with $1 million in prize money ($500K for first place), Tom made extensive preparations both logically and intuitively. On the first day, he combined logical, intuitive and ARV to make predictions. That proved to be too much over the day's 15 races, so on the second day he used intuition and ARV to make predictions. He did much better on the second day (three contest winners and one second, $75 in contest points) than the first (zero contest winners, five seconds, $34 in contest points). Tom finished 91st of 301 entries.

In addition to this, Tom, Teresa and others participated in other NHC qualifying events, without successfully qualifying for the year-end championship.

Tom led a group of six viewers/judges (including Teresa and himself) to take down second prize in the $5,000 Grade One Racing 10-race contest with 166 contestants in 2010, winning a cash prize of $750. The viewers (two per race) were assigned races at random, did a session on all the horses in the race, then self-judged their session against all of the associated photos without knowing which photo was associated with which horse. Tom also performed an independent judging in parallel, then made a decision based on both judgings to predict the contest winner for that race. Tom did his usual emotional preparation for the contest. Elections were based on ARV data only (no logical handicapping). The odds on the team's three winners on the 10 races were 44 to 1, 9 to 1 and 2 to 1; longshots were the key winners here as before. Winning viewers included Joanie Sullivan, Loraine Connon and Patsy Posey.

The last contest race was a good example of the prediction protocol. Of the two viewers for this race, Loraine self-judged her session and got horse #7 as her prediction, with #2 ranked second, while Tom ranked #7 and #2 as tied for the top rating; Joanie self-judged #4 on top and #7 second, and Tom agreed with her assessment. So it was close between #2 and #7; Tom used his intuition to select #7, which won at 9 to 1 and elevated the team to finish second in the contest.

Tom, Teresa and others participated in various horse racing contests, the best result being in a Derby Wars contest in 2012, where Tom, Teresa and Loraine finished eleventh and earned $50.

These contests all demonstrate ARV's success in picking longshots, which one has to do in order to win – the most logical horses never provide enough contest points to finish high up in the standings. Tom always stated his intent beforehand to hit at least three longshots (defined as win odds greater than 8 to 1), as this was usually sufficient for a successful contest outcome.

Combining non-ARV intuitive with logical horse race handicapping predictions

Tom, Teresa and others used intuitive methods other than ARV to predict the outcomes of horse races. From 2015-2016, Tom made predictions for the winner of 964 races using a combination of his intuition directly, in combination with logical horse race handicapping predictions. Instead of performing time-consuming manual logical handicapping, logical predictions were obtained using the output of Handicapping Technology and Research (HTR) software and data from Handicappers Data Warehouse (HDW) fine-tuned to produce horses of all odds, including longshots. HTR predictions by themselves show very little or no profit over time but served here as a starting point to combine with intuitive predictions.

Protocol. The software analyzed its data to produce "spot plays" for all North American tracks, based on suggestions from the HTR developer and fine-tuned using Tom's savvy from long-term experience betting horses. This resulted in more than 200 logical predictions per month.

For each of these logical predictions, Tom used his intuition directly to predict whether the horse would win the race. He did this by ranking his confidence in the prediction on an Intuition Rating (IR) scale of 5 to -5. If his intuition said YES, he placed a bet; if NO, he passed the race.

Tom's main intent for this experiment was to achieve an ROI of +20% or greater. Only YES predictions are legally bettable in North America, where Tom is based, so ROI is primarily reported here for those predictions.

Results. There are two ways to look at the results: one, focus on the ROI, since that is what indicates financial success or not, and that was the primary intention here; two, to analyze the hit rate.

ROI results. For 961 intuitive YES predictions, the hit rate was 20% at average odds of 4.4 to 1, producing an ROI of +9% profit per unit bet. In contrast, the ROI for 1,806 intuitive NO predictions was -2% loss per unit bet. The logical result for the software predictions alone without intuition for 2,767 predictions was a hit rate of 19% and ROI of +4%. So these Intuition+Logic predictions did slightly better than the logical software predictions alone.

Hit rate results. For YES predictions, a hit occurs when the prediction *wins* the race. For NO predictions, a hit occurs when the prediction *loses* the race. For all of the 2,781 intuitive predictions – YES plus NO – the hit rate was 60.2% +/- 1.1%. This is well above the random expected value for the binary YES/NO choice of 50%, with a p-value of p<0.000001.

Early success. Most of the intuitive YES prediction success came in the first three months (297 predictions) of the experiment, where the ROI was +42% profit per unit bet; for the last 11 months of the experiment (664 predictions), the ROI fell to -15% loss per unit bet. This may indicate the well-known decline effect in parapsychology.

Longshots. All of the profits came from betting longshots, defined here as horses with post time odds of 8 to 1 or greater. The 25 longshot winners from 291 predictions produced an ROI of +39% profit, while the 670 lower-odds predictions produced a loss of -3% per unit bet. Betting only such longshots sounds great in theory, but with only about two winners out of 20 predictions per month, frequent losing streaks are likely to erode viewer confidence.

Real money bets. The above ROIs are reported as per unit bet – assuming the same amount of money was bet on every prediction. Tom wagered real money on his selections, with the bet size varying during the experiment, in accordance with prudent betting strategies (e.g., never betting more than 5% of the wagering bankroll). Over the course of 14 months, Tom bet on every YES prediction, for a total of $132,000 wagered; the bets yielded a net profit of $3,700, or +3%, less than the +9% unit bet profit reported above because he wagered more proportionally on losers than winners. He showed a profit of $20,000 after three months but lost $16,000 of it over the next 11 months.

Intuitive feeling ratings. For part of the experiment (1,875 of the predictions), Tom assigned each prediction an Intuition Rating (IR) of -5 to 5 based on his simple intuitive feeling about the possibility the logically predicted horse would win. Over the entire experiment's length, the ROI as a function of IR showed a slight decline with decreasing IR. This effect was most pronounced in the first three months that IRs were assigned – for IRs of 1 to 5, the ROI was +11% profit, for 0 to -5, -1% loss. In contrast, for the rest of the experiment, the situation reversed – IRs 1 to 5 had -6% loss, while IRs 0 to -5 had +9% profit. This may be another indication of the decline effect in parapsychology.

The initial success was most pronounced for the highest ratings – predictions with IRs of 4 to 5 for the first three months had an ROI of +26% profit for their 231 predictions. This led Tom to use these predictions as the basis for real money wagers through the rest of the experiment. Unfortunately, thereafter the ROI for predictions with IRs of 4 to 5 dipped to -5% loss for their 388 predictions. This led to the losses detailed above in the real money bankroll.

Over the entire experiment, IRs of 4 to 5 showed +7% profit, as opposed to all lower IRs combined, which showed a loss of -1%. These highest IRs constituted 33% of the 1,875 predictions. Tom concluded higher IRs do have some predictive value for ROI success.

Conclusion. This experiment indicated that financial profits from horse racing may be possible using non-ARV intuitive methods in conjunction with logical data. Results were best at the beginning of the experiment, especially for predictions that were self-rated with a high Intuition Rating score. The ROI for YES predictions was well above random betting and the hit rate for the binary YES/NO choice was well above chance, strongly indicating the presence of anomalous cognition.

Applying intuition directly to predict saddlecloth colors

From 2017-2020, Teresa led a group of 26 viewers who used the Direct Intuition (DI) protocol to predict horse race results. This was a form of ARV using colors rather than photos, but it's discussed here as DI because of the protocol's simplicity and to avoid confusion with the binary ARV outcomes.

These races were the same ones predicted by the APP binary ARV group Teresa managed. As discussed earlier, the First Groove viewers predicted whether the Morning Line Favorite (MLF) would win, whereas this DI group's predictions drew from the whole field of horses in a race. Viewers submitted their individual ARV and DI sessions separately, prior to Teresa sharing any predictions with the groups. For those who placed bets, the DI predictions were especially helpful when First Groove predicted the MLF would not win. Some viewers participated in only one of the groups.

Protocol. In all North American races, each horse is assigned a saddlecloth, whose color corresponds with the horse's program number: #1 is always Red, #2 is always White and so on. For this experiment, viewers were emailed a list of the possible saddlecloth colors for a given race and asked to intuit the saddlecloth color of the winning horse from among the five to 20 entries/colors per race. These races had an average field of 10 horses, so the expected random hit rate was around 10%. For feedback, project manager Teresa sent viewers an email with the color word (Red, White, etc.) associated with the winning horse's saddlecloth color.

Results. Direct Intuition (DI) protocol predictions had a hit rate of 11.2 +/- 1.2% over 694 trials, compared to a randomly expected value of 10.4% (averaged over the field sizes for each viewer). This result was not statistically significant, so anomalous cognition was not indicated over all viewers. However, viewers having the most experience with the DI protocol did better than those without much experience with this protocol – the top half of viewers in number of trials had average hit rates of 12.7 +/- 2.0%, which is above the random value of 10.3%.

Conclusions. Results indicated the more experience a Direct Intuition viewer had, the higher the success rate. Overall, the more experienced viewers showed possible anomalous cognition with scores above random, whereas no such effect was seen for overall results of viewers using intuition directly to predict the winners of horse races.

Application. The name Applied Precognition Project implies "applying" ARV results. In this APP group, the DI viewers bet individually, if at all, so ROI data was anecdotal. Unlike Tom, Teresa is not an experienced logical handicapper. Generally, she placed a $2 bet based on her personal color prediction in a race as an affirmation of confidence in her viewing. Also, it was more difficult to place a group bet because often the viewers chose different colors. However, for a race on Sept. 26, 2020, all four DI viewers selected either Green or Yellow from a field of nine horses. She bet on the associated horses (one of which was a longshot) and got a $26 return on a $2 Exacta Box bet. For Teresa, the best part was applying their precognitive predictions for a team hit!

Tom Atwater's method

Tom employed a regular method by steps, which he has permitted us to present in summary form:

There were eight steps for each race, using a program written in PHP and a random number generator.

1. Photosite ID generation: Software generates a five-digit number when viewer is ready to do a sketch. Each time the viewer is ready to do a sketch, the software is run to generate a five-digit number. Midway through the experiment, a sixth digit was added to the random number, for reasons explained below. (This number is known as the "coordinate" or "target reference number" in most RV studies.)

2. Viewer undertakes a cooldown period of relaxation throughout the week prior to the races. The viewer's intent is to draw the feedback image that will be sent to him after the event. That image will be the one associated with the winning horse. Viewer makes a sketch, which is labeled with an ID number.

3. A race is selected. Races were selected at random from among the races at major North American thoroughbred tracks. The number of races selected was the same as the number of sketches available. Each race was randomly associated with a photosite ID number.

4. Image selection: The five-digit number generated in step 1 was used as the search string in Google Images. A sixth digit was added to the random number to find a unique image from the Google searches. Images were then manually filtered to remove images used before, computer graphics, etc. One image for each horse in a race was chosen, labeled with a letter.

5. Association of images with horses: The software randomly assigned a letter to the name of each horse in a race. The viewer did not look at this file till after the judging.

6. Judging: The viewer compared the sketches from Step 2 with the images and assigned a SRI/Targ score (range is 0 to 7). This was done both before and after Step 5 "in the spirit of the timelessness of psychic phenomena."

7. Race result: The horse with the highest SRI score was the selection. The race took place and the results and all other data were recorded on a web page and on a summary spreadsheet.

8. Feedback: The viewer noted the image associated with the winning horse and sent that image via email to himself. This step reinforced successes "for the intuitive/RV self."

Alongside the ARV process, Tom used standard handicapping techniques to make a "logical selection." This was based on data about past performances by horses, together with his 35 years' experience in handicapping. This was almost always performed before judging (Step 6).

Tom believes he has an open mind about most subjects, believes strongly in his "own intuitive self" and also in the "reality of extrasensory methods of accessing information." He placed great emphasis on intention and at times used a list of 38 positive intentions about horse racing events and his life generally. As examples, the final three in the list are:

- I have had GREAT success in the past at ARVing for horse races – and I know it is in me to do it again.
- I am really getting good at doing ARV sessions and judging.
- I am doing great!

A note from Debra and Jon

We very much appreciate the contribution from Teresa and Tom given their extensive experience and knowledge in this area.

Over the years we have occasionally participated ourselves in horseracing events, some at conferences where we learned innovative approaches. One of these was an event led by Teresa. Debra was excited because she didn't know Teresa was about to ask the audience to describe a color that she had pre-associated with a horse. This had never been raised as a possible method or alternative to ARV using photos before. However, suddenly Debra became aware she was seeing a color for no apparent reason in her mind's eye, with her eyes wide open.

About 30 seconds later, Teresa invited the group to picture a color. While we can't recall the exact number of horses, it was at least 10 for this prediction, hence a color for each horse. Of course, Debra didn't have to tune in any further. After all who were present wrote down a color, polls were taken to determine the top repeating color. The group color was not the same as Debra's and the horse associated with her color was not considered a favorite, but she went and wagered on this color only. The horse came in first place, but unfortunately, not being a big spender, she had only wagered $20 and her earnings weren't as impressive as her precognitive psi hit.

Catterick Race Course – Tunde Atunrase

In Chapter 24, Tunde Atunrase presents his Primary Pool RV method, which his team used with great success to predict the outcome of football (soccer) tournaments. Please refer to that chapter for details of the PPRV method. Tunde has kindly shared one of his horse racing successes with the same method, betting at the Catterick Track in the United Kingdom.

The horses were divided into two pools, as shown below. Two photos (each a potential target) were used, each corresponding to one of the two pools. Tunde self-judged his session against the two photos. The full session is shown below, followed by the track tickets. Tunde uses low-level data such as colors and basic shapes as criteria to decide matches between transcript and photos.

For this target, "Wheels" was a strong indicator and others were "outdoors," "road" and "movement," as well as the color green (though here the print is in B&W). Below the images is Tunde's complete ARV session.

Target 410-002

POOL 3	POOL 4
INGLEBY HOLLOW	FLOWER POWER
BOLLIN JOAN	LADY KYRIA
GYLO	TOMMY HALLINAN
EURO IMPLOSION	JAN DE HEEM

DISPLACEMENT ALERT MEDIUM
Group photo pick HR3 Truck with Green building in background

Figure 18.2

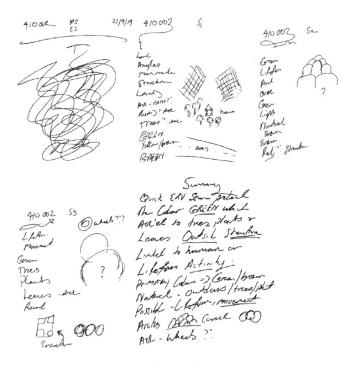

Figure 18.3

Text from summary: Quick ERV scan protocol. the color GREEN, which AOLed to trees, plants & leaves. Outside. Structure Linked to human or Lifeform Activity. Primary colors -> Green/brown. Natural-Outdoors/trees/plants. Possible – Lifeform, movement. Arches, circles. AOL: Wheels?

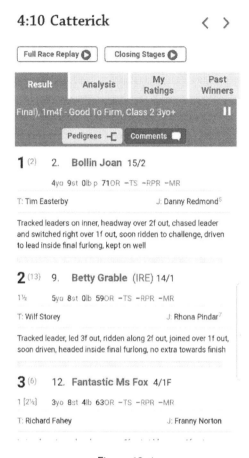

Figure 18.4

Figure 18.5

CHAPTER 19

Election Predictions: Did ARV Get It Right?

Who will be our next president? While this question has been one that many citizens of any country with a democratically elected leader will ask throughout their lifetime, the significance of the question has taken on epic proportions recently given the devastation caused by the global pandemic. Whoever is at the apex of power does in fact have the ability to tremendously impact, for better or for worse, the health, financial stability and cohesiveness of a nation's people.

While predicting the outcome of a presidential election is a casual past time for the average person, it is serious business for pollsters, statisticians and investors who make financial decisions based on these predictions, although the extent of the impact of these predictions is often debated in the literature. In a note to clients, the Goldman Sachs investment firm suggested:

the political stakes in presidential, parliamentary or legislative elections often translate into changes in policies that can reshape the economic environment. Second, the regularity with which elections take place in most countries may give place to cyclical patterns in government and investment behavior. And third, elections can markedly increase political and social uncertainty. These three factors have the potential to affect all asset classes, especially equities, given their strong sensitivity to changes in the economic outlook.[421]

Traditionally, presidential forecasters have made predictions based on complex algorithms combining polling, demographics and sophisticated analysis of swing states. Allan Lichtman, an American political history professor at American University, has successfully predicted the outcome of every presidential election since 1984, often months ahead, by using a process he and Russian scientist Volodia Keilis-Borok developed. The method employs a series of true/false statements addressing variables such as party mandate, incumbent party, third party, short-term economy, long-term economy, policy changes, scandal, social unrest and charisma of incumbent vs. challenger.[422]

Statistician and predictive analytics expert Nate Silver successfully predicted the outcome of the presidential election in 49 of 50 states in 2008 and all 50 states in 2012 using big data methods. These included the analysis of multiple factors such as past election results and current polling data. However, three days prior to the election on Nov. 4, 2016, between Donald Trump and Hillary Clinton, Silver was not confident about his overall prediction due to what he termed "the uncertainty factor," which consisted of harder-to-predict variables such as voter turnout in particular states and the impact of one state's outcome on others in the final hours. He speculated "while Clinton's a 76 percent favorite to win the popular vote according to our polls-only forecast, her odds are more tenuous – 64 percent – to win the Electoral College. (Her chances in the polls-plus forecast are identical.) It would not necessarily require a major polling error for Trump to be elected, though he would have to do so with an extremely narrow majority in the Electoral College."[423]

Silver further compared his own predictive model to other polling-based models giving Clinton a 77 to 99 percent chance of winning. Actual results of the election as of Dec. 22, 2016, as reported by the Associated Press, showed Hillary Clinton surpassed Donald Trump in the national popular vote by nearly 2.9 million votes with 65,844,610 votes across all 50 states and Washington, D.C. This was 48.2 percent of all votes cast. Trump received 62,979,636 votes, which was 46.1 percent of all votes cast. Clinton therefore had 2,864,974 votes more than Trump, the largest popular vote margin of any losing presidential candidate in US history. Trump won the presidency by clinching 304 electoral votes, whereas Clinton won 227 electoral votes.[424] From the above, it's clear that predicting elections is tricky business, even with the best analytic models and tools.

So, what if there was a way to combat the "electoral uncertainty principle," to essentially leap over the unknowns, bypass all surprises, and latch onto the final outcome, no matter what unexpected twists and turns happened in the days or even hours leading up to the election? The projects discussed below were designed to do just this, working on the premise that if there is a way, it is likely not purely an analytic approach but one grounded in intuitive-based processes.

2012 elections

Remote viewers predict outcome of the 2012 elections by Debra Katz & Michelle Bulgatz

With the above in mind, Debra and Michelle, later assisted by T.W. (Teresa) Fendley,[425] designed a project to determine whether remote viewers using a double-blind protocol could describe a human subject in enough detail so raters could choose between two potential candidates and predict the outcome of the 2012 United States Presidential election. Remote viewers use intuition and structured protocols to obtain information that lies outside their analytic minds and conscious knowledge. Unlike other intuitive disciplines that focus on human subjects, in Associative Remote Viewing, human subjects are among the lesser-used targets. The three researchers set out to answer the following:

1.   How strongly did the viewer's candidate preference affect their session?
2.   How does a project involving a human target differ from one that targets objects and locations?
3.   Should project managers consider the use of human targets in remote viewing projects or research which involve a binary outcome?
4.   Which session rating method/system is the most helpful with human subject targets?
5.   Why are human subjects/targets typically not used in formal RV research studies when they are quite often the main focus for intuitive practitioners?

Initially, 11 remote viewers were given a target number and no frontloading. However, after a few transcripts were turned in with no mention of people, the researchers made the decision to start the project over and task the viewers with the cue "The target is a person." Sessions were turned in one week prior to the election. Each word and sketch was input into a spreadsheet and compared to both candidate photos using both the SRI/Targ Scale and the more sensitive Dung Beetle system.

The researchers used a consensus team judging approach to score every word and sketch. This required them to meet online and agree on each item, which they were surprised to find was exceedingly difficult. Even items that seemed obvious became items for debate depending on the news sources the raters tended to watch, as well as on definitions of words. For example, one viewer stated the person had multiple brothers and gave a specific number. An online search showed this could apply to Obama but the viewer had one too many people. The judges debated whether half-brothers from another parent or ones he hadn't lived with should count. From all of this, they realized if one has to argue or research a response for more than 15 minutes, it should just not be counted.

Once the scoring was completed, the results were sent to Alexis Poquiz (originator of the Dung Beetle scoring system), who calculated the percent that matched (Correct), did not match (Wrong) or that were unknown for each candidate. The results showed of 11 sessions, eight were deemed a match for Obama and three for Romney. The "Lower Q%" score also yielded an overall group prediction for Obama.

In summing up the experiment, the researchers came to a few conclusions about the challenges posed by rating human targets. First, humans have too many inherent similarities. For example, they generally have the same number of body parts, a few different skin tones, hair and eye colors, etc. So, while it might seem as if Mitt Romney, a Caucasian man, and Barack Obama, a Black man, would be seen as very different, they are both male, tall, slender, have brown eyes and hair, are the same age and have the same number of body parts. The only difference is skin tone.

The next problem was that of the viewer's "subjective relational descriptors." These could be defined as words that mean different things to different people, usually because the word's meaning is already comparable with something else. For example, when a viewer says a subject in a remote viewing session is "tall," what does that mean? How tall is tall? A viewer that is 5-feet tall is going to describe someone who is 6-feet tall as tall, whereas a 6-foot-tall viewer may not mention or even notice someone else is "tall" unless that person was 6'6". This is true of lots of words, such as "attractive," "intelligent," "friendly" and "quiet."

In this project, a couple of viewers mentioned the subject was "darker skinned," while another mentioned "lighter skinned." The viewers who mentioned "darker skinned" were Caucasian, while the viewer who mentioned "lighter skinned" was African American. Barack Obama is darker skinned than Caucasian people but for someone of African American descent (his father's side of the family) he is fairer complexioned. So what did the viewers mean? This finding was rather surprising – that it could be so challenging to tell the difference between a Caucasian and a black man. We think this drawing by R.E. though, demonstrates the challenges. Did he draw Mitt Romney or Barack Obama?

Sketch by Viewer 7 - Session pointed to Romney, but there was a high number of "Q"s.

Figure 19.1

The third challenge in rating humans was a judge's inability to perceive a subject's inner life in the way a remote viewer can. Or to put it another way, viewers will observe many factors about a person – their thoughts, emotions and overall private life – that a rater would never be able to verify. For example, several viewers indicated this person had a public life but was actually a very private person who preferred to be alone. We could confirm that both people had a public life since they were politicians (which remember the viewers did not know – all they knew was the target was a human). However, we could not confirm which, if either, prefers being alone. All the judges could go by was what has been revealed in the media, which brings us back to the problem noted above – one judge primarily watched one news channel (Fox News) and the other watched another (CNN), meaning even things considered a "known fact" to one judge were often questioned by the other.

Based on the above discoveries, it was recommended that human targets not be used in remote viewing research projects or applied precognition projects involving binary outcomes unless only one of the photos is of a human. It was also found that Poquiz's Dung Beetle scale was a useful rating tool as it did allow for breaking down and evaluating every item on its own terms.

One other factor tested in this project was whether a remote viewer's preference for a particular candidate could affect their session, even when they were unaware what the project entailed. To test this, after all transcripts were turned in, viewers were polled on their preferences. Viewer preferences for a particular candidate were compared to their judged prediction. Seven of 11 viewers indicated a preference toward a particular candidate. All seven voiced a preference for the candidate their session pointed to, including one whose session pointed toward the wrong (losing) candidate. Still the sample size was not large enough to make any definitive statement about this other than that the concept of a viewer's preference toward a particular outcome warranted further investigation.

2016 elections
ARV predictions of the presidential elections
by Debra Katz & Michelle Bulgatz

For the 2016 elections, Debra and Michelle undertook another experiment, this one using an ARV protocol. After the project was completed, Debra's classmate, Nancy McLaughlin-Walter, who was also a student at UWG finishing up her Ph.D., joined the project. Nancy served as their statistician. This project was carried out by Debra and her research partner, Michelle Bulgatz, and was originally published in *Eight Martinis* magazine.[426]

Forty-one moderately to highly experienced remote viewers were tasked with describing a feedback photo they would see at a future date. Viewers were kept blind to the nature of the project. The undisclosed tasking was "Describe the photo that you'll see after the elections as your feedback that is associated with the actual winner of the 2016 US presidential election."

The hypothesis was that most of the remote viewers would have descriptors and sketches that would strongly match the photo associated with the winning candidate and have lower correspondence or no correspondence with the photos associated with the competitors. Viewers were given a target number and told their target was the feedback photo. They had approximately one week to turn in their transcripts. Once these were turned in, researchers compared the remote viewers' transcripts to a set of four photos – two associated with the Republican and Democratic frontrunners, one with a third-party candidate and one with an impossible option that served as a control. A session was considered predictive only if it gained a score of at least four on the SRI/Targ 7-point scale, meaning there was close correspondence and little incorrect data. If this score was not reached, then it would not count toward a prediction.

Once all transcripts were evaluated against each photo option and scored, the scores were added together and a formal prediction was issued. The prediction favored Hilary Clinton and a third-party candidate over Trump.

Out of 41 participants, 19 had scores that were high enough toward one photo and low enough for all others to issue a prediction toward one choice. Twenty-two passes were issued either because of low scores across the board or conflicting scores indicating more than one photo may have been described. Of the 19 predictions, eight pointed toward Hillary Clinton, eight toward an independent candidate and only three toward Trump.

No predictions favored the control photo, which was attached to an option we knew would not win – that the researchers (Debra and Michelle) would win the election. As a result, the formal prediction was for Clinton, since it seemed unlikely that a third-party candidate would win.

The formal prediction was shared on a remote viewing forum from which several of the remote viewers had been recruited. However, the researchers changed their minds by the end of the day about exposing viewers to the prediction (even though it is not unusual for predictions to be shared with viewers in applied projects). This resulted in some viewers seeing the prediction and some not. The researchers decided since this had happened that they would take advantage of the disclosure and do a test to assess whether exposure to a potentially wrong prediction might result in displacement to the wrong photo. They polled all viewers to find out who had been exposed to the prediction. Other variables such as viewer preferences and voting behaviors were also assessed.

After the election results were in, it was clear the group prediction had been wrong. Results indicated that rather than describing the photo the remote viewers would see at the future date, they instead tuned in to photos they would not see.[427]

Why did this happen? What went wrong?

These researchers were certainly not the only ones asking this question, as just about all the national polls had made the wrong prediction in the days leading up to the election. For example, according to an article published by the Pew Research Center on Nov. 9, 2016, by Mercer, Deane and McGeeney:

> The results of Tuesday's presidential election came as a surprise to nearly everyone who had been following the national and state election polling, which consistently projected Hillary Clinton as defeating Donald Trump. Relying largely on opinion polls, election forecasters put Clinton's chance of winning at anywhere from 70% to as high as 99%, and pegged her as the heavy favorite to win a number of states such as Pennsylvania and Wisconsin that in the end were taken by Trump… The fact that so many forecasts were off-target was particularly notable given the increasingly wide variety of methodologies being tested and reported via the mainstream media and other channels. The traditional telephone polls of recent decades are now joined by increasing numbers of high profile, online probability and nonprobability sample surveys, as well as prediction markets, all of which showed similar errors.

One theory was that perhaps the polls hadn't been so off. According to a *Chicago Tribune* article: "Clinton had clinched 47.7 percent of the votes cast, while Trump sat at 47.5 percent." According to a CNN report issued on Dec. 12, 2016, "The Democrat outpaced President-elect Donald Trump by almost 2.9 million votes, with 65,844,954 (48.2%) to his 62,979,879 (46.1%), according to revised and certified final election results from all 50 states and the District of Columbia. However, Trump won the electoral vote, which is determined by a process conducted by the electoral college, based on the original Founding Fathers' mandate in the Constitution that

> Each state shall appoint, in such manner as its legislature may direct, a number of electors equal to the whole number of senators and members of the House of Representatives to which the state may be entitled in the legislature.

Still, this theoretically should not have affected the ARV process, since Trump officially won the election and became president. This was very frustrating to the researchers.

From the start, based on their extensive experience with ARV as remote viewers, judges, project managers and researchers, the researchers were already cognizant of the possibility of displacement – and determined to do what they could to guard against it. They attempted to mitigate this in a variety of ways. They made sure the photos to be paired with each outcome were different from each other in every aspect, but well balanced in their attractiveness, entropy, numinosity, emotionality, etc. They wanted to make sure they both equally liked each one – to lessen what Charles Tart found in a project in which taskers had preferences for target choices. They also used a consensus approach in both photo selection and subsequent judging to reduce subjective preferences with regard to photos, viewers, appearance of transcripts, etc. Further, to lessen distractions and noise, they decided to keep the viewers blind to the project's purpose, meaning that potential participants were informed only that this was a remote viewing project. They were told they simply needed to do a good job describing their final feedback photo, which would be about a real location and anything occurring or found there. Viewers were not told this was an ARV-related project or one that had to do with the presidential election.

Some might suggest it wasn't too hard for the viewers to figure out this might pertain to the elections, which was just a month away, and the researchers had done the above-mentioned ARV project for the elections four years earlier.[428] However, given that viewers were not directly tuning in to the candidates, but instead viewing photos they were blind to, the authors did not think this would call into question the validity of the experimental design. In fact, in many ARV projects the viewers do know the project involves ARV and are aware it is for a stock market prediction or a particular sports event. It should be noted in any analysis of the causes of displacement that a few viewers did later say they suspected this project might involve using ARV for the election.

Once the election happened and the world was surprised by the outcome, viewers were sent the feedback photo. Analysis showed those who had been exposed to the prediction had a higher rate of displacement to the wrong photo than those who hadn't. However, given that these groups were not large (not everyone responded to whether they had been exposed), results can't be seen as definitive. Still, given that groups today are often formed around being able to offer predictions to its members, the question of whether there could be a displacement effect due to knowing the prediction itself is one that needs more exploration.

Additionally, the researchers polled the remote viewers for their preferred candidate, thinking perhaps they had described their own preferences; however, no trend in the preferences was observed. In fact, the viewer who clearly did the best was former TransDimensional Systems project manager and viewer John Vivanco. The photo associated with Trump was a decorative hanging light fixture made of broken plates and cups strung together (which was randomly chosen, but now seems quite symbolic). John essentially named this highly unusual target. In his summary he wrote, "Subject didn't feel so solid…biggest gestalts were plates clamping, moving together & over each other." We questioned Vivanco about his preference prior to results of the election being known and he stated he was strongly for any candidate besides Trump.

Finally – the question had to be asked – could the researchers' preferences for one candidate over the other have affected results? (Especially since they knew the goal of the project and served in all roles – target selection, rating, and predicting?) At least one of the researchers had a very strong aversion to Trump as a candidate and a belief he would not win. The other researcher was more neutral but also definitely not a fan. While we can't be sure, it is possible that the preferences of a tasker or analyst may impact the overall outcome – in this case, not a preference regarding the photo choice but the outcome itself. This would not be surprising given early researchers' beliefs that the only way psychic transmission could happen was between two individuals having a telepathic link through intention, attention and relationship.

Figure 19.2

John Vivanco's session pointed to Trump being elected: "Subject didn't feel so solid...biggest gestalts were plates clamping, moving together & over each other."

2016 election prediction Facebook poll conducted by Daz Smith

Daz Smith conducted a poll about the 2016 Presidential race on his Facebook group, Remote Viewers and Remote Viewing. The poll generated intense interest and the thread ran to 250 comments. Daz asked viewers to post who would win based on their use of ARV for this election. There is no way of knowing if everyone used ARV, but the poll showed 30 viewers predicted a Trump win, 17 a Clinton win and 6 a Sanders win. Four predicted a win by other candidates and two other votes may have been submitted as jokes. Those who preferred Trump predicted he would win and the same for Clinton. However, this was not formally polled.

In the aftermath, the project administrator, Daz Smith, was very candid about his session:

OK so just looked at my ARV feedback image again and re-looked at my ARV of the session for this and yes, probably all the signs were actually there in the data for a trump win but my NON blind analysis of my own data was clouded by my wants in my final decision.

So this is my first off target session in a very lonnng time.

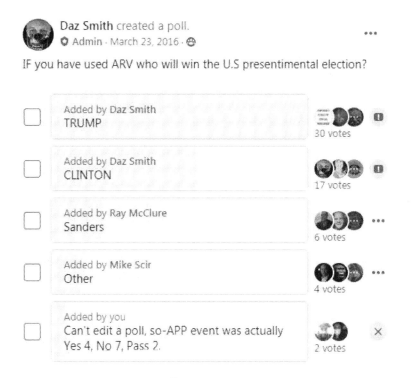

Figure 19.3

Long-time viewer and author Tunde Atunrase commented as follows:

I think if anything this election with regards to ARV has demonstrated several things.

1. Multiple viewers does not guarantee an accurate prediction.
2. Having an emotionally invested interested in an outcome will skewer your results.
3. Be blind to the target and stick to your best viewers.
4. Trust your instincts. If you are not confident about your session even if you did an 8m hit...don't take it too seriously.

Number 1 and 4 are very important points to hammer home. For example, despite two sessions pointing towards a Hillary win I never placed any bets or put too much stock in it. Those that know me know I bet heavily when I'm confident and have seen the winnings but for some reason…I really had zero faith the data would make money by placing a bet on Hillary. Listening to that hunch saved me a whole heap of heartache, pain money loss. So it wasn't all bad. I learned to curb the urge and just have fun with it despite the outcome. Will be interesting to hear what other viewers who predicted the wrong outcome have to say about their experience. By the way well done to everyone who contributed 🙂 hopefully we can predict more events this way. Overall this group poll was a success. The correct prediction was reached overall. Well done guys.

2016 APP conference prediction

We don't have much information about this project but during the 2016 APP conference, a trial to predict the outcome using ARV was undertaken. The question was: Will Hillary be the Next President? The result was Yes 4, No 7, Pass 2, meaning the APP also predicted a Trump win.

Summary of 2016 predictions

All told, this meant that the predictions for 2016 were:

Candidate	Katz	Facebook	APP	Total	Percent
Trump	3	30	7	40	45.9
Clinton	8	17	4	29	33.3
Other	8	10	n/a	18	20.6

Figure 19.4 – Candidate Table

That is to say: for 2016, ARV had it right.

2020 elections

2020 Applied Precognition Project APPFest U. S. presidential election prediction

The following write up is a collaborative effort by Crystal Hope Reed and Debra Katz, with assistance from other APP members.

The 2020 Applied Precognition "APPFest" Conference was held the weekend of Oct. 23-25. Rather than their usual meeting in Las Vegas, due to the Coronavirus pandemic it was held on Zoom, in which participants met via audio and visual conferencing. As usual, attendees participated in several collaborative ARV trials, with some acting as viewers and some as judges, the latter of whom had gone through a training to learn Tom McNear's judging method (see Chapter 4). Given the proximity of the November election, on the final day of the conference organizers decided to task the group with an election-related question.

The tasking question was "Will Donald Trump be sworn in as president on Jan. 20, 2021?" One photo was to be paired with "Trump will be sworn in as president" while the other photo would be attached to the option, "No, he will not be sworn in as president."

This seemed like a fail-safe tasking. The term "sworn in" refers to inauguration day, which didn't happen until Jan. 20, 2021, almost two months after the election to allow all state election offices to formally verify their results and any court proceedings to be resolved. Regardless of how the candidate makes it there, one way or another a candidate will be sworn in on this day, barring a catastrophe. This was especially important in 2020 given that incumbent Trump had given every indication he would contest the outcome if he didn't win. (That's exactly what happened, and uncertainty remained after Election Day about what would transpire on Jan. 20, 2021.)

The remote viewers were told the tasking question. All judging was done by independent judges, so the remote viewers were not exposed to the photo options at judging time. The only photo they would ever see was the one associated with the winning outcome, shared with them after the swearing in. It was decided that another webinar would be held in which viewers would receive this feedback photo sometime after Jan. 20.

Viewers did their remote viewing sessions from the privacy of their own homes while logged on to the conference. They quickly scanned their transcripts and emailed them to the judging manager (Crystal). She distributed transcripts to the judges, who were in a separate Zoom breakout room. The entire process took about 45 minutes, at which time the judges issued a prediction, which was shared with all conference participants. The verdict was in: The overall group tally indicated President Trump would grace us with his presence for another four years. Of 21 scores, 17 pointed to Trump and only four to "Not Trump" – a landslide prediction for the incumbent.

At this point, Debra suggested (as usual) that data be tracked so a future formal assessment could be made. Then Marty Rosenblatt (as usual) pointed out this was not set up as a formal research project, but just a group activity that might help establish the usefulness of remote viewing should the results be positive. Then Debra countered (as usual) that without proper reporting and communication of methods and results, no one would ever know whether a project is successful, at which point Crystal stepped in and offered assistance (as usual).

Crystal, who had overseen the judging and other conference organizational tasks, agreed to collect all the data and biographical characteristics of the participants, while maintaining a delicate balance of confidentiality and consent. First, we tried to identify who was an experienced viewer and who was brand new. We were able to ascertain at least half the viewers were highly experienced, but we weren't sure about the other half. The next thing we tried to determine was the "presidential preference" of each viewer and judge. This is obviously a sensitive issue and one not everyone was comfortable discussing, but the majority responded on the condition their responses remain confidential.

Prediction 5: 894377 – Actual Scores		
Viewer	**CR PhotoSite A** (Trump will not win)	**CR Photosite B** (Trump will win)
Viewer 1	0.00	5.25
Viewer 2	1.43	3.22
Viewer 3	2.45	2.96
Viewer 4	1.61	3.01
Viewer 5	2.15	4.78
Viewer 6	5.00	4.50
Viewer 7	1.93	5.36
Viewer 8	2.91	5.27
Viewer 9	1.14	4.99
Viewer 10	1.17	4.20
Viewer 11	0.78	1.78
Viewer 12	2.14	4.52
Viewer 13	4.30	0.58
Viewer 14	0.47	5.25
Viewer 15	1.56	5.42
Viewer 16	2.45	5.42
Viewer 17	1.93	5.08
Viewer 18	4.06	2.24
Viewer 19	4.38	3.15
Viewer 20	1.11	2.10
Viewer 21	0.72	5.48
CR Summation	43.69	84.56
Mean CR	2.08	4.02
Prediction	No	Yes

| 21 viewers | 4 transcripts with 3.5 or higher for Trump will not win | 17 with a score of 3.5 or higher for Trump will win |

3.5 or higher – we added up all scores that were at least 3.5 on a 7-point scale.

Figure 19.5 – Prediction 894377

Results: The actual outcome of the election was that Joe Biden won, not Trump. The group prediction of Trump was a clear miss. The Mean CR for the viewers was 4.02 for Trump and 2.08 for Not Trump. Seventeen votes for Trump were incorrect, 4 for Not Trump were correct.

Scores: From this table, it is clear that scores were substantially higher for the photo associated with Trump, with a wide spread between scores for each photo. In fact, the only time the difference was not wide was for the two transcripts that had higher scores for the Not Trump option (viewer 6 and viewer 19).

It appears the scoring did not allow for passes in cases where there was a small spread or scores were very low. However, we reevaluated those, looking to see how this might have affected the final results, and found it would not have. Of the 21 predictions, five scores were quite low and would have been discarded if there had there been a passing choice. Even if a rule had been established to include only scores of 4.0 or higher, and to allow a pass when scores for each option were of approximately equal value, this would have resulted in 13 votes for Trump, and 3 votes for Not Trump.

In order to understand what went wrong with the prediction, we looked at two measures: viewer and judge preference (see Table # below).

Viewer preference: Of the 21 viewers, 17 shared their preferences. Of the 17, 10 candidate preferences matched their individual prediction and seven did not. While slightly more (three) did than didn't, this is not enough to establish that viewer preference was correlated with predictions.

Judges' preference: Further, the eight judges were also polled on their preferences. Seven responded and of these, five had preferences that did not match the overall prediction, while two did match the prediction. Judges who did not prefer Trump still had a majority of their viewers' predictions favoring Trump. Only two of these judges had a single viewer whose sessions pointed away from Trump, with 10 toward Trump, which contradicts any hypothesis suggesting judging preferences could have played a role.

However, for the two judges who preferred Trump, of their seven viewers' predictions, five matched the judges' preferences (toward Trump). Still, this is much too small a sample size to draw a conclusion. To make that assessment, we would have needed many more judges preferring Trump and more viewers assigned to them who picked Trump.

Of the four correct predictions: Two of the four predictions correlated with the viewers' own preference, one did not and one was unknown. Two of the correct sessions correlated with judges' preferences, and two did not. Two of the viewers were experienced and the other two had unknown levels of experience.

Table # showing APPFest Predictions – "Yes" refers to a vote for Trump as either the prediction or the preference; "No" refers to a preference for any candidate other than Trump. "Same" refers to whether the viewer's preference for the candidate is the same as the prediction their transcript yielded. New/Exp refers to viewer's level. The ("no" or "yes") next to the judge's name reflects the judge's preference for the candidate.

Prediction 5 894377	Analyzed by & Judge's Preference	Y/N Pref	Y/N Score	Same?	New/ Exp
Viewer 1	Judge D (no)	Yes	Yes	Same	Exp
Viewer 2	Judge F (no)	Yes	Yes	Same	
Viewer 3	Judge C (yes)	No	No	Same	
Viewer 4	Judge E (no)	No	Yes	Diff	
Viewer 5	Judge G(yes)	No	Yes	Diff	New
Viewer 6	Judge E (no)	?	Yes	?	
Viewer 7	Judge E (no)	No	Yes	Diff	
Viewer 8	Judge G (yes)	No	Yes	Diff	Exp
Viewer 9	Judge F (no)	No	Yes	Diff	New
Viewer 10	Judge H (no)	N/A	Yes	N/A	Exp
Viewer 11	Judge G (yes)	No	No	Same	Exp
Viewer12	Judge C(yes)	No	Yes	Diff	
Viewer13	Judge C (yes)	Yes	Yes	Same	

Viewer14	Judge D (no)	Yes	No	Diff	Exp
Viewer15	Judge A (no)	N/A	No	N/A	
Viewer16	Judge F (no)	Yes	Yes	Same	Exp
Viewer 17	Judge G (yes)	No	Yes	Diff	Exp
Viewer 18	Judge E (no)	Yes	Yes	Same	
Viewer 19	Judge F(no)	N/A	Yes	N/A	Exp
Viewer 20	Judge B (?)	No	Yes	Diff	Exp
Viewer 21	Judge B (?)	No	Yes	Diff	
		Yes: 6	Yes: 17	Same: 7	
* A=No, B=Yes		No: 11	No: 4	Diff: 10	
		N/A: 4		N/A: 4	

Figure 19.6

Analysts (& did their preference match outcome they judged?)	New/Exp	Pref	Judged
(1 match) Judge A	Unknown	No	Viewer 15
(N/A) Judge B	Exp	N/A	Viewer 19, Viewer 21
(2 out of 3 trump predictions) Judge C	Exp	Yes	Viewer 3, Viewer 12, Viewer 13
(1 match) Judge D	Exp	No	Viewer 1, Viewer 14
(0 match) Judge E	Exp	No	Viewer 4, Viewer 6, Viewer 7, Viewer 18
(0 match) Judge F	New	No	Viewer 2, Viewer 9, Viewer 16, Viewer 20
(3 out of 4 yes predictions) Judge G		Yes	Viewer 5, Viewer 8, Viewer 11, Viewer 17
(0 match) Judge H	Exp	No	Viewer 10

Figure 19.7 – Predictions 5

Discussion

We cannot say for sure why the overall group predictions and the majority of predictions failed. Many sessions were scored for the outcome that did not occur, raising the question of whether the judges rated one photo higher because they preferred that photo or whether the viewers' subconscious went for the photo that was more interesting. Different viewers and judges will obviously have different responses to photos. The ARV process is predicated on the theory that viewers will and can focus only on the photo attached to the winning outcome, but as we know, displacement frequently occurs in binary ARV.

Co-author Jon served as a judge and he feels strongly that the photo associated with a Trump win was far more interesting than the other. For one thing, the content of the winning photo was very dynamic, of the type that Star Gate viewers always got.[429] As Jon wrote, "It was striking and numinous while the alternative photo was boredom personified." Two of the APPFest managers, one of whom was also a judge, did not agree with Jon's assessment that the mismatch in the photos could have been responsible. Jon was surprised at this because he feels disproportion between two photos is one of the primary causes of ARV displacement and one of judges had agreed with this point in the past. In Jon's view, these two photos were as clear an example as one could ask for.

This was the feedback photo corresponding to the win by Biden:

Figure 19.8

Upon seeing the actual feedback photo, Debra concurs with Jon – very boring. (Readers, however, can decide for themselves). Also, one thing we've observed in other projects is if you give remote viewers a scene that is mostly empty, they will assume there has to be more, that they must be missing something and will fill in the blanks.

Another possible reason for failure for viewers to access the correct feedback photo could be that some viewers felt they did not get to see their feedback photo in a timely manner, or at all. The feedback photo was emailed to the viewers when the outcome was known months later, but some expected to come together as a group to receive the feedback. Others said they didn't see the email until they looked in their spam folder. An audit of whether viewers all saw the feedback has not been conducted.

What else could have gone wrong in this ARV prediction? One possibility is that the prediction was announced to the group at the conclusion of the viewing and judging, prior to the outcome being known. (Which is kind of the point of doing such a project – making a prediction, except the viewers themselves don't necessarily need to know what that is.) As noted in Debra and Michelle's 2016 Presidential prediction, researchers questioned if issuing a prediction and announcing it ahead of the outcome being known could lead viewers to describe the photo associated with the *candidate whose winning was predicted* rather than the candidate who actually does win. This is a similar question to what they explored in their 2012 election project – could a viewer describe a photo associated with the *candidate who is their preference*, as opposed to the candidate who turns out to be the winner.

A future project might seek to control this variable by dividing viewers (and judges) into two groups, those who receive the prediction and those who don't.

2020 presidential election prediction project by "Frank"
Does proximity in time to the event produce better results in precognitive projects?

"Frank&Friends" (hereafter Frank) is a moderator of the Subreddit RV Group. While he's done research in other areas, this was his first formal RV research project. Beginning in December 2019, Frank initiated an ARV experiment to see if viewers' predictions could be used to successfully predict the 2020 US presidential election. He wrote:

> This would also serve as an experiment that would look at 1) the reliability of ARV predictions relative to time from the event in question, and 2) better understand how to predict the outcome of an event using crowdsourced remote viewing data obtained from online communities. The election was chosen as it would be an emotionally and psychically "noisy" binary outcome event, and because polling

and reporting leading up to the result would provide a proxy counterfactual. The experiment used nine ARV targets/sessions for the same event. 18 Remote viewers were self-selecting members of the Reddit /r/RemoteViewing community who submitted their RV session data via the forum. The main hypothesis for this experiment was that an economic concept known as "discounting" would provide a reasonable and acceptable reduction in prediction strength when applied to ARV binary outcome remote viewing data. Discounting reduces the perceived value of something right now the farther in the future it will be realized – in this case the reliability of binary ARV predictions today on an event in the future. This is intended to provide easier interpretation of remote viewing prediction data, and more accurate predictions when looking at large data sets over time (p. 1).

Methodology

The methodology we will describe is based on the paper Frank wrote. We've used some of his wording and interjected some of our own.

Tasking

Frank created two targets at the start of the project. One had the undisclosed wording: Target Number ####-####: "If the incumbent wins the November 3 US Presidential election, the viewer will describe _____. ONLY. The second one had the undisclosed wording: Target Number ####-####: If the challenger wins the November 3 US Presidential election, the viewer will describe _____ ONLY".

He used the terms terms "incumbent" and "challenger" since it was unknown whether the people in these roles might change. The first target was assigned right before impeachment hearings started, from Dec. 22, 2019, and Nov. 3, 2020.

Because Frank was using the already established reddit.com /r/RemoteViewing community forum, he had to follow certain rules, which led him to making procedural choices he might not otherwise have made. The Reddit rules required that he post the tasking to a target in a section entitled "REAL TARGET" to demonstrate this was for an actual project rather than for practice. It also required him to post that there would be "FEEDBACK" within seven days of the tasking.

Since he had to post something, he chose to post the feedback photos associated with both target photos. He was not requiring the viewers to self-judge since he was doing the judging, but he did occasionally ask the viewers which photo would be the correct one. Therefore, we assume viewers were exposed to and at times looked at both photos.

Frank noted that for the final prediction, which took place on Nov. 3, he provided only the photo attached to the winning option:

When revealing targets, the author posted at least one image related to each target, and where applicable a website describing the target for greater detail. For the Nov. 3 target, feedback was posted on Nov. 6 and only images related to the challenger outcome were posted as every major media outlet had called the election for the challenger.

This was a break in protocol since all the other trials had included photos that were something other than the feedback photo (since the outcome wasn't known before Nov. 3, 2020). Frank described the scoring system he created:

This adapted point scoring system offers a streamlined way to make use of remote viewing online communities as pools of amateur and semi-professional viewers. Crowdsourcing remote viewing data from online sources allows for a far greater number of participants while maintaining a blind nature target. However, the fidelity of the data received are often lacking as compared to more experienced viewers who take their time to their sessions. This method attempts to bridge this gap and allow easier access to a wide, albeit it shallow, pool of remote viewers.

Adapting a point scoring method for widespread internet use

User-submitted session data posted to /r/RemoteViewing is varied, and can be everything from brief, text-only descriptions, to scanned images of notes with additional write-ups. To better accommodate the use of a discounting factor and inclusion of low-fidelity data, session data were evaluated using a 0, 1, or 3-point score: 0 points for a mixed hit or inconclusive but relevant, 1 point for a weak hit, and 3 points for a clear hit. The author scored hits for the incumbent as positive points since incumbents statistically retain seats during re-elections, and recorded hits for the challenger as negative points. A sum of the total points would form a basic prediction akin to the more standard ARV point scoring methods. The author gave even moderately mixed results a zero score, presuming that inconclusive data would add noise and should be eliminated, rather than included.

Confidence ratio

The researcher also sought a way to assess their relative confidence of a prediction using the point system. In the absence of robust data, this would control for mixed predictions, or data points where few viewers score strong hits to one side and several score weak hits to the other side. The author weighed point totals with the ratio of sessions agreeing with the predicted outcome, out of all of data submitted (including mixed results scoring zero points). This allows up to 100% of the sum to remain as the prediction strength for unanimous results and deflates mixed results.

$$C = S_p / S_n$$

C=Confidence Ratio, Sp = Count of Sessions agreeing with winning prediction, Sn – Total Sessions

For example: Target 1135-29H4 had three sessions reported, scoring -3 points, 1 point, and 0 points. The total of -2 points is a prediction for the challenger, albeit a weak one. The confidence ratio is 33.3% as only one out of three sessions agreed with the predicted outcome.

Discounting factor

Discounting, also called delayed discounting, or temporal discounting, is a method of assessing the perceived value right now of something in the future. Economists, accountants, and financial planners use discounting in cost/benefit analysis calculations or in behavioral economics to evaluate what something with a delayed payout is worth today, such as investment in climate change policy. A discounting rate is a percentage of how fast the value drops over a specific period of time: a percentage per day, per week, per month, per year, etc. A discounting factor is then a multiplier between 1 and 0 that is based on the discounting rate and time from the event. The discounting factor is then applied to the original value of something and reduces that original value to what the value actually seems to be worth.

Frank noted a 2% discounting rate is typically used for a number of applications when no established discounting rate is known:

Though, rates of three, five and seven percent have their place, such as if the item under discussion is perishable…Discounting applied to remote viewing predictions using a point system smooth large data variation in early predictions by reducing the value of a prediction. Using the previously mentioned point system, this means reducing the point total of a prediction based on how far the prediction took place from the event…For this experiment a standard discounting formula was used to find the discounting factor for the date of each target ID. A discounting rate of five percent (5%) and two percent (2%) per day were used for initial analysis. The resulting discounting factor for each prediction was then applied to the product of the raw point sum multiplied by the confidence ratio. The resulting figure for each target ID were then charted to provide an adjusted graph of predictions over time.

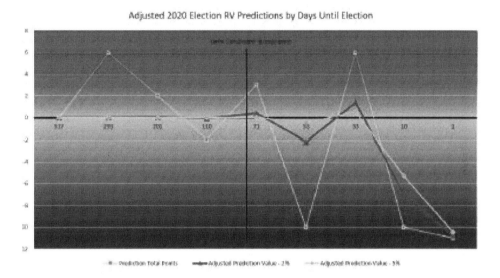

Figure 19.9

Chart showing raw prediction point totals, point totals after confidence ratio and a 2% daily discounting factor applied, and point totals after confidence ratio and a 5% daily discounting factor applied.

Results

Frank noted that "Predictions varied over the course of the data recorded, with the viewers providing more aligned data as the election approached:

> After applying the confidence ratio and discounting factor to point totals for each target ID, the adjusted prediction value for both a 5% and 2% per day discounting rate was plotted against the raw point scores. The adjusted values suggest that early predictions prior to the challenger primary process concluding were simply not trustworthy as point scores dropped to near zero, even for accurate predictions. Sessions from after the primary process, including a relatively strong prediction for the challenger in mid-September, become stronger and lean more accurate over time, whereas a moderate-strength prediction for the incumbent is tempered

by both the discounting factor and confidence ratios. The final two sessions provided fewer mixed results, which in addition to being close to election day, gave significantly higher scores.

His assessment was:

In terms of simply charting points over time, this would lead to the conclusion on the surface that either 1) the early ARV predictions were not reliable in the first place, 2) or that the outcome of the election changed as time progressed and the early predictions were accurate and tracked this change...

Feedback was provided to the November 3 target, 018-A724, based on the calling of the election by AP News and Reuters, on November 6. This was the only target with both a unanimous prediction and no mixed results. It seems likely that providing feedback for only one outcome may have led to a stronger prediction for that target, in addition to the target being posted and most data being returned after all votes were cast.

The true discount rate or appropriate unit of time for it to be applied is impossible to know for certain due to the unknown nature of remote viewing. Both a two percent (often used as the standard for inflation and interest calculations unless another established rate is known) and five percent daily rate were used for initial analysis. While subjective in this application, this rate is frequently applicable in most situations to provide a reliable discounting factors across a number of industries and disciplines. A 5% per day discounting rate seems to have diminished the scores considerably, and may be too large a rate to be used practically as any predictions more than 90 days before an event would drop to a near zero score.

An assessment of the project

The main goal of this project was to test whether the proximity of viewing to the actual event can produce better results. We (Debra and Jon) believe this is one of the most important questions one can ask about a precognitive task because it speaks to the nature of time itself – whether the future is set or can change and what effect that might have on predictions. (There is also the more "out there" question: what effect do predictions have on events themselves?)

Unfortunately, the methodology as it was carried out had procedural weaknesses, which cloud confidence that results were due to the proximity to feedback. This is because for the first eight trials, viewers received both photo options as their feedback, whereas for the final one their feedback included only the photo attached to the winning outcome. Further, viewers were given their feedback under a heading called "real targets" and "feedback." Despite the researcher's intention for viewers to describe only the photo associated with the winning option, this setup, which was partially dictated by Reddit group rules, may have inadvertently encouraged viewers to focus on both photos.

Future recommendations: We hope this researcher or others will try to replicate this study under conditions in which feedback does not need to be given prior to the outcome being known. This might require using a platform other than Reddit or changing the Reddit system to allow projects to provide feedback much further out.

While we understand the desire to use crowdsourcing to get more data points, not many viewers participated in each trial (compared to RV Tournament, APP or IONS psi games, which attract thousands of participants). However, Reddit is growing rapidly (36K members of the RV Subreddit) and future experiments might well attract a much larger number of participants. One option to consider is to use a pool of more experienced viewers to ensure that results are less likely due to learning curves by participants. Learning curves would be much more pronounced for newer viewers and this could account for improvements in sessions as the trials progressed.

We also suggest testing the distance between when the remote viewing sessions are done and the event itself, on the one hand, and RV sessions and the moment of feedback, on the other.

The three-point scoring method the researcher used shows promise; it seems like a simplified version of the University of Colorado rating scale, which isn't overly complex but does require some getting used to (see Chapter 4).

The discounting method is quite interesting and deserves further exploration. We provided only a summarized version of it here. We suspect we will hear more about this approach in the future and encourage readers who are interested in this concept to reach out to the author directly.[430]

In closing, this was the researcher's first attempt at conducting a complex, formal RV project, under the restrictions of an existing platform, and his innovative approach was impressive. He probably would have benefited from a proposal review before starting to work out the feedback issue. This is an example of how a slight change in protocol during the trials can have substantial effects on a project. Such a review can be conducted with the aid of the International Remote Viewing Association's Research Unit (IRU), which is presently overseen by Dale Graff and Debra, or through a review by the Rhine Research Center. Also, the Parapsychology Association recently started a new mentorship program that might serve as a good resource.

Further, it's our hope that in providing a summary of his write up, other publications will also follow suit and summarize projects that might have methodological weaknesses but nevertheless offer important concepts and tools such as the important premise, scoring and prediction tools this project offered.

If nothing else, this experiment shows that for trials sharing the correct feedback photo close to the event, successful predicting of a presidential election may be possible. It also supports the body of evidence indicating that providing only the photo associated with the outcome is preferable to exposing participants to both photos in advance.

US presidential elections 2020 using ARV: What went wrong?

The following essay was written by Daz Smith, editor of *Eight Martinis* magazine, author of six books about RV and psi, founder of a very large remote viewing Facebook site and a key member of the *CryptoViewing* team.

Notes & thoughts – by Daz Smith, December 2020

After two years of focused ARV research, more than a decade of sporadic ARV participation and over 24 years of combined RV project management and participation, Daz recently arrived at a few conclusions, which are presented here:

ARV, for some reason or as I may show (many reasons), is less accurate than a traditional or normal Remote Viewing target. For a standard remote viewing target, I myself (and others I work with) with some regularity achieve accuracies of well over 75%. These are of targets that are usually present or past targets.

There seems to be something within the process of doing ARV targets that seems to affect the overall feel and accuracy in a negative way bringing the accuracy down to something like 55-68% *(this is my approximation and is not based on data analysis, but on my research and discussions)*. *BUT* clearly everyone seems to see/have less accuracy on ARV-style RV projects.

For two years I have run a series of ARV/Unitary ARV projects for CryptoViewing as a tasker and project manager and here are some thoughts that I have formulated.

As with all remote viewing, we still do not know how it works. Therefore, imo, anything has the potential to influence the data and hence its accuracy. I have gathered and I feel, shown, that there seems to be a communication channel between taskers and viewers. Shown in projects whereby the target only exists in the mind of the tasker – yet still can be accurately recorded by remote viewers. But there are many other factors that I feel CAN hinder any remote viewing project but probably more so on ARV style targets that involve typically both the future/forecasting, money and invested intent from people involved.

As an example of my thinking I will use two projects I ran from March 2019431 – (ongoing) to help diagram my thoughts.

Public ARV Project 1 – to predict the outcome of the U.S elections in November 2020.

Start – March 2019. End – December 2020.

The project was a binary ARV to determine if Donald Trump would be re-elected US President. The target was set for over nineteen months in the future and the viewers were to sketch and describe the image given as feedback – only.

I first analyzed the data. I had full knowledge of the target as far as which image was appointed to what outcome. I also asked colleagues Jon Knowles and Tunde Atunrase (both seasoned ARV/RV experts) to also BLINDLY review the data and to match the data to one of two images – they had no knowledge of what images represented what outcome. All three of us selected the same outcome target image as a match. This image represented the B target in the binary set – to be shown as FEEDBACK if Donald Trump WAS re-elected. Although still in play – it's looking more likely each day that this is not going to be the case and the Biden is going to be the newly elected next US President.[432] This was the A target in the binary set. So what went wrong?

On reviewing this project I can see several main factors that may have caused the inaccuracy:

1.   Project setup
2.   Time to the event
3.   Intent/over time
4.   Errors in the analysis

So let's look at these

First, the project setup. On reviewing I can see no real issues that would have caused viewers to report more of target B than A. I was careful to select two targets of the same approximate age, size, form/function and interest. On reviewing, I have to admit that target B DID have a slightly more interesting shape/form than A, but I don't believe that this alone is enough to cause any major displacement.

The actual target cue was good for both of these:

- **A** - The remote viewer is to move to the optimum position/location to describe the ACTUAL structure focused upon in the feedback image if Donald Trump is **NOT** re-elected president and this target is given as feedback. ONLY.
- **B** - The remote viewer is to move to the optimum position/ location to describe the ACTUAL structure focused upon in the feedback image if Donald Trump **IS** re-elected president and this target is given as feedback. ONLY.

Point 1 – Setup I feel is OK.

Point 2 - Time to the event. Now, in this first project the time to the actual event and feedback was twenty-two months. (It is commonly thought within the RV/ARV community that the further out a prediction is from the predictive event, the less accurate it seems to be. This is based on the theory that over time and moving closer to events, the options for it decrease coalescing over time into a single route.) I'm not sure I have read any scientific projects that validate this theory, but it is common thought. As this project was quite some time away from the prediction event, If the theory holds true, then this would impact the accuracy of the remote viewing data – So this COULD have been an effect.

Point 3 – Intent /over time. Now, this is a complex part of RV. Intent. It's known that the intent of the people involved in the project, especially the client, tasker, project manager, analysts and viewers, CAN have an effect on the results and the data presented. It's known within Remote Viewing research that a level of telepathic communication CAN possibly be involved.

In this project I was the client, tasker, project manager and one of the analysts. My intent on this project is known and can be computed. In March 2019, I did not like TRUMP, did not want him to be re-elected. Therefore at this date, IF my intent were to influence the final RV data it would have created an A target selection – Trump NOT to be re-elected.

But hold on there – it gets more complicated than this. My intent over those 22 months dramatically changed. This is for two main reasons. The first is that I did a second ARV project for a client, CryptoViewing. And secondly, both projects were public, so over time I had a personal interest for the predictions to be correct to validate themselves for myself and to satisfy the client, CryptoViewing.

I think it's safe to say that in November 2020, My intent had now morphed into one somewhat schizophrenic in that I still did not like Trump, but that I also had a need for the two public predictions to be correct predictions. Over time my intent had dramatically shifted – this has to be listed as a potential cause of an effect on the remote viewing data. Especially as I was in this first project client, tasker, project manager and part analyst. If this is the case though, then it has to be conceded that my future intent MAY have influenced past data from the viewers.

Point 4 – A miss in the analysis. In this project I knew the targets and analysed them knowing this and I chose the B target as the best choice. My analysis shows that although there was some displacement in three of the seven viewers, three were also clearly B target descriptions and only one viewer outright seems to be describing the A target.

The second person to analyse the RV sessions was Jon Knowles. Jon is a very knowledgeable person in remote viewing, one who has spent well over a decade looking at ARV. Jon did NOT know what target represented what outcome – he was BLIND. Jon's analysis was: *"Three passes and four sessions favoring B suggests a moderate to strong pick for B."*

The third analyst – Tunde Atunrase – is also a very knowledgeable person and longtime practitioner of ARV projects with great successes. Tunde also reviewed the ARV data and was blind as to which represented what outcome. Tunde reported: *"For me the overwhelming favorite is the Atomium structure in Brussels B Target."*

In conclusion, one unblind analyst and two blind analysts ALL picked the B target – a TRUMP re-election as the prediction. I can't find any issues of bias in the analysis of this project.

Therefore the main factors that I feel are the cause of this ARV inaccuracy must be:
1. Project setup ✓
2. Time to the event ✗
3. Intent over time ✗
4. Errors in the analysis ✓

My further thoughts on this… On many thoughts about this and the other ARV projects I have worked in and managed I have come up with this structure that may both help explain where things go awry, but also may be used to calculate future probabilities for the accuracy.

Cofactors.

This is how I'm rating this and the cofactors I feel MAY influence each stage of the RV/ARV process.

- C – Client intent
- T – Tasker intent
- Pm – Project manager intent
- Ps– Project setup (numinosity and values)
- V – Viewers intent
- F – Feedback
- Ti – Time
- S – Social
- Fi – Future Intent
- Fa – Fatigue

I'm giving each of these either a 1 or 0 rating, 0 being unbiased, 1 being biased or influenced. I feel the best case scenario score would be O, but in this case a score given in March was 20. It could be more but viewer's score is unknown. In November this very much changed to be even more negative at 24. More on why later.

I feel this Facebook ARV project had this algorithm.

March 2019: $C0 + T0 + Pm0 + Ps0 + V? + F0 + Ti22 + S1 + Fi0 + Fa0 = 23$

November 2020: $C\ 1 + T1 + Pm1 + Ps0 + V? + F1 + Ti22 + S1 + Fi1 + Fa0 = 28$

So first:

C – Client. The client has an intent, an expectation and a want from the project, this will have an effect. In this case above, the client was myself. I am not a US citizen and my personal intent at this stage was to know my thoughts on Trump (at the start of the project were that I didn't really like him – but I had no investment either way.) This of course VERY much changed in the later months. So: **C0 +**

T- Tasker. I was also the tasker of the target and again my intent and/or influence in March was imo, 0. So: **C0 + T0 +**

Pm - Project Manager. Again. I was also the Pm of the target and again my intent and/or influence in March was imo, 0. So: **C0 + T0 + Pm0 +**

V –Viewers. We had seven remote viewers in this project and we do not know their thoughts on this project in March and they were blind to the target at this stage, so I feel it's safe to give this a score of 0. **C0 + T0 + Pm0 + V0 +**

F-Feedback. This is a target that WILL have (imo) real/solid feedback, so I gave this a March score of 0. **C0 + T0 + Pm0 + V0 + F0 +**

Ti – Time.

Now this is a calculation based on how far in months between the target time and the viewing time. It seems that targets further into the future MAY have more probabilities or possibilities that MAY lessen the closer the viewing is to the target time. So, in this case I scored 1 per month between viewing and target time. Imo, targets within a month or so, seem to be way more accurate that far-out predictions. In CryptoViewing our monthly predictions of the next thirty+ days seems to be scoring an approx. 75%+ accuracy month on month. So I added the 22 for the 20 months between viewing and target time.

C0 + T0 + Pm0 + V0 + F0 + Ti22 +

S- Social.

This effect I feel is necessary because high profile and global effect targets like the US elections will/does get a lot of global social interaction and noise. Knowing that time within remote viewing isn't linear then future social noise probably has an effect on the target accuracy. With this election – there has been a huge amount of noise. **C0 + T0 + Pm0 + V0 + F0 + Ti20 + S1 +**

Fi – Future intent

With some targets like the US elections and the events surrounding myself and project manager, tasker and more, it's probable that with far-out predictions of this magnitude my future intent will/did change and that this MAY have affected the project data.

In this example I also ran a secondary US elections Unitary ARV project for CryptoViewing. This was started on 13 September 2019. This project using a single photo image tasked to me to project manage by my client, CryptoViewing, had its own set of calculations. But as things progressed over time towards the actual outcome, it's sure that my needs and intent also changed. With now two predictions that TRUMP would win, my intent had obviously changed because being agnostic before, now I had two ARV projects in the public domain, and a client in CryptoViewing to please, my intent was now conflicted, but MAYBE wanting a TRUMP win to appease the RV community, fans and my client at CryptoViewing.

This Future intent change – has to be factored into any calculations that may have affect both project outcomes.

March: C0 + T0 + Pm0 + Ps0 + V? + F0 + Ti22 + S1 + Fi0 + Fa0 = 23

On reflection I would say the score in November probably would change to something more like:
November: C 1 + T1 + Pm1 + Ps0 + V? + F1 + Ti22 + S1+ Fi1+ Fa0 = 28
*The viewers intent and possible knowledge they had been involved in the project would also have an unknown effect on their data.

This November score shows the Client's intent (me) to have changed because I was now invested in both predictions being accurate within the RV community and for the client that came on the scene for the second ARV project on US elections in March 2020.

If a single predictive project had been done in say October of 2020 with me as the client, tasker and Pm, then it may have created a score like:
C0 + T0 + Pm0 + Ps0 + V? + F0 + Ti1 + S1 + Fi0 + Fa0 = 2
If I would have a second project in play for a client then this may be a score of:
C 1 + T1 + Pm1 + Ps0 + V? + F0 + Ti1 + S1+ Fi1+ Fa0 = 6

Conclusion: In my ARV projects I feel that both the time between RV data/ prediction and the prediction event coupled with a second project and with the projects being public. This may have influenced the accuracy of the data in these predictions.

Future projects should be done:
- As close as possible to the prediction event to decrease TIME options
- Probably noted that public projects may change the INTENT from those involved due to wanting the predictions to be accurate and to please the RV community

- Stick to one project or prediction. Additional projects for other clients may impact ALL projects due to intent and wanting to please clients.

Jon's email response to Daz's comments:

Daz, Thanx for sharing this. That's an innovative approach – to give a multi-factorial algorithm to try to estimate factors in an ARV prediction.

The first thing it reminded me of is "Drake's Equation." Wiki says: "The Drake equation is a probabilistic argument used to estimate the number of active, communicative extraterrestrial civilizations in the Milky Way galaxy." It was apparently designed to stimulate research rather than produce a meaningful calculation. It may be that your algorithm will stimulate inquiry into the relative importance of these factors in ARV as Drake's equation did (e.g. for Carl Sagan).

For my part, first thought, is that in the public project in which I was an analyst, that second photo is way numinous compared with the first photo and so plugged into your algorithm I would give that a very large value. (I may have mentioned that Marty and Tom McNear disagree with my opinion that one photo in a tasking that we did at APP was also way more numinous than the other and IMO it was a determinative factor. Debra was a viewer there so I won't say what the two photos were. But there is historical support for my opinion in that case – not so for the two photos in the 2020 election trial that I judged for).

Intent. This is so hard to parse. Conscious intent. Unconscious intent. If there is a higher self or other entity "overseeing" intent. I need to think more about this one.

Time to the event: You may have seen that Frankandfriends posted his 15-page paper on "discounting" in his experiment over 11 months to predict the outcome of the 2020 election. His goal was to see if a discount rate for such a prediction could be derived or at least estimated. Over time, the crowdsourced self-selecting viewers converged on a 98% chance of success of "the challenger." I haven't seen this approach before, one in which you try to establish the "value" of the prediction for a given time slot in the time between the first prediction and the event. I pointed out to Frank that some may prefer to bet big at the outset when the odds are best. e.g. the 2006 or so predictions that Liz Ruse made and that we bet on…But it is true some may like this way of approaching ARV betting or purchasing and may make repeated bets over time based on the "value" of the prediction. The value is arrived at by scoring sessions in binary ARV. He used a 2% and a 5% discount rate, which is apparently used regularly in economic forecasting.

Probability of event – Tom Atwater and Marty have discussed it a lot. Does the probability of an event happening change from moment to moment, hour to hour, day to day, month to month, year to year? It would SEEM that it does. Trump gets Covid and so he would not be a member of the "Incumbents" in Frank's experiment. Of course he did get it but it wasn't serious enough. Another angle on this is that Yes, the event appears probabilistic to us humans, but if the universe or multiverse is a "block multiverse" as some physicists hold, then in some way the outcome exists and is known and there is the possibility that the viewers go directly to that outcome. (I raised this with Frank too. He said he didn't share that ontology. I said I'm open to the block universe but as with so much else with RV and ARV, just don't know.) We get results galore but our explanations lag badly.

Those are my first impressions. Thanks for coming up with this innovative way of looking at the issues. Best, Jon.

Sean McNamara 2020 election group prediction

We heard that Sean McNamara, author of *The Telekinesis Training Method*[433] and *Signal and Noise: Advanced Psychic Training for Remote Viewing, Clairvoyance, and ESP,*[434] had carried out a group prediction for the 2020 presidential election. We asked if he could outline his approach and findings. He confirmed that on Oct. 11, 2020, he had hosted an online remote viewing public group event. They had just finished discussing the results of a lottery prediction and one of the participants suggested doing an election prediction before they finished. Sean provided the following account:

I felt like I'd been put on the spot because judging by everyone's reaction to her request, I already knew this prediction would contradict a couple of important principles:

- The viewers wouldn't be blind, since they all knew this was a prediction about Joe Biden and Donald Trump.
- They were emotionally involved, since this was such a touchy subject, and it was likely that each viewer's bias could affect which potential target their subconscious focused on.

Yet I decided to do it in the spirit of fun. This was the most slapdash prediction I ever put together, since I did it in just a couple of minutes with the viewers waiting on the Zoom call.

Because the political climate has been erratic, especially the personal behavior of President Trump, the prediction was formulated this way, "Who will speak at the Inauguration in January as the recognized president?" This would give some time for the storm to pass after the vote tally was complete following the November deadline.

I quickly browsed my video target database and selected three targets.

To represent Trump as the president in January, I chose a video taken inside a state park visitor center in Colorado, featuring a display of a stuffed elk and other forest animals. For Biden I chose a video taken inside of a store featuring Christmas ornaments. I also selected a third target, representing the possibility that neither candidate became president. We knew this was a possibility due to age, illness or other conditions. For that prediction, I chose a video of a dam located in the mountains just west of Denver, in Evergreen.

I asked everyone to spend a few minutes sending their mind to the future, perceiving the feedback they'd receive the day after the inauguration, and jot some notes down to serve as their transcript. We didn't use coordinates. They had no idea what the potential feedback images could be.

Then, I asked them to describe their transcripts to me right there on the Zoom call. As each person spoke, I noted which target they connected to the most. I was very familiar with the target videos, so this part was easy.

There were 13 transcripts. 8 of them connected to the Christmas store (Biden). 0 of them connected to the Elk display (Trump). Curiously, 5 of them connected to the Evergreen Dam. I had used the dam video one time before, with a different group.

That video seemed particularly powerful at attracting the viewers' subconscious attention, which led to a lot of "misses." I decided to use it again for this election prediction knowing it could have a powerful draw once again. My intention was for it to serve as a "strength-tester," drawing the viewers with less "subconscious conviction" toward it. This way, only the strongest, or "most certain" transcripts would lean toward one of the candidates.

I think afterward they were left wondering, "Was it really that probable that neither candidate would be recognized as president on Inauguration Day?"

Now, several weeks remain until Inauguration Day. At that point, we'll know if the prediction was correct or not. Regardless of the results and of the obvious holes in this prediction's methodology, everyone had a good time doing it together. Sometimes doing an RV prediction the wrong way is a good way to learn how to do it the right way.[435]

Does it need to be so complex? Direct clairvoyance for elections – A few thoughts from Debra

Since I'm trained in a variety of psi-related disciplines, which operate within different psi collectives, I quite frequently operate under less-than-blind conditions and tune in to topics I'm already somewhat familiar with to see what I might learn that I don't already know.

Ingo Swann sometimes knew the nature of his targets, such as when he and Harold Sherman set sights on Jupiter and conducted the "Jupiter probe experiment" in 1973. He knew he was exploring Jupiter and his aim was to access information no one knew about but that would likely come to light in the months ahead when images came back from the Pioneer 10 and 11 "flybys" of 1973 and 1974. That was how he discovered the rings of Jupiter. He also accessed information that scientists knew but he did not and some that he himself likely knew but experienced in a new way.

While knowing more up front about a topic (like the potential candidates in an election are) makes it harder to verify later, being blind for operational projects can often waste a lot of time. As we have found out with the above-mentioned election projects, if you ask viewers who are used to describing locations to do a blind session, they will usually give you location-based information with perhaps a few mentions of people. They will then need to be retasked and may not have time to do more viewing or another interfering factor may have arisen. Having them start with the tasking, "The target is a person, describe the person" is the way to go if you want to keep them blind to the essentials of the task but, as mentioned elsewhere, a viewer is never without assumptions.[436] Keeping them blind ensures their logic is less likely to kick in for the person who is the target, but this in no way ensures they won't form an opinion, most likely an incorrect one, and then start making all sorts of logic-based assumptions about it.

I often do an exercise with my clairvoyant class in which I give them a variety of famous historical figures to tune in to. Sometimes I just call these "mystery people" and assign a number to them. Sometimes I provide their name. For the latter, yes, of course, they go into the session with many preconceived notions and biases, to the point they wonder themselves if they can even do the exercise. Then something interesting happens – they start accessing information, some of which they didn't know before. At the end of the exercise, I ask if their perceptions have changed about that person. About 75 percent of the time, they respond "yes." Overwhelmingly, they respond with having more compassion from this intimate perspective – although on occasion, it's just the opposite.

Speaking of which, if someone wants to know who the next president will be, it might be more economical to say, "Here is President Trump – Tune into him on Jan. 20 and look at what he is doing or feeling," or "Move to Inauguration Day, we want to know anything about who is giving the inauguration speech." Not all viewers can work this way, but many can, especially those already used to reading people. During my career, I've looked at hundreds of relationships for clients – often love interests, but also family members, co-workers, pets, etc. It's not that difficult. Unlike ARV, it requires no project set up, no blinding or randomization protocols, just a couple minutes deciding what is the best way to proceed on an internal level (e.g., where to direct one's own attention).

2020 election – Two months before the election, I decided I would personally tune in to Inauguration Day and see what was happening. I lay down to get comfortable and imagined I was looking at my TV screen at the time the inauguration speech was being given. I didn't feel like I was getting much, and I kept getting distracted and having to refocus. I started to drift off to sleep. I was almost unconscious when I suddenly had a flash of two people coming out of what looked like a tunnel sports figures emerge from at the start of the game. I could see one was a darker-skin woman and the other a white-haired Caucasian man. It took me a few seconds to even know what this was connected to as I had forgotten what I was doing. Then I realized I had just gotten some very clear information about the future outcome of the election!

Given that Biden's running mate, Kamala Harris, is an African American/Indian woman and he is an older white man with white hair, I felt I had my answer. Trump would not likely be giving his speech with a darker-skinned woman nearby and his hair wasn't white.

Having dozed off and forgotten what I was doing gave me higher confidence that it was coming from an intuitive place and not my logical mind. At any rate, I feel that if I put together another precognitive-based project for a future election, this is the way I will personally go – full frontloading – and see where that leads us.

CHAPTER 20

The Psi Frontier: Alphanumerics

While some accounts describe early parapsychological experiments that test whether psychics could psychically "read" letters and numbers, the literature is not extensive. In his groundbreaking book *Extra-Sensory Perception*, published in 1934, J. B. Rhine[437] summarized the parapsychological research that began in the late 1880s. Tens of thousands of trials were conducted by about a dozen researchers interested in moving psychical research from its earlier focus on case studies of mediumship to a statistical laboratory approach. In the latter, participants underwent telepathic and clairvoyant exercises starting with the use of numbers (digits 0-9) and playing cards (numbers and suits), then just the suits of the cards and eventually Zener (ESP) cards.

Rhine and his colleagues[438] did not go into great detail about the success of the experiments with numbers except to say that above-chance levels were sometimes achieved and some individuals did better than others. They also did not provide details about the decision to move from numbers to symbols. In the Rhine collection at Duke University are documents entitled "Alpha-numerics." Unfortunately, we were not able to access those prior to the publication of this book due to COVID restrictions. It's our hope the files contain clues as to why American parapsychologists moved away from tasks related to describing numbers and letters.

Experiments at the Stanford Research Institute

As we noted earlier, SRI pioneered free-response "outbounder" experiments. Researchers observed that viewers were far better at describing and drawing outdoor scenes than they were in discerning words and numbers. There were exceptions, however. In one of the most (in)famous and well-documented accounts, former Police Commissioner Pat Price undertook a tasking given by the CIA and spelled out highly classified project names inside file cabinets at a secret NSA installation. Although some of the specifics were not correct, the report alarmed the authorities so much that a security investigation was immediately undertaken. Price's feat is credited as a major factor in the CIA offering consulting contracts to SRI.[439,440,441]

In the same vein, Ingo's students (such as Tom McNear) were occasionally able to provide the names of target locations in their training sessions (e.g., Stanford Linear Accelerator and Oral Roberts University). It's unknown if the information was derived from direct perception (such as clairvoyant reading of letters) or if elements at the location were recognized, enabling the viewer to deduce the name of the location. McNear states that he just knew the name of the site though he says he isn't sure how, or by what mechanism he knew.[442]

Joe McMoneagle's success in counting has also been noted in the literature. He psychically entered a large building and observed construction of a submarine, which had a number of missile tubes. His monitor pressed him to count the tubes. Rather than providing a number, Joe sketched 18 to 20 of them. Months later, it became known a new type of Russian submarine (Typhoon) was indeed launched from the building and it sported 20 missile tubes.[443]

Joe reported one of the alphanumeric SRI experiments he was involved with:

Yes, there were some preliminary experiments done regarding words and could they be remote viewed. The subject was me, and Russell Targ was the one who ran the experiment. They used a random number generator to generate a seven-digit number. They used this in conjunction with a very large Webster's Dictionary.

Digits 1 through 4 represented the page number. The pages went well up into the second thousands – two thousand and a couple hundred. Digits 5 through 6 represented how many words they counted down from the beginning of the page in number of words. Digit 7 was the number of four-letter words they then counted from there to the word they would use.

They then created the four-digit word by creating the letters, using black tape on a 1 x 1-foot white card. They then put the card with the letter on a board mounted on a wall within the location they were in. Once they had done this, they would click the transmit mike-handle on a small walkie talkie twice to let us know in another location one mile away that it was time to target the word. Only the two clicks would tell us it was time to do the remote viewing. No one would speak during the procedure and the double click on the mike was all that was done.

I can't remember all the words, but we did seven of them and I was able to get five of the seven words. I got two of the words wrong. I cannot remember which words were wrong, and I cannot remember all five of the words that I got right. But I do remember getting the word "FLYS." This is important because someone said it proved nothing since anyone could guess a four-letter word since all words had vowels – A, E, I, O, U. What was interesting, however, is that none of the five words we got right had any vowels except for the "sometimes Y." Now go and compute how difficult it would have been to randomly select five words in a row with only the "sometimes vowel" Y? It's off the charts. Also, of the five words I got right, the last two they put up, they scrambled the letters. In other words, they would have put the word FLYS up like; L-F-S-Y, but I still got the word FLYS. Which means I'm not actually seeing the words, I'm understanding what they represent in terms of meaning.[444]

SRI carried out a few other experiments to see if letters and words could be recognized.[445] In "Remote Viewing of Alphabet Letters," three pilot experiments are described.

In the first experiment "Subjects H1 and I1 were located in their respective homes in Los Angeles and New York City, while the targets were posted daily, one per day, in a laboratory at SRI in Menlo Park, California." Twenty forced-choice letter-guessing trials were undertaken. "Both subjects submitted their lists of 20 letters. The results were not found to depart from chance expectation" (p. 161).

In the second pilot experiment testing free responses, the subject was told to treat the task (identifying a letter) as if it was a remote viewing target and to draw it in non-analytic fashion. Another person would judge what letter they thought was indicated by the transcript. The viewer (S1) was told not to try to guess what the letter was, but instead the viewer thought it was a U and said so. The judge, Hal Puthoff, judged it a U, which was correct. The next day the letter V was similarly identified.

In another series of three free-response experiments, viewer H1 correctly identified three letters out of five, odds of less than one in 1,500.

This suggests that the way to increase the analytical capability to include written material is to arrange to separate the perception from the analysis, to encourage the subject to describe only his or her perceptions, and to follow up by having a different person do the final analysis on a blind basis (p. 161-62).

The third pilot experiment involved a machine-generated target. The task was to identify letters based on form and shape. Slides (35 mm) of black letters on a white background were prepared and shown on a screen, four slides at a time. The target slide was chosen in random fashion; after the viewer chose one of the four, the target slide was lit up. The machine chose the letter, to which both viewer and experimenter were blind. If the subject wanted to pass, he could hit the Pass button. Encouraging phrases were shown at the top of the machine if the viewer had increasingly correct choices. The letter combinations ABIO, CDGQ, EHLT and KWZY were employed. The viewer succeeded in this alphabet reading at odds of p = 1 and 10.[446]

> …the results obtained in the remote viewing and machine approaches to reading remote alphabet characters do indicate a potential for developing acceptable levels of reliability in reading text for operational purposes. Further study is required to determine whether this reliability can be achieved with a reasonable effort (p. 162).

Another of the pilot experiments at SRI featured Viewer 372 (Joe McMoneagle). In 12 trials, the target was a randomly chosen three-letter word in red, block letters eight-inches high on white cardboard. Words were chosen from a 1,700-page college dictionary using a random process to direct the experimenter to a word in the dictionary. On the first try, the viewer failed to get the first letter but, when told what it was, he got the entire word (GUN) right away. In the same way he identified the word VAT. In the bulk of the series (trials 3 through 12), feedback was given only after all three letters were named. The letters were displayed one at a time in the order in which they came in the word (with one exception). In trying to discern nine three-letter words, the viewer was correct on six of the 23 letters. "This pilot result was therefore encouraging" (p. 278).

Experiments by Ingo Swann

In the 1970s and 1980s, Ingo Swann took part in thousands of experiments in what he called "Analytics," which we have referred to previously. A great number of these were done with Hal Puthoff in the 1970s and later with an associate or solo in the 1980s. The archives at the University of West Georgia contain many pages of Ingo's raw data and notes and they are worth devoting some space to.

One reason is that, according to Ingo's long-time and best student Tom McNear, his goal was to be able to determine Yes or No answers without AOL (analytic overlay) occurring, and thereby identify or "get" numbers and letters using CRV. He wanted to make a breakthrough on this most difficult challenge for remote viewing.[447]

A second is that Ingo was one of the most innovative (co-)experimenters anywhere in parapsychology and his zeal in trying to perceive future or distant symbols, letters, numbers and words is unmatched. In the process of exploring these Analytics, he constantly monitored his mental and bodily impressions. His notes enable us to gain insight into what those processes were, as well as what worked and what did not.

By contract, Ingo had to regularly write a report to SRI about these experiments on Analytics and he dutifully submitted them to Dr. Ed May, head of the program from 1985 to 1995. His reports are clear but the raw data and notes are often difficult to make out. As he proceeded, Ingo would jot down ideas, hypotheses and initial conclusions. He would note some small gains, then setbacks, and this cycle would be repeated. He did claim a few breakthroughs, which he noted in his reports to SRI.

Here are some of the symbols and letters Ingo tested – sometimes by himself, sometimes with an associate. In one series, he tried to determine if he could distinguish between whether the next target was a 0, 1 or A.

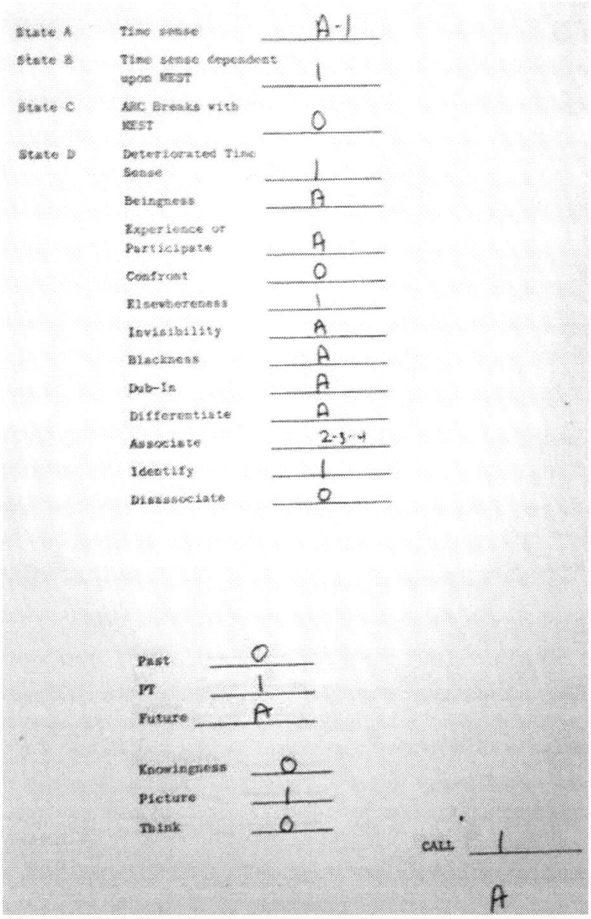

State A	Time sense	A-1
State B	Time sense dependent upon REST	1
State C	ARC Breaks with REST	0
State D	Deteriorated Time Sense	1
	Beingness	A
	Experience or Participate	A
	Confront	0
	Elsewhereness	1
	Invisibility	A
	Blackness	A
	Dub-In	A
	Differentiate	A
	Associate	2-3-4
	Identify	1
	Disassociate	0

Past	0
PT	1
Future	A
Knowingness	0
Picture	1
Think	0
CALL	1

A

Figure 20.1

He did a series to see if he could tell if one of four letters was curved or "oblique." For example, if he got the letter S and the target was 0, he considered this a hit because both are curved (while the other two were not). On a tally sheet of 100 such attempts, 49 were correct.

He tried to identify whether the target was red or black, to discern if a sequence was 011, 110, 111, etc. In another experiment he drew squiggles (his word) until the marks resembled the target. In yet another series, he tried to discriminate between the numbers from 0 to 9 and in some way his notes indicate the concept "blackness" was important.

Drawing on his background in Scientology in which he would explore "isness," Ingo drew panels like the following for gymness, cobness, boxness, hipness and other "nesses."

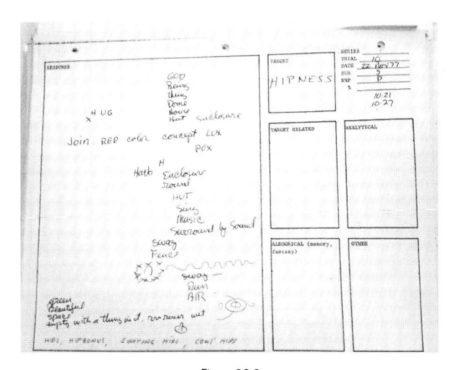

Figure 20.2

Associative Remote Viewing in the series

Another series appeared to be using yellow, blue and black to try to distinguish between O, H, I and A. Using colors to discern letters would seem to be a form of ARV.

A different set consisted of 40 pages of drawings using the same panels shown above in which the "target" is entered as 10, 18, 27, 35 or 38. There is no indication why he chose those numbers. The drawings did not appear to attempt to resemble numbers; they are simple shapes and forms. We don't know how Ingo scored the results, but this appears to be an attempt to recognize numbers through drawings, which is also a form of ARV. Below is one of the most detailed.

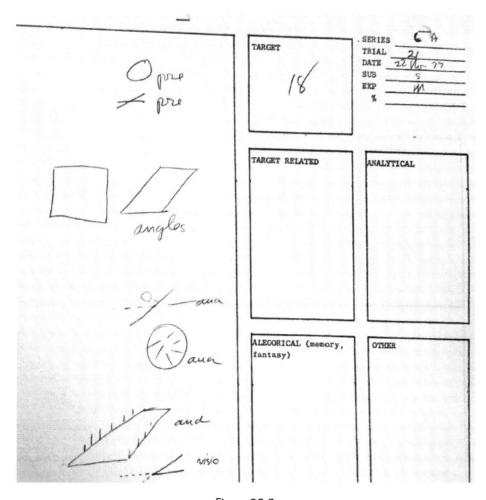

Figure 20.3

Ingo also tried his hand at predicting lottery numbers. A few sheets show a miss on a Pick 3 and a Pick 6. It appears he simply wrote down the numbers he came up with. As we noted elsewhere, Ingo admitted in public that he never won a lottery and, in a private 2012 conversation, he drolly told Tom McNear he was an "abject failure" at it.[448]

During many of the above experiments, Ingo wrote down ideas and speculations, many of them intriguing and thought-provoking:

- Numbers are not physical, but do they have a subset phoneme system?
- Psi is a whole-body process, and a bidirectional one at that.

- There is a signaling system in the gut.
- There are probably two or more psychic channels.

Ingo also suggested our psi faculty seems unable to use deductive or inductive reasoning on its own. It has to develop its own analytical ability. For this reason, Ingo reported, it was a great breakthrough when he was able to identify the condition of a target across the country – for example, whether a candle was burning or not. This was significant, he said, because to do so he had to use an analytical capability during a psi experience.

Another of Ingo's observations was that psi is not exact with regard to time.[449] Hal Puthoff and Ingo observed that sometimes Ingo would get the next target in a series rather than the current one. He also had the impression that ordinary time would break down while he was attempting to access targets and the targets would "line up." He called this phenomenon the "Analytical Summation Pool" (SUMP) and described its purported characteristics.

Swann reported that he experienced learning curves with letters and numbers, but results continued to fluctuate. He had many successful shorter runs interspersed with setbacks.[450] Ingo did informally offer a figure of one to 1 million (or one to 2 million), with reference to his successes in these experiments. However, despite this remark, Ingo did not succeed in using analytics to be able to get numbers or letters, according to Tom McNear, and Ingo "gave up" on Analytics.[451]

As noted above, apart from Ingo Swann's extensive Analytics explorations, the reports in the *Star Gate Archives* (Volume 1)[452] describing alphanumeric experiments done at SRI indicate positive results and hopeful signs:

> The data indicating that a viewer can describe an individual slide as it is shown on a screen shows that targeting on high-resolution transient targets (charts, maps, etc.) is not out of the question. This, coupled with our findings that a viewer may be able to describe and identify alphabet letters is a most encouraging development, and one deserving of further work. Extension of the RV process to include high-resolution material, especially with a reading ability, would constitute a significant breakthrough for operational applications.[453]

But in the end, SRI directors and researchers concluded viewers and psi participants just aren't that good at describing letters and numbers compared with their abilities to describe objects, structures, places and people.[454] They theorized that different brain structures might be involved in identifying and labeling alphanumerics as opposed to the structures or networks employed in visual recognition of colors, shapes and patterns.

As a result of these pioneering researchers' conclusions, many remote viewers and psychics do not try to develop alphanumeric skills. While we don't have formal data to support this inference, in discussions over the years, remote viewers usually say letters and numbers are something we just aren't good at. When asked how much experience or training they have in this area, they usually indicate little or none. Of course, very few viewers these days have a monitor sitting across the table prompting them to count the number of objects they see, as Joe McMoneagle did.

There are outliers, though, some of whom are featured in this book (particularly in Chapter 21 on the lottery). One is Richard Ireland.

Ireland is known for his "X-Ray Clairvoyance," which was documented in a 1969 video of his appearance on the Steve Martin show. Dr. Ireland was a psychic with celebrity clients such as Mae West, Amanda Blake and Glenn Ford. Ireland could call out long strings of numbers found in people's pockets. He correctly provided first and last names of friends and relatives (both living and deceased), furnished accurate dates of pending births and identified phone numbers and addresses. He apparently became aware of these gifts following eye surgery at age five when medical staff became aware he could see with eye bandages on.[455] We mention Ireland because he is reported to have collaborated with a US Army Special Forces unit involved with remote viewing.

How much information can a remote viewer hold in conscious awareness?

One of the characteristics that makes perception of letters and numbers so challenging is how "sensory-poor" and similar in shape they are to each other. Another is their absence from the natural environment humans have existed in over millennia, so the historical "meaning quotient" of letters and numbers is low. These inhibiting factors naturally lead to the desire to learn more about how we process information, how ordinary perception and memory work, and how our knowledge of these might apply to psi.

Cognitive science research has demonstrated that the average human can remember only a few numbers at once – the figure usually given is seven. However, research has shown that mnemonic techniques can greatly expand this capacity.[456] The existence of successful memory courses and programs demonstrates this, as well.

To learn more about this, researchers at SRI explored how much information a remote viewer or psi participant can receive at any given time. These experiments have been interpreted as showing the "channel capacity" is probably too low to win a very large (six-number) lottery. According to Ed May:

From information theory, the guestimate for the RV channel capacity is only about 0.6 millibits/second. This crude number emerged from 20 years of research—that does not make it correct, but to ignore it is a mistake.

In another example, to win a lottery of 6 two-digit numbers requires 6 x log2 (100) = 39.9 bits of information. So if the above channel capacity is correct – a mighty big IF – then it would take continuous RV for 18.5 hours straight! This is the geeky answer to the question we are all asked: If psi is real how come you ain't rich??[457]

As we note in Chapter 21, while we are not aware of any six-digit lottery win using psi, there have been at least three five-digit lottery winners.

People who demonstrated exceptional acuity with letters and numbers

Despite negative indications about our abilities with numbers, there are several well-documented case studies of individuals – savants – who were able to hold an unfathomable number of items in their mind such as phone numbers, addresses, dates, mathematical calculations and words in multiple languages, with hardly any study time involved.[458] One of these was Kim Peek, the man who inspired the 1988 movie *Rain Man*.[459] He was said to have memorized at least 6,000 books, reading and memorizing two pages at once, one eye on each page as he scanned down the pages. If someone gave him their address, he could tell them who their neighbors were. He memorized the maps in the front of telephone books so he could immediately state how to get from one US city to another, and then how to get around in that city street by street.[460]

Autistic savants who may be using telepathy to access numbers

While most of those who have studied these savants have not – publicly, at least – proposed that a psi element is involved, there are exceptions. Diane Hennacy Powell, M.D., a neuroscientist and psychologist, has studied psychic abilities in autistic children. She is author of *The ESP Enigma: The Scientific Case for Psychic Phenomena*.[461]

Dr. Powell noted in an article for *Edge Science #23* (2015)[462] that her research was initially inspired by anecdotal reports by psychologist Bernard Rimland:[463]

Rimland studied over 5,400 autistic children, 119 of whom were savants. Four reportedly exhibited ESP, which Rimland listed as a savant skill. These children routinely predicted events in advance, especially concerning their caregivers and provided specific information that only these caregivers could have known…After stating my hypothesis that savants might be the most likely to demonstrate ESP, I found many parents and clinicians in the autism community who believe their children are precognitive and/or clairvoyant, and even more say their children are telepathic (p. 13).

In 2013, Powell began evaluating child savants in India. She recounted: "One was a six-year-old boy with an encyclopedic knowledge of science, reportedly without having studied. Another was a girl who always knew exactly how many potato chips her father had reserved for later" (p. 13). Her first case in the US was that of Hayley, a nine-year-old autistic and mute girl who could only communicate "by either pointing at letters and numbers on thick plastic stencils or typing into a device called a 'talker' that converts text to speech." Although she could not do simple math, she could "give answers to increasingly complex problems involving several-digit numbers" (p. 14).

One day she typed her answer in an exponential format for the first time. She hadn't been asked to, but the therapist's calculator had just accidentally been switched to displaying results in that notation. The shocked therapist asked how she knew. Hayley typed, "I see the numerators and denominators in your head." Hayley then accurately answered questions for her therapist that she shouldn't have known the answers to, such as her landlord's name, "Helmut." Hayley also could type the exact words her therapist was thinking to describe pictures hidden from view. She even typed prose, word-for-word, including several foreign languages, but only when her therapist knows or reads it.

Powell[464] set out to do a series of telepathy experiments with the girl and her therapists, having them look at and write numbers while apart from each other. For the first one, feedback was intentionally withheld in an attempt to control a number of variables. This was a sharp departure from the protocols she was used to using with two therapists, who had frequently played telepathy games with her. Powell reported "she was 100% accurate on three of 20 images with descriptions containing up to nine letters, 60% to 100% accurate on three nonsense words, and 100% accurate on two random numbers: one eight digits, the other nine" (p. 15).

During the second experiments, feedback was allowed between trials. "Random numbers between one and one billion were used to create 12 equations (multiplication, addition, division and cube roots) with solutions of up to 12 digits." She writes:

These were determined and written with their equations on individual slips of paper, stacked face down, and given to the therapist at the beginning of the next session. Data from day two with Therapist A included 100% accuracy on six of 12 equations with 15 to 19 digits, 100% accuracy on seven of 20 image descriptions containing up to six letters, and 81% to 100% accuracy on sentences of between 18 and 35 letters. Data from the session with Therapist B showed 100% accuracy on five out of 20 random numbers of up to six digits, and 100% accuracy on five of 12 image descriptions containing up to six letters.

Powell points to other anecdotal examples provided by herself and doctors as early as the 1930s of children with autistic-like symptoms (they didn't have the diagnosis yet) performing successful telepathy with numbers, words and secret information of their caregivers.[465] The consensus was this was telepathic (mind-to-mind communication) because the children seemed to be successful only when specific caregivers or relatives were present (p. 16).

Powell's results were first publicly reported at the 57th Annual Convention of the Parapsychology Association in 2014. Her impressive presentation included videotaped sessions of her experiments with a girl who spelled out a long string of digits, which her tutor had just spelled out and hidden behind a divider in a separate room. While Powell was only able to conduct a modest number of trials with each child, she went to great lengths to perform rigorous tests, even knowing some of the protocols she put in place would likely lessen results since autistic children don't respond well to change.

Sadly, at the above-mentioned conference, two longtime parapsychologists sitting next to Debra mumbled they didn't believe it. Thinking they had noticed something flawed with the study's protocols, Debra asked them to explain their skepticism. While they admitted they couldn't find anything wrong with the methods as reported, they insisted, "This is just too impressive; no results can be this good."

We reached out to Powell to see if she had further insights or updates to her work with autistic kids and numbers. She wrote in an email in December 2020:

I am pretty selective about whom I have formally tested, so each of the children has scored between 95% and 100% accuracy. Ramses was the youngest child I tested and had the shortest attention span. Nonetheless, he was able to accurately vocalize 4-digit randomized numbers both times we tried. Akhil can type independently. I tested him for randomized 6-digit numbers in groups of 10 on three occasions and his speed and accuracy were as remarkable as Hayley.

TransDimensional Systems

From Jon: When I was first involved in remote viewing around 2000 in TransDimensional Systems, the leader, Pru Calabrese, suggested we try the lottery as a team. She said the team previously had been very successful. (The success is detailed in Chapter 21 on the lottery.) We used a finger-testing method, one digit for each number, and Pru applied algorithms to our results. We did not win any lotteries. Though we didn't succeed, this attempt (and hearing about the previous success) whetted my appetite and over the years since TDS closed down in 2003, I have experimented with and played various lotteries. I explored all the methods I could find, devised some of my own and wrote a piece for the TenThousandRoads website titled *Lottery Methods Rounds Up*. Our next chapter explores attempts by remote viewers to win lotteries, some of which have been successful.

Viewing numbers and letters – "the final frontier" of psi

The ability to "read" letters and numbers has implications that go far beyond winning the lottery. This ability has been useful in applications such as finding missing people or pets, solving crimes and gathering intelligence information. The ability has been useful in a few instances in medicine, science and engineering applications. Imagine if remote viewers could identify chromosomes or genes involved in disease replication or suppression, or identify chemical compounds that prove so useful in the many fields dependent on chemistry in our technological society.

As it stands right now, in the practical world of locating missing people or objects, viewers most often rely on psychically "visiting" nearby landmarks to get the job done. Sometimes this is sufficient. Psychic detective Pam Coronado, former host of the hit TV show *Sensing Murder*,[466] recently shared with Debra that she worked with a law enforcement officer whose address was unknown to her. He gave her the challenge of finding his house, which for security reasons was unlisted.

> I was training with this department to track fugitives in real time. A detective would spend the day running around and I would track him in real time. I ended up locating his house because he went home.
>
> I knew what general area but not what town he lived in. I was working with a monitor, another detective I have worked with for years. She found the town based on my description of a library next to city hall and police station. Then she found his street based on my description of a creek behind his house and an independent (weird name) church that looked like a house, down the street. I actually landed on his next door neighbor's house because I had the wrong color shutters.[467]

As another example, Debra located keys that were missing on a 2.5-acre desert property. They were underground inside a sprinkler system and she found them using simple clairvoyance, which was only possible because of recognizable landmarks on top of the sprinkler. Even her son Manny, at about six years old, found a missing game by closing his eyes and seeing a map in his head that led him right to it under a sofa. Still, the sofa was in his own home, not on the other side of the city, country or world at a completely unknown location, which is often the challenge a remote viewer faces.

Jon experienced the difficulty of using landmarks to pinpoint a location in a murder case.[468] One viewer (Daz Smith) described an iconic tower five miles away and another viewer (Don Walker) described an observatory just one mile away from what was eventually found to be the location. Finally, Daz Smith described a very winding road (Snake Road), which came within 250 feet of the spot the victim was buried in (and even drew a strong likeness of the hole she was found in). But:

> As we have shown in this report, the Aurora team developed a large amount of data congruent with the location – from the surrounding region, to the salient landmarks nearby, to what was in the immediate vicinity, down to the specific spot – that there was a hole, it was near a path, near a road, on the edge of town, wooded area, in a park, etc. However, as we have mentioned, finding a specific geographical location with remote viewing is extremely difficult…the police reported that even when they were in the immediate vicinity, they would not have been able to pinpoint the site without Reiser being there to show them. It was in their opinion a "clever place" to hide a body. In short, getting this vital piece of information was an extremely difficult task which no one was able to accomplish, searchers, police or our RV team.

In other words, the practice of looking for landmarks is often just not practical. It can be very time consuming and may fail to yield results, as in the situation above. Also, too many elements at the location may match other things in the vicinity, or a person or vehicle may be on the move. Imagine how inconvenient it would be if every time you needed directions, you were given only landmarks, without street names or an address. Then imagine how useful it would be if you could intuitively get the phone number or email address.

This is why Debra and one of her research partners, Michelle Bulgatz, feel strongly that the field of intentional psi will not progress until this final frontier of psi reading of letters and numbers is conquered. They are in the process of devising an experimental design they hope to implement in the near future. This will task remote viewers, psychics and mediums with license plates and street signs, trying out a variety of approaches to see what works best. In advance of this, Debra and her pal Natalie Cormier have created license plates, street signs, three-digit number and lotto-number targets on their practice target site www.remoteviewing.net. (If readers find they are having success, do please let us know so we can include you in future studies.)

Debra also has conducted some preliminary and very informal tests with her clairvoyant students. During one of these, she wrote the nonsensical letters VROOM on a dry-erase board in her office, keeping the students blind to the letters. Her students were at their homes at locations around the globe connecting via tele-seminar. With just a few minutes of concentration, one student reported the letters "ROOM" and other students had three letters correct.

In a more recent test with a remote viewing class, Debra wrote down a three-digit number. None of the students had a strong match beyond a single digit. However, Debra was startled when one student wrote "8439" in the chat box, stating she didn't know why she got those numbers but they did come in clearly. The reason Debra was so surprised is that her house number is 8439! (Any readers wondering why our book contains so much emphasis on displacement, wonder no more!)

In summary, we believe one of the reasons more progress has not been made with alphanumerics is the same reason many people don't have the slightest idea they have the ability to use remote viewing (or any intuitive faculty) with any degree of control – they have been programmed by society to believe it's not possible. They don't know where to begin and therefore have never tried. A further complication is that many intuitives have artistic temperaments and simply don't like numbers or don't feel competent doing math, so their relationship with numbers is an adversarial one to begin with.

Still, even in our optimism, we do not disagree with the early SRI researchers who suggested there are specific challenges in psychic tasks involving numbers because we don't understand how the mind and brain function with regard to psi. Theories and some supporting data about hemispheric differences suggest similarities in the way remote viewers and those with certain brain injuries function. For example, some people with brain injuries develop a condition known as aphasia; they can draw something, but not name what they are drawing, or they can portray correct elements of a picture or object in their drawing, but will place these elements in the wrong spatial orientation to one another.[469]

From our observations, there is some similarity with results remote viewers produce. However, the study of psi processes and brain functioning is still in its infancy and theories about left/right brain functioning are oversimplified. Problems also arise when trying to discern when a viewer is actually engaging their intuitive faculties vs. logical ones. They may think they know when they are doing this, but no one else can be certain. Still, current research with remote viewers at the Monroe Institute and the University of Virginia gives promise that challenges in understanding brain functions in the psychic perception of numbers can be overcome. Perhaps it will be in a way similar to workarounds neuroscientists have found for seriously disabled people, which enable them to do things they couldn't do before.

If nothing else, it is our hope that this chapter inspires readers to start to practice skills related to numbers and letters, especially those remote viewers, clairvoyants and mediums who already have mastered other types of targets and are ready to stretch themselves. The lotto is only one practical application of this ability, albeit a pretty cool one.

CHAPTER 21

Remote Viewers Tackle the Lottery

"Hey! If remote viewing is real, how come you haven't won the lottery??"

Psychics hear this all the time. Well, as it turns out, contrary to what the general public and even many in the remote viewing community think, some viewers have in fact won lotteries – many times! Even a few relatively big lotteries.

Early on, researchers at the Stanford Research Institute conducted forced-choice trials involving numbers and letters. As we reported in earlier chapters, Ingo Swann engaged in thousands of such trials, which he referred to as "Analytics."

These began during his work in the psychoenergetics program at SRI in the early 1970s and were very much the focus of his work there in his final year. These trials were some of his least-publicized work, but absorbed much of his time and attention. During these trials, Swann documented that he experienced "learning curves" with letters and numbers, yet results continued to fluctuate.

While Swann later tried his hand at the lottery, but he was not successful. His niece, Elly Flippen, who lived with him as his assistant and is now the executor of his estate, personally shared with Debra that "Ingo could not use his psychic abilities to win money if his life depended on it." She felt he had "money issues," noting that others who followed in his footsteps or engaged in similar ideas or projects always seemed to be more financially successful than he was. His longest-running student and favorite viewer, Tom McNear, recounted during the October 2020 Applied Precognition Project conference that in his later years, Ingo would send a friend running to the corner store to put down a few dollars on the lottery – but he never won.

Swann and others offer various reasons for the difficulty in using psi to describe numbers. Explanations include that it is because numbers are "not real" or are abstract concepts or are manmade. Another explanation goes that because the viewer knows the entire set of targets, as in a Pick 3 lottery (the numbers run from 0 to 9), the conscious mind becomes active and selects numbers instead of letting the subconscious mind gather the information.

When asked to explain his lack of success with the lottery, Swann joked:

> Either God says no, or it's a different process altogether. Remember, the lottery is a very quick thing, and it involves numbers, which are mental. The human being makes these numerical sequences. They're not found in nature. Nature doesn't say already there's nine eggs here. Humans have to learn to count to say there's nine eggs here. It's known in research with innate abilities that each child is born with a general mathematical hardware program in them, and they can count up to five. And estimate the numbers thereafter. That's innate in all kids. I haven't learned how to do that then. But the minute I do it, I'll never be seen here again. (laughter)[470]

As we noted earlier, a related line of research carried out at SRI came to the conclusion that the psi "channel capacity" is probably too low to win a very large lottery. As Ed May wrote to Debra:

> In another example, to win a lottery of 6 two digit numbers requires 6 x log2 (100) = 39.9 bits of information. So if the above channel capacity is correct—a mighty big IF—then it would take continuous RV for 18.5 hours straight! This is the geeky answer to the question we are all asked: If psi is real how come you ain't rich??

Successes and failures at the lottery

We turn now to attempts by several individuals and groups to use psi to get winning lottery numbers. Some are little known or unknown, and a few have been successful.

The most well-known attempt was by parapsychology researcher James Spottiswoode, who in 1996-97 conducted an experiment to try to win the California Fantasy Five lottery.[471] In this lottery, you pick five numbers from 1 to 39: e.g., 15, 18, 24, 31, 38. Odds are 575.757 to 1. The method was explained in detail by Stephan Schwartz, who was one of 17 viewers.[472] Consider a grid with 575,757 squares, which contains all the number combinations possible in this pick 5 lottery. There will be 256 blocks, with each block having 2,249 squares. Four teams of four viewers worked on their own, deciding which block the winning number was in. Four trial efforts were done and each trial was successful, but with no purchases made. The correct block was selected by consensus each time. James Spottiswoode was prepared to spend $2,249 to win the lottery, for which the payoff would have been between about $50K and $250K, depending on the number of players, number of winners and carryover if the lottery had not been won recently. Unfortunately, he left himself only three hours to purchase 2,249 tickets and he was unable to purchase all the tickets required. The procedure did lead to the block with the winning number, but logistics foiled the effort.

Milan Ryzl

Milan Ryzl was a Czech biochemist and researcher who was the first experimenter to transmit a long series of numbers under laboratory conditions using psi, and by report he also had considerable success with lottery predictions. He first worked in Czechoslovakia but the government frowned on his experiments and eventually he left the country, crossing the border in a hair-raising journey, and eventually sought refuge in the United States.[473]

In Czechoslovakia, Ryzl had decided to use hypnosis in his explorations of psi. He established a training protocol in stages. In stage one, the goal was to give the participant confidence as they visualized images, which would contain extrasensory information. The emphasis was on getting clear images, not fuzzy ones (which runs counter to the advice given nowadays by remote viewing instructors). In the second stage, immediate feedback was provided so participants could gain a sense of their mental state when receiving correct psi information. In the third stage, Ryzl would guide the viewer to be able to perform on their own, without his assistance.

Ryzl tested 500 student volunteers between the ages of 16 and 30, of whom about 50 had some success. He chose his best subject (Pavel Stepanek) for an experiment transmitting numbers. The numbers were converted from base 10 to base 2 to permit a binary choice of colors. Eight sequences of colors were used to determine each digit. The procedure was cumbersome – it took 19,350 single color calls (green or white), 400 per hour, with two participants – a grueling undertaking. The data was analyzed mathematically and, in the end, five three-digit numbers had been transmitted without a single error.

The experiment was designed as proof of principle, not something that would have a practical application. However, Beverly Humphries of the Defense Intelligence Agency suggested the model might have relevance to the Star Gate remote viewing program and it would be worthwhile to try to replicate aspects of Ryzl's design.[474]

The following piece was written by a person who evidently had detailed knowledge of Ryzl's experiments with hypnosis. Particularly noteworthy is the reference to synesthesia at the end of the second paragraph, since – based on his many years of experimentation – Ed May (among others) has offered the opinion that psi may manifest through synesthetic impressions.

In a review of the total experiment, Ryzl concluded that there had been a number of obstacles to be overcome. The first of these obstacles occurred during the initial phase of the experiment, when the subject was first brought to a hypnotic trance corresponding to the proper level of consciousness in which ESP manifests. At this stage the subject was in an extremely suggestible state. Unfortunately, the maintenance of such a state requires the suspension of critical thinking. Without this discriminatory aid the subject makes mistakes, as he or she is unable to determine the difference between true impressions and other sensory impressions. To overcome this difficulty, Ryzl juggled the different levels of hypnosis. Thus, while the subject was in deep sleep, he was more receptive to extrasensory impressions, and while in the lighter stages, he could use his critical faculties and memory. In this way the subject was able to progress by correcting his own mistakes and by learning to rely upon, and trust, his own judgment.[475]

An interesting difficulty that arose concerned the resistant aspect of psychic impressions. Psi impressions do not seem to occur in the same set patterns and symbology as do sensory impressions. Extrasensory perceptions are usually perceived subjectively and manifest most frequently through the physical senses as hallucinatory experiences. This means that a color may manifest itself as a texture, sound or temperature.

Ryzl found one of the difficulties in testing for ESP is that psychically received impressions, manifesting as false sensory hallucinations, are frequently indistinguishable from conventional hypnotic hallucinations. ESP subjects must double their energy for they must constantly be assessing their impressions against what they know to be reality.

Several reports describe Ryzl's attempts to win the lottery. According to Jeffrey Mishlove, Ryzl's viewers predicted the winning number in a Czech lottery several weeks in a row, winning several thousand dollars.[476] Remote viewer Martin Wszolek provided more details, noting that Ryzl worked with 47 intuitive typists in Prague who had been trained in hypnosis, meditation and remote viewing techniques. Two methods were used: one concentrated on typing "regular repeated numbers" and the second focused on "repeating archetypes and symbols associated with specific numbers."[477] As far as we know, this is the first use of psi/ESP to win the lottery. An exact date is hard to come by, but it appears to have been in the late 1960s or early 1970s.

Recent examples of lotto wins

We have been in touch with remote viewers who have won either a large lottery or had many wins on a Pick 3 or Pick 4. While all these folks are remote viewers, they did not use ARV in each instance in their wins, but they did use psi. The wins have been mostly box wins – the numbers can be in any order. Because of the difficulty of getting a win in a Pick 3, Pick 4 or higher lottery, a box win pays quite well. For example, an online bet of 50 cents on Betanysports wins $187.50 for a Pick 4 box win. A straight win – having the numbers in the exact order – pays considerably more (50 cents wins $450.)

We will take a look at a few of these winners. Unfortunately, lottery winners generally prefer to keep as low a profile as possible and do not post online. They may be concerned about being targeted by hackers or criminals, accosted by random strangers or nagged by friends and relatives. Stories abound of people whose lives have been upset or even ruined following a large lottery win.

Against this background, two of the remote viewers with major wins have kindly agreed to provide some details. A third has declined, in part because he is writing his own book on psi and the lottery. Because of their understandable reticence, we are not able to provide documentary proof of these wins. However, we have seen the winning ticket for one of the large wins, and we have email and other evidence for a second large win. The desire for anonymity or a very low profile is similar for people who have had multiple small wins. We have seen details provided by these folks and, in addition, Jon has had multiple small wins himself.

A Pick 5 lottery win of more than $100,000

Twenty-three years after Ryzl achieved success in the lottery, Muspsi Arte – a remote viewer who lives on the East Coast of the US – won more than $100K in the Pick Five, using a far simpler procedure than those noted above. She has studied many psi modalities and is a trained remote viewer. For this event, she was experimenting and used internal body sensing (interoception), which resulted in this amazing multi-number hit, as she explains below. Muspsi kindly permitted her method to be posted on Daz Smith's Remote Viewer and Remote Viewing Facebook group (March 27, 2019), about which Daz wrote:

A few weeks back a remote viewer approached some of us with some details about how they had successfully used remote viewing to win over $100K on the lottery. Since this time they have taken some time out to acclimatize but we have been given the go ahead to share a few details with the community at large…So well done, thanks for sharing and let this give the rest of you ARV, RV lottery and other money-generating ideas and projects some much needed inspiration – it can be done.

These are Muspsi's comments:

I was partially blind. I knew to get numbers, however, I was not aware they were going to be played or the game. I was just trying to have a fun exploring exercise. In my case, the number range was 35 numbers. I was to pick 5 numbers but instead had 6, one was not certain about but since my throat was a "yes" – I coughed and mistook it as a positive. The method I used was something I did not practice or was exploring during this time – so, it wasn't something I was "training" to do. It is something I felt like doing in the moment.

My partner was an equal contributor because it was they who picked the game and actually decided to play the numbers. How they treated the "6th" number was to just alternate the number on 5 rows on the ticket. To play 5 rows it costs $5 ($1 per line). Instead of picking the exact date of the game they played the same numbers for a week. So, $5 X 7 days was a total of $35 spent on the betting. The first day of playing won on the whole ticket. 1 row (non-coughing number) won $100,000 and the 4 other rows won $250 each (the next amount one could win). So, all total the winning ticket was $101K and ended up after taxes (and a long wait to release funds due to security) $72,000. The check was written out to my partner who picked it up on the way to their work.

This is a quick explanation of how I approached it. It's a mixture of body (kinesthetic) dowsing, manifestation block elimination, hypnosis and neuroscience concepts, and remote viewing concepts…The concepts I put together are based on a variety of borrowed tools from other fields (no one person's invention or ownership but they may be put together uniquely by someone, which will be considered "their" approach). I was trained in a good number of things prior to remote viewing being publicly released and some tools were used in teaching music. So, I believe in any and all approaches should be experimented around on. Good luck for those who may be out there working on this stuff. Yes, it can be done.

Daz concluded with,

> Please don't ask who this is – the person for now wants to remain nameless, but ask any questions here and we will see that they get to see them. Please respect their privacy – it's all true, I have seen the lottery slip and winning check.

Jon and Debra know this talented remote viewer and we'd like to add the following. We have also seen the winning ticket. Muspsi has made the following additional points: she recommends asking your body first what the signal for yes and no will be for each number. They may not be the same each time. She said the signals for her win were a falling feeling in her stomach area and a tightening of the throat.

A very large lottery win of $325,000

The largest remote viewing lottery winner we know of did so as a member of a team effort. The team wishes anonymity so, unfortunately, we can't reveal their name, location or other defining details. The win was confirmed by a viewer we know who was aware of the win shortly after it happened. This viewer vouches for the winner's absolute integrity and indeed the person whose session generated the win is known and respected in the RV community. The winning team used ARV to get the winning numbers, and this was their method (quoted with permission):

> Perhaps the toughest job in any lottery remote viewing attempt may be that of the pool person. It can be quite difficult to come up with 49 photos that are uniquely different, so that the pool person can correctly match the drawings of the remote viewer with the appropriate photo image. It may require someone who also has remote viewing training and skills to put together a good pool of targets.

> Let's quickly review some of the best practices in building an ideal pool of photos. The pool manager would put together a pool of 49 images and label them with numbers 1 to 49. But, rather than cutting photographs and images out of magazines and using actual envelopes, it is much easier to just use photos (digital images) on your computer and number them 1 to 49. You can find a lot of good photos on the internet. It's a lot easier to deal with and easier to make. That way, when the draw is over, the pool person can just e-mail the correct images to the remote viewer.

> In addition to this, photos cut out of magazines, etc. can be extra confusing because there may often be a photo or image on the reverse side of the page as

well. The remote viewing mind may not realize which the correct image is. The associated numbers don't need to be written on the photographs, however, if they are, they should be put in the upper left hand corner of the photo page.

All of the photos should be a fairly large size as the image is more easily detected by the remote viewer if it is large and/or fills the screen. (Thumbnail size photos are too small.)

Only photos of real objects, buildings, etc. seem to work. Clip-art images that come with computer software, while simple and easy, don't seem to work well because they don't carry the same energy as a photograph of a real object, etc.

There shouldn't be any similar photographs. For example, you wouldn't want a photo of a ball, the moon or a brass ring in the same pool, because the remote viewer probably can't tell the difference. He might only say, "I get something round." And the pool operator wouldn't know which one it pertained to if there is more than one photo of something round. For example, the pool shouldn't have any more than a couple of photos of buildings, and if there are two of them, they should be distinctly different. You wouldn't want to have more than two photographs of food, and they would have to be decidedly different as well.

The remote viewer envisions (remote views) what he/she will see when he/she receives the e-mail containing attachments with the correct images associated with the winning numbers after the draw is made. The remote viewer then makes a drawing of his/her impressions, and scans and sends the drawings to the pool person who then attempts to match each of the 6 drawings to 6 of the photographs in the 49 number pool, thereby identifying which numbers need to be purchased for the lottery.

When the 6 (or 7) winning lottery numbers are known, the pool person sends (e-mails) copies of the 6 correct images associated with the 6 winning numbers to the remote viewer.

This is very important. The pool person must send only the 6 images corresponding to the winning lottery numbers. Only the 6 Correct Images, nothing else. Sending the remote viewer anything but the correct winning images will only guarantee that he/she will perceive the wrong ones. Even if the remote viewer didn't correctly identify a single correct image, the remote viewer must never be shown anything other than the correct images. The remote viewer should never see the misses, because it will only reinforce the incorrect image in the viewer's mind.

After each lottery draw, the pool person must remove the 6 (or 7) images that the remote viewers have seen from the pool, and replace them with 6 new photos. It is only necessary to replace the 6 (or 7) images seen by the viewers, not the remaining 43 images. Once the remote viewer has seen the images, they can never be put back in the same pool for the same remote viewer.

ARV for horse races, auto races or other events
Horse races are much easier to do, as usually there are only eight horses in a race, so the pool person only has to assemble eight photographs. In the beginning, it's probably easier for the remote viewer to only try to pick the one winning image (horse). Don't worry about the "Place" and "Show" horses. If you wish you can add these to the mix after a bit of practice.

To do this properly, the remote viewer must separate the first remote viewing from the rest in each subsequent remote viewing. Remove the photo and horse that was selected in the "Win" round of RV. The subsequent remote viewings would be performed for the "Place" horse, and then for the "Show" horse. The thing to remember here is that once the "Winner" image is remote viewed and selected, it must then be removed from the subsequent remote viewings for the "Place" horse, and then the selected "Place" horse image must also be removed from the pool for the final "Show" horse remote viewing selection.

So, the process would actually require three separate remote viewings – first remote viewing the "Winner" out of eight images, then remote viewing the "Place" out of the remaining seven images, and then finally remote viewing the "Show" horse out of the remaining six possible images.

Since this person and the team won such a large amount and has had other wins as well, their practices are worth paying close attention to. We add the following point, based on observing and taking part in many lotteries, horse races and other ARV predictions. The above recommendation is to use photographs as targets, but targets that are drawings by hand have also succeeded, including in the third largest lottery win that we are aware of by a remote viewer.

As we mentioned, we have been in touch with several viewers who have had multiple small wins (winning the Pick 3 or Pick 4 from 20 to 100 times). One of these is Jon, whose results we turn to now.

Multiple Pick 3 and Pick 4 wins

For a long time I have been interested in seeing if we can "get" alphanumerics using remote viewing. If we could identify numbers such as dates, telephone numbers, scientific formulas, street addresses, numbers on charts and diagrams, remote viewing would be a much more powerful tool than it is now. It is unfortunately the case that it could and would be used for unethical/repressive as well as ethical/liberating purposes. One of the most popular uses would be to win lotteries, big or small. Looking on the upside, with enough success in the lottery the reality of remote viewing could come into focus in society at large.

When I was first involved in remote viewing (1999-2000), after being intensively trained in the Bananaslam program I became a member of the TransDimensional Systems team. The year previously (1998), TDS had remarkable success with the lottery:

We have started a lottery group which has been RVing the "Georgia Lotto" and the multi-state "Big Game" every week. The participants are using the "fingers" technique (which utilizes kinesthesiology – biomuscular feedback) to remote view the correct picks. We are compiling the numbers and statistically choosing the most likely combination. Any winnings are split evenly between everyone participating that week. We buy only one group ticket per lottery drawing.

The group consists of viewers trained at TDS, CRV-trained viewers, Farsight-trained viewers, Psi-Tech-trained viewers and the people who learned the fingers technique in the weekend workshops. The more people that join the group, the better the results have been! Group energy seems to help the process. So far, our winnings have far outweighed our investment. (Our investment is a whopping $1 each lottery drawing!) In the month of February 1998, our earnings have been in excess of $338 per participating viewer.[478]

Group	Members	Winnings	Cost	Net
1	24	$304	$13	$291
2	22	$250	$13	$237
3	10	$1103	$13	$1099

Figure 21.1 – Lottery Group Winnings During March 1998

Each group purchased three tickets per week – one for each of the lotteries targeted.

Pru Calabrese, the founder and head of TDS, suggested we try the lottery as a team. We would use the same finger-testing method that had had success and she would devise an algorithm to select numbers from the team's picks. So we finger-tested the numbers, submitted our results and Pru ran her algorithm. However, we did not win any lotteries. Though we didn't succeed, these attempts whetted my appetite and over the years since TDS closed (March 2003) I have often played with and played the lottery.

I explored all the methods I could find, including finger testing. I devised some of my own and posted a list of many methods on the TenThousandRoads website. The post is titled *Lottery Methods Rounds Up* and here are the first 14 of the 33 in the list.

Ways and methods of selecting digits (in no particular order):

1. **Language method:** Use a language other than your native language. e.g., write in German if the databox refers to a winning number. Write in Hebrew if a triple digit, Spanish if a double digit. The idea is that perhaps a lesser used language channel will be less overloaded with associations for the viewer. (Jon K tried this, no particular success.)

2. **Letter the data boxes:** For the Daily 3 lottery, letter some boxes for data with A through J. Assign a number for each box (1 through 9, 0). Assignment can be made in variety of ways.

3. **Notch method:** Count as you would by drawing lines in the sand till you got to a certain number. e.g., 1111 = 4. (Idea is knosomatic – body-related.)

4. **Word count:** Write words in data box to correspond to the number that the box represents, e.g., if you write 5 words in box, 5 is the digit sought.

5. **Feel it:** For example, using an actual ticket or a ticket model, use your fingers to "feel" the winning numbers.

6. **Blanks method**: Write digits 1 through 0 into 10 databoxes (not in order). Then do the session, going over the databoxes one by one. Where there is a non-winning number, write in any kind of data (or in a specific language or

use a particular symbol or sign). Otherwise leave the box blank. The blank boxes will be the winning numbers. Alternatively: Put a symbol or sign or specific language data in the boxes with winning numbers.

7. **Use ideograms, pictograms, archetypes or elements from the stages of visual processing:** For example, make an ideogram for each number and practice that ideogram. Make up an ideogram or use hypnosis or self-hypnosis to find out appropriate personal ideograms (or symbols or pictograms or signs). Then when you do the session, write the ideograms for the 3 winning numbers. Or vice versa, write ideograms for 7 non-winning numbers. Alternative: Use shapes and patterns characteristic of processing in one or more of the several stages of the complex visual process, e.g., a corner, cross, star, edge, etc.

8. **Music association:** Designate an association between a piece of music or title with each number (for example, assign Tea for Two to number 2). Drill on this. Then hear the tune in your head or see the title as you do the session.

9. **Visualize the numbers and draw them**: Visualize them in your mind or as they will appear in the paper the next day or as they will appear on a computer screen when checking results. Perhaps the most extensively used method. Does not appear to produce results over time. VSW posted some winning tickets using one form of this method.

10. **Model the numbers:** Model the winning numbers with clay. Use either Arabic numerals or sand counting method. For example, / / / / / = 5. Supplement: As you model the numbers, be aware of any databits occurring in your mind that relate to the method you are using.

11. **Associate numbers with body parts and/or movements**: Assign a number to each finger. Then let the fingers tingle or otherwise attract your attention to indicate the winning numbers. Pru Calabrese used this "knosomatic" digit method in TDS. Did not get any substantial results. One could also use and move elbows, ears, nose, trunk; make facial gestures, etc.

12. **Use a viewing team and an algorithm**: Use a team of viewers, have each come up with the 3 numbers. Use an algorithm to produce best picks based on past records of team. TDS tried this with SuperLotto – 6 digits – with no positive results. However, it was not done long enough to track individual's records and hence weight their picks. Algorithm used is not known.

13. **Divide up the set of numbers**: Divide up the numbers among the viewers and have each viewer responsible for the numbers in his or her section, e.g., Viewer A is given 1-4, Viewer B 5-7, Viewer C 8-0. Assignments are

rotated (or not) for each draw. Tasker tasks the viewers and collates the results. Viewers say whether they have one of the winning numbers in their slot for the day. Aurora tried this briefly without a win.

14. **Dowse the numbers:** Many ways to do this. e.g., prepare a piece of paper with all the numbers on it, either free form or in a grid, written or printed. Dowse, cast your fingers on or use a pendulum to pick the winning numbers.

The list continues and since then I have come across many other methods. Chances are, if you come up with a method, someone has probably tried it – but it may be one that few know about.

In the years since 2003, I had been remote viewing numbers for the lottery, betting some, and I had a few Pick 3 hits over the years. However, not many hits and I was not coming out ahead. The method that delivered the most hits was what I called California Dreamin' – using intention to get numbers in dreams, then the next morning doing a little simple addition with them, using intuition. In about 2016 I began experimenting with two other remote viewers to see if we could get more hits.

The Method of Advanced Picks (MAP)

After having only a few hits in pilot runs (and not betting), one of the viewers suggested we check to see if numbers hit on days following the designated day. This was not a new idea in lottery predictions (nor in parapsychology research), but it was new to me at the time and I felt it was definitely worth trying.[479] I learned afterward when I explored Lottery Post and other lottery sites that some zealots bet 10 or 20 numbers for an entire week on the Pick 3 in many states, all at once. It's a scattershot approach and – from what the statistics on Lottery Post indicate – not a successful one.

Since I had kept records of my lottery experiments over the years, I decided to reexamine the stats to see how many wins there would have been if I had placed bets on the days immediately following the designated date.

CA Daily 3 Same Day Hits					
Year	Dates	Picks	Same Day Hits	Hit%	Picks Per Day
2003	Mar-Nov	992	6	.6	24.0
2005	May 3-31	50	3	6.0	.54
2005	Jun 9-30	40	1	2.5	.52
2005	Jul 1-31	54	1	1.9	.57
2006	Dec 17-31	114	2	1.8	7.2
2007	Dec 28-Jan 10	107	1	.9	7.6
2007	Jan 11-29	61	0	0	3.3
2008	Jan 9-31	64	2	3.1	2.9
2008	Feb 2-29	92	1	1.1	3.4
2008	Apr 25-May 31	351	0	0	12.0
2008	Jun 4-9	37	0	0	7.4
2008	Jun 30-Jul 16	71	0	0	5.4
2010	May 10 & Jun 22	13	0	0	6.5
2010	Oct 22-29	52	0	0	6.5
2012	Jun 15-19	5	0	0	1.0
2013	Jan19	2	0	0	2.0
2015	Jul 20-Aug 1	11	0	0	.9
2015	Aug 6-15	5	0	0	.5
2016	Jul 18-20	3	1	33.	1.0
Totals		2124	18	.85	NA

Figure 21.2a

CA Daily 3 MAP Hits 2003-2016					
		MAP hits	Days in future		
Year	Dates	Total	0-4	5-6	7-8
2003	Mar-Nov				
2005	May 3-31	1			1
2005	Jun 9-30	1	1		
2005	Jul 1-31	2	2		
2006	Dec 17-31	1	1		
2007	Dec 28-Jan 10	0			
2007	Jan 11-29	0			
2008	Jan 9-31	6	4	2	
2008	Feb 2-29	11	3	7	1
2008	Apr 25-May 31	18	11	7	
2008	Jun 4-9	0			
2008	Jun 30-Jul 16	4	1	3	
2010	May 10& Jun 22	NA			
2010	Oct 22-29	NA			
2012	Jun 15-19	1	1		
2013	Jan19	NA			
2015	Jul 20-Aug 1	1	1		
2015	Aug 6-15	0			
2016	Jul 18-20	0			
Totals		46	25	19	2

Figure 21.2b

Key: CA Daily 3 solo efforts by Jon 2003-2016. Lottery trials with groups and my solo trials betting the same numbers in multiple states have been omitted. CA has two draws each day, Day and Evening. MAP = Method of Advanced Picks, which includes same-day hits plus hits on days immediately following.

I was surprised by what I found. There were only **18 same-day hits** out of 2,124 picks from 2003 to 2016, for a very low 0.85% hit rate. Figuring the exact odds for all these trials would be complex. I am not trained in statistics and, in any case, I do not have all the data from those 13 years of trials. Odds for a same-day box hit (three non-repeating numbers in any order per pick are 1/167= .0059 or 0.59%). California has two draws and I bet different numbers of picks per day and sometimes per draw, which complicates figuring the odds.

In contrast to the same-day hits, there would have been 46 MAP hits. This struck me as a lot, especially since I did not have data for four periods (marked NA above) when presumably there were other MAP hits.

Naturally, if you bet the same number on multiple days (and/or more than one draw per day), you will have more hits. The question is, Will you have a lot more? My data seemed to show I did. I also noted the hits tailed off over the succeeding days/draws, which could indicate the first day – the designated day for the bet – was a focal point in time but "not the whole story." There were more hits on days 0 through 4 than on days 5 through 8 (or later days). All this suggested there might be something to this approach.

Based on the above, I made lottery predictions from 2016 to 2019 working side-by-side with the two other remote viewers while viewing and betting on my own.

Methods: Automatic writing, visualization, "thot"

The first method I used was automatic writing (AW) with both hands grasping a pencil (or pen) and making "numerograms," which is what I call attempts to draw a number automatically but without trying to make it look precisely like a number. That is, it is done with the general intention of drawing a number but not letting the conscious mind guide the hand to try to make sure it is recognizable as a number. Hence, the term "numerogram" by analogy with the ideograms in remote viewing.

Since the resulting marks are not always recognizable (and sometimes not even close to resembling a digit), I sometimes repeat the numerogram and may do so a third time – till I can recognize a digit. My numerograms are almost always in the proper orientation; that is, straight up, not rotated 90 or 180 degrees, except for 6 and 9 on occasion. Nor is there a left-right reversal. There are few substitutions of numbers with visual similarity like 0 and 8, 1 and 7. (Some people in lotteryland get these substitutions, track them and seek patterns.)

I grasp the pencil with both hands rather than with my dominant (left) hand, because as soon as I started using both hands (Oct. 7, 2018), I got an exact hit the next day and more hits as time went on. Bilateral holding of the pencil or pen may not be fruitful for others, but perhaps it will be for other "kinaesthetic numeronauts."

I feel it's important to use more than one method during each session. The purpose is to avoid relying on one method, which perhaps "me-myself-and-I" would get bored with and stop playing the game.

After doing the automatic writing, which takes only a minute or two, I visualize a trip starting in my backyard, walking into a large opening in the ground and passing down a short twisting staircase made of bamboo and wood. The staircase ends and I come out onto a warm sunny beach. My mental eye takes me across the beach, with blue water on my right, and I come to an old TV set resting on a small table at the edge of the water. I draw an outline of the TV set in my session and when ready, I push the button on the TV with my pencil. I note what image, if any, appears on the screen. I call this method Viz (for visualization). Sometimes a three- or four-digit number will pop into my mind when doing these visualizations and I call that method "Thot."

I use all three methods and do very short sessions. Here is an example of a four-minute transcript from June 20, 2019. It's from the Texas Daily 4 lottery rather than the CA Daily 3. Texas has four draws per day, which enables you to amass data quickly (and is perhaps a factor in the success I've had with that lottery).

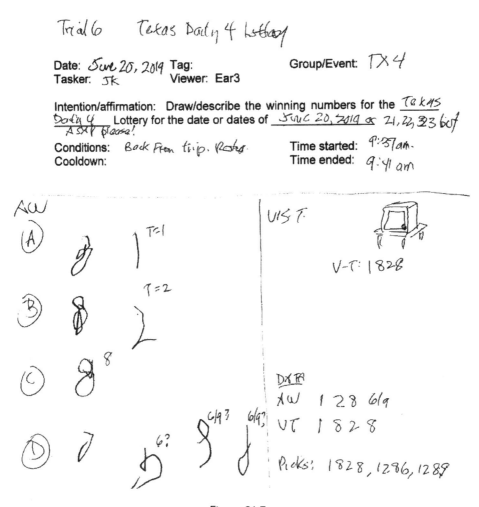

Figure 21.3

Key: Trial Six refers to the sixth lottery event in the series. Ear3 is my long-time viewer name. I sometimes think of it as an alternate identity. The intention is to get the winning numbers on one of the days listed but ASAP if possible.

I do just one session and check results for the next several draws or days for the state lottery I have bet. (You can bet all the states online.) The number 1982 hit the next day (June 21) in one of the four draws and my box bet of $.50 won $187.50. Five days later, on June 26 the number 8128 won. Because it was a "double" (one digit is repeated), my bet of $.50 won $374.50.

Texas Pick 4 Lottery win June 26, 2019							
Date	Amount	Balance	Description	Ticket	Risk/Loss	To Win/ Won	Status
6/26/19 2:00 pm	$375.00	$878.88	**Wager Won**	298.00	$0.50	$374.50	**Win**

Figure 21.4 Texas Pick 4 Lottery win June 26, 2019

Here is another example of a win, actually a win for the next day followed by a win on the day after that (that is, +1 and +2 wins). Remarkably, both wins were exact hits (correct order of digits), which needless to say are hard to get.

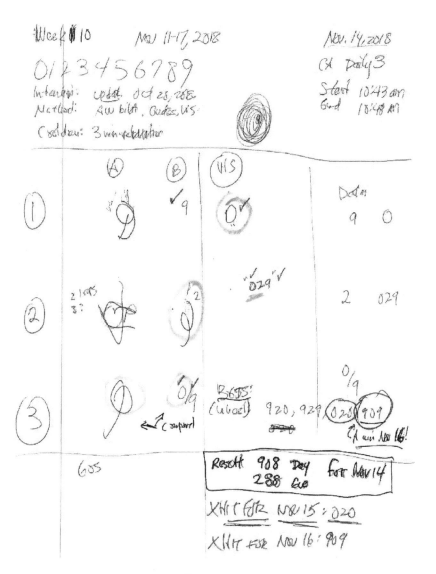

Figure 21.5

Key: *The session was done with a green pen. The red and yellow markings are comments, partly for excitement/"reinforcement" after I learned the winning numbers.*

Some people advocate celebrating your wins since if you are sending information from yourself in the future, you want emotion during feedback (which will act as an "attractor"). Maybe that is the case, and maybe it isn't. I feel good when I get a hit, but I don't try to enhance the feeling. I "go with the flow" and "it is what it is."

These bets were based on a wheel, as were the Texas Pick 4 bets above. As an example of a wheel: the numbers I got were 9, 2 and 0/9 from AW and Viz. I made box bets with combinations of those three numbers: 920, 929, 020, 909, doubling the 0 and the 9. Both

020 and 909 were winners, one on Nov. 15 and the other on Nov. 16. This session was rare in that I generated only a few numbers, yet they resulted in two hits. Sometimes I generate too many numbers and so I don't bet them. The wheels become too large.[480]

Note that the handwriting reads "AW bilat, Guides, Vis." This refers to bilateral Automatic Writing and Visualization. "Guides" refers to my "Higher Self" and/or spirit guides that I asked to assist, if they wanted to. One such potential guide was philosopher William James, whom I chose based on my reading of Jane Roberts, *The Afterdeath Journal of an American Philosopher: The World View of William James* (1979).[481] One of the persons I was working with on the lottery, Dr. Elisa Lagana, has familiarity with guides and she helped guide me in this unfamiliar territory. I don't know whether William James or the other guides helped, but the trial did result in those two exact hits within two days.

The other viewer I had been working with suggested tracing whether hits occurred in the past since time may be symmetrical. If there were hits in the future, perhaps there were hits in the past. I examined a few sets of past numbers but did not find the substantial increase in hits that looking ahead produced. However, more work on this might show there are "behind" hits and while they would not be useful in the lottery, they could be for addresses, dates and other numbers one would like to know. Being able to do this would have very strong practical consequences.

CA Daily 3 Map Hits 2018									
Year	Dates	Picks	Same Day Hits	Hit%	Picks per day	Total Map hits	0-4	5-6	7-8
2018	Sep16-Nov17	194	0	4.1%	4.6	8	6	2	0

Figure 21.6

The above table summarizes the results of trials from two months in 2018. There were no same-day hits but there were eight MAP hits, with a hit rate of 4.1%, considerably above the rate from a MAP analysis of my solo trials from 2003-2016. In this set, I undertook 12 trials in the California Daily 3 lottery (two draws per day) with a single trial lasting from four to seven days. I made an average of about eight picks each time. Overall, this set had a negative ROI (Return on Investment) because I did not bet all the hits.

TX Daily 4 – 2019 – Jon's Bets and ROI								
Trial	**Dates**	**$Bet**	**$Won**	**$Lost**	**Hits**	**#Days**	**Total Bets**	
1	Mar 4-9	$42.45	$93.75		1	6	28	
2	Mar 13-19	$43.00	$93.75		1	7	32	
3	Mar 20-28	$40.00		-$40.00	0	9	23	
4	Mar 29-Apr 4	$51.50	$93.75	$42.25	1	7	23	
5	Apr 8-15	$52.25		-$52.25	0	8	40	
6	Jun 20-29	$128.00	$562.00		2	10	46	
7	Jul 2-6	$66.00		-$66.00	0	5	8	
8	Jul 8-13	$106.00		-$106.00	0	6	10	
9	Jul 16-20	$30.00		-$30.00	0	5	3	
10	Aug 26-31	$43.00		-$43.00	0	6	27	
11	Sep 24-28	$17.50	$93.50		1	4	15	
12	Oct 10-17	$51.00		-$51.00	0	7	41	
	Mar 4-Oct 17	$670.70	$936.75	-$388.25	6	80	296	
	ROI (Net/Bet)	39.7%			**Average**	6.7	24.7	

Figure 21.7

Key: *UniPicks refers to the number of unique four-digit numbers bet. The number of four-digit numbers I bet varied from day to day. The average number of bets for the 12 trials was 24.7 and the average trial lasted 6.7 days.*

The above is the longest series I undertook, 130 picks over 12 trials in an eight-month span in 2019. I took breaks of a few days between most of the trials. Taking such breaks has been widely recommended in the field and perhaps they contributed to the success of the series. A positive ROI of 39.7% was achieved and would have been higher but I did not bet several numbers that hit. That is because when betting ahead several days, you may (and I did) decide to end some trials just before a number hit. Overall success was achieved with a positive ROI in the minority of trials (five) and negative ROI in the majority (seven) because of the very large amounts you win for a very inexpensive purchase.

It is rare to have a positive ROI in the lottery after even one trial, no matter what method you use. The above shows it is possible to use a psi-method and get repeated hits on something as difficult as the Pick 4 over a period as long as eight months. This and later trials lead me to think there is indeed something to these methods. They may not be transferable, but on the other hand, maybe they are.

I've found that hits tend to clump close to the day of the first bet. That includes lotteries with multiple draws each day. For the 2003-2016 data, there were 31 MAP hits

(57%) for days one, two, three and four; 21 MAP hits for days five and six (39%) and only two MAP hits for days seven and eight (3.7%).

After doing these series, I learned about Ingo Swann's and Joe McMoneagle's opinion that when you remote view, you don't get just one moment in time (see Chapter 7). You get information from the near future and the near past, or other distortions of time (Ingo's "fields"). My lottery results would seem to be confirmation of that, a practical application of "fuzzy psi time." Two other viewers with multiple lottery hits have mentioned experiencing the same "advance hits," citing wins on the third day. One of these viewers says he has won the Georgia Pick 3 "hundreds of times" and a few came after the designated day.

Below is a table showing my Pick 3 and Pick 4 hits from Sept. 16, 2018, to Sept. 25, 2019, indicating the methods that generated the hits.

September 16, 2018 – October 17, 2019		
Method	**Wins**	**Percent**
AW	13	35.1%
AW-V	2	5.4%
VIS	5	13.5%
Thot	5	13.5%
Dream	5	13.5%
Body	2	5.4%
V-T	5	13.5%
Total	37	100.0%
All V-T	15.5	41.9%
Lotteries: CA3, TX3, CA4, TX4		

Figure 21.8

Marty Rosenblatt   Jon Knowles

Debra Lynne Katz

Figure 21.9

This table shows the number of lottery hits from each method, both singly (i.e., AW or VIS) and combined (AW-V and All V-T) over the course of a year (Sept. 2018-Oct. 2019). The Automatic Writing total was 35% and the Visual-Thot total was 41.9%. These methods had the most hits, but they were also the methods used most often. The table shows a variety of methods can achieve Pick 3 and Pick 4 hits and, as mentioned earlier, it may be that using a variety of methods keeps things fresh in the kinesthetic and unconscious realms. If you choose to have fun and experiment with lottery methods, you will likely find your own unique profile with your methods.

As this book was nearing completion, I wished I had a hit to show a screen shot of. I hadn't yet found the ticket shown above and I was unable to scan others because the online betting site had closed down. So I decided to go back to California Dreamin' and combine it with the MAP method to see if I could get a hit. I intended that numbers for CA Pick 4 would come in a dream. Based on a tip from another viewer who has had success, I decided to ask my ancestors to put the numbers into my dream. I had never tried to be in touch with my ancestors before. I had a dream on March 13, 2021, with four people in it, standing on a field. There was some sort of event and a twisted cord to work with. The dream was in fragments and obscure. But the number four was impressed on me. I forget how two other numbers (5 and 2) figured in the dream, but I wrote 452 on the pad next to my bed during the night.

In the morning I had only three numbers, but the dream was supposed to be for a Pick 4, so I bet 4452, 4552, 4522. I also made a box bet on the CA Pick 3, since I got only three numbers in the dream. As per the method of advanced picks, I bet the two draws for March 14 but also the draws for March 15 and 16. My Pick 4 numbers did not come in between March 14-16 – the most I had was two digits correct: 6541 on March 14. But the CA Pick 3 results produced a win: March 14 - 894 and 038; March 15 - 550 and 388; and March 16 - 315 and 254.

Based on a dream, I bet a total of $21 and won $298. Not bad!

677354360-7	3/16/2021	LOTTONC		1.00	Lotto CA 9:15PM Pick 4 03/16/2021 (4-5-2-2) BOX $1.00 ticket for $750.00 prize
677354360-8	3/16/2021	LOTTONC		1.00	Lotto CA 9:15PM Pick 4 03/16/2021 (4-5-5-2) BOX $1.00 ticket for $750.00 prize
677354360-9	3/16/2021	LOTTONC		1.00	Lotto CA 9:15PM Pick 4 03/16/2021 (4-4-5-2) BOX $1.00 ticket for $750.00 prize
677355426-2	3/15/2021	LOTTONC		2.00	Lotto CA Midday 3:45PM Pick 3 03/15/2021 (4-5-2) BOX $2.00 ticket for $300.00 prize
677355426-5	3/15/2021	LOTTONC		2.00	Lotto CA 9:15PM Pick 3 03/15/2021 (4-5-2) BOX $2.00 ticket for $300.00 prize
677355426-6	3/16/2021	LOTTONC	298.00		Lotto CA 9:15PM Pick 3 03/16/2021 (4-5-2) BOX $2.00 ticket for $300.00 prize
677355426-3	3/16/2021	LOTTONC		2.00	Lotto CA Midday 3:45PM Pick 3 03/16/2021 (4-5-2) BOX $2.00 ticket for $300.00 prize
677487004-1	3/15/2021	LOTTONC		0.50	Lotto CA Midday 3:45PM Pick 3 03/15/2021 8-2-2 $0.50 ticket for $450.00 prize
677487004-2	3/15/2021	LOTTONC		0.50	Lotto CA Midday 3:45PM Pick 3 03/15/2021 2-8-2 $0.50 ticket for $450.00 prize
677487004-3	3/15/2021	LOTTONC		0.50	Lotto CA Midday 3:45PM Pick 3 03/15/2021 2-2-8 $0.50 ticket for $450.00 prize
677487004-7	3/15/2021	LOTTONC		0.50	Lotto CA 9:15PM Pick 3 03/15/2021 8-2-2 $0.50 ticket for $450.00 prize
677487004-8	3/15/2021	LOTTONC		0.50	Lotto CA 9:15PM Pick 3 03/15/2021 2-8-2 $0.50 ticket for $450.00 prize

Figure 21.10

In addition to the above accounts, we have been in touch with a few other viewers who have had multiple small lottery wins using a variety of analogous psi methods. For a change of pace, we end this chapter with one scientist's highly unusual attempt to predict the lottery: mimicking intuition through a radioactive device and taking geomagnetic factors into account.

An Artificial Intuition Device (AID)

Mark Zilberman, a Russian astrophysicist living in Canada, examined what he called the "True Predictions Density" (TPD) – the fraction of lottery ticket winners vs. tickets sold in the French and Russian state lotteries. This is a massive source of information since the TPD in Russia alone in a year is based on three to four billion predictions. Zilberman found, surprisingly, that the number of winners varied with the season in both Russia and France. Given his training, Zilberman happened to notice the variations correlated with sun cycles (11 years). As others have noted, when solar weather is "high," psi declines. Specifically, Zilberman's hypothesis is that when the Ap index is less than or equal to 5, psi is more likely to be successful. The Ap index is "the average planetary index of geomagnetic activity."[482]

We mention Zilberman's work for a second reason. He constructed what he calls an Artificial Intuition Device (AID). The device predicts six numbers for a Pick 3 and suggests 15 wheels for the top numbers. Zilberman's explanation of the device is not clear to us, but this is what he seems to be attempting. He wants to measure deviations in frequencies of microevents close (in time) to a macro event with "significant entropic potentials." His source is radioactive decay of an enclosed substance. As he explains:

Every five minutes, recording software displays the next number from the sequence [0,1,...9] in the application text box. Each impulse coming from the detector was recorded into an SQL database with the number displayed at the moment when the impulse was detected and timestamped.

Each daily series started on 11:00 PM of preceding day and finished on 10:00 PM of draw day. Ontario "Pick 3" lottery draws happen usually at 9:10 PM (EST) and have a cut-off time of 9:00 PM (EST) sharp for draw entries. At 5:49 PM (EST) of each day, the recorded data is analyzed, and then sorted in accordance to AID algorithms.

Amazingly enough, the device generated a positive ROI (Return on Investment) in transparent public testing in 2008-2009, with a range of success from 100% to .4% depending upon the betting configuration. However, a later document by Zilberman says the method fails to generate a positive ROI for individual lottery players.

As noted throughout this chapter, remote viewers have used quite a variety of methods, including ARV, to win lotteries both large and small, and repeatedly for the smaller ones. So, yes, you may be able to win the lottery. You may not get rich, but you may be able to get enough wins to stay ahead, at least in some periods, and have some fun.

As far as we know, this is the first time successes using remote viewing for the lottery have been collected and shared – some of these results are unknown even within the remote viewing community. We hope the above accounts stimulate others to look at the possibility not only of winning lotteries but of using ARV and remote viewing to explore numbers and words – signs, addresses, dates, records, formulas, etc. The development of RV capability with numbers or words would have enormous consequences.

CHAPTER 22

A Fresh Look at the Lottery – Sean McNamara

Sean McNamara is the author of several books related to meditation, telekinesis and remote viewing. He is a licensed massage therapist and energy healer, has a bachelor's degree in computer science and is working on a graduate degree. He frequently teaches classes in which he conducts "experiments" with his students.

In 2019 and early 2020, Sean McNamara led a group of remote viewers through a series of sessions focused on predicting Colorado's Pick 3 lottery. During their sessions, they experimented with techniques involving enhanced physiological relaxation and hyperarousal of the sympathetic nervous system. Their efforts led to two box wins for the Pick 3 lottery, with potential numbers 0 through 9, with the possibility of any number being drawn twice.

We interviewed Sean on these wins, and what follows are his responses:

What were the winnings? A $1 ticket yielded $80. Some of my group members may have paid more (a person can buy a $2 or even a $5 ticket), but if so, they didn't tell me. Also, all our predictions have been to buy an "any order" (box) ticket, *not* an «exact order» (straight) ticket, which has a higher jackpot. However, coming up with an «exact order» prediction would require a larger team of viewers than I had available to me.

Was there verification? For a lottery with such a small jackpot, the lottery commission makes no formal record of winning. Rather any lottery retailer, such as the local gas station or grocery store, can pay the winnings. However, I've attached a photo of my winning ticket from Nov. 10.

The winning numbers for the mid-day drawing on Nov. 10, 2019 were
5-8-2
(see set "D" on my ticket)

Figure 22.1 – Pick 3 Winning Ticket

What approach was taken? We used a "team style" ARV approach. Each viewer was assigned their own unique coordinate that was created by using an online random number generator. Each coordinate was associated with one of the lotto numbers. Each coordinate was then also associated with two potential feedback images. One image was associated with the outcome of that number being "drawn" while the other was associated with that number being "not drawn."

All the viewers knew we were doing lotto predictions, but none of them had ever seen the potential target images before and they didn't know which lotto number their particular coordinate was associated with.

Each transcript was judged, and then the prediction for each lottery number (0-9) was shared with the whole group so everyone could go buy their own tickets before the lottery was drawn at 1:30 p.m. (for t he midday drawing).

After the drawing's results were posted on the Colorado lottery's website, each viewer was shown their feedback image, which had been printed on paper and kept in its own envelope.

"Nov. 10 Pick 3 Coordinate Assignment" attached below. This is the same principle as that shown in the attached photo "Sept. 15 judging sheet."

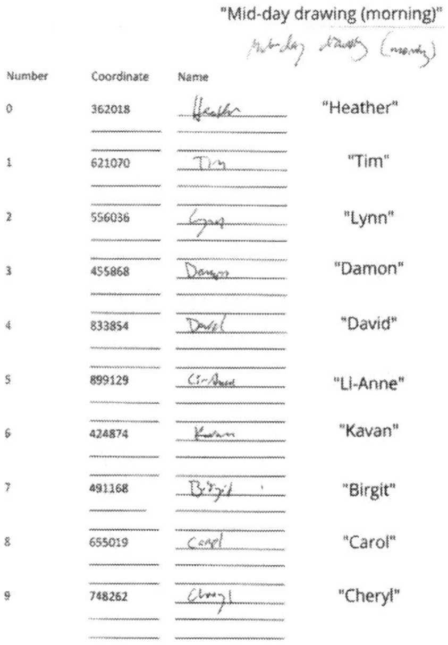

"Mid-day drawing (morning)"

Number	Coordinate	Name	
0	362018	*Heather*	"Heather"
1	621070	*Tim*	"Tim"
2	556036	*Lynn*	"Lynn"
3	455868	*Damon*	"Damon"
4	833854	*David*	"David"
5	899129	*Li-Anne*	"Li-Anne"
6	424874	*Kavan*	"Kavan"
7	491168	*Birgit*	"Birgit"
8	655019	*Carol*	"Carol"
9	748262	*Cheryl*	"Cheryl"

Figure 22.2 – Midday Drawing Morning

A special note about our approaches in September and November

Sept. 15, 2019, Prediction – When my group met for our lottery predictions, we also experimented with techniques to boost each viewer's psychic receptivity, as well as their ability to "send back" the feedback image to themselves at the time of their viewing session. This is why the book is titled *Signal and Noise*. We wanted to enhance the psychic signal and reduce the noise, which includes *analytical overlay*, *displacement* and *bleed-through*.

On Sept. 15, we experimented with heightening a viewer's transmission of psychic information from the future (the feedback session) to the past (the viewing session), making the viewers better senders of their own feedback images. For example, imagine a mother peacefully reading a book when suddenly the image of her son getting into a car accident enters her mind. People usually credit her relaxed state of mind for allowing the psychic event to occur. We should also pay attention to the fact that her son's state of mind was opposite to hers – he was in a heightened state of stress, which made him a powerful *sender*.

We sought to artificially, and safely, recreate those conditions. We did this by stimulating the viewers' sympathetic nervous systems to a high degree during their feedback sessions. I had them dunk one of their arms into a pail of ice water while looking at their feedback image. The consequent pain from the cold water was enough to produce the desired state of stress.

Below are images of one viewer's (Stacy) transcript along with the feedback photo. This technique seemed to have a similarly positive effect on most of the viewers that day.

Figure 22.3 – Stacy's Sept. 15 Transcript, Close-ups & Feedback Photo

Figure 22.4 – Fan-like Motion

November 10 Prediction – On Nov. 10, we tried an approach I refer to as "remote influencing." Below is an excerpt from the chapter titled "Experiment 3" from *Signal and Noise*:

The viewers' task was to send the information about their feedback image (the correct target photo) back in time to themselves at the time they did their viewing, several hours earlier. But this time, they had an additional task. I handed each of them a piece of translucent tracing paper, with instructions to lay it over the feedback picture and slowly trace the significant lines. At the same time, they were to intend to send two experiences back in time. 1) The imagery of the lines they drew on the tracing paper. 2) The sensation of their arm and hand movements (of the limb being used to draw).

Now, let's go back in time, a few hours earlier, when the viewers began their viewing sessions. They'd been given their coordinates and have spent some time in their "cool down" process, then putting their impressions down on paper as usual. This time, though, I added a new stage to the process. When they were finished, I handed each of them a piece of thick 9"x12" drawing paper. Then, I repeated an induction similar to ones I'd used before, returning them to a more receptive state.

Since the interruption of introducing a new piece of paper and explaining we'd be doing something new had shifted their minds to a more "beta"/alert state), they needed to relax again before proceeding. Then, I asked them to simply allow themselves to feel into their future selves at their "feedback time." More than that, they were to feel the sensation of their "future arm" moving over the tracing paper and to allow their present arm to move accordingly, while putting pen to paper. My aim was to help them establish a telepathic link with their future selves, and use *remote influencing*, not only across space, but across *time*, to put lines to paper.

I've attached photos of transcripts and target images from Heather's Nov. 10 session using the tracing technique. The target was a green rice terrace. Readers can compare the "psychic tracing" portion of her transcript compared to her actual tracing of the feedback image. We can see how helpful the "psychic tracing" can be in helping to match a transcript with a potential feedback image.

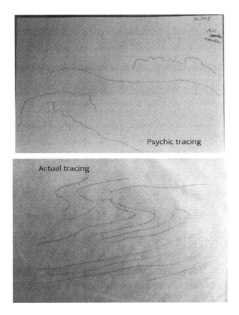

Figure 22.5 – Heather Nov 10 Transcript

Prediction: 0 is NOT drawn - hit

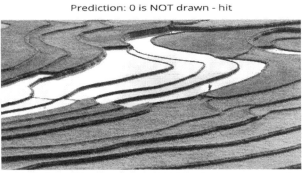

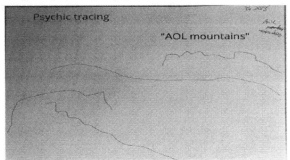

Figure 22.6 – Rice Paddies

461

How many times did we win? – Followed by my thoughts on statistics (in the excerpt below)

For our Sept. 15 prediction, two members of our group had winning tickets. For the Nov. 10 prediction, most, if not all, of the group members had winning tickets. In both cases, I say that "we won" since it was a group effort to produce the predicted numbers. The problem was, we usually came up with more than just three potential numbers. For example, on Sept. 15, we predicted 1, **3**, 5, 7, **8** and **9**. The actual winning numbers were 9, 3, 8.

I've attached three photos from Sept. 15. One is of my own ticket, which was not a winning ticket, but came close. A second photo is of me with the two winners, enjoying their winning glow. The third photo for that day is the "judging sheet" I used to track the prediction of "drawn" or "not drawn" (noted as "N" and "Y") for each potential lottery number.

On Nov. 10, we did better, since we only came up with 1, **2**, **8** and **5** as potential winning numbers. The actual winning numbers were 5, 8, 2. Each individual viewer had to decide for him/herself what combination of numbers to buy for themselves. Hence, not all viewers had winning tickets.

Schedule of Sean's Groups attempts to predict the Pick 3 lottery:	
Date	Outcome
Dec. 29, 2018	Lost
August 3, 2019	Lost
Sept. 15, 2019	Won
Oct. 12, 2019	Lost
Nov. 10, 2019	Won
Jan. 5, 2020	Lost but got 2 out of 3
Feb. 16, 2020	but got 1 out of 3

Figure 22.7 – Schedule of Sean's Group's Attempts

At this point, some readers may be wondering about the "odds" of winning the Pick 3 lottery. After all, to randomly pick the three winning numbers out of a set of 10 shouldn't be astronomic. It's not. The odds of winning by selecting the three numbers *in their exact order* of being drawn are 1 in 1,000 according to the Colorado lottery site.

However, with only 10 viewers, we could only do an "any order" prediction, meaning we'd win a lesser jackpot for selecting the numbers out of the order they were drawn. Those odds are 1 in 167. *(Note: 1:167 if there is no repeating number; 1:333 if there is a repeating number, e.g., 223.)*

But to think about our success in terms of "odds" is short-sighted, because it strips away the human capacity and effort, the very *life*, which went into making these predictions, replacing it with a cold, hard number.

What are the odds a person can close their eyes, see some wispy shapes enter their awareness, then write them down without allowing the conscious mind to filter or deform those images, then match them up to a target photo which they'll be shown several hours in the future?

And what are the odds of 10 people doing the same exercise and being sufficiently correct *at the same time* in order to win the Pick 3 later that day?

The lottery "odds" don't reflect the quality of a viewer's transcript or the quantity of psychic information recorded on it sufficient to correctly choose the target image. Yes, one could say a transcript only has to be "good enough" to make a prediction and beat the 50/50 chances of choosing the wrong image.

But this wasn't really about winning a lottery. So frankly, I don't care about the statistics. The lottery simply served as a third-party random number generator – an impartial event we could use to test our precognitive abilities against. Thankfully, nobody did it "for the money," because the prize on an "any order" bet with a $1 ticket is only $80.

I assure you, the time and energy each viewer put into a full day's work to produce a lottery prediction was worth far more than $80. It was invaluable.

A comment about our attempts since then

In 2020, the Covid epidemic made it impossible for my RV group to meet in person. However, we have made several attempts this year, and I've also led some public groups in attempting to predict the Pick 3. Neither my regular group nor the public groups have won in 2020, although several attempts yielded "1 out of 3" and one yielded "2 out of 3."

Since publishing my book in early 2020, I stopped formally keeping track of our attempts largely because I didn't have the energy to continue doing so. We're also beginning a series of experiments using other types of perception for rendering predictions, and I'm sure we'll disclose them at some point in the future if it seems appropriate.

A final, personal note

The two times my group succeeded, our viewers were together in the same room while they were doing their viewing, as well as when they received their feedback images (they did not look at each other's images during the feedback session). Also, their relationships and bonds had already strengthened by the time we had our first win. I believe the strong sense of connectedness and friendship shared among our group members had a great deal to do with our success. Of course, there's no way I can prove it. But I think our success occurred because some sense of "meaning" wanted to be born into our reality, and our situation was ripe for it to happen with us, particularly on Sept. 15 and Nov. 10. I think this is why I've had lesser success with my public groups, and even with our own group since then. It's not about talent, but about what kind of meaning can be created from the situation. What does it mean for us as individuals, and as a group, for something like this to happen?

When the event (of winning) has occurred once or twice to the same group of people, is there really a need for it to happen again, to make whatever point *reality* needed to make? I think not. This might explain the infamous *decline effect* reported in so many other experiments done by other viewers on projects like these. This isn't a cold, mechanical process. It's about consciousness, and it's alive. Strangely, in a recent experiment with my group a couple months ago, nobody had a winning ticket, except for one viewer. After looking over our group's prediction set, she spontaneously, and *seemingly randomly*, chose her own set of numbers and won. To say these experiments vex the conscious mind would be an understatement.

The details of these sessions, including the instructions for how anyone can replicate them at home, are recorded in my book *Signal and Noise: Advanced Psychic Training for Remote Viewing, Clairvoyance, and ESP.*[483]

Photo from Nov. 10, with the group celebrating our win. Four of them have their faces obscured for privacy per their request. Another photo is of my winning ticket for Nov. 10.

Ffigure 22.8 – Photo from November 10

Sean is with viewers Stacy & Michelle, who selected the correct combination of numbers from their "prediction set" to win the Pick 3 lottery on Sept. 15, 2019.

Figure 22.9 – Sean s with Viewers Stacy & Michelle

Predicting the MegaMillions Lottery

In early January 2021, Sean conducted another lottery experiment. He solicited volunteers online to attempt to predict the huge MegaMillions lottery for Jan 19, 2021: the pot had grown to $865 million. He used the same approach as with the earlier lotteries: each viewer on the team would do a session, which would be evaluated to make a prediction whether a specific number would be drawn. For example, Abel's binary ARV session was used to determine if the number 1 was a winning number. Baker's session was used for the number 2, and so on. This approach is similar to that of Muspsi, who evaluated each number to decide whether it was a winning number, using internal body sensing (interoception).

Using this method for MegaMillions requires 95 predictions because there are five winning numbers ranging from 1 to 70 and one winning number (the Mega) from 1 to 25. Sean wanted to evaluate each of the 95 numbers but "only" 76 people volunteered. He was able to evaluate each number from 1 to 70 to see if it was a hit and for the Mega one viewer's session was a weak indicator that 18 would be the Mega number, so Sean used that for the Mega on the tickets he purchased.

Sean received sessions one or two days before the draw and compared the sessions with two videos (45 seconds and just over two minutes), one associated with the number not being drawn and the other with it being drawn.

To undertake the large amount of work required, Sean charged each viewer $2, which he used as a screen so people couldn't just sign up without reading the instructions. Participants bought their own tickets.

In a video about the experiment, Sean said he does not consider the experiment to be about the money at all. **"It's not about the money!"** It's about feeling connected with others and the universe; there is a spiritual aspect to it. It's also to meet people and have fun. It would have been nice to have won, but if they had won, Sean said he would probably be on a beach somewhere now.

The viewers ranged from newbies to very experienced viewers. The instructions were specific and designed to prevent displacement. For example, viewers would only see one feedback video, not the other, and viewers were asked not to discuss their sessions with other viewers after the drawing. There was no general meeting summing up the experiment. Matching each number with a viewer was done randomly by selecting a strip of paper with a number on it from a bowl, entering the numbers into an Excel sheet in the order drawn, and then filling in a viewer's name next to a number as registrations came in.

Two viewers did not submit a pick and Sean had to pass on making a prediction for four others for various reasons. Sean's analysis indicated that seven numbers were being chosen, which meant participants would have to choose five of the seven numbers for the main field. He had hoped there would be only five numbers predicted to be winners.

Of the numbers from 1 to 70, scoring of the sessions projected that the following numbers would hit: 40, 13, 63 and 39 ("moderate" strength predictions) and 70, 47, 10 would hit as well ("weak" strength predictions). Only one of these numbers did in fact hit – 10. The winning numbers were 10, 19, 26, 28 and 50. The Mega was 16.

What Sean found remarkable about the experiment was that of *65 numbers the team predicted would **not** be drawn, the team was correct on 58 of them.* That is, 58 viewers predicted their number would not be a winning number, and it was not. Only 12 misses occurred – when a viewer predicted a number would not be a winner, but it was (four instances) or they predicted a number would be a winning number but was not (eight instances). However, as we discussed with Sean, we have no other data from experiments with this setup, so we don't know what an average or high number of such "negative hits" would be. Still, 89% does seem high.

In some other lottery-prediction methods, you don't make a prediction about each number – you don't assess whether each number is a winning number or not. (For example, you can try to draw or get a psychic hit on just the winning numbers, as in Jon's multiple small lottery wins.) But that is the method Sean chose, as did remote viewers who had the two largest lottery wins of which we are aware – Muspsi won more than $100K using interoception (body sensing) to evaluate 35 numbers and the team that won $325K successfully matched six sessions from among 49 photos, each of which represented a potential winning number. Given that these two large lottery wins used this method, Sean's approach makes good sense.

CHAPTER 23

Riding the Cryptocurrency Roller Coaster

We now take a look at two groups in the remote viewing community that try to predict the dizzying rises and falls of Bitcoin and other cryptocurrencies. The groups have made other financial predictions as well, albeit "for entertainment purposes only." The groups are CryptoViewing spearheaded by Daz Smith, Dick Allgire (and others) and the Precognitive Trading Group with Igor Grgić and associates.

Before taking a look at these ventures, let's jump into the exciting and culture-changing world of cryptocurrency and the blockchain, which appear to be forming a radical new core for monetary, business, scientific, legal, cultural and many other transactions challenging the existing worldwide banking and monetary systems.

Bitcoin. Ethereum. Blockchains. Mining. Digital currency. Exchanges. Wallets. Altcoins. HODL. Proof of Work vs. Proof of Stake. The Tangle. Even the Byzantine Generals Problem! What's it all about?

First, about that strange term "HODL." This is simply the crypto way of referring to holding coins. That is, not trading them – "Hold On for Dear Life" – hold them till they rise, then sell them.

Second, why the term "crypto?" Because you can encrypt the transactions so they are secret and not hackable (in theory anyway).

Cryptocurrency is a movement to bring money into the computer age by enabling payments and storage of money in digital form. Does it make sense to go to a bank, withdraw pieces of paper, carry them around in your wallet and exchange the paper for groceries at the supermarket when you could pay by inserting a plastic card into a machine at checkout? No. And now people in many countries are making purchases primarily with credit and debit cards. It's the wave of…the present.

That's the first step: payment with a card instead of cash (the latter is "fiat currency").

Now, what backs up that currency? How is the value of it stored? Since the Bretton Woods Conference and the US going off the gold standard, money is "backed" by the faith and credit of nations and the banking systems within them. This varies with the perceived strength and reach of the economy (and the military might) of each country. This system has developed to the point that the wealthy nations completely dominate the developing countries. And with the majority of people in the world in poverty, climate change already severely affecting these areas and with uneven development across the globe resulting in grotesque differences in wealth, the system is increasingly perceived as unjust – and, with its booms and busts, doomed.

After the crash of 2008, a person or persons named Satoshi Nakamoto published a paper outlining a system for digital currency. They were of a libertarian mindset and wanted people around the world to be able to buy and sell commodities, locally and worldwide, without being beholden to the dollar, the yuan or any other national currency. Person-to-person transactions. No middlemen, no banks. The amazing advances in computers and digital technology made this vision realizable for the first time in history.

With cryptocurrency, transactions are conducted on a "blockchain." This is a ledger that is validated by users in complex ways we will not go into here. The ledger may have contracts and other documents of all kinds attached to it.

Then, you need a coin to be a store of value. Bitcoin (BTC) was one of the first such coins. It is still dominant, having risen from practically nothing to a peak of around $60K (as of this writing in April 2021), albeit with huge rises and scary falls. Since BTC is digital, the coin does not have a physical form. It is a coin in name only. You must store the value of it somewhere. The solution? A digital wallet of course. An amount like "15 BTC" exists as an entry in a digital wallet. The wallet may be online, on a desktop computer or in a stand-alone physical device (e.g., Trezor, Ledger Nano).

Once Bitcoin came into being, others said, Why just Bitcoin? Let's have other coins. And so thousands of digital coins were created – they are called altcoins. *Bitcoin, Ethereum, Binance Coin, XRP and Tether* – to take just the top five coins as of April 2021. These were sold in ICOs (Initial Coin Offerings) along with hundreds of other altcoins.

Mining – coins? What's that? Bitcoin came up with the idea of creating a total of only 21 million Bitcoins. There will never be any more. They also said let's have "miners" receive bitcoins for validating the blockchain. Miners can receive some Bitcoin if they are the first to solve a mathematical puzzle. An odd idea, but it took off.

There are now thousands of "miners" on the Bitcoin blockchain and they validate their claims by Proof of Work (POW) using lots of electricity for their rigs. Others said, that's nuts! and devised an alternative method called Proof of Stake. Still others said the

BTC and Ethereum blockchains are too slow! We need something faster like the EOS blockchains. Or something entirely different – "The Tangle," which uses a different approach to connect its elements.

About those "Byzantine Generals?" The term derives from a parable about the dilemma generals would face when they must launch a coordinated attack but lack complete information about what other generals are going to do. Some of the generals may be corrupt or disseminating false information. This problem arises when trying to develop architecture to validate transactions in a blockchain. A thorny problem, and solutions are being worked out in the intense competition between blockchains in many countries, each of which is creating their own "national crypto stack" of assets (blockchains, coins, exchanges, regulations, etc.).

There is a lot more, but we hope this is helpful in laying out a few of the basic ideas about cryptocurrency.

2014 – Remote Viewing community discovers Bitcoin

On August 6, 2014, a would-be remote viewer who goes by Galderma posted:

Some of you might have heard of a digital currency called Bitcoin. Just as in Forex trading it is possible to trade Bitcoin against traditional currencies, such as the Dollar and the Euro. The interesting thing about Bitcoin is that the price fluctuates much stronger, allowing much more profit to be gained than traditional Forex trading. Besides Bitcoin, there are also alternative digital currencies, so called "Altcoins." Some of these "Altcoins" fluctuate so much that 200% profit a day is not an exception! I am trading altcoins myself and lately I'm very interested in Remote Viewing. This brought me to the idea of combining trading and Remote Viewing in a new and very interesting way. I would like to know if there are any Remote Viewers interested in this and see if we can start some kind of project together. If there are people interested or want more information (even if you don't understand anything about Bitcoin at all), please feel free to ask any question! Would love to explain more.

The next day one of the present authors (Jon) replied:

Synchronicity lives! I say that because someone has posted recently in APP (Applied Precognition Project) discussion suggesting we seriously consider RV and bitcoin. I felt it was likely the same person posting here since those are the first two mentions of bitcoin and RV in the same sentences that I can recall seeing

– and within a week or so of each other. But it isn't the same person. Maybe he will join you in this thread. I'd like to learn more. I've read a little about it and there is a lot of disagreement about the nature, purpose, stability, etc. of bitcoin, litecoin and the others. Even as to whether they are actual virtual currencies. There are some advantages, no doubt. Viewers in different countries can use the same (virtual) currency, if currency it be. No middlemen. It's not illegal, etc.

Unfortunately, Jon did not follow up to purchase Bitcoin at that time and missed out on having an early stake. He did get into the market with altcoins in October 2017 just before the huge rise in 2018 and he remains an expectant "HODLer" of BTC and half dozen altcoins. In 2018, Jon started a FB page called CryptoViewing, but when Daz and his team got into the game, Jon withdrew his since he was not planning to undertake crypto right away nor in the big way Daz and the team were.

CryptoViewing

Figure 23.1 – Daz S. and Ed R.

CryptoViewing is the biggest success story among RV/ARV businesses in recent years – the largest in the US since Pru Calabrese's TransDimensional System (1997-2003). By building up a large subscriber base on Patreon (a site where subscribers pay for content), they have been able to support about a dozen full-time and part-time staff. They also pay selected contract viewers from time to time (Jon has on occasion been one of them). Key viewer and CryptoViewing partner Daz Smith provided us with the following description of the group (personal communication):

CryptoViewing is a US-based business that employs full and part-time paid staff that includes consultants, editors, managers, marketers, I.T. and both full and part-time remote viewers from all over the globe. CryptoViewing as of 2021 is in its fourth year of operations and growth.

Unlike most other remote viewing operations that focus upon a purely business path, CryptoViewing's model is based around creating informative and fun consumable media that is accessed by subscribers in a tier system.

The team at CryptoViewing has centered around using remote viewing to look at and report around the growing interest in Crypto currencies, but they also look at other financial markets such as Gold, Silver and other financials. Every month the viewing team also release predictions of the following month's top world news and occasionally there is the odd mystery target thrown into the mix. CryptoViewing has a solid base of clients and subscribers who access the digital content through many channels like email, social media, YouTube, Patreon and now a private members website with discussion forums. Every month CryptoViewing also hosts online video chats with the remote viewers where the subscribers can ask questions.

We spent some time looking into CryptoViewing and found the following information.

CryptoViewing provides information and entertainment to three tiers of subscribers. At the time of publication, these were Supporters ($9.97/month), Viewers ($19.97/month) and Visionaries ($38.88/month). Depending on the tier, subscribers get some or all of the following: Dick Allgire's World Views, CryptoViewing Interviews, Crypto Sessions (two per month), Crypto Session Debrief, WooWoo Sessions, Health Related Sessions, World Events, World Events Debriefs.

Daz Smith has a very large base of followers based on his top-notch remote viewing expertise demonstrated in live videos and in sessions posted on his site, his FB group with about 5,000 members and the hundreds of thousands of people who have viewed his UFO and photography sites. He is the author of several books.

Viewer Edward Riordan's online diary includes 344 videos showing him doing and discussing remote viewing sessions. He has 9.77K subscribers.

Four of the viewers had built up moderate to large followings before CryptoViewing came together. Dick Allgire was a TV reporter in Hawaii for many years and is skilled at making videos. His guitar licks and a hot-shot character named Barry Schmelling flavor the CryptoViewing videos. Allgire's "Dick's World" YouTube channel has 9.7K subscribers, while his "Dallgire" channel, which includes music and remote viewing, has 4.26K.

The CryptoViewing YouTube channel itself has 52K subscribers. Both Dick and Daz gained considerable recognition on YouTube and other social media by taking part in several "far-out" projects by the Farsight Institute (53.2K subscribers). The topics include crowd pleasers like how the pyramids were made and hot topics like the supposed truth about Jesus.

With these social media bases and networks, the CryptoViewing team has generated more income from its thousands of Patreon subscribers and has provided more jobs than any other RV group in recent decades. This is a great accomplishment in a field where a good many commercial applications of remote viewing may exist, but we supposedly don't and can't know anything about them, primarily because of nondisclosure agreements. It is true more is going on behind the scenes than is known to the public – or to many in the RV community, for that matter. However, it would surprise us if any of these efforts have resulted in the (modest, but outstanding!) 10 or 12 full-time and part-time jobs created by the CryptoViewing team.

Much of the prolific content CryptoViewing delivers is for paying subscribers only, which makes it difficult to get an accurate overview of their work. But it is clear they have developed a substantial paying audience for remote viewing, one of the few RV companies to accomplish this in the 25 years since remote viewing came into the public domain.

Predictions

The CryptoViewing group makes a variety of predictions, from finance to world events, as noted above. This is done by evaluating the work of the viewers, most of whom use CRV, while Dick Allgire uses the HRVG method. In addition, the group has ventured into experimental work using ARV to make financial predictions "for entertainment purposes only." Their record from April 2019 through December 2020 for (mainly) financial predictions using ARV was seven correct and four incorrect. Of the seven correct predictions, five used Unitary ARV and two binary ARV. Of the four incorrect, two used Unitary ARV and two binary ARV.

Three of the four incorrect predictions were financial, and one was about the US Presidential election. They are awaiting the results of two long-term financial predictions (silver, gold) and they passed on two financial predictions because viewers' data was split.

In other words, for this period of financial predictions, using ARV CryptoViewing had seven correct and three incorrect predictions, for a very solid 70% accuracy. Of the Unitary ARV predictions, five were correct and two were incorrect, for a slightly higher success rate of 71.4%. The ARV taskings had wording like "84% probability the DOW moves below $30K" and "Gold will not rise above its all-time high," both within periods specified in the tasking.

CryptoViewing is a "large" company by remote viewing standards, with both group and individual predictions being offered, but there is no central tracking of predictions and their level of accuracy. The group covers a wide range of targets, including predictions of events all over the world. Many of these are open-ended, prognosticating the subject of major newspaper articles that will appear the following month. These run the gamut from the COVID-19 virus, presidential elections, catastrophic weather or geographical events, accidents, and political and financial developments. Many of these predictions involve remote viewing of newspaper articles, while others use Associative Remote Viewing protocols.

Crypto predictions have been issued as many as four times a month and approximately 40 to 50 new predictions are issued each month. Members gain access to these features based on their level of membership.

Jon speaking: For major event predictions, CryptoViewing often posts an image with a few headlines and a few lines from the prediction and declares it a "Hit." I wondered how many of CryptoViewing's predictions of major events for the next month were, in fact, hits and. if so, how strong they were. I had no doubt it is possible to predict events ahead of time. In TransDimensional Systems, we often did this in training sessions and Pru Calabrese, the head of TDS, often assigned "the major story in the local paper tomorrow" in introductory presentations. In fact, she did so the first time I heard her in 1999. On checking the papers the next day, many times a session will indeed foreshadow the lead story.

While predictions of all the cryptoviewers have not been posted, Daz Smith has posted about 50 of his predictions that he considers hits. Among them are very strong hits, for sure. For example, in January one of Daz's predictions was about an airplane disaster in Turkey due to technical issues. Sure enough, on Wednesday, Feb. 4, a Boeing Pegasus Airlines aircraft carrying 170 people failed to stop before the end of the runway at Sabiha Gokcen airport and fell 30 meters into a ditch, splitting into three pieces. It appears mechanical failure was the cause. Three passengers died, according to news reports. The plane sustained serious damage and from the pictures and video online, it's amazing casualties weren't higher.

Figure 23.2 – Daz Top News Turkey Plane

The text reads: A vehicle crash. A long linear vehicle, AOL: Plane. Feels like a crash event with lots of deaths/all dead. AOL: Turkey. A feeling of a vehicle crash event, very messy, lots of wreckage, feels like over land/land base/recovery. Feels like it's over a border or line or mountains or similar? Remote land not an urban/city. Feels like a technical/engineering issue causes loss of altitude and a crash event.

To get a sense of the accuracy of Daz's predictions from February 2019 through December 2020, I reviewed 45 of them, which included crypto and financial predictions. To be clear, this was a sample of predictions made by Daz only, not the entire team. Since the full data from the session is not included, I had to make a subjective estimate based on images and text posted after the event: the few lines or occasional drawing from the transcript alongside a single image of image or text about the event predicted. In *my subjective estimate* of this very limited feedback data on Daz's site, of the 45 predictions (including crypto predictions), 12 were strong hits, 10 were moderate, and 21 were weak, mostly by not being specific enough about the event to be considered a hit. I rated two as "weak-moderate." Excluding the crypto predictions, the historical predictions were rated as eight strong, six moderate and 21 weak. None of the crypto predictions were rated as "weak."

Returning to CryptoViewing's event predictions as a whole, some have been useful in predicting the impact of events on financial markets, which is what some subscribers are most interested in. However, many of these predictions were general, not naming a particular country, person or institution, or exact day of the month. The world is a vast place and many of the types of events forecast (floods, rebellions, plane crashes) occur in multiple places every day.

Decades ago, when Ingo Swann first started a yearlong precognition project, one of the scoring criteria for whether a prediction should count as a hit was that the event had to be one that was not foreseeable.[484] The downside is that this criterion could potentially disqualify valid predictions through no fault of the person making them. For example, a viewer might legitimately get lots of details using their psi abilities about a President's upcoming harsh words toward a specific person or about pending earthquake in a certain country. However, since this President tended to use harsh words toward others and earthquakes are very common, from a scientific perspective these would have to be ruled out as strong predictions, unless there were more specific details such as the date and a specific area in the country.

Despite the challenges, from the selection of predictions informally assessed by Jon (and Debra a bit, too), specifically made by Daz, we found a number of additional predictions that would meet Ingo's strict criteria and others that would not.

For example, another prediction issued by CryptoViewing viewers focused on a crypto investment called "Libra." The prediction was made in February 2020 for March 2020. The viewer/prediction stated: "Libra. I feel that Facebook/Libra will restart of re-ignite in the media in March. A new release/new information of similar. A trial product… Media, event/demo. They take advantage of the global situation." The feedback image stated, "Facebook latest update: Facebook plans to incorporate the launch of Libra in a new wallet product." Now in assessing this, one issue is that we don't know whether the viewer was already familiar with Libra nor how likely it was for the viewer to anticipate that an announcement would be made by March, especially since the feedback indicates Facebook had "last year" announced its intention.

The next three examples all involve the COVID-19 pandemic. Readers will recall that in January 2020 there was mention of COVID emerging from China, yet the US government and the UK governments seriously downplayed this. US President Trump advised on Jan. 22, 2020, "We have it totally under control. It's one person coming in from China. It's going to be just fine." This was not the message, however, from the CryptoViewing prediction, which recognized the scope and depth of the pandemic. The data was "more deaths, a feeling of an outbreak in a major urban location, the word 'London'." A "Controlled Spread." Numerous newspaper reports confirmed there were more deaths in London, thought to be coming from Italy, but that they were still keeping the virus under control through contact tracing.

Another example of a strong COVID-related prediction was issued in September about an important COVID-related development that would occur the following month. Daz wrote, "A discovery of breakthru; A treatment that is discovered; a reaction at a very small level." The essential item here was mention of "enzyme: T Cell." In October, an article came out in *Science Daily* entitled: "T-cells from recovered COVID-19 patients show promise to protect vulnerable patients from infection." Here the transcript includes explicit mention of an important development with T-cells in "Covid news" and the news story from *Science Daily* refers to the same thing. It is easy see why this could be considered a hit.

This next example was judged to be weak by Jon, while Debra judged it to be stronger. This demonstrates the variability in judging, as we discuss elsewhere throughout the book. In March 2020, Daz wrote there would be a news item in April that involved "Life in, or on top of structures." He said people were separated by a wall or structure, yet were still communicating despite this separation. He said this was a "feel good" story.

Jon's assessment noted the viewer didn't mention tennis or Italy and therefore did not provide enough specifics. However, while it's hard to see in the feedback photo, another person is in the picture and the two are "still communicating" even though they are standing on two different rooftops. This is a specific detail and a unique situation. Further, there is mention of "a good feeling story" and the headlines state "good news, deaths, patient decline." We leave it to you, the reader, to make your own assessment.

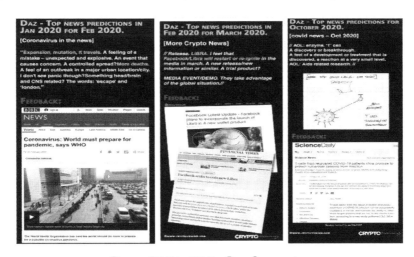

Figures 23.3 to 23.5 – Daz Coronavirus

Figure 23.6 – Daz Coronavirus

Jon and others have posed the question, "How is a prediction like this useful?" Debra's response is that people want to know if things are going to ease up so they can make plans for their future and also find some comfort, hence a large percentage of daily news items attempting to predict and report on whether COVID patients and deaths will increase or decrease. Jon replies: Yes, of course people want to know if conditions are going to improve, but a single "feel good" story like this does not provide the generality that would support the conclusion that things are going to ease up.

From the information we were able to glean, reviewers' and patrons' opinions of CryptoViewing's predictions, both worldwide and crypto, are generally very favorable. Many indicate they really appreciate the entire project and have saved or made money, while a few are critical of both types of predictions. Overall, the team is providing a voluminous prediction and entertainment service, which has resulted in a very solid user base, a breakthrough in what so far has been the miniscule development of the business side of remote viewing.

CryptoViewing states they have completed about 10% complete of their goals, which are to create a safe haven for remote viewing and the advancement of the art and science of remote viewing. "We plan to connect remote viewers on every continent throughout the world."

Precognitive Trading Group - Igor Grgić

Figure 23.7 – Precog Trading Group Logo

A recent entry into the commercial remote viewing field is Igor Grgić's Precognitive Trading Group, also a subscription-based service. Igor is the creator of ARV Studio, which we discussed in Chapter 13. He is the co-experimenter in several formal RV-related research projects, such as the ARV rejudging project, as well as informal projects like Project Firefly (see the Appendix). He is the founder of the Precognitive Trading Group (PTG). As with some other accounts in this book, we begin with what Igor says about his group and himself:

> Precognitive Trading Group is comprised of a think tank of 14 dedicated and passionate Remote Viewers with a goal to successfully predict future outcomes for financial markets. Their main focus is on Cryptocurrencies and the major Forex currency pairs. On the team are Igor Grgić, Anita de Lange, Tom Atwater, Jason Webb, Mike Myers, Karsten H and others.
>
> The team delivers precognitive predictions for daily price movements of various financial markets, including Cryptocurrencies (Bitcoin/BTC, Litecoin/ LTC, Ethereum/ETH) and Forex (EUR/USD).

Igor traces the origin of the method used by the Precognitive Trading Group:

> I and group of remote viewers made an effort to directly remote view future movements for Forex charts and currency pairs. I wanted to test this new approach since after many years of doing by ARV method, I wanted to explore other

possibilities. A team of 11 remote viewers was assembled, some I knew and worked with years before, some completely new people that were interested to join my public call on this experiment. My role was a tasker, analyst and feedback/outcome provider for the viewers. There were total of 6 RV taskings, once per week.

The task for the viewers was to locate the start of the most significant up or down movement of a financial instrument for the next week. A sample wording of the tasking would be: "Locate the start of the most significant DOWN move on Forex EUR/USD 1-hour chart for the next week." (Forex is the Foreign Exchange market. EUR is the Euro and USD is the US dollar.)

Viewers were given a template with a blank rectangular area and were asked to view/ sense/dowse the most significant move, being blind to the pair of Forex currencies involved. Igor as analyst was blind, as well. After viewers submitted their sessions, Igor analyzed the data and marked the predicted area(s) for the next week's Forex chart.

Week #	Prediction outcome	Subject
Week 1	spot-on correct	week's most significant 70 pips move.
Week 2	spot-on correct	week's most significant down move of 90 pips. Any trade entry at predicted & marked chart areas would produce significant profit in desired direction.
Week 3	correct	week's most significant up move 60 pips
Week 4	most significant	63 pip move predicted with two additional moves (45 + 50 pips) in correct direction. Interesting, all these three moves were marked precisely at areas for the most profitable trade entry.
Week 5	mixed result with 3 predicted areas, but 3rd area was an exact trade entry	mixed result with 3 predicted areas, but 3rd area was an exact trade entry for the most significant 70 pips down move.
Week 6	three predicted areas – one wrong, one ok, and one/third was an exact and correct	correct entry predicted for 115 pip move.

Figure 23.8 – Weekly Predictions Outcomes

These predictions were not traded but were analyzed to see what could happen if they were traded. The evaluation was that after six weeks, the profit would have been about

500 pips. The estimate was conservative. In addition to trade winners, failed trade entries exited with a small loss were also calculated.

A few years later, after major tweaks and modifications, the Precognitive Trading Group implemented the protocol in a formal manner.

Igor has stated he does not consider the PTG method to be Associative Remote Viewing. Originally the method was dowsing the most significant move on a blank rectangular grid area. It has evolved to using direct viewing of graphs in some cases, as discussed in Chapter 16.

Igor intends for PTG to be entirely transparent. Predictions will be maintained on a page available to the public. The predictions are time-stamped through an online open-source page.

List of all PTG predictions since launch in December, 2020			
Date	Type	Prediction	Result
01-28-21	Bitcoin	up	Correct
01-25-21	Ethereum	down	Correct
01-21-21	Bitcoin	up	Correct
01-18-21	Bitcoin	down	Incorrect
01-14-21	EUR/USD	down	correct
01-11-21	Litecoin	down	correct
01-07-21	Litecoin	up	correct
01-04-21	EUR/USD	up	Incorrect
12-31-20	Ethereum	up	correct
12-28-20	Ethereum	down	correct
12-24-20	Ethereum	up	correct
12-21-20	EUR/USD	down	correct
		Hit Rate: 83.3%	

Figure 23.9 – PTG Results After One Month of Predictions

That is, 10 correct, two incorrect, for an excellent hit rate of 83.3%! These are predictions of an Up or Down movement within a specific period, but they do not quantify the amount of the rise or fall, which is something others in remote viewing are trying to do. Even without quantifying, however, 83% is truly exceptional and could be the basis for cryptocurrency predictions, should one venture to use the information for more than "entertainment purposes."

At the time of this writing, the PTG offers two levels of subscriptions: PTG Silver at $25 per month, allowing subscribers to receive one prediction emailed per week (four per month) and PTG Gold at $50 per month, allowing for access to all available predictions per week (eight per month) sent via email. Given its newness, its location overseas and the lower-key approach to marketing and social media presence, PTG does not have the same level of social media commentary as other groups currently have.

In conclusion

What is evident from the work of the innovative groups discussed in this chapter is – even though they are distinct entities and even, one could say, competitors – it's clear that long-term cooperation, an open learning environment and open sharing of information between remote viewers and project managers within applied and formal projects has made it possible for these companies to launch and thrive.

We know many of the viewers who have been involved in these two companies, as well as those taking part in remote viewing networks and one or two other companies. It is clear that many viewers choose to work for multiple companies, which is wise given that hardly anyone is supporting themselves on income from any one group. However, the fact that some remote viewers are being paid at all on a regular basis is certainly a sign of progress. Many viewers and groups or project managers have participated for years or even decades in various RV-related projects in a variety of roles without financial compensation. We are heartened, too, that these companies are demonstrating that remote viewing and psi can be useful for a mainstream purpose such as currency trading. As we have noted in this book, remote viewing does not yet have legitimacy in society. Many of the subscribers and clients of these companies are professionals who would never "come out" to their family, friends and coworkers that they were involved in something as "dodgy" as psychic activity. The situation appears to be changing and groups like these two are leading the way.

CHAPTER 24

ARVing the NFL, MLB, NBA and European Soccer
Lounsbury, White and Atunrase

In this chapter, we present the approaches, methods and results of ARV efforts in sports by three very experienced Associative Remote Viewers. All three have had considerable success in sports betting using binary ARV.

We start with Lincoln Lounsbury, a professional writer who has successfully predicted several thousand NFL and college football games, basketball, baseball and hockey games, horse races, and currency, equity and commodity markets. He'll share how he tracked the state of the earth's horizontal and vertical components of the geomagnetic field to find out how it might impact his own ARV performance.

Then we move to Mark White, designer of ARV Creator. Mark has participated in a whopping 10,000 ARV trials. These were for many different types of events, including the lottery. Jon will add additional commentary and examples of sessions when he participated in Mark's ARV Club.

The third contributor is Tunde Atunrase, who focuses on large-scale projects like predicting the 2019 Soccer Champions League Final. He has used ARV successfully in several big tournaments. He is the author of *Remote Viewing UFOs and the Visitors: Where Do They Come From? What Are They? Who Are They? Why Are They Here?*[485]

Lincoln Lounsbury

(The following was written by Lincoln Lounsbury.)

Being a successful ARVer can be quite fulfilling, just plain fun and financially rewarding, too. Aside from using ARV for investing, ARV can be practiced casually, as a regular hobby or even with a commitment to train and develop others. If one of these latter possibilities is how you choose to participate, congratulations. The personal growth and development, the validation the statistics bring and the accomplishment of powerfully sharing your expertise could not be more gratifying.

If you choose to use ARV as an investment tool, the potential for financial reward needs no explanation. Happily, the step to financial success from mere statistical success is nothing more than following a mechanical process – a mere "set it and forget it" undertaking once you have navigated around a few pitfalls associated with the money-making route.

In 2011, after seven years of ARV practice, research and development, I passed the tipping point in a series of breakthrough discoveries which summarily and sufficiently explained why traditional ARV technology wasn't working for me. Importantly, the breakthroughs made my practice of ARV successful. Below I will be sharing some of these discoveries, committed and confident they will make a difference for you.

Immediately following this series of breakthroughs, I purchased the domain name AssociativeRemoteViewing.com. The price was ten dollars annually and the name was available for sale to anyone at that price. It is telling that anyone could have purchased the domain name at all considering ARV predates the World Wide Web, and thousands of posts on the subject had been viewed hundreds of thousands of times over the 16 years prior to the time I purchased the domain name.

In my own practice of ARV since 2011, I have excelled statistically, made money, and trained others to do the same. I have successfully predicted several thousand NFL and college football games, basketball, baseball and hockey games, horse races, and currency, equity and commodity markets. My intention here is to briefly share one piece of my ARV history with you, then get to some of the breakthroughs I have made, discoveries which should help propel *your* ARV practice forward.

Specifically, I will be sharing about my experience betting on NFL football, one of the best opportunities I have found for both honing skills and investing with ARV or honing skills and investing elsewhere. The final outcome of individual NFL games can have a significant impact on playoff opportunities, and the intense energy transmitted by the players prior to the start of a game is practically like working with an ARV Outbounder.

Below I will begin with a simple explanation of NFL betting 'against the spread,' followed by a review of my personal betting statistics having bet this way. Next, we will discuss a few pitfalls when doing ARV followed by a powerful session technique I use. After that we will discuss the current state of NFL sports betting and how you can cash in on this lucrative opportunity. Finally, on a different note, I will share about the powerful effect of the earth's magnetic field on ARV performance and how you might benefit greatly by following the geomagnetic field yourself.

NFL games

When betting on an NFL football game, there is an option, of course, to simply bet on the team you believe will win the game. Another more popular betting option is betting 'against the spread' where points (the spread) are either added to the 'Underdog's' final score or subtracted from the 'Favorite's' final score depending on whether one bets on the Favorite or the Underdog.

For example, let's say a Sportsbook (the part of a casino dedicated to sports betting) advertises a spread of 3.5 points for a game. If a Bettor places a wager on the Underdog, the Sportsbook will add 3.5 points to the Underdog's final score at the end of the game, and the Bettor will win if the underdog's final score plus the spread (3.5 points) is greater than the Favorite's final score.

If the final score of the game is, say, Underdog 10, Favorite 13, the Underdog Bettor wins the bet because 10 plus 3.5 is greater than 13. If the final score of the game was Underdog 7, Favorite 13, the Underdog Bettor loses because 7 plus 3.5 is less than 13.

If a spread Bettor places a wager on the 'Favorite,' the Sportsbook subtracts 3.5 points from the Favorite's final score, and the Bettor will win if the Favorite's final score minus the spread is still greater than the 'Underdog's' final score.

If the final score of the game is, say, Underdog 10, Favorite 17, the Favorite Bettor wins because 17 minus 3.5 is greater than 10. If the final score of the game was Underdog 7 Favorite 10, the Favorite Bettor loses because 10 minus 3.5 is less than 7.

Betting against the spread using Associative Remote Viewing

Session by Lincoln Lounsbury

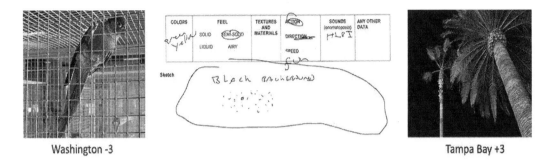

Washington -3   Tampa Bay +3

Figure 24.1 – Buccaneers vs Redskins Session

'Black background' and a sketch of the hub of the tree – successful session predicting the Tampa Bay Buccaneers (+3) would cover the spread against the Washington Redskins on 10/25/15. The final score was Tampa Bay Buccaneers 30, Washington Redskins 31.

When betting in Las Vegas, or through most venues where sports betting is now legal in the United States, one must accurately pick winners against the spread, on average, 52.38% of the time to break even financially, yet successfully predicting winners against the spread 'just' 54% of the time in the long term would be good enough for those starting with larger bankrolls to earn a comfortable living.

The potential earnings from such low success rates may seem incredible, but there may be no statistics more powerful than these figures to show precisely how difficult it must be to predict winners against the spread in the NFL. It is estimated that more than 200 million people bet money on the 2020 Super Bowl, yet in the entire world, today, there may not be even a dozen non-psychic individuals, cadres and syndicates left who successfully pick NFL winners against the spread 54% of the time or more in the long term. The bookmakers employ the best analysts, too.

An internet search for such success turns up only a handful of sports bettors who have publicly documented this kind of achievement, yet all the achievement above a 54% success rate (still better than virtually anyone) is ultimately only success in recent history or a short history at best, while long-term success at 500 picks or more, if they have even made that many picks, still runs below 54%. By 'publicly documented' I mean 100% of NFL predictions are publicly posted no later than immediately prior to kickoff and show a long-term success rate against the spread above 54%.

To be sure, there are countless websites built by dishonest touts who claim far greater, mostly absurd levels of success, all to cheat people out of their money. These touts, when not just plain lying about their results, regularly advertise success rates after as few as 100, 50, or even 10 picks as great accomplishment when, by accepted statistical measures, the results are based on such a low number of predictions success can't even be measured.

Success

The success of even a single practice of ARV has profound implications that eclipse its value as a tool to merely generate income, but when you're using ARV to bet on sports, invest in financial markets or show on a public website, for starters, that a technology has been developed that actually works to predict what is popularly called 'the future,' it's indisputable numbers that spell crystal clear success that matter, and nothing spells success, here, like a positive Return on Investment.

Between 2013 and 2018 using Associative Remote Viewing, I successfully predicted winners against the spread in 277 out of 502 NFL games, a 55.2% success rate. Betting $100, for example, on each of these 502 games (flat betting) at average Vegas odds of 10-11 (.91 to 1) would have produced a gross profit of $2,707, a 5.4% Return On Investment. $1,000 bets would have grossed $27K, and for a professional sports betting enterprise using my predictions, bets of $20,000 – the typical limit Las Vegas casinos happily accepted on individual NFL games at the time – would have generated $541,400 in gross profit per employee placing the bets.

Investment professionals will recognize these numbers greatly understate the potential for profit, even for those investing with smaller bankrolls because the numbers only represent one outcome that's possible from placing a mere 502 bets over a six-year period using flat betting. Just by employing financially sound money-management strategies like the Kelly Criterion, for example, investors making the 502 bets might easily have realized twice the gross profit of their flat-betting counterparts, or $1M *per employee* making the bets.

Importantly, sports betting has changed drastically in the past few years, although we will see in a moment that much of this change for the worse for professional sports bettors does not apply to ARVers. No matter how good any sports bettor may be today at traditional quantitative analysis and picking NFL winners, the big challenge in sports betting now is 'getting your money down' or simply being able to consistently place large wagers or even modest wagers. Contrary to their advertising, casinos do not like winners nor do they like bettors who appear to be professionals or are working for professionals. They are weeding them out by not taking their bets.

At a recent MIT Sloan Sports Analytics conference, moderator Jeff Ma questioned one of the panelists, a major bookmaker, whether it had participated in a practice of aggressively limiting betting and banning winning players. The bookmaker responded by saying, "We have a business. We're not a not-for-profit." The already small pool of professional sports bettors in the world has been shrinking for more than a decade, now, but this accelerating trend of not accepting larger bets from most anyone represents the final nails in the coffin for non-psychic, professional sports bettors.

The good news for ARVers

The above-noted trend does not affect us. In fact, it works to our advantage. Using ARV to bet on NFL games consistently calls for betting on games that professional, non-psychic handicappers do *not* bet on. While sportsbooks quickly limit betting for the best bettors, or 'sharps' as they are known, for betting on both the 'right' games *and* picking the 'right' team to win against the spread, they will still take money, happily, from the 'squares' or amateurs who either pick the 'wrong' games to bet on or the 'wrong' team to win in the 'right' game.

Sharps bet on games where their analysis shows the biggest discrepancies between their own calculated odds of a team winning against the spread and the odds the House has calculated for a game. Games with the larger spreads generally represent more opportunity for Sharps to find an edge. For ARVers, it is different. NFL teams generally want to just win the game and they don't care about winning against the spread. Ideally, then, ARVers should really bet on outright winners rather than winners against the spread, except for one reason. Betting on outright winners means incredible volatility in one's bankroll, and ARVers, like most people, should simply avoid it. The next best alternative for ARVers is only betting on games with the *smallest* spreads and, as fortune has it, these are the games the Pros generally don't bet on. This works out beautifully for ARVers and you will look like a Square.

As it has turned out over the past seven years as I bet the smallest spreads, I picked Favorites to win more often than Underdogs, possibly making me the most successful NFL spread bettor in the world who has done this, if not the *only* one – that's public or private, psychic or non-psychic. A popular adage in the world of NFL betting is this: no professional bettor has ever succeeded by picking mostly Favorites. Well, not only have I mostly picked Favorites to win, my Favorites have a higher success rate than my Dogs! By the way, in case you are wondering, I know next to nothing about traditional NFL analytics, but I have, of course, learned about ARV and betting, and I do enjoy a good football game.

Lastly, and to add an additional opportunity for ARVers, there will always be a place online for exceptional ARV touts to sell their picks, and this may ultimately be one of the best ways you can cash in as a successful ARVer. You will have to educate bettors on the definition of a good success rate but achieving a good success rate is solely sufficient to thrive as a tout. Meanwhile, you can work from the comfort of your home, and your customers should not have so much trouble getting their money down.

In all my NFL ARV sessions, I simultaneously ask all players on both teams how the game will go immediately before I select a random photoset to work with. I am not asking for an opinion, here. Rather, speaking out loud, I start by matter-of-factly acknowledging the spread, the context being the small spread is real, but irrelevant. Next, since the players know how the game is *really* going to go, I simply ask them, out loud, who is going win given the spread? The context of my question is that the question itself is a game and I am asking they cough up the truth. In a fun, yet taunting way, I ask something like, "Okay, who is going to win the game you guys – *really?*" I am not looking for an immediate answer, but rather the answer within the session. Still, sometimes a response will come immediately, and I say "okay" and set the response aside, still committed to gathering data by doing the session. I train myself to access data when and how I want the data.

Should you choose to proceed here as I do, you will need to sort out for yourself how you want to deal with things should the immediate response not match the session result. My no-exception rule (I have made up plenty of no-exception rules and you should, too) is I always go with the session data solely on its own merit or I call the session a Pass. By creating a no-exception rule under which I always proceed, I train myself to access and receive psychic data consistently, and how and when I want it. If you choose, for example, to go ahead, instead, with the immediate answer and even stop the session on the spot, that is perfectly valid as well. If this is how you want to proceed, make it a no-exception practice.

At every decision point when doing ARV, choose a protocol that you will use consistently throughout your practice and *never* give yourself the "benefit" of the doubt. For example, you determine in a session that Team A will win, but you mistakenly bet on Team B and Team B wins. The first time something like this happens you should always call the session a Miss. Remember, you are training yourself, here, and training yourself to stop making mistakes is good. Next, you should immediately create a rule for how you will permanently handle this issue should it happen again. You either decide for the future that what you bet is always the final prediction or what you predicted in your session will always count as the final prediction. Do not cheat yourself, here, and be sure to accurately capture all your misses in your statistics. Use your Misses to develop yourself.

Practice tip: Consider maintaining a high ratio of non-psychic to psychic time in your life. A high ratio, here, will have psychic data occur more sharply distinct from the rest of life experiences. Develop a skill for accessing psychic data and stopping access to psychic data as cleanly as turning a light switch on and off. Begin by using your session start and stop times. And when I say stop, I mean stop. Do not casually or semiconsciously drift off throughout the day to access psychic data. Wait until you write down your next session's start time to once, again, access psychic data. Be a professional, in charge and on purpose with your psychic ability.

If you are like most ARVers, you probably learned remote viewing before you learned ARV. As a matter of pedagogy, it could be that learning remote viewing first is ultimately the best way to be taught the two practices yet the current teaching of ARV from the protocols of remote viewing has been fraught with problems, problems that would not have existed if ARV was learned first.

While we don't have the space here to discuss the many concerns, there is one matter we must address before moving any further. The issue is the specific remote viewing training that says being successful as a remote viewer means you must strictly and always follow tried-and-true remote viewing protocols. Without giving the notion of protocol any further thought, this "always follow the remote viewing protocol" rule was taught to students learning ARV, and it instantly destroyed for most people the possibility of ever achieving long-term success from an ARV practice.

Remote viewing protocols often don't work for ARV. Just the sheer length of time it takes to follow traditional protocols and complete a proper remote viewing session, even only half-way through the stages, can take from half an hour to days. Well, you can forget about being a successful ARVer if you plan on spending this kind of time doing your ARV sessions. Here's one reason why: Long-term success as an ARVer is determined *only* through the use of accepted statistical measures and, first and foremost, this means having results to measure. And having results to measure means doing *lots* of ARV sessions, and doing *lots* of ARV sessions means doing lots of *short* ARV sessions, and this means *not* following traditional remote viewing protocols and doing long sessions.

Never mind that doing lots of sessions, a.k.a. practice, is a big stretch of the path to ARV success. And how willing are you, really, to call a session a Pass at the slightest presence of uncertainty when you just spent half an hour or more doing the session, even fifteen minutes? Sometimes, I don't want to pass a session when I only spent one minute doing it. For whatever reasons, and there are many, being foolishly attached to using a session to make a prediction is a success-rate killer, and nothing compels you to use a session like spending considerable energy following protocols and lots of time doing the session.

My current sessions last around two minutes but most of that time is spent doing administrative work. I make videos of my sessions for documentation and post them to my website and YouTube. The time I actually spend being psychic in a single session is typically only 5 to 15 seconds. Now, I am not suggesting that you should suddenly do what I do. While it took me quite a bit of time to naturally shorten my sessions without negatively affecting my success rate, you can at least speed up the process by working at it much more intentionally. While we are on the subject of making sessions shorter, take a look at the following ARV session and consider how you might create photosets given the lesson here.

Associative Remote Viewing Session

by Lincoln Lounsbury

Figure 24.2 – Syracuse Will Win

'flat sheet of metal' - successful session predicting Syracuse would beat Boston College in the football game between these two teams on 11/28/2015. Syracuse won 20-17 with a last-second field goal.

Unlike the extensive detail that is a desirable product of a good <u>remote</u> viewing session, an <u>Associative</u> Remote Viewing session only requires enough information to distinguish one image from another. Taking half an hour or longer, here, to produce a detailed description and sketch of the entire bus, and all that is going on inside and outside the bus, would have produced a session having no more value than this actual session which took about 10 seconds.

Again, in the light of this lesson, give some thought to how you might now create ARV photosets.

Geomagnetic field

Over the last 1,528 ARV sessions I completed, I recorded the state of the earth's horizontal and vertical components of the geomagnetic field at the time I did my sessions. I found the earth's magnetic field had a statistically significant impact on my ARV performance.

When I first began capturing geomagnetic field data, I had no reason to believe the earth's magnetic field would have any effect on my ARV performance, but an increasing interest in the topic on the Web had piqued my interest enough to start collecting the data. All data was collected and recorded only at the very end of every session, after predictions had been made. Until just a couple hundred sessions ago, I had not attempted to analyze or even casually look at this data.

Below, I will discuss only generally what I have found for several reasons. The impact of geomagnetic field on psychic performance needs more study and I do not want to undermine the potential for further unbiased research by precisely detailing my findings to the tenth of a nanotesla. Geomagnetic field is also a localized phenomenon and there is no reason to think that the impact of geomagnetic field conditions on one individual's success rate in one location should be fully realized by someone else in another location. Further, although you can loosely attempt to forecast geomagnetic field conditions in which you do your sessions, you can only learn of the actual conditions approximately three minutes after your session is complete, and if you haven't already reduced ARV session time to less than, say, 5 minutes, you may not be able to identify the actual minute in which you accessed psychic data. For these reasons and others, it is prudent to not share the detailed parameters of the geomagnetic fields under which I have produced results.

The purpose of my research has been to simply determine whether any specific combinations of the vertical and horizontal components of earth's magnetic field had affected my ARV success rate. For each of the 1,528 two- to three-minute sessions, I recorded horizontal component and vertical component data of the geomagnetic field by the minute as published from the sources below.

For Horizontal Component data I used the K-index data from NOAA's Fredericksburg, Va., magnetometer as it is located less than 50 miles from my home. NOAA's K-index website shows fluctuations in the magnetic field, tied to specific geographic locations. The index ranges from 0 to 9 and is directly related to the maximum amount of fluctuation (relative to a quiet day) in the earth's geomagnetic field over a three-hour interval.

Those interested in collecting their own data but are not close to one of the three NOAA magnetometers in the U.S. can use other magnetometers around the world, or the estimated planetary data on NOAA's website.

For vertical Component information I used the Bz data compiled by NOAA from Nasa's Advanced Composition Explorer (ACE) satellite which provides a three-minute delay of the earth's magnetic field data by the minute.

Data

My data show my success rate repeatedly improved then declined in a wave-like pattern over linear changes in geomagnetic field conditions. Using a minimum success rate of 60% as a threshold, I identified nine specific ranges of vertical geomagnetic field conditions within each of which I produced statistically significant results. Forty-two percent of my 1,528 sessions fit within these ranges. My lowest success rate in a range was 63.8% (224 sessions, $P = 0.00002$), my highest success rate was 85.3% (34 sessions, $P = 0.00004$), and my average success rate was 69.6% (642 sessions, $P < 0.000001$).

Oddly, I also identified a single point in the vertical component of the geomagnetic field where my success rate was 30% over 30 sessions, $P = 0.021$, regardless of the state of the horizontal component of the geomagnetic field. Even more curious, this single point, measured to the tenth of a nanotesla, is in the middle of the three ranges where I achieved my highest success rates.

If I have learned anything about the impact of geomagnetic field on ARV performance it is this: At any time, under any of the vast geomagnetic field conditions in which I have completed sessions, it is possible to do an extraordinary ARV session for a Hit. Of course, any number of extraordinary sessions resulting in Hits is abject failure if the binary success rate is 50%. With regard, then, to distinguishing the impact geomagnetic field may have on your ARV performance, it is important to collect sufficient data for analysis and not jump to premature conclusions about having identified ideal geomagnetic field conditions as we collect data.

To be clear, using the measures as I have laid them out here, it is unlikely you will have captured enough data to determine whether the geomagnetic field has had any impact on your personal ARV results until you have completed as many as 500 to 700 sessions or more. It is imperative you identify and record the precise minute and geomagnetic field in which you gathered the actual psychic data on which your prediction was based. Sessions two to three minutes in length will help greatly here and, if you are good, you might complete this task in a single year. Until you have amassed sufficient data, simply continue collecting data and doing sessions when the best times are for you as your schedule allows. Once you have produced statistically significant results, you will be able to make use of NOAA's geomagnetic forecast.

To scientifically and properly advance geomagnetic research as it applies to ARV, an appropriate party or parties will need to step up. Data can come from ARVers willing to offer up their own historical session data that shows session location with start and stop times and/or, ARVers who now begin recording their session location with start and stop times for future analysis. Smaller data sets that are not yet sufficient in size for individual use may still be useful when combined with data sets from others. Publicly documenting sessions, even using YouTube to effectively time-stamp predictions, would be helpful. Shorter sessions are a necessity if we want to identify effect sooner than later; longer sessions will take more time to identify effect. Please make recording geomagnetic data part of your ARV process.

Feel free to use videos of me doing NFL ARV sessions to grow and develop yourself as an ARVer. Borrow any of the techniques or protocols for your personal use. (For links, please visit http://www.arvbook.com)

ARV Club & ARV Creator – Mark White

Mark White is a longtime remote viewer and ARVer who developed *ARV Creator*, a scripted Excel file along with a pool of photos he created, which can be used in tandem to conduct solo or group ARV trials. Ten years ago, Mark led a group called *ARV Club*, which coauthor Jon was part of. Mark is one of the very few remote viewers who has done more than 10,000 ARV sessions.

In 2012, Mark White and Jon were members of an ARV group run by another person and they got in touch. That group was not doing well. Mark already had another team of three viewers and after discussing how to proceed, Jon joined the group. The viewers in ARV Club did sessions to predict the outcome of sports events and shared their picks from April through December 2012, using Mark's ARV Creator. The results will be discussed below.

Mark's methods

While each participant in ARV Club used their own preferred method for their sessions, Mark's method was similar to Greg K's: do many short ARV sessions for each event and evaluate them to see which photo the sessions point to as a group. As Mark put it, the method "involved nesting multiple trials until you get a clear consensus. This is the only way I've found to maintain 80% accuracy, and it still takes a lot of work."

Mark explained:

Each seemingly "dead on" hit in one direction is about 6% accurate. Even Ed (Dames) has stated that he can only achieve 60% accuracy using HARV for future events. So, it's still very possible to get a great hit in the wrong direction. However, if you stack enough trials, you can push your probability of success up to 90% or higher. I've done this many times. At home, in Las Vegas, trading Stocks, Commodities, etc. It's all about putting in enough work to make darn sure you're making up for those 35% to 40% "chance" hits in the wrong direction. Sometimes it's easy and the sessions line right up in the right direction, and other times it's a little more difficult and you have to be careful not to jump to conclusions...I don't think 1, 2 or 3 sessions will ever lead to long-term success. 10 to 20 is more like it. I do fast sessions, so even 20 sessions at 5 minutes each is still less than a 2-hour time investment for me. These sessions are easy for me to space out over the course of a week leading up to a single event, and if I get good alignment on my data I can sometimes achieve an extremely high confidence in a prediction, allowing me to risk thousands of dollars. I would never risk that kind of money with our current group predictions.[486]

Further:

I don't get overly hung up on the cue or the theory of displacement or any of that. I've heard it all. I used to work without a cue and it works the exact same way. What Ed and others fail to recognize is that intent is more powerful than a cue. Yes, a cue is good if you are working a large project and you want to organize into work groups, but for predicting the future, you have to set your intent to the "correct" target. There may be other ways that people have developed that work for them, but from past experience, I don't trust anyone's work as much as I trust my own.

Greg K understands the numbers game and has the persistence (he's a world record holder for human power distance). I think the things that might be causing him problems are: he doesn't use protocols and he believes too much in this solar wind speed and sidereal time effect. Personally, I pay no attention to external factors, I believe it's all internal. I also believe the protocol is very beneficial. He closes his eyes and does more of an ERV. Problem with that is takes more time to cool down, and you don't learn to read your sessions. For every session hit that I have which is obvious, there is data coming through like ideograms or rough sketches that I know are hits that other people would not be able to read. Also, the protocol makes doing the work easy, and you don't have to try to make conscious target contact. (I think that would be very tiring.) I know Greg has had some streaks of 80%+ accuracy, but he told me last year he did a lot of viewing and was only at 60% (May 22, 2012).

When I trained at the Farsight, we used to do ideogram drills where the instructor would call out the ideogram and you would quickly produce it. We trained on land, water, mountain, movement, energetics, structure, subject, I think. When I did that training, my ideograms were more likely to be on target. I still get some of the gestalts. Usually a person is a loop, water is a wavy line, structure is a right angle, mountain is angled up. Lately I kind of ignore my ideograms, they just get me to S2 and S3, but if they are consistently producing accurate data for you, by all means I would use them (June 3, 2012).

By August of 2012, the ARV Club results had fallen off a bit and Mark wrote:

It sounds like we all need a break. That is totally understood. I'm still well ahead, but don't want to give too much back. We even saw this with _____ group. They had pretty good success, but week after week the results started fading and getting inconsistent. After a break it appears things were somewhat back on track, but only time will tell.

In early October, Mark invited three other viewers he knew to join the club, and they did. The reason was to get more picks, which could help us "grow our accounts faster." However, results did not improve by taking a consensus of the enlarged group of seven viewers.

Jon's first results in the ARV Club

On Oct. 10, 2012, Jon wrote about the first set of games he did using ARV Creator (slightly edited):

I've completed 24 games in the ARV Club and I decided to see what can be learned from my viewing and scoring of them. 24 is a bit below what is usually considered a threshold for statistical meaningfulness; 30 instances is a figure often cited. Nine of the games were passes so only 15 were hits or misses. Passes do play a role though so they need to be included in drawing conclusions from the process. Of the 24 games, I had a correct prediction in 10, incorrect in 5, and I passed on 9; 67% accuracy of those in which a pick was made. I had a relatively low efficiency in that I had so many passes. Mark and Vicki had far fewer passes.

We used a 4-point scale Mark devised. Generally I used scores of 3, 3.5 or 4 as the indicators that the session matched the photosite of the associated team. We each use the scale a bit differently – I give very few 0s or 4s, for example. Not that many 3.5s either. When I made a pick, 9 of 15 times the winner had more scores of 3-4 than the loser did. In a couple of games I don't know how many scores of 3-4 the loser had since I looked only at the photosite for the winner. Those sessions (two games) scored high enough that I didn't look at the photosites for the other team (the team that ended up losing). I did this so as to reduce the possibility of displacement, which has been a big problem in regular binary ARV.

I don't see any fine-grained lessons so far in the data. I do see that when you have a preponderance of scores of 3-4 for one team and fewer for the other then that generally should be the pick. It doesn't always turn out that way, of course!

It doesn't seem to matter how many sessions I've done per game. The range in number of sessions I've done per game is from one session to 17 sessions. The one-session effort was a win while the 17-session effort was a miss. I find it tedious to do 10+ sessions per game. I'm now doing around 5 sessions per game. Others may feel more comfortable doing more. One downside of doing so many, as Mark has mentioned, is that occasionally the high scores will swing over to the other team. Bummer! We each have to figure our own optimum number of sessions per game.

Another point is that sometimes you will do a session or a couple and feel like that locks it up for one team. You have great confidence that the data points to the winning team. I've had a couple like that and gone with them, and they have been correct. There is no rule here; just intuition. Of course, naming or

497

drawing the target exactly gets high marks and if it's exact enough, that may be convincing all by itself.

I had one session like that where I displaced to the *next* target in the series that I had set up for that game. It was not displacement to the other photosite for that target but the next one. In other words, it was displacement in time.[487]

Figure 24.3 – Image of Dam

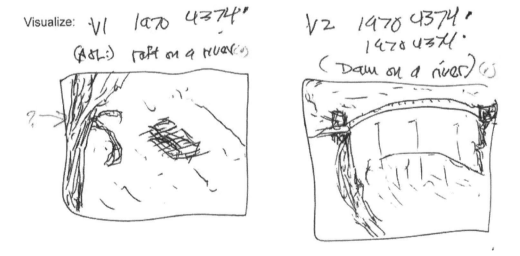

Figure 24.4 – V1& V2 Sketches of Dam on a River

I had been experimenting with visualizing and hadn't had great success at that point with "Viz." However, sketch V1 seemed relevant and V2 nailed it. The sketch is clearly of a dam and it is called a dam. The type, shape, curvature and proportions of the dam were caught, as well as the embankment on the left and the structures on either end of the dam. The lower part of the dam was missed, but this is as good a simple sketch as you get in an ARV transcript.

Overall ARV Club results

The overall hit rate for the four active viewers was a very respectable 61.9%, with a decline to 57% for all seven viewers. The three viewers who joined after the group had formed did not do much viewing (a total of 12 sessions) and their results are not tallied in the tables below. Nearly all of the picks by the four active viewers were bet. Jon had the highest hit rate but also the most passes (41%). Many ARVers have found that passing is an important component in having a high hit rate, while others refuse to pass, feeling they should be able to make a pick each time.

ARV CLUB TOTALS			
Predictions	107	All	Mar-Jon-Sue-Vic
Win	61	57.0%	61.9%
Loss	46	43.0%	38.1%
Pass	34	24.1%	23.0%

Figure 24.5 – ARV Club Totals

Viewer	Hit	Miss	Pass	Hit%	Miss%	Pass%
Mark	26	17	8	60.5%	39.5%	15.7%
Jon	12	5	12	70.6%	29.4%	41.4%
Sue	5	4	5	55.6%	44.4%	35.7%
Vic	17	11	7	60.7%	39.3%	20.0%

Figure 24.6 – Viewer Hit-Miss Pass

Jon's overall betting records in the ARV Club is as follows. Bets were made on MLB and NFL games between May 28 and Dec. 30, 2012. Both single games (28 of them) and parlays (multiple games – 58 parlays) were bet. Of the 86 bets, 28 cashed in. The amount risked was $1,524 and the net gain was $141.50. A very modest amount but it was a 9.2% gain.

Mark and Jon corresponded as 2012 wound down and discussed what they would be doing in 2013. We offer excerpts from a few emails to convey Mark and Jon's observations and conclusions at the time and to provide a feeling for the tenor of ARV discussions.

Dec. 12, 2012 – Jon to Mark

There is something about this 60+% range. The 8 separate (but related) 1ARV groups (in APP) are at about 62% with an N over 130. Marty just sent out the latest figures. We are at 63% in the ARV Club (not counting the new folks). Greg K too had 60% success with actual trades made during those 12 or so years, as per the draft paper he wrote. So how to get to 70-79% is the next goal, as I see it.

Dec. 13, 2012 – Mark to Jon

Interesting, Good job with that stats. My goal has been to break out of that 60% zone into higher success, but so far that remains elusive. I am getting better at knowing when I don't have it locked, just need to stop fighting those scenarios where it's obviously inconclusive. Take your time on break, I'll probably view through the NFL playoffs and SB, and then take a break. It gives me something to do in the cold Winter. I've had great success streaks out of the blue, but a break probably does help.

Dec. 26, 2012 – Jon to Mark

I'm forwarding an email from Marty (and Tom Atwater and Chris Georges) about the new project they are putting together - **the APP**. It's going to be or is an LLC, I understand. They now have 2 to 4 people who are knowledgeable about statistics, including one who is putting together a MySQL database to better analyze the statistics.

They are opening an Options group, which I thought you might be interested in. I'm going to find out more about it and perhaps join. I'm told it's not quite as complex as Forex.

Dec. 26, 2012 – Mark to Jon

I may be interested. I'm kind of mulling over what I want to do with RV this year. In all honesty I've had the most success working alone so I'm wondering if that is my best path. Other options are trading earnings and FDA approvals (since they are already known before announcements)…I think Marty is the reigning king right now, he's been at it the longest and hasn't given up. A real statistician might be good for spotting trends. I'd like to do more of that but just don't have the time.

In March 2013, Jon wrote to Mark about an upcoming APP Conference saying that he would be doing a presentation and would like to use examples by Mark and mention the ARV Creator. Mark said that was fine and Jon carried through on that at the conference (which took place in June 2013).

Jon later wrote to Mark asking whether he would approve circulating ARV Creator for use in the RV community and Mark said yes. They agreed that Jon would handle the distribution and ask for a $30 degree donation to be given to APP or another group.

Mark's ARV Creator

Mark wrote this about his creation of a pool of photos for the Creator:

> My targets are filed into these categories. Activity & Subjects, Animals, Machines, Nature – Landscapes, Objects, Structures, Vegetation and ALL (everything in one folder for some pure random selections) So, I randomly choose between dissimilar pools. Machines vs Landscapes. Activity vs Structures. Objects vs Animals, etc. I have thousands of pictures. I prefer targets with lots of activity, so there is more to zero in on. I hate still life, unless it's something impressive, but I do mix in some boring targets from time to time.

To help circulate the ARV Creator, Jon wrote an explanation titled "What the ARV Creator Does":

> For the two choices in a binary event (e.g. two teams or the Over/Under), the ARV Creator automatically selects and presents to you two random TRNs (Target Reference Numbers) and two randomly selected photos corresponding to the two outcomes. It does this for however many TRN's and photo pairs you would like to use for the event.
>
> You can select which categories of photos you want to be paired (e.g. Vegetation and Machines) or use the default settings. The ARV Creator automates the time-consuming process of finding suitable photos, pairing them, and arranging to have them presented "blind" to you. The ARV Creator is ideal for solo work but can also be used by a Group Manager with one or more viewers. You can customize the ARV Creator in several ways described below. The ARV Creator creates a project; it does not have features for scoring sessions.

Although ARV Club was no longer active, Mark and Jon kept in touch over the succeeding years.

May 29, 2013 – Mark to Jon

I could possibly go through all my folders over the past 10 years and estimate how many sessions I have done, but it has to be over 10k. I don't keep detailed statistics on everything I do like Greg K did, although I'm not sure that all of his recordkeeping has led him to anything that increased his accuracy. He admits that he had more success in the beginning.

I think overall I'm probably in the low 60% long-term. I have had extended runs of 80% (over 30 trials) but that always seems to be followed by a stretch that is closer to 50/50. The 60 % includes a lot of experimentation and some sloppiness. About 10% of the time I get a golden round that bumps that particular pick into the 85% to 95% accuracy range, but these are definitely not the norm. Most of the time when I lose I conclude that it was my own doing (poor judgment or rushing and not doing enough work) and not the evil matrix. I do get the occasional switches, but those typically happen when I am rushing to try to get to an answer, rather than just working steady and letting the answer come to me, and I don't allow myself to do enough good work to get around those switches. I'll see if I can go back through the past year and determine my actual win rate.

July 27, 2013 – Mark to Marty Rosenblatt

I have a few different theories on why we go through slumps. We may never know since this process will probably always be a mystery. Unfortunately I haven't kept perfect records like Greg K. So I can't pour through my data and determine that during full moon I hit 75% and new moon is 50%. I think the data can easily mislead as well, since this is not a science that can easily be quantified.

My main thought is the streaky nature of this process mostly has to do with rigor and statistics/probability. The effect size (of seeing the future) may not be as strong as we like to believe, so a really good hit on a target may only be 60% chance of being correct. So it takes several of those really strong hits stacked together to get the win rate at 80%. And there is always the 40% chance of hitting in the wrong direction, so you have to work extra hard to get around those chance hits in the wrong direction. Even if a process is good 70% of the time, statistically you are going to have some losing streaks. I think if you shoot 100 free throws at 70%, statistically you will miss five or six in a row at some point. That is the time when normal people declare the process doesn't work and quit. I think this is what Greg K realized and he would give it the Iron Man dedication and work to get a really large sample size. I've pretty much used that method to try to maintain 70% or better. It seems to work for a while. I think I have lightened up in the

R=rigor at times. I have a job and a family and I can't dedicate more than an hour a day to this process. I have pulled the trigger on trades or wagers when I knew things were not aligned as well as possible. The thing I like about the multi-trial approach, I do on occasion get a data set that lines up so well that an outcome is a 90% probability or better. That maybe happens 1 in every 10 projects.

I guess (there are) other theories on why we go through the slumps. Maybe all events are not yet decided? Maybe there are times when there is cosmic fog in the 4th dimension and we can't see through it? The other aspect about the multi-trial approach, in these scenarios I would think the results would be inconclusive, so at least you can pass rather than going in big on one or two false positive hits.

The following year, Mark remarked that he had gone back to his earlier method:

Jan. 5, 2014

I've been toying with my cue a little bit and I went back to my original ARV cue, which seems to be working fantastic at the moment. I think the Hybrid ARV cue was producing inconsistent results personally and my original cue produced results that were 70% accurate. I may test this theory over the next few weeks before I decide what I'm going to do.

So, in theory I think HARV and _____ 's methods were a diversion. I was doing pretty well on my own, but I thought there might be some magical method to get from 70% to 100%, but it really went from 70% to 55% using HARV and _____. I'm back to doing fewer sessions and trying to get one or two quality sessions in and over the past few weeks I've been at 80%. Curious to see if it can last this time around

Mark took part in some APP Groups and was one of the many viewers in the Firefly Investment Club in 2014-2015. Mark and Jon reconnected in 2019 and Mark wrote:

June 13, 2019 – Mark to Jon

I started a new ARV effort. I've changed my philosophy somewhat only giving minimal effort to each game and going with the first good data that comes up. Generally three or fewer quick sessions. Here are my results over the past six weeks. Betting relatively small to keep emotions out of the process but I think this is promising.

Wins	Losses	Percent	Gross Win	Gross Loss	Net Profit
39	22	63.9%	$2991.96	$1,486.96	$1,505.00

Figure 24.7 – Win Loss Percent Gross Win

The above provides an example of the thoughts and methods of ARVers, Mark being one of the most experienced around. Like others, he has tried a variety of methods, sometimes returning to earlier ones, going through the ups and downs that all practitioners who keep at it experience, and in Mark's case, ending with a success rate in that range that keeps recurring (60-65%), and with an excellent Return on Investment, too – 50% for his most recent efforts!

2020: More recently, Mark has traded in pencil and paper for an iPad Pro 11-inch model. He recommends using Notability software with an Apple Pencil 2 for best results:

The benefit of working with an iPad is you can perform sessions in a variety of positions (sitting at desk, lying in bed, riding on a train) and it reduces paper. Also, sessions can quickly be uploaded to a PDF in OneDrive or Google Drive and organized into project folders, which provides better access to sessions for scoring/judging, comparing sketches and targets side by side on screen, as well as providing more opportunity for feedback. I can view feedback on my desktop computer, my iPad or my phone, which gives me more time to make sure I review the correct target after the event. I don't have enough long-term data to be certain this new configuration is allowing me to maintain 70% accuracy, but so far the results are promising and I have switched a few would-be losses to wins by having better access to my data during judging. This plus the new availability of in-person legal sports betting on the East Coast has re-ignited the passion and yielded some great success. I went on a 6 out of 7 run through August and September 2020 and profited $2,500.

Examples[488] of a session by Mark White and winning tickets:

Figures 24.8 & 24.9 – Examples by Mark – Saw and Sketch

Figure 24.10 to 24.13 – Capital One Arena Tickets

Primary Pool Remote Viewing (PPRV) – Tunde Atunrase

Tunde Atunrase is a very experienced remote viewer based in the UK. Tunde trained in the TDS method with Pru Calabrese and then with the Bananaslam program, was a member of the Aurora Remote Viewing Group and has been successful in predicting sports events both solo and with a team using ARV. His team of three viewers predicted the winner of the World Cup of 2018 before the event began. He wrote about the feat in *Eight Martinis*.[489]

2018 World Cup

To pick the winning team out of 32 teams in the 2018 World Cup, Tunde decided to view the pools of the teams that made it to each stage. This meant predicting which pool the winning team would fall into, starting with 32 teams, then 16, eight, four and the two finalists. As Tunde notes, if the viewing had been incorrect for the first pool, the entire setup would have collapsed. Five viewers took part (Tunde, Elisa Lagana, Glyn Flyers, Daz Smith and coauthor Jon). The protocol requires viewers to do self-judging, and fortunately all five of us picked the pool that contained the eventual World Cup winner.

From that point on, as the events took place, Tunde and Elisa continued to choose the correct pool, ending up with both forecasting a win by France. A "bonus" was

> because I had reliable RV information that France would most likely win the WC, it gave me the confidence to wager and ultimately win every qualifying game France played in the tournament right up to the final itself. We had beaten the football experts and city wiz kids who produced all kinds of sophisticated algorithms and statistics to predict the winner and we did it using nothing but pure psi and a simple ARV process.[490]

Tunde calls his method PPRV – Primary Pool Remote Viewing. For full details of its successful use for the World Cup, please see his article in *Eight Martinis*. Here we present elements of Tunde's account and offer our comments as well.

In *Eight Martinis*, Tunde included the photos that corresponded to the winning team and showed extracts from Elisa's and his sessions. These extracts have correspondences with an exploding volcano (which was the target) with Tunde even naming the target. Tunde's ARV experience has led him to conclude that it is best to rely on "low-level" data: gestalts, basic sensory impressions like colors, movement and shapes whether in words or drawings and not to rely on higher-level data like names.

Moreover, as in most ARV sessions (especially if the session is four or more pages, as these were), the viewer may start off with material that seems "off" but then narrow their focus on the target. For example, one of Tunde's sessions for which the target was a volcano starts out with a person but quickly moves toward heat, bright colors, exploding, coughing, etc. Coughing is relevant because it may be that a person nearby was affected by the smoke and ash from the eruption. Or because no person was nearby but Tunde's subconscious used coughing (and running away) to convey information about the volcano. As he self-judged, Tunde easily concluded the volcano was the target.

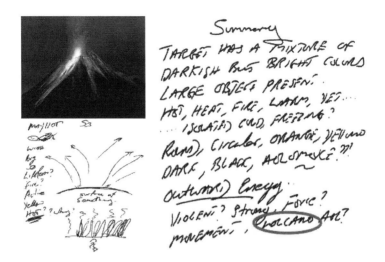

Figure 24.14 – Volcano and Summary

Summary: *Target was a mixture of darkish but bright colored / large objects present / hot, heat, fire, warm yet / isolated, cold freezing? / round, circular organic, yellow, dark, black, AOL smoke??? / Outward energy / violent? strong, force? / Movement, Volcano AOL?*

Elisa's session starts with a manmade structure and something flowing, which she feels is water. The low-level gestalts of falling and wavelike, emphasized in both drawings and words, are repeated throughout the transcript but are linked with water. Water serves as an analogy for what is falling and wavelike – correct low-level content. A manmade structure recurs as well and elements of it suggest the volcano. One of her drawings captures the volcano shape.

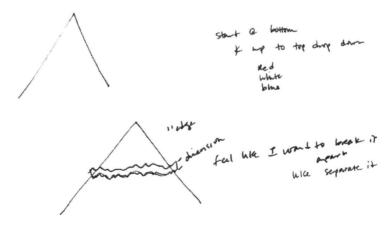

Figure 24.15 – Volcano Drawings by Elisa

Like Tunde, Elisa has considerable experience in ARV and with self-judging and easily chose the correct photo.

2019 Champions League Final 2019
(Tunde's account:)

This one again followed the same format as the highly successful 2018 World Cup. I decided the team would view the 2019 Champions League Final in March of that year. The final was due to be played in May, three months into the future.

I used my regular team – Glyn and Elisa and decided to see if others would like to participate in the last 16 round of games of the Champions League. Viewer X joined, boosting the number of viewers to four (including myself),

As with all our PPRV projects including the 2018 World Cup, I use Igor Grgić's ARV Studio software to blindly pick the photo pairs, which also enables me to remain blind to the photos and participate as an active remote viewer. This can be used for solo or in this case, group PPRV projects. With everything in place and target IDs sent out, we began viewing the four PPRV targets.

The 16 remaining teams were split into TWO groups and based on my analysis we ended up with Tottenham or Liverpool as the two finalists. The data based on the chart heavily favoured Liverpool.

Bets were subsequently placed on these two teams in March, who by the way were NOT the favourites to win. In fact as the competition progressed to the semifinals, both teams strangely and inexplicably LOST their first-round games (Liverpool lost away 3-0 to a much fancied Barcelona), yet somehow miraculously the two teams overturned their huge deficit scores in their respective return legs and together both reach a historic Final, against all odds.

As predicted three months earlier and despite the IMPOSSIBLE task, Liverpool reached the Final… Liverpool beat Tottenham 2-0 in the final to lift the trophy at the Wanda Metropolitano Stadium in Madrid on the 1st June, 2019.

The low odds of either team reaching the final vastly increased the return of investment plus numerous cash out opportunities on bets made on those teams not just winning the trophy but also by simply reaching the final itself. It remains ARV Sport's most significant prediction to date.

One of half a dozen winning payout tickets:

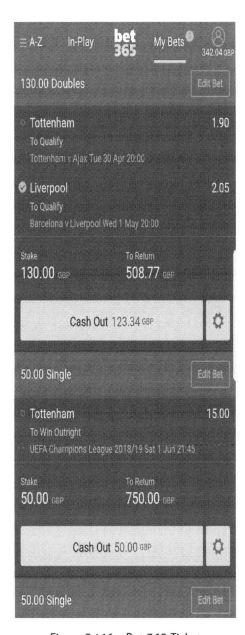

Figure 24.16 – Bet 365 Ticket

Lessons learned and ARV advice

With Tunde's permission, we repeat the lessons he offers after many years of using ARV for sports events:[491]

> Practice, Practice and keep Practicing. Learn how to describe the target in detail with just a few pages. Pay attention to low-level gestalts over high-data descriptors

in your session. Show coherent and detailed data in your final summary report to help your analyst or during self-judging.

Pick your ARV photo pairs with great caution and care – Photos should have an even balance of entropy within them. Not too high and not too low but also remain completely different from one another. For example, do NOT use photos that have a space shuttle rocket launch against a photo of flower in a green field. NEVER use photoshopped images under any circumstances. Try to use 'live' real world images.

Do not use photos of animals unless you know your viewers are good at describing different types of lifeforms.

POOL based ARV works!!! Try it. Modify it to suit your events, forecasts or targets. Prepare the pool carefully in advance if you are a tasker and ensure everything is as clear as possible.

Use ONLY experienced, confident, trusted and reliable viewers/psychics for your ARV projects and ensure they have a clear focus and are fully committed to the project just as you are (the tasker/project manager).

In my experience, it makes no difference if the viewer never sees one of the other photos – displacement still occurs. Likewise, keeping the ARV tasks separate (analyst, tasker and viewer etc.) does not guarantee a displacement-free ARV result.

Self-Judging Works! – The myth that viewers cannot be trusted to judge their own sessions is just not true. ALL viewers in this project judged their own work.

Take responsibility for failure and learn from it – it's easy to get discouraged after a miss but use that disappointment to increase your intent, focus and determination to do better.

Absolutely 100% focus on your feedback results regardless of whether you missed or hit the target. Experience the event, observe every detail and try imagining sending yourself the most important part, feature or description of each winning target photo for every session you did.

If you are an ARV project Tasker/Analyst/Manager, know your viewers' weaknesses and strengths.

Encourage positivity and keep it fun.

Encourage confidentiality. Unless you are in the business of selling predictions, try not to brag about your prediction until AFTER the event. Believe it or not, some people would love to see you fail or miss and the effects of this can be quite demoralizing for some viewers.

I see no reason why this cannot be replicated for future projects. We managed to replicate the success of the World Cup producing excellent predictions months in advance such as: The 2019 NFL Super Bowl finalists (LA RAMS reaching the Super Bowl – this project was started before the season even began!!!

In closing

Lincoln and Mark represent viewers who believe more is better (as does Marty Rosenblatt). It is important to note this is the opposite approach from others we mention in this book. The latter have found having short runs with larger wagers and taking more time for each session yields earnings and may be much more feasible in terms of time and effort – all provided someone can risk larger wagers. Still, much of this is about personal characteristics, skill level, interest in statistics and wagering, and willingness to explore changing protocols to adjust to fluctuations in success and apply lessons learned from one's own projects and those of others.

PART FOUR

HOW TO AND OTHER TOPICS

CHAPTER 25

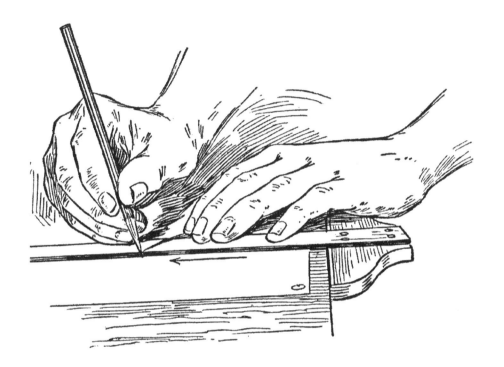

How to Remote View for ARV Projects

One can set up a remote viewing session or an Associative Remote Viewing session in many ways. Some are highly structured "method" approaches, such as those used in Controlled Remote Viewing (CRV) and in offshoots such as TRV, TDRV, SRV and Resonant RV. These proceed in specific stages and you do best if you "stay in structure." This is the dominant approach nowadays to doing remote viewing both for practice and for operational work.

The other basic approach is "simple," "natural," "generic" remote viewing, which is what was done at the Stanford Research Institute. This is free form and informal, and viewers generally produce less data using this approach. However, Joe McMoneagle and Gary Langford, both highly skilled viewers, have produced many pages of data using the particular "non-method" procedures they developed over the decades.[492]

To do an ARV session, one can use any approach. Most ARV sessions are shorter, often much shorter, than sessions done with "method" approaches like CRV. In fact, many who use CRV for Associative Remote Viewing stop after the first three stages, which can take about 10 to 20 minutes. Others like Greg K, Mark White and Lincoln Lounsbury do a great many very quick sessions for each event to be predicted.

Debra has trained with many different mentors using a variety of psi modalities (including several who teach CRV and its many derivatives), more freestyle forms of remote viewing and another psi modality that focuses on clairvoyance (a visual orientation of psi originally conceived by Louis Bostwick). In this chapter, she will use the breadth and depth of her experience, including lessons learned as a longtime teacher herself, to offer a detailed explanation of how to do a structured session, with certain adaptations for ARV. She will also present suggestions for disconnecting from a target when the session is over.

Debra calls her approach Multi-dimensional Remote Viewing for ARV (MDRV-ARV) because it is based on maintaining awareness of whether a viewer is operating in 2-D or 3-D space and modifying one's focus to adjust to whatever the project calls for. Finding this level of detail outside a course is rare, and we feel readers will benefit from Debra's fine-grained account.

Jon's remote viewing training was different (primarily from TransDimensional Systems). After Debra's presentation, he will offer his perspectives, including how he does ARV. Jon will also touch on the most common practices of ARVers, including those of successful viewers who are featured in our book. We begin with Debra's presentation.

MultiDimensional Remote Viewing for ARV (MDRV-ARV)

These guidelines are for ARV in which a) a photograph will be shown to the viewer as feedback after an event and b) judging is done by the viewer or someone else (a "judge").

I (Debra) will first give my views on how to set up your paper and establish some rules to follow. These instructions may be useful for other remote viewing or psi-based projects that require detailed information (such as missing person cases), although these will require staying in session longer. More complete summaries than are typical for ARV sessions are also required in this approach. Modifications may be needed for such projects depending on the situation: for example, whether the work is undertaken to confirm information already suspected to be true or to locate persons at unknown locations, which requires complex mapping and descriptions of nearby landmarks.

I have found that remote viewers who work only on ARV-related projects often don't provide enough information when tasked with these "regular" standard RV projects, including for classroom assignments. New viewers don't realize how much more they could get out of their sessions if they used simple techniques that build on information already gleaned. However, if they have patience and aren't addicted to immediate feedback, they can usually get up to speed fairly quickly. Later I will discuss two of these techniques – primary and secondary probing. The latter is what really develops a session.[493]

For those who wish to dive more deeply into the art and practice of remote viewing, Jon and I recommend taking a class in Controlled Remote Viewing or one of its derivatives such as TDRV[494] from an experienced remote viewing instructor. That being said, so many free or nearly free resources are available that any highly motivated person can find what they need to progress. Becoming involved with a group of remote viewers can help bring you up to speed, in addition to self-study.

What I am presenting in this chapter is a consolidated and modified version of what I've found to be the most important components of "method" approaches over the past couple of decades. However, Ingo Swann, who was the prime developer of CRV, would certainly object to any abbreviated version of his method. He'd probably refer to the term "abbreviated" as a misnomer of sorts since his method requires the viewer to move from stage 1 to stage 6 in a very disciplined way. The stages approach aims to achieve deeper site contact and a greater flow of information as the viewer progresses through the stages (although the later three were designed so the viewer could move back and forth with fluidity). I know from experience that his original approach has merit. I hope he will forgive me posthumously here.

The reason I'm willing to break all rules is that my very abbreviated version seems to work well, especially for those who are willing to adhere to structure but don't have the patience for the rigor required by CRV in its pure form. There is a learning curve, although once one practices for a while, the procedure becomes second nature. Further, even those who are originalists when it comes to CRV report that they don't stay completely in CRV structure when doing ARV.[495] This is another reason I feel comfortable sharing my approach.

For those who wish to read more about CRV methodology in its entirely, I recommend books such as Paul Smith (2015) *The Essential Guide to Remote Viewing,* Jon Noble *Natural Remote Viewing;* Daz Smith *CRV - Controlled Remote Viewing: Manuals, collected papers & information to help you learn Controlled Remote Viewing;* David Morehouse *(2011) Remote Viewing: The Complete User's Manual for Coordinate Remote Viewing;* and Lori Williams (2019) *Boundless: Your How To Guide to Practical Remote Viewing – Phase One.*[496,497,498,499,500] Also, there are plenty of comprehensive home-study video courses, including one by Lyn Buchanan. Additionally, the original military remote viewing manual is still available.[501]

Simple Clairvoyant-Somatic Approach

You can certainly do a remote viewing session without following the structure below and sometimes you may be quite successful, but other times not so much. The "simple" approach has been dubbed "natural RV." It was used by the researchers at the Stanford Research Institute and is still recommended by pioneers like Russell Targ, Joe McMoneagle and Ed May. In this approach, you generally do most or all of the following: Relax. Clear your mind. Write your current thoughts on a piece of paper and toss it away. Write your name and the target number on the paper used for the session and enter any other "log-in" information. Then close your eyes and see what comes into your mind or visualize a screen. Touch the target number or say "Target!" Wait for an impression to come into your mind and sketch anything you see inwardly. Write down words or phrases that pop in. And that's it! We both recommend trying this basic approach since you will likely learn a lot from what works and what doesn't. Some people prefer simple, natural RV to any of the "method" approaches.

What I describe below may help you overcome common difficulties and challenges at the outset. I explain the overall set up in this approach, then establish some ground rules. I follow that with a brief review of Analytic Overlay and then go to step-by-step instructions.

Structure the page

Your writing surface serves multiple purposes. I will refer to paper although more and more viewers are turning to tablets, as well as white boards. Your paper is a *recording* device since it becomes your RV transcript. Whatever is on it counts and whatever isn't does not count. I encourage you to put down absolutely everything that comes to you because if you don't, you will usually regret it.

Your paper is *also a communication device*. Others may end up looking at it to decide which photo you were describing. They will rate your responses, not to judge you as a viewer (although they may form opinions about that), but to judge how well your response matches the photo options.

Like an artist's canvas, your paper is *an interactive tool* that allows you to interface with the target. The physical aspects of your paper such as size, shape or color can affect your sessions. Likewise, what you do with the paper – where you place your impressions, the size of your sketches, the positioning and even interactions, such as where you touch the paper – can make a difference. This means your paper is also part of your method. It can be used to access the intuition-based information.

Paper set up

Through organizing your words and sketches in different parts of the paper, you will learn how to decrease (not eliminate) interpretational errors. You'll also create a transcript your project manager or a rater will have an easier time understanding. This will increase the likelihood they will invite you back for future projects. You can think of your paper as split into three sections. This will initially feel awkward, difficult or even impossible. Just push through the discomfort. You'll get the hang of it after a few attempts.

- **On the right.** Your record keeping is on the right (name, date, time). Also this is where you declare and release your AOLs (Analytic Overlay – nouns or higher-level descriptors such as proper nouns, which could potentially derail your session).

- *To the far left.* This is where you will record your sensory descriptors, adjectives and low-level components related to textures, shapes, color, luminosity (brightness, shadows), dimensions (shapes, sizes, weights), sounds, tastes, temperatures, action words (falling, running, leaping) and concepts (historical, old-fashioned, religious), etc.

- *In the middle.* Here you write the target number, tasking/frontloading (e.g., "The target is a location. Describe the location."). This is also where you place your sketches and commands/prompts to yourself (such as Move to the right, Move up high, Move to something I've missed, etc.)

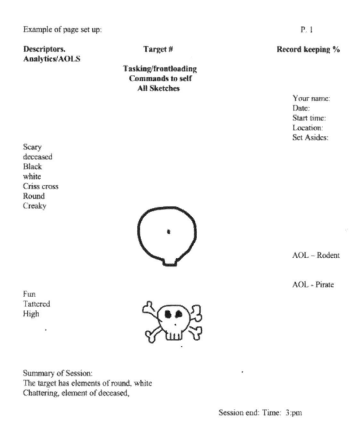

Example of page set up: P. 1

Descriptors.   Target #   Record keeping %
Analytics/AOLS

                                        Tasking/frontloading
                                          Commands to self
                                             All Sketches

                                                                                            Your name:
                                                                                            Date:
                                                                                            Start time:
                                                                                            Location:
                                                                                            Set Asides:

Scary
deceased
Black
white
Criss cross
Round
Creaky

                                                                                            AOL – Rodent

                                                                                            AOL - Pirate

Fun
Tattered
High

Summary of Session:
The target has elements of round, white
Chattering, element of deceased,

                                        Session end: Time: 3:pm

Figure 25.1 – Examples of Page Setup

What is Analytic Overlay (AOL)?

In normal perception, a thing does not stand in isolation in the environment or in meaning. We look at it and that act may stimulate a memory or an association with other things or ideas. In perception with our everyday physical senses, this is not a problem because we know the difference between our thoughts and what we perceive in front of us – that is what enables us to function and survive. For example, if I see a dog, I may be reminded of my dog I had in the past. I know the difference between my thoughts and the dog I see, although my memories may color my reaction to and sentiments about the dog in front of me. However, in remote viewing we can't as easily distinguish between the target and the thoughts that arise in association with the target.

An example of Analytic Overlay: A round orange pillow is in the photograph. As you remote view, you might visualize the image of a pumpkin and be reminded of trick-or-treaters on Halloween. The round aspect and orange color are correct, and imagining is useful in helping you get there, but if you start to describe little kids in costumes, the session could be derailed. Far too often it is difficult to know which aspects are correct and which are incorrect when these types of impressions come up, so a viewer should want to minimize their occurrence rather than welcome them.

If they do come up, they should be reported and recorded – for example, AOL: pumpkin/Halloween, on the right side of the paper – and then write "set aside," which means setting the intention to discard the impression from your mind. Take a momentary break if you want. Setting aside such impressions is one of the most important parts of remote viewing and if you practice any form of meditation, you will find it easier to do this. (Many experienced remote viewers believe the single most important practice you can engage in to become a better remote viewer is to meditate regularly.)

Also, as you get more used to remote viewing, you will have an easier time extracting what is useful. Sometimes you will get several AOLs with a common theme – perhaps the names of tropical islands come to mind, for example. Most likely the target will not be any of those places, but it may involve a distant tropical location. AOLs can arise in two ways – spontaneously, when you didn't go through any thought processes and "the US Capitol building" comes to mind – or through deductive reasoning – you get an image of a white building. It's large, has columns and what appears to be a flag outside, so you decide it's the US Capitol. The challenge is that an AOL will sometimes be correct, but most often it won't be. In this example, the words "building" or "flag" could be AOLs, or not. It is up to you to practice enough to learn whether these types of impressions tend to be accurate or misleading.

Difference between Sensory Descriptors and AOLs

While many viewers are taught that sensory descriptors are adjectives and AOLs are nouns, I like Courtney Brown's usage in his advanced Scientific Remote Viewing training course, which distinguishes between lower-level and higher-level descriptors. The low-level descriptors are sensories like brown, moist, soft, and salty; they tend to be adjectives. The higher-level descriptors are proper nouns (names of people or places) or ideas/

concepts. Many "middle-level words" are trickier – some viewers might want to put them in the left (descriptor) columns and others favor the right (AOLs). Building, flag, street, windows and doors are nouns and could be considered "mid-range" terms. The actual target might be similar to one of these, e.g., it looks like a door but it's a wall. While we don't have room for a full discussion of AOLs, beginning students, as a general rule, should log them as AOLs. As you get more practice, you'll understand how you can "break the rules."

Simple Rules

Let go of your need to know
Your job is not to know or understand the target. It is simply to connect, experience, observe and report. Nothing else. This is really important, and it's somewhat of a paradox. As the *National Inquirer* used to say of its fictitious gossip magazine articles, "Inquiring minds want to know!" Of course, we all are curious about what the target is. Part of what makes remote viewing so much fun is getting to see the feedback, particularly when a target is a photograph. That's when we get to see what the darn thing was and how well we did. More importantly, we can try to understand the relation between what went on in our head and body and the photograph we now see before us.

During a session, a focus on "What IS this?" can easily lead to an AOL "storm." No viewers are totally immune to AOLs because of the nature of human perception and cognition. No remote viewers can totally let go of the desire to understand what it is they are dealing with, but some have an easier time than others. Overall, I suggest avoiding the question of "what" since that leads to wanting to attach a name to something. Naming things leads to AOLs. Even sounds can lead to AOLs if you ask yourself, "What do I hear?" Instead, if you want to know if sounds are present, prompt yourself with just the word: "Sounds?"

Keep every word and every sketch on its own line
Put one word or sketch on the paper, one under the other. This means you don't write all over the place but instead in an organized fashion. This won't be easy at first, but the idea is to force your mind to stay busy and arrange things; eventually this will seem like second nature.

Exception #1 – "intuitive statements." Occasionally a group of words that wouldn't make sense if separated pops into your mind. Keep them together as a phrase or sentence.

Exception #2 – "Cognitron."[502] Occasionally a sudden stream of information about the target will come in very fast as if it's poured into your awareness. When this happens, you'll feel inclined to put it all into a single paragraph, as if you can't even write fast enough. In that case, it's important to get it down on the paper. It will be up to you where this goes. If you think there could be Analytic Overlay involved, put it to the right side of your paper; if you feel it contains more raw and solid data, put it on the left.

Be cognizant of where you place your pen and hand

Wherever you touch your pen on the paper, you may get streams of impressions. Since lower-level descriptors tend to be more accurate than AOLs, you want to keep your hand and pen in that column rather than letting it rest in the right column where you drop your AOLs. You can also spend as much time as you like in the middle column for your sketches.

Paper tips

No need to limit yourself to one piece of paper. (Sorry, trees!) While you don't want hundreds of descriptors or many pages for an ARV session, you'll find if you restrict yourself to one page, your session may be limited. Also make sure your sketches are not too small. New students often do this because they are trying to save space and because they don't feel confident. The only time a sketch should be tiny is if there is something small in the photo and you want to emphasize this impression. Otherwise sketches should be medium size, while allowing them to be large if something really big is in the photo or at the location. If the sketches are too small, they will often be overlooked by judges and they are also hard to probe.

Sketch!

You are sketching not to display artistic talent, but because it is very useful. If you are self-conscious or embarrassed about your sketching abilities, just get over it. Stick figures are fine. Take an art class. Slow down. Go to therapy. Do whatever it takes, but sketch. Your subconscious has a way of figuring out how to best communicate with you to produce sketches that are great matches to the target, even when all else is incorrect. Sketches can also be user-friendly for judges since they are trying match noticeable elements in your transcript to a photo. ("A picture is worth 1,000 words.") Sketches flow from your subconscious and your body, although they can certainly be distorted by your logical mind (try to leave your AOLs out of them).

Let your subconscious and body do their job by adding sketches. Any shape can be sketched. Motion can be sketched, too. Sketches can consist of ideograms (symbolic representations of the feel and motion of a target or an aspect of one), very basic drawings or developed drawings. It's possible you will do better with word descriptors than with graphic ones, but we urge you to explore sketches since for many viewers they provide very good or even their best data.

The Steps

Pre-session meditation/ warm up/cool down
Technology prep
Arrange beforehand that when your session is done you can scan your transcript. Many free apps for smart phones allow you to batch scan and create a single pdf of your entire session with a few clicks. You can then easily email the session. Also create folders on your desktop for your sessions. Create a folder for each trial or project as it will be a handy way to look back at your work.

Your room, self and paper prep
Find a quiet room and set aside or turn off your cell phone. Tell your family or others not to disturb you. Check to see if your body needs anything before you start (Water? Bathroom? Nap?) Some people RV better when sleepy, while others don't do well if they are tired. You won't know until you practice in different states.

Prepare a clear and clean writing surface. Even paper on a clipboard while sitting in bed will do. Get out several sheets of blank paper (nothing on the back). Have a black pen ready (and optionally colored pens). The imprint or mark of whatever writing device you use should be easily visible on your transcript and on the scanned pdf. Use a pen instead of a pencil – you won't want to ever erase anything. Close your eyes.

Meditate Before Beginning
Meditate until you have the feeling you really want to get started. Many viewers don't bother to meditate in advance while some do so for as long as an hour. Meditation has cumulative effects so experienced meditators don't find it necessary to meditate in advance of each session. Joe McMoneagle takes about three minutes to reach his desired state while Daz Smith meditates for 30 minutes or longer. Tom McNear doesn't meditate at all. He simply takes a few seconds to relax and then begins.[503]

Those who are brand new to both meditation and psi-based practices might need more time as they begin a session. In my book *You Are Psychic: The Art of Clairvoyant*

Reading and Healing,[504] I have outlined meditation exercises designed to assist with clairvoyance, centering, setting energetic boundaries, making separations from targets, using a mental reading screen, etc. While the book outlines methods for doing readings of people (which will not interest all readers of this book), many will find the "psychic tools" chapters very useful.

Many remote viewers like the audio programs offered by the Monroe Institute that feature binaural beats and white noise. Other practitioners suggest pranayama yogic breathing exercises, such as alternate nostril breathing (Nadi Shodhana). Another breathing exercise can be thought of as full body breathing – you imagine every pore of your body is soaking in oxygen and light and then releasing stress or any energy not conducive to the upcoming session. Some remote viewers simply like to listen to music prior to a session.

Connect mentally with the target

Pick up your pen. This is a sign to your subconscious you are ready to get started. You wouldn't just shoot at arrow or rifle without first scoping out the target. You wouldn't try to hit a ball on a pool table without focusing on the pocket or roll. In the same way you don't want to start without aligning yourself with the target. So I suggest that you first think through what you are about to do.

What is your target? Is it the feedback photo? If it's the feedback photo, where will the photo be when you see it? Where will you be? Visualize yourself and that spot at that time. If you won't get a feedback photo and it's simply the photo associated with a winning outcome, where do you imagine that photo existing in space after the outcome is known? Once you imagine where it will be, send some attention toward it. The attention could be in the form of giving yourself a grounding cord and giving the photo a matching one. You can also create an anchor for it, or you can put a bubble around it. Think of it like you are saying hello to it. For example, "Hey there target, I'm here, you are somewhere else in time and space but we are going to now connect. I'm going to interact with you and you are going to interact with me, and I just need to experience you enough to get a depiction of you on my paper." You might even thank it and yourself for the great relationship you are forming. You can think of the target like an excited puppy running around across the street. You need to look at it, call to it, wave your arms and capture its attention. In doing so, it will capture yours.

Now it's time to really get started.

At the top right corner, you do your record keeping

This will let you and others know who you are and the circumstances under which you did your session. Different project managers may ask you to set this up in different ways. Basically, though, you can't go wrong by providing your name, date, time, location, whether or not you are monitored for this session and your "set asides."

Set asides

These are anything that might be bothering you – sometimes called "personal inclemencies." Also, if you are already getting images or words you think could be related to the target, but aren't sure, write them down here. They are called "set asides" because you want to let people know you are dealing with them in case they have an impact on the session, but you also really don't want them to bother you. The act of writing them down should serve as a means to flush them from your space. Lyn Buchanan suggests speaking these out loud while you write them down. Don't feel like you have to get too specific with them on the paper and don't feel like you have to come up with something.

Declare frontloading, tasking and write the target number in the center

Write down any frontloading you have been given about the target. Then write the target number, which is usually referred to as the TRN – Target Reference Number.

Create an ideogram based on your target number

While this step is not essential, it is standard practice and can be quite useful. While I won't go into detail about it here, think of an ideogram as a somatic motion that is reflective of a major aspect (aka "gestalt") of the target. This is made on a subconscious level, meaning your logical mind has nothing to do with it other than to prompt yourself to make it.

It is recommended you write your target number in the middle of your paper. As soon as you write the final number, without lifting your hand from the paper close your eyes (or at least look away) and let your hand make a quick scribble. Your pen shouldn't come off the paper and it should be done very quickly, a second or two at most. There are several ways to do ideograms, some of which can ultimately lead to a viewer being able to identify gestalts and continue them on a conscious level. However, for the sake of this lesson, which is primarily on ARV, I will keep the discussion to the most basic way to work with the ideogram, which is the way Ingo Swann intended (according to his best student, Tom McNear, who spent about three and a half years with Ingo).

Work with the ideogram

Now you have two things to probe with your pen or the finger of your off hand: the target number and the ideogram. Feel these, probe them. Imagine you can touch the target through them. Or imagine that they open up like a portal and you can slide right through and come out on the other side to the target.

Start your viewing!

Then wait and when you get your first impression, put it down on paper in the correct spot – again, to the left if it's a descriptor and to the right if it's an AOL. You really don't want AOLs so remember to keep your hand positioned close to the middle or left to make it more likely you'll generate lower-level descriptors or sketches. You can then keep probing the ideogram (or create another one), or you can put your pen on the left side and tell yourself to start listing words that come to mind. What you are now doing, too, is waiting until you feel the urge or inkling to draw, at which time you do so. I'll discuss below what to do after that.

Alternative or in addition to ideogram – Visual Screen

First, decide what and when you want to visualize, such as the screen at the moment you see your feedback. You can simply intend to visualize something and let it happen or you can intend for it to appear on a "screen" in your mind. Joe McMoneagle suggests a screen in the middle of your head. In this case, you can imagine you are turning your screen on and the moment it lights up, you'll get an image. In Chapter 21, Jon describes an imaginary TV set he turns on, and he has had lottery hits with this method of visualizing. Don't try to get a clear, precise high-level image. Just ask or intend to see something you can sketch. You can also choose to see a black screen and let your images light up on it.

Somatic-psychometric alternative: Use another piece of paper as a 2D representative of the target

Take out another piece of paper and let this be the 2D representative of the photograph. Think of it as a magical paper screen. Wherever you touch the paper, an impression (word or image, thought or feeling) will emerge from that exact spot. You can either take a free-form approach and just touch wherever you feel inclined to and then make notes, or you can be more organized by sectioning it off and numbering the sections. For example, the far-left upper corner could be labeled 1, the top middle 2, the far right 3, far left row below that 4, etc. It's totally up to you how you wish to proceed and you can revisit different sections. It will be very important as you do this to record your impressions on your transcript. In addition, you might take notes on what you are probing. Make sure you continue to sketch with this approach, too.

3D real targets – Locations and movement commands

If you know your target exists in time and space, such as a photo of a real location or object, you will have a lot more to work with. This is because you can now use your intuitive sense ("imagination") to move to the location or object or to bring it to you, and then explore it with all your bodily senses and the use of movement commands (explained below). The important thing for ARV projects or any project where judging is involved is to limit yourself to the photo itself – that which is within the frame. This is basically the main difference between doing remote viewing for ARV projects vs. other sorts of projects when there is a need to use the information obtained for further exploration. In the latter case, there may not be photographic feedback or any feedback.

Movement commands can be extremely useful and productive. You can do them from the very start of a session. You can write your target number, make an ideogram and immediately think of it as if you are moving through the ideogram straight to the location (or to the object). Once there, decide where you will go first. Starting at "base level" can be useful as it creates a grounded connection and orientation. If you start there, write in the center of your paper: "Move at base level." You can look down at your feet and notice what impressions you get. You can jump up and down and feel what that's like. As your impressions come, just make sure they are placed in the appropriate place on your paper, working in downward fashion.

If you find something, you can move to it immediately. For example, if you saw something triangular, you could write the word "triangular" on the left side of the paper, then sketch the triangle and then move to that triangle and see what happens next. This is what we refer to as "secondary probing." Or you could wait to do that later and instead do a movement command to go somewhere else such as to the right 10 feet. Some viewers have a routine about where they go while others use whatever feels right at the time. Moving above the target to different heights and vantage points is also helpful, as is looking down from a bird's eye view. If the target is a photograph, you have to be careful about going too far up since this, more than anything else, could expose you to too much information outside the frame.

A few other things to keep in mind with movement commands.

- Always indicate on your paper (in the middle) where you are going.
- Touch the part of the paper you want to move to.
- Think of it like you are really moving there.

- Use your body as much as possible. Don't just depend on visuals. Listen, Smell, Inhale. You can go through different senses whenever you move.
- Don't worry if you listen and get a color, or you smell and hear a sound. Celebrate any impression.
- Take moments to orient yourself to the new movement command – you could imagine you are coalescing your energy (or spirit or astral body). It doesn't matter if this is really happening (it could be, but we can't prove it). But the more you imagine you are doing this and gather yourself at the place, at surface level, or above, etc., the more likely you are to get impressions.
- Move around frequently. Don't just hang out if nothing is coming to you. After 30 seconds, move again.
- Revisit the same area at least twice, if not more.
- Let new information come in. Don't discount something just because you didn't notice it before or if it doesn't match up. To even have the thought that "this doesn't make sense with what I saw already here" is a sign you are now operating from logic and very much in a danger zone – your job is not to make sense of anything.
- Don't make assumptions about the target! For example, the reason I say "base level" (where your feet would be) is because the ground may or may not be there! It might be water, a hole, an incline or even air if the target is in the sky.
- You can use a separate piece of paper to represent areas to move to such as top, bottom, sides, etc.
- Again, only do this if you know the target will be a real location or object!

Secondary probing

In the above example, if you become aware of a structure, a vehicle or a person, move to it and probe. You can do this with any impression. It's always great to probe any shape or drawing you have already made. You can work with these in a 2D manner by putting your pen on the sketch or in a 3D manner (bring your consciousness, your spirit, your energy body to what you just sketched). In all of these approaches, use your sense of touch. For a 2D sketch, feel the paper under your pen and think of it as if there is a magical element there and whatever you touch, you'll get a flow of information. For 3D, think of it as if your hand is reaching out to touch the object, the person or whatever element you got. For example, let's say that you were 100 feet above the target, looked down and got a sense of red. You can just say, "I will now move to the red."

You are not trying to name it or understand it; you are simply moving with the single intention to make contact, connect, explore, observe and report. Once there, you can reach out, you can move into it or even hug it. Then just breathe, relax and wait. You'll get an impression soon. If you don't, you can move somewhere else or explore something you got earlier. Or you could move back to primary probing if you need to get refocused. Rewrite the target number and draw an ideogram.

Secondary probing of objects and structures. Mapping around an object or structure

This is essentially what the latter stages of CRV teach. This is how you can really start to get more and more details and redraw sketches. Don't add to earlier ones, but rather redraw them as new information develops. For example, let's say that right away you had a sense of being on the water and something that seemed manmade was floating on the water. Your mind is going to immediately generate the idea of a boat. This has to be treated as an AOL. So, you write down "AOL Boat." Take a break by writing "AOL break" right under it, go do something for a few minutes and return to your session. When you come back, you can now think of it as if you have surgically removed the idea of boat. (This doesn't mean it's not a boat; it just means it is way, way too early to accept that assumption). Then start exploring. You'll want to explore the right side of this. Move your body there. Grasp or hug it. This will give you an impression of the size in relation to yourself. If there is exhaust or water coming out or a tail (maybe it's a sea creature, not a manmade object), you'll sense something is there. Then move to the object's opposite side and make physical contact with it. Try to stand on it. What happens? Can you move above it slightly? What impressions come? Then explore the top, the middle and the bottom. Don't forget to move inside. Also don't forget to continue to use your other senses. What happens when you breathe in? Again, you can map it in 2D or 3D. Make sure you continue to develop your sketches.

Secondary probing of people

People can be probed by touching them. (Appropriately!) Talk to them, interact with them. Feel free to interview them. Just keep in mind, the purpose is not to learn who they are or their personal issues. They are serving a purpose at the target. They are there to show you something about the location or what's happening in the photo. So interview them about that. Observe what they are wearing. Sketch body positions, clothing, accessories or relation to others. Make a note of how many people there are. Often you'll find groups of people in a target and this can be indicative of what's happening at the location.

I have developed a target pool that helps viewers practice secondary probing of people and animals. The targets are labeled "people targets" and "animal targets," so one is already frontloaded and can proceed as if they had appeared in a session and now need to be worked with further. For animals, you will learn a lot by probing their backside, front side and listening to them, even smelling them! (For links, please visit http://www. arvbook.com)

Playing with different mediums (Forgive the pun!)

Clay modeling – You can use clay or Play-Doh to model.

Using large canvas – Again, what you use as your physical tools has an impact on what you experience and express. The Hawaii Remote Viewers Group, The Farsight Institute and now the CryptoViewing Patreon group have used white boards for some sessions. Ed Riordan, a CryptoViewing viewer, uses a white board. Quite frequently you can see his entire body pantomiming and expressing elements of the target. Also, John Vivanco, who teaches the TDS methodology, has great demonstrations in his classes of what viewers come up with in bodily expression when they are freed from the limitations of working with a small canvas.

Final prompts

A remote viewer can use a variety of prompts at the end of a session to make sure they haven't missed anything important. These are basically instructions to one's subconscious to get additional information. The first two could be used at the start of the session or as ways to task oneself.

- "Move next to the photographer (whoever took the photo) – touch his/her shoulder, look through the camera lens or look at what the photographer is looking at."
- "Move to whatever is most important about the target." This can be used at the start or at the end of the session.
- "Move to whatever I have missed about the target."
- "Describe the purpose of this target." It is recommended you ask about the purpose. This may generate more conceptual-based information, which could otherwise be confused with physical attributes.

Summary

You may or may not want to do a summary for an ARV session, but you would for a regular remote viewing session. If you know your handwriting is hard to read, do a typed summary. Summaries are very much appreciated by project managers and group managers. This is where things can really come together. When you summarize, you are including information that you got throughout your session and sometimes you may still be in an intuitive mode.

The different kinds of summaries include narrative and outline. Each has its advantages. A narrative may consist of just a few sentences or a paragraph that describes the overall sense of the target. I suggest leaving out your AOLs. An outline summary is a brief list of main and minor points or it can be detailed. For the latter, use an Excel spreadsheet and go through your entire session, breaking it down into themes and subthemes. These are very useful. I suggest moving from the first page to the last and considering every word, but again leaving out your AOLs. At the very bottom of your summary, you could include the AOLs separately.

An example of a remote viewing session and Excel spreadsheet summary can be found at http://www.arvbook.com

Overall compilation sketch

A compilation sketch is a pictorial summary where *you force yourself* to bring all sketches of objects, structures, elements, people, etc., into a single drawing with the hope of demonstrating the relationships of the elements. At times you'll have a real sense where the sketches should be positioned, especially if you did movement commands and secondary probing. But more often, you will feel you are guessing, taking a stab in the dark. The task may feel daunting and even intimidating. This happens whenever your analytic mind is trying to make sense of the intuitive impressions. (Intuitive functioning is never stressful). No matter how unconfident you feel about creating a compilation sketch, just do it anyway. The beauty of compilation sketches is that quite often you find they are pretty accurate.

Here is one Debra created that felt much like this. This was for a practice target assigned as part of the International Remote Viewing Association's monthly "Focal Point" target practice. You can see it's not perfect but it gives an overall idea of the location.

Figure 25.2 – Debra Katz RV Telescope Sketch and Photo

Ending your session

When you are done with your session, make sure you write "Session End" or "EOS" (End of Session) in the far-right bottom corner of your page. Write down the time, so the project manager can assess how long you were in session. As you write EOS and the time, tell yourself you must now disconnect and EOS is a sign you are disconnecting. Be disciplined and really end your session at this point.

Meditation: Call your energy back.

You are an energetic being. When you connect with a target, your energy or "life force" may become strained or scattered. While we can't prove this scientifically, we can attest to how many people are amazed at how different they feel when they learn it's possible for their energy to extend to and connect with sources outside themselves and then consciously call it back thorough visualization.

To do this, simply ask, "What does my own energy look like? What color is it? What is its consistency? Its luminosity? Is it more like light, liquid, a cloud, molecules? Where is it right now? Then pull it back to you. Make sure you visualize it coming back completely. Let it, and only it, come back into you and fill up wherever you are lacking.

If you want, you can run it through a "filtration system" first to separate out other energies. The filtration system can be a light (as in a brilliant sun – see my *You Are Psychic:*

The Art of Clairvoyant Reading and Healing[505]) or it can be an advanced water filter-looking device or something you invent. Your thoughts and emotions at the time may be signs and indicators of where your energy is. Just as you can separate from the target, you can do this with anything you've been obsessing about, stressing about, worrying about or trying to change, fix or heal. You can repeat this exercise if you find yourself thinking about the target.

Meditation: Break the grounding cord you may have established with the target

If you initially connected with the target with a grounding cord and created a matching one for the target, it is time to break them. See them exploding and evaporating or imagine you are dropping them into the earth. You can then give yourself a brand new one that connects your body (at the base of the spine) with the earth.

Meditation: Release pent up emotions.

Many times emotions will emerge after a session, having been stimulated during it. Sometimes they are emotions we encounter at a location. Other times, they are emotions we have by association with a location or target. Grounding cords stabilize your energy and give you a means of release. Create a new grounding cord that represents you at the present time. Take some deep breaths now that you have your energy in your body and allow the emotions to release down your cord. What color are the emotions or sensations? Force yourself to see these as colors and watch as they seep out of your body and melt into the ground. If there are any emotions that need to get reintegrated into other parts of your body, let that happen now.

And that's how you do a remote viewing session!

Remember, sessions for ARV don't have to be long. What you've learned above may be a lot more than you need for many sessions.

A few more concepts

Breaks

Occasionally you may wish to take a break from your session, knowing you will resume work on it later or even on another day. In this case, write "Break" and add the time you break and, when you come back, the time you resume. Be sure to separate from the target during your break. This separation should happen in your mind and energetically. Don't let yourself tune back in until you are back in front of your paper. You can also do the exercise for calling back your energy at the start of your break or repeat it if you are having trouble taking a break, then reconnect with the target when you resume.

Some people find they can't properly separate or stay disciplined during a break. In this case, they should use whatever time they have for a session and be done with it. This tends to happen more with remote viewing than with other kinds of psi practices (such as doing readings for people) because often we know we will have definite feedback and are waiting for it, but we are anxious about how we did and have questions like "Did I do enough?" For those who notice they are having trouble with self-discipline, this is going to be key.

Retasking

Retasking happens when your project manager sees something suggestive or promising in your session. Retasking is quite common in application-based projects. The tasker or project manager may not know what the photo or target is but feels you should explore it further. Either way, the rule of thumb is the tasker isn't supposed to frontload you, but just say, "I see you mentioned a structure on page 3. Please go back into your session and probe further." Or "You refer to a group of females on page 6. Go back and get as many details as you can about this group. They may or may not be females. We don't know, so just let that idea go and see if you get more about them."

You may be asked to move somewhere you didn't move to. It's not a bad idea if you are working with others to invite them to do this if you feel they have not provided enough information. Eventually though, part of being an advanced remote viewer is being able to retask yourself. This means being able to look at your transcript and ask yourself, "What have I not probed enough?" "Did I sketch a shape or write 'movement' without going back in to get more?"

Aesthetic Impact (AI)

Aesthetic Impact is another term Ingo Swann is credited with. This happens as viewers experience themselves in relation to a target. They may make a statement such as "I have a sense this structure is much larger than me." "The sunlight is so bright it hurts my eyes." "There is such a sense of sadness here I want to cry." From these examples, you can see AIs are relational, can be dimensional (pertaining to size and shape) or can pertain to luminosity or emotionality.

I recall once when I gave myself the command to move next to the photographer and look through the camera. Suddenly I felt as if I was stumbling off a cliff. The photo was taken from the top of a mountain. Another time, I had a spinning feeling, like I was suddenly suspended in the air getting a 360-degree view of the Earth, and that was exactly what the photo was.

I had a strong AI during Extended Remote Viewing training with John Herlosky. I had the bizarre sense of lying on the ground in a suit of armor with a sword sticking out of my chest. The target turned out to be an ancient castle. I still don't know how I ended up in that soldier's body. I didn't feel pain or distress, just the sense of a weapon embedded in my chest. These kinds of experiences are what keep some remote viewers coming back for more.

An AI will usually only be felt when the target is a real location. This will happen to viewers whether or not they know the phrase (AI) or what it means. Some believe this is a precursor to *bilocation* – the sense of being separated from one's body and being at the target. It usually lasts for just a split second and often it's so subtle, unless one is looking for it, they won't really catch on to what happened. Often however, a quality makes the experience impactful. Many have noted the information tends to be quite accurate, although bilocation can be accompanied by emotions and bodily sensations that distract a viewer from exploring other or more important aspects of a target. That said, for most viewers, intense bilocation ("I'm there!") is a very rare occurrence and for some it never happens.

Jon's perspective

In my book *Remote Viewing from the Ground Up*,[506] I discussed in detail the TransDimensional Systems method I was taught by Pru Calabrese and John Vivanco in a long and intensive program. I later ended up teaching it myself as Training Coordinator in the TDS Bananaslam program. I included extracts and transcripts from sessions I did and examples from client work so I will not go into detail about the TDS method here and instead refer you to my book. (I recommend it also in case you would like to learn more about the most successful remote viewing company in those early years, which was TransDimensional Systems, 1998-2003.)

The TDS method involves the entire body. A term was even coined for it by Don Walker, a member of the TDS public demonstration team[507] – "knosomatics" – to emphasize that the approach is based on "body knowledge." Don was intimately familiar with the human body from his decades as a chiropractor. Many of the particulars Debra incorporates into her MDRV are similar to those in TDS. TDS employs other techniques, as well, such as probing a "gingerbread man" outline of an object or person or doing a "consciousness map" of sentient creatures.

In doing ARV sessions, I'd like to repeat the point Debra made earlier that you don't need to do a long session to be successful. The great majority of ARVers do short, quick sessions. One or two trials doesn't work for some people, though, and successful viewers like Greg K, Mark White and Lincoln Lounsbury use a consensus from many sessions to decide which way their psi-ber sense is pointing. Few ARV practitioners go through all the steps Debra details, but these steps are very useful to know for practice or operational work.

Photograph or that which the photograph is depicting?

One point on which Debra and I differ is whether an ARVer must confine themselves to what's in the photograph and not include what is in the vicinity. My take on this is that the tasker and the viewer should agree beforehand on this issue. If the tasker wants the viewer to focus only on the photograph, that can be part of the methodology, so long as the viewer agrees. On the other hand, the tasker and viewer may agree the target is a "photosite" rather than just what's in the photo. Marty Rosenblatt has long used the term "photosite," indicating the viewer may pick up information outside the photo and can include it in the transcript. Joe McMoneagle has commented that there's no law which says the viewer is limited by what's in the photo. However, Joe works closely with Ed May, whose approach in his CAS software is that only data in the transcript that matches elements visible in the photo is credited.

A sample session by Jon

Below is an example of a session I did based on the TDS method but shortened for ARV. It is neither the best nor worst ARV transcript of the many hundreds I have done by now.

My ARV sessions average around seven to 10 minutes. I generally do three "scans." The idea in the TDS method is that you will get different aspects of the target in each scan. (You can do more scans but we generally did only three in regular remote viewing and I do three for ARV.) In the "Collector," you probe the three scans and harvest the data by probing with your offhand or pen. At this point, I usually end my ARV transcript. However, I sometimes go on to make a "General Sketch" that combines the scans and attempts to put the elements in the orientation found in the target photo. Sometimes (but rarely) I go on to do an elementary Matrix, which is a set of categories found in many "method" approaches. You go through the categories one by one, moving down the page and recording impressions.

Very often only one of my scans will have strong identifying information. The others will have less or even no relevant data. (This isn't rare in ARV transcripts.) This comes to a point Debra and I want to make – a point which may not shine through our emphasis on very good and even amazing extracts of transcripts that match the target.

The truth is in both regular RV and ARV, the judge or analyst often has it tougher than the viewer. This is especially true in operational remote viewing when the project manager or analyst may receive many transcripts containing a cornucopia of impressions from viewers who have different styles and strengths. This is a challenge even if all the viewers use the same methodology. The analyst must go through all the data, paying close attention to each element to see what patterns emerge and possible likenesses in the visual elements that correspond to whatever is known about the target or can be surmised as a likely match.

Gail Husick confirmed this point quite strongly when she discussed the work of her successful RV company.[508] She noted how important it was to her as an analyst that the viewers all use the same methodology (which is CRV for her company). The same factors are at play in ARV as well.

In my session below, as an example, there is one significant matching element. It's in Scan 1, the drawing of the structure. Although the drawing certainly looks like a building, I questioned whether the target was a building. I wasn't sure, and indeed on first glance the target resembles a basket more than it does a building. Note that the windows of the sketch are a strong match for those in the "basket building." My sketch of the building would be the main element extracted in a composite of winning sessions (as we will see in the next chapter on "Pictolanguage." It is true a few other databits are relevant to the target, such as "Seldom seen." Leafy, twig and the Y-shaped drawings suggest the scrawny trees in the foreground. But there is much in Scans 2 and 3 that does not seem relevant.

Further, in the second quick session (not shown) done for the other potential target in the pair in this binary ARV, the first data I got was "sense of a building!" So how would the analyst know which to choose? If the analyst was familiar with my sessions, they would likely choose the building depicted in the sketch because my drawings, though artistically weak, often contain my best data.

Figures 25.3 and 25.4 – Photo of Basket Building and Jon's RV Sketch

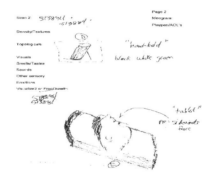

Figures 25.5 – Sketch Page 2

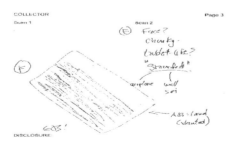

Figures 25.6 – Sketch Page 2

In sum, the TDS method has worked for me in hundreds of ARV trials. I was one of the four viewers with the highest hit rates that the Firefly project (over 50 viewers) turned to when it ran into poor results. Extracts from my transcripts appeared in many compilations of successful predictions during the years I was active in APP (2010-2016) and my ARV results have ranked high in other ARV groups, as well.

To conclude, let's illustrate the point Debra made about the differences between simple sketches and more developed sketches with two concrete examples.

The first is my session above in which the drawings are simple, with few details. Compare them with the artistic rendering of a building made by one of the very best viewers ever, Joe McMoneagle. Joe has had training in art and drawing and even in a five-minute session such as this one from the 2014 APP conference, he produced a highly detailed drawing with specific relevant descriptors. To begin with, there is no doubt he drew a building, and the target was a building (whereas in my session, I was not sure about my building).

He writes: "Gestalt: castle. Manmade, lots of stone, fortress like, old, impressive, ice & snow, mountains in background, animal storage, medieval, 1500's-1600's, Germanic."

Most of the words (especially, castle, medieval, animal storage and fortress like) and the fact that the drawing appears to show a drawbridge outside the "Castle" indicate the transcript represents a building from medieval times, whereas the building in the photo is from a much later period. However, the descriptors in the left column match the photo, with the possible exception of "old." Not only that, but the windows and the columns in the drawing resemble those in the photo and both of these elements are very prominent in the photo.

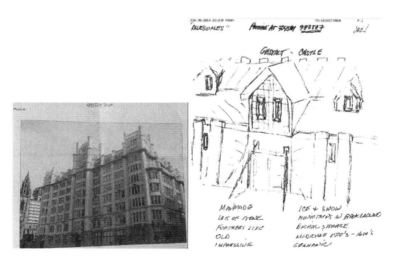

Figures 25.7 (left) – Large Building and 25.8 (right) – Building Sketch by Joe McMoneagle

CHAPTER 26

The Pictolanguage of ARV Sketches

Remote viewers receive impressions in the form of words, images, sensations and feelings. They may make sketches to convey the images, or they may "just draw" while tuning into the target and see what emerges on the paper. In this chapter, we focus on these drawings and what Ingo Swann called the "pictolanguage" of psi.[509]

In November 2013, Jon wrote an article for *Eight Martinis* #10 on *The "Pictolanguage" of Psi Sketches*[510]. and the same year presented the topic at the APP conference.[511] A second presentation was titled *ARV Sketches from Six Viewers in Relation to Photosite Attributes*. Here we will draw on ideas and images from those presentations and add a great deal more as well. We present examples from about a dozen remote viewers illustrative of themes about analyzing and improving sketching.

Jon says: We feature many of Debra's examples to give a profile of a talented viewer working different targets over the years and because she deftly illustrates from her own sketches how to probe and turn simple drawings into more informative ones. This is something new viewers may be not aware of and it is difficult to learn about outside of a remote viewing class.

Ingo Swann summed up his view on picture drawings in his book *Natural ESP*.[512]

The relative ease by which picture drawings can be produced by non-artists, together with the striking similarity of all picture drawings, suggest that the drawings are not the product of an individual's artistic processes, but are a kind of basic psychic language in themselves. A language that has gone totally unnoticed by all parapsychologists. This psychic pictolanguage has one element in common among all picture drawers. It translates the incoming psychic information into basic forms and shapes which are then recognized by the individual's psychic system and consciousness. The picture drawing mechanism seldom goes beyond this specific task, and it is unusual to find picture drawings fleshed out into highly artistic renderings. When the drawing is fleshed out, we are likely to discover that it has been done so by consciousness trying to fill in the holes and that what has been filled in is erroneous.[513]

While we'd agree with Swann, he did write the above just as he was getting his own training program off the ground, delivered to fewer than a dozen trainees in the years just before the publication of his book. Swann developed his Controlled Remote Viewing method to help viewers develop their initial impressions and sketches with ever-increasing levels of detail. As it turns out, it is possible to develop sketches by spending time on them with continued "probing" of the target, executing movements in space, mapping all sides of an object and combining earlier sketches into a detailed compilation.

Let's look at historical precedents and then at themes that Swann explored. We present and comment on many examples from present-day viewing, ranging from extremely simple to complex detailed sketches. Simple sketches – enough to get the main "gestalt" of a potential target – are sufficient for ARV but more developed sketches are useful as well and are highly valuable during operational remote viewing. We offer our take these drawings based on the thousands of sketches we have seen over the years in a variety of remote viewing projects.

In all of this, we bear in mind that sketches represent the way the conscious mind and the unconscious work,[514] and they bear the stamp of what viewers have learned since childhood, including instruction in art or drafting and from remote viewing instruction.

Picture drawings in parapsychological experiments began around the year 1882. The targets were generally very simple drawings. The following appeared originally in Rene Warcollier's book, *Experiments in Telepathy* (1948):[515]

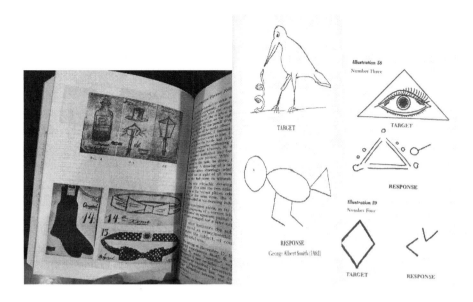

Figures 26.1 (left) – Photo of Book and 26.2 & 26.3 – Simple Drawings

Next are a few examples from the famous American novelist Upton Sinclair's *Mental Radio* (1930):[516] The artist is Mary Craig Sinclair.

Many of these early examples can also be found in Ingo Swann's *Natural ESP.*[517]

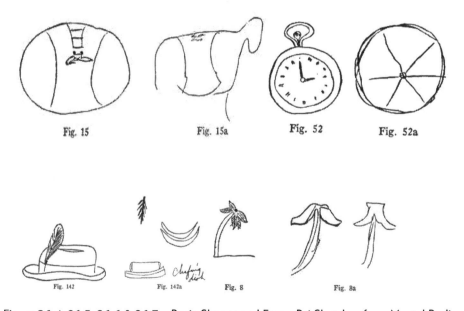

Figure 26.4, 26.5, 26.6 & 26.7 – Basic Shapes and Forms Psi Sketches from Mental Radio

The sketches often represent "basic shapes and forms" and they tend to omit detail. Psi drawings have other characteristics in common. For example, the viewer often takes the drawing to be something other than what it is. A football (Fig. 15, the target) becomes a calf (Fig. 15a, the sketch by the viewer). Mary Craig noted she was not a sports fan and so she drew something she was familiar with. The viewer's conscious mind will often mislabel what the unconscious has produced (whether a drawing, word, or both).

After analyzing the examples in Warcollier's and Sinclair's books, Ingo summed up the characteristics of these drawings:[518]

Accuracies	Lack of Fusion	Error Contributions	Associations
Correct, in all aspects	All parts are correctly perceived, but will not connect to form a whole	Thoughts that have nothing to do with the target	Associations of feelings, etc.
Correct, but some distortion	Some parts are fused; others are not	No contact or correspondence at all (barriers)	Things that are or might be expected to be associated with the object
Correct, but something else added Correct, but some information missing	Fusion is only approximate	Illusion or imagination	Something the object or location reminds you of
Details preserved, but not the whole, only part or parts perceived Correct relationships perceived	Parts are correctly fused; all parts are there; but put together in such a way as to falsely create another image	False guesses, or just guessing	An image of something similar to the object
General idea correct, but overdeveloped Correct but elements reversed			

Figure 26.8 – Summed-up Characteristics of Drawings

We largely agree with Ingo's categorizations. It should be noted, however, that the sessions in both Sinclair's and Warcollier's books (and even in Ingo Swann's earlier experimental attempts) are premised on telepathy – they use a sender/receiver design. The psi participants were not simply mentally accessing a book page or a picture on a wall but instead connecting with the mind of someone focusing on the target. If a mind-to-mind connection was actually occurring, it is possible different effects could be produced compared with efforts when this person was not aware of the target. How telepathic influence, overlay or connection might lead to distortions or to improved accuracy is not clear. Perhaps Upton Sinclair was thinking about a football rather than a calf and Mary Sinclair read his mind. We just don't know.

Further observations about pictolanguage elements

Supplementing and expanding Ingo's observations, here are points that stand out for us. Some of these are seen primarily in experienced viewers.

- The most prominent or significant feature is often drawn, but not always
- Sometimes only one part of the target is drawn
- The viewer may add details that are not in the original
- The basic "shape-form" of an object is what is most often drawn
- A simple outlined whole or part of an object, thing or person is generally what is drawn. The surface, details, texture, etc. are often not depicted
- Drawings focus on simply rendered edges of parts or wholes of persons, places and things
- Viewers vary in levels of complexity both in conceptualization and in drawing styles
- The artistic traits and personality of the viewer are often present
- Horizontal orientation (something facing to the left or the right) is often correct
- Relations among components different in size are often correct
- Body positions of people and animals are specific and often correct
- Eye positions and eye qualities are often included in targets containing people
- Hand, arm and feet positions are often included and correct
- Interactions of objects or placement between one or more is often included
- Entities are sometimes depicted by themselves but sometimes are integrated into settings
- Some viewers label aspects of their sketches while some omit words entirely. There is a mixture of correct and incorrect descriptors. Sometimes the viewer will correctly "name the target"
- Parts of objects may be over- or underemphasized and not be in a correct position
- Through intent, training and taking more time, simple sketches can be compiled into full and more complex depictions

- Viewers will often use a word to convey the color but not use color in their sketches. Some of this may be because the military remote viewers advised against coloring, which did not come through in copies because of limited technology. Also colors have the potential to lead to greater Analytic Overlay once introduced into a sketch. If wrong, colors may overwhelm the rest of the sketch. Some, however, do use color very effectively, such as Dale Graff and Patricia Cyrus in their dream images. (See Chapter 17)

Differences in sketching styles – simple to artistic

Let's look at examples of sketches by viewers from recent years to see if they bear out Ingo's and our analyses. We don't have feedback photos for some of the first few – they simply demonstrate features common to many sketches. One way to approach the points Ingo makes about accuracies is, What simplifications does the psychic make in producing picture drawings? Does the viewer use basic geometric forms like circles, squares, triangles? Does the viewer draw the entire figure or just part?

Here is one of the simplest sketches possible. Jon drew this at an online site where you use your mouse to sketch on the screen. It is clear he got the target, but his "sub" extrapolated from what is shown to the part we can't see. The orange may or may not be missing a slice on the far side.

Figures 26.9 & 26.10 – Orange and Jon's RV Sketch

Here we have a simple sketch of two figures (rider and horse) conveyed with a few basic shapes and lines.

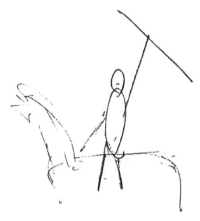

Figure 26.11 – Rider and Horse

Next a simple sketch by Debra Katz: outlines, not to scale, simplified, no shading.

Figure 26.12 – Woman's Head

This is a sketch from Natalie C. – apparently this guy was once on the planet Mars. He's got no hands but some interesting strands of hair on his head.

Figure 26.13 – Man with Weird Hair on Top of Head

Next is a sketch by RE, a portraiture attempt in a missing person case.

Figure 26.14 – Sketch by Russ E. – Head and Upper Torso

Another example of a portrait, also a simple sketch, is this one with feedback.

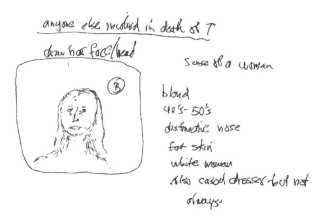

Figure 26.15 – Simple RV sketch of a Portrait by Jon

Figure 26.16 & 26.17 – Woman Arrested in 2019 for Being at the Scene and Inciting the Murder

The viewer (Jon) knew there had been a murder but did not know who the suspect(s) were. His simple sketch is a suggestive likeness and the physical descriptors are accurate (blond, distinctive nose, fair skin, white woman, long hair shown), but her age is off. Note: Jon drew her as she looked in a media photo at the time of the arrest (left), the same time the session was done, not as she looked after two years in jail (right). This is consistent with the observation that if the viewer understands the target is in the present, as was the case here, they will tend to go to the approximate time the events were taking place, not to the distant past, nor the future. One suspects, too, that Jon picked up on the media photo itself.

More artistic sketches

Below is sketch using shading by Daz Smith – an artist, photographer and highly experienced viewer. Daz participated in a remote viewing project tasked by a UFO group wanting to know if and when future contact might be made. Viewers were blind to the target question. Notice Daz provides written descriptions of facial features. He did not draw the entire face, perhaps because that is all he saw. He provides a close-up version of the eye as well.

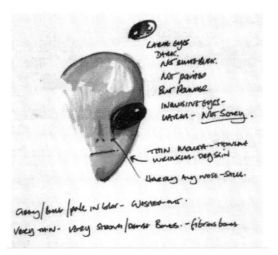

Figure 26.18 – Shaded RV Drawing of Head by Daz Smith – Big Eye

Here is an image of a hand Daz saw during the same project. Notice that the hand appeared to him as separate from the rest of the body, unattached. Similar to the above photo with the eye, Daz drew an entire hand and then he redrew the finger to show details on the underside. He even then drew one part of the finger, which he labeled "mind suckers." It is quite common for viewers to draw single body parts when humans or animals are in the target (whether a photo or real life).

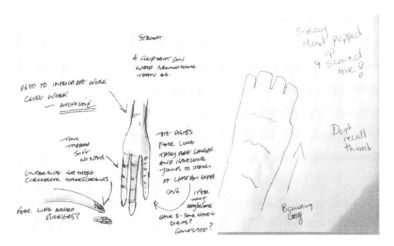

Figures 26.19 (left) Daz – Alien Hand & 26.20 (right) – Debra – Alien Hand

For this same project, an alien hand unattached to the rest of a body popped up unexpectedly for Debra. She described it as smelling like mushrooms. It was so real in her mind's eye she ran out of the room in fear and did not want to return. As you can see, her sketch, similar to the woman's face above, is less shaded and detailed than Daz's. The hand also has an additional digit; neither of their sketches included thumbs. Viewers often vary in details like this. Both wrote notes about what they drew – these can be very helpful for the tasker or a judge. Debra wrote down the color "browning grey" but didn't attempt to recreate the color.

Viewers for this same project (for which there is no feedback photo) also drew flying crafts, depicted below. These were drawn by four viewers, all working independently in the same project. They illustrate the different levels of complexity that viewers proffer. All seem to convey both inside and outside aspects of the object. Again, Debra's is relatively simple, while the other viewers add artistic detail.

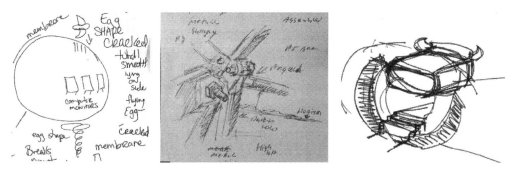

Figure 26.21 (left) Debra Katz – Egg Shape Sketch
Figure 26.22 (middle) Michael Ash – Shaded Drawing
Figure 23 (right) Coral Carte – Sketch of Structure

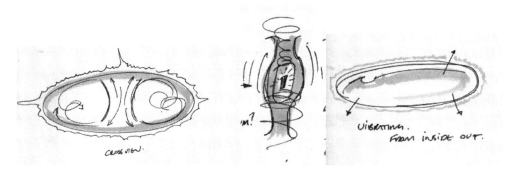

Figures 26.24 – 26.26 – Daz Smith – Cross View of Bottle Shape from Inside Out

Many sketches display body positions

Another common feature of sketches is depiction of body positions. Often these will be specific and correct, but sometimes reversed or with an incorrect orientation, as in the following example.

Figure 26.27 – Soccer Player Midair

In this example by Jon, the extreme splay and relative length of the player's limbs are caught well but the upended position was missed. This sketch confirms Swann's observation that viewers get shapes, but they often mistake what the shape is – here one arm becomes a leg and vice versa. Jon's descriptors are soft, flesh, rounded, humanoid with arms outstretched and motions/vectors. Jon didn't convey this was a soccer game, but the main gestalt (marked D) and unusual movement are well depicted, more than enough for an ARV match.

Figures 26.28 & 26.29 – Jon's Sketches B & D

Here is another body-position sketch, by Debra. She has drawn this figure facing to the right with a bent-over body, which is a correct depiction (perhaps a bit exaggerated). This is not typically the position in which one sees a human body, giving more certainty the image was not drawn by chance.

Many times, if someone is holding something in their hands, this will be incorporated into the sketch. Debra has correctly drawn the rider's hands grasping the bars. While the central figure doesn't have on glasses, the figure next to him does and she has depicted glasses or goggles, as well. She conveys the sense of movement by writing the word, by the lines drawn and even by his hair. The sketch is representative of the central feature of the photo yet still is simplistic and with little shading.

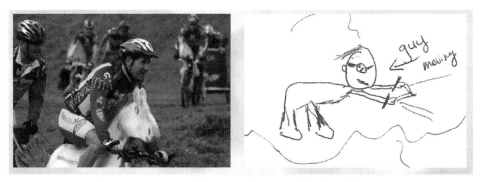

Figure 26.30 – Hobby Horse Riders and Debra's Sketch

In the next sketch of Dorothy interacting with the Wizard of Oz, there is mention of a female facing to the left. She is leaning slightly forward. The viewer (Debra) has sketched a female and the look in her eyes is "pensive." She is described as "short," which compared to the wizard, she is. Also, she is depicted as holding something, which was intended to be a cup and not a dog or basket, but the holding part is correct. Further, the sketch includes an image of a shelf with little items to the left of the female, which mirrors the shelves in the photo. The drawing is simple but conveys the correct posture of Dorothy and other information from the scene. There are distortions, however, and details missing. It's interesting that only one person was drawn, not two.

Figure 26.31 (left) Dorothy in the Wizard of Oz and
Figure 26.32 (right) Sketch of Person Leaning Forward

Experienced viewer Mark White has done thousands of sessions, as many as 25 per game. In the sketch below, Mark draws the net and adds a ball and player. One might ask, is he adding these because they are associated with it, or is he actually accessing the photo when a player was there? Could there be a ball and player on the other side of the field that we don't see in the photo? Impossible to say. Still, what makes a soccer net a net and not just a piece of material suspended between two poles? The activity happening there. Also note Mark's use of basic geometric forms.

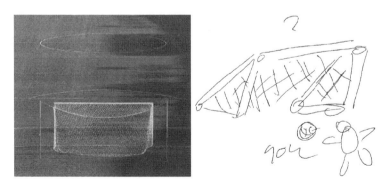

Figure 26.33 (left) Soccer Net and Field and
Figure 26.34 (right) Soccer Net and Field Drawing – Mark White

The next sketch is also by Mark White. Here Mark has drawn a few basic shapes depicting a person and he names what the person in the photo is holding. The pole is in the correct position, slanted to the left, and the scale of pole and person is correct. He

551

also conveys wavy motion and the line coming down from the pole. There are no details about the net nor the water, yet the few basic lines somewhat express them.

Figure 26.35 & 26.36 – Man Fishing Photo and RV Sketch by Mark White

Similar to Mark's sketches using basic shapes is a simple yet expressive and highly accurate ARV sketch by longtime remote viewer T.W. (Teresa) Fendley. The target is a girl on a swing and Teresa labels her sketch "swing." She conveys a sense of movement and the angle of the swing is the same, though the orientation is reversed. She has embellished the scene by adding more of the swing set than is shown in the picture, but there is little doubt that the swing would have the components Teresa added. The child's left arm and leg positions are shown and are correct, despite it being such a simple sketch.

Figure 26.37 & 26.38 – Girl on a Swing Photo and RV Sketch by T.W. (Teresa) Fendley

From the above examples we see that viewers often add more than there is in the photo. We don't know for sure in any of these cases if the viewer perceives more of what was going on in the photo or if they subconsciously made associations with things one would likely find at the location. Asking the viewers about their state of mind at the time they were sketching could provide insight about this (and is one argument for self-judging).

The next sketch is from a very experienced viewer, JFK. He has drawn two basic but full humanoid figures with a double line behind them, as in this photo of the moon walk. While the photo only shows one figure, we know two were there. He has correctly drawn the orientation of the figure (facing left), and also drawn their arms at their sides.

Figures 26.39 & 26.40 – Moon Walk Photo and Sketch by JFK

The next photo is interesting because it is of a close-up of what one would think was a simple object – a rope knot. This photo was assigned by Debra to two different RV classes. The first sketch is by Aala, a young college student from Pakistan. All her sketches from the start to finish of the 12-week class tended to be remarkably simple, yet were an accurate depiction of the target site. The second, by Lyrysa Smith, conveys the crisscrossing motion in her first sketch, then further develops the gestalt/shape by sketching an interconnected center of two circles and then rectangles on either side. She explained in class that she had a sense this was "woven fabric" and not plastic after she tuned in and attempted to "feel" the target using various methods.

 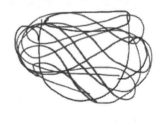

Figures 26.41 & 26.42 – Knot Photo and Drawing by Aala

The third sketch is by Igor Schwartzman, another new RV student, who provides a very simple sketch but representative nevertheless. These convey the detail of the intertwined knots. These three demonstrate how viewers will convey relationships of parts to a whole in varied ways.

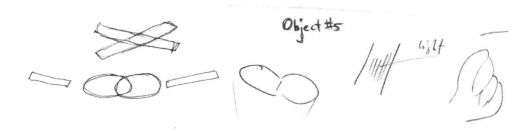

Figure 26.43 (left) Sketch of Knot by Lyrysa and
Figures 26.44 & 26.45 (right) Sketches of Knots by Schwartzmann

Here is an example of a compilation of sketches from an Applied Precognition Project group tasking, which led to a correct prediction. This can be a difficult task since viewers in a group have different styles and strengths. If the transcript is two or three pages, which it often is, there will very likely be a mixture of matching and non-matching data for one or both photos.

Again, the sketches here are simple and represent basic shapes. Extracts from Jon's session are on the left. His descriptors include "structure" and "just sitting there," and he draws one large circle and three small ones, representative of the car's wheels. The most distinctive drawing is the "vertical surface with rounded edges coming forward," which resembles the cowling of the headlights. The image from another viewer (top right) is labeled "window" and "clear" and the drawing resembles the windshield of the car. This sort of "division of labor" is common in ARV and also in operational remote viewing.

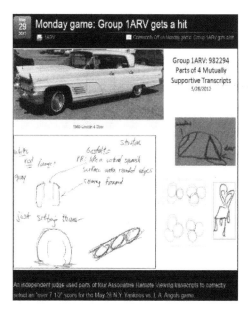

Figure 26.46 – Monday Game Group 1ARV Hit

Here is another simple sketch from Aala. She has drawn two cans, shading the top to depict the different colors near the tops of the cans. She also drew an exclamation point, which is a perfect match of a symbol on one of the cans.

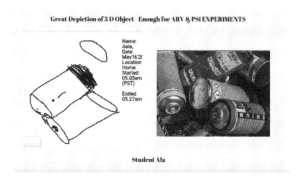

Figure 26.47 – Great Depiction of Can by Aala

Here Aala sketches an image of a trumpet. It is facing in the right direction and she depicts the correct number of valves. She uses basic shapes (oval, circle, rectangle and lines) to convey the idea of a trumpet.

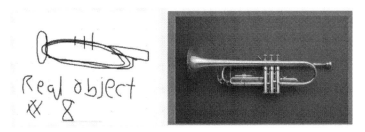

Figure 26.48 – Trumpet by Aala

Minimalist sketches and complex targets

This next example exemplifies a minimalist approach despite the complexity of a target. This sketch is also provided by Aala. The target was complex – Beyonce's Coachella concert. The students were given just a target number and the frontloading that the target involved a location, activity and event. They were told to take their time and that they would receive video feedback. Much of the concert took place on a stage with bleachers. Beyonce and her huge array of musicians, backup singers and dancers interacted with the bleachers. Aala drew a "minimalist" sketch of a figure at the bottom of the stairs. The feedback photo from the video is representative of the main setting. This is quite interesting because, in keeping with her style, Aala provided very little data in her hour-long session other than this sketch – yet if there were any two images representative of the target, a human figure and some steps would be it.

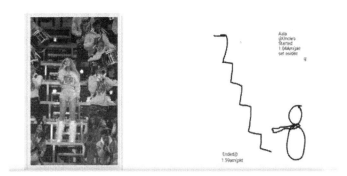

Figure 26.49 – Beyonce in Stands and Drawing

These next three sketches are from Lyrysa Smith, who is a journalist, writer and editor. They were provided to Debra by Garret Moddel of University of Colorado, who is friends with Lyrysa and began running her through a variety of informal ARV exercises, suspecting she would be good at them. These are from her early transcripts. In the first one, she has depicted a metal ball with spikes coming out of it, an extraordinarily accurate rendering. She has just it turned in the opposite direction. She writes the word "metal," "speed" and "vastness." This demonstrates that not only does her drawing match the image, but her words pick up on the conceptual aspect since the target was Sputnik moving through space.

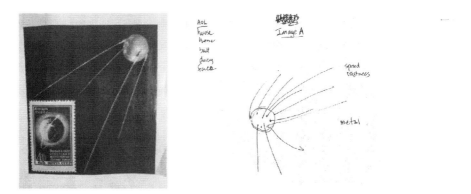

Figure 26.50 & 26.51 – Sputnik by Lyrysa

Lyrysa's next sketch is highly representative of the target photo. She has conveyed a close-up of a face in the hollow of a tree. She has added shading to show the roughness or darker color of the tree bark. Lyrysa captures the prominence of the eyes and recalls feeling the stare. She wrote "forest with many trees" as an AOL and "scent of a forest" (she smelled pine), but she knew the image was not a forest. This shows her ability to recognize something as an AOL and thereby not include it in her drawing.

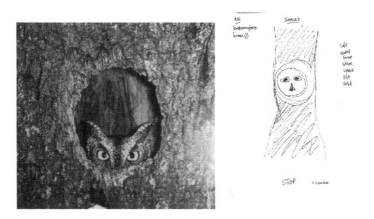

Figure 26.52 & 26.53 – Owl in Tree and Sketch by Lyrysa

About the idea of a face, Lyrysa writes:

In the transcript about the owl, my descriptor of "face" is in the AOL column, where I added the "null" symbol as an emphasis to further help myself in the midst

557

of the session to firmly keep "face" out of the drawing and my thoughts and only have it in the AOL column, only. I knew it wasn't a human face I was "seeing." I didn't know what it was, but it was clear to me that it wasn't a human face.[519]

Below is a final sketch by Lyrysa. She has drawn simple shapes in good relation to each other, with the exception that the railing around the top of the building is placed higher than in the photo. She sketched a door where there may not actually be one – a reflection on the building in the shape of a door at first glance looks just like a door. The sketch places the "door" in approximately the location where, in fact, one might expect to see a door. This shows remote viewers can pick up on "tricks of light" the camera catches and that someone at the scene would see. Further, she sketched what looks like an antenna, although out of position. Her descriptive terms – snow, cold, big, searching and hard surface – all are good matches for the photo. Altogether, this is another stunning sketch.

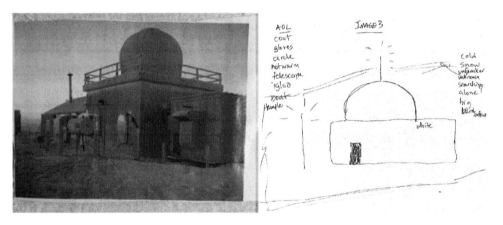

Figures 26.54 & 26.55 – Observatory and Sketch by Lyrysa

Animal targets

Moving on from simple sketches, noted researcher/viewer Angela T. Smith provides an example from a project with Marty Rosenblatt going back to the year 2000. Note that Angela has shaded the drawings of the shark and has tried to convey something of the surface. This is rare – most ARV sketches are outlines. Another common characteristic: a viewer may make several sketches of the target, but only a few or just one truly captures the graphic essence.

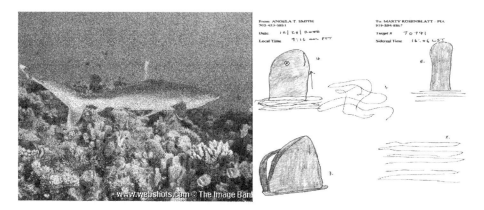

Figures 26.56 & 26.57 – Shark and Sketch by Angela Smith

These next two sessions were contributed by Carl McLelland, a talented viewer from the UK, when he was practicing with a target pool of photos of animals. He was frontloaded with "the target is an animal" (unlike Angela's example, whose pool was not animal-specific). Many times viewers will get a sense of an animal but not provide enough details to distinguish it from other animals. For this reason, Debra and her friend Natalie Cormier developed a target pool (www.remoteviewer.net) setting up categories of targets with the idea that if a viewer knows the pool consists only of animals, they will then have to both get past the frontloading (which tends to generate false information) and be as specific as possible. This is considered an advanced and challenging task for any viewer. While Carl didn't correctly identify the animal, he did get much of the body shape and did describe it accurately as a "predator."

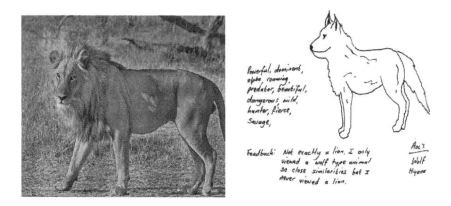

Figures 26.58 & 26.59 – Lion and Sketch by Carl McLelland

This next target was testing a student's ability to tune in to gestalts at physical locations. There happened to be a duck in the water and Carl perceived a bird flying over water. He indicated he didn't know it was a duck; he just got a picture of a bird in his mind and drew it.

Figure 26.60 – Duck Photo and Birds RV Sketch by Carl McLelland

Carl also provided this example during his time as a beginning student. He was tasked with an outbounder exercise: to describe a windmill formerly located at Debra's desert home in the Mojave desert. He did a terrific job of getting the shape of the spinning top and also the triangular bottom. While we are focusing on sketches here, from his words "space shuttle" it is clear images and ideas coming to him about the target were not correct, but still his sketches are highly representative, though simple. This again illustrates another important theme – remote viewing sketches tend to be "efficient"; the most important aspects are drawn.

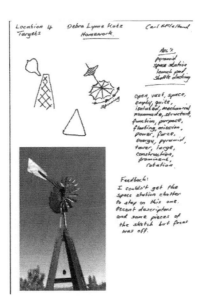

Figure 26.61 – Windmill Photo and RV Sketch (561) by Carl McLelland

The next example is by Alexis Poquiz, when he too was fairly new to remote viewing. When the target is a transportation device, it is common to misidentify it. Here Alexis turns a mountain bike into a "motorcycle." What is particularly interesting is the level of detail of the bike. While his wheels are too thick, the bike in the photo does have thicker wheels than other types of bikes (such as racing bikes). He draws what is clearly a bike of some sort and notice how he correctly shows the turn of the handlebars and wheels. This is further evidence that viewers can capture the orientation not just of human bodies but of physical objects, as well. In this case, he did not include the rider in the sketch, and it is common to omit vital features from a photograph and focus on another vital one, or even something less prominent.

561

Here is an example by a newer viewer, Alexis Poquiz:

Figure 26.62 – Man on Bike – Alexis Poquiz

Here is another example in which a central feature of the image is omitted but other essential elements are depicted.

Figure 26.63 – Skiing in the Desert Photo

The viewer is Jon and his sketch captures some, but not all, significant elements of the picture.

Figure 26.64 – Sketch by Jon

Jon gets the prominent horizontal lines and the small figures in the distance but omits the man looming in the foreground! He also correctly notes something "big" on the left, but the shape is not correct.

This three-page session also exemplifies another aspect of ARV. In the TDS methodology Jon uses, you do three "scans," which represent three different aspects of the target. In this case, it was the third scan (page 3) that had by far the most accurate information, although "humanoid shapes" are mentioned and drawn in Scan 1, the *top half* of a person is drawn in Scan 2 (but showing a man's face), and "two humans or humanoids" are mentioned in Scan 3.

As in Debra's sketch of the hair of a man, here Jon's sub may be trying to capture the agal (cord) of the keffiyeh (yes, we had to look that up), and he had never drawn something on top of a head like that before.

Figure 26.65 – Man with Hair on Top of Head by Jon

When a group manager gets sketches and descriptors from several viewers, he is almost certain to have to evaluate a mixture of correct and incorrect elements. Some may match one photo, some the other, and some both photos. Piecing together the correct descriptors is an art. (Anecdotally, when Jon was in TransDimensional Systems, the watchword was "Everyone wants to be a viewer and no one wants to be an analyst!" Honing skills as an analyst is very important because proficiency in analyzing is needed in operational RV and also in ARV.)

Viewer correctly describes elements present, though they do not appear in the photo itself

Russian Submarine Session by David Silverstein, as we have noted above, viewers' sketches sometimes depict aspects of objects or indeed entire objects that are not in the photo. This speaks to the question: Does the viewer describe only the 2D-photo itself or what is actually present at the location when the photo was taken? In this example, the feedback photo sent to Dave showed gauges aboard a submarine.

Although it was judged a miss, Dave felt certain he was on target, searched for other photos of this seagoing vessel and found it was indeed a submarine – a Russian one. In essence, by doing so, Dave added these photos to his feedback. This is one of the photos he found.

Figure 26.66 – Submarine Photo

In the page from Dave's session (shown below), he draws a large vessel and writes "military ship, lots of blue water, high seas, radar antenna rotating, skipper" and on page 1 (not shown) he labeled the object "submarine." It is clear Dave accessed the submarine, although some of his impressions are off ("cargo ship").

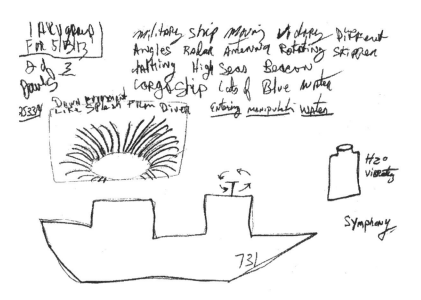

Figure 26.67 – Sketch by Dave Silverstein

His sketch mentions movement of the vessel: "down movement like splash from diver" and "entering manipulate(ing) water." We do not know if the photo of the gauges was taken during a dive or not. We don't know if Dave viewed the sub in dock or if the sub was actually diving when the photo was taken. What we do know is that during his session, Dave extrapolated from the part (the gauges) to the whole (the entire vessel) – that is, to elements not found in the feedback photo, but implied.

For this next target, which was a man flying through the air on a motorcycle, Dave produced a simple sketch of man turned upside down next to a downward pointing arrow. He writes "free falling," "dare devil" and "White Helmit." While we don't see the motorcycle coming down in the photo, we know for sure gravity will bring it down in just a moment in a "free fall." This description shows Dave accessed the situation and got impressions of the event itself, but not the exact position of the person in the photo. Meanwhile, what is incorrect is the manmade item, related to transportation. It looks like he has sketched the outline of an airplane, which one would expect to find in the air. One wouldn't expect to find a motorcycle in the air.

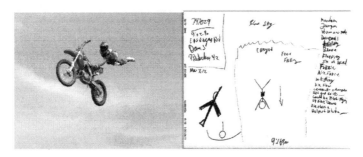

Figures 26.68 & 26.69 – Motorcycle Midair and RV Sketch by Dave Silverstein

Example of details that are correct, with very slight distortions

Next is a final example of a sketch from Dave Silverstein, and one of our favorites. It is a compilation of an entire scene. The photograph shows a single wake boarder and a person suspended in the air, to the left of a boat filled with people. The airborne person is attached to a boat with a rope.

Dave's sketch is very accurate, with some minor and somewhat amusing inaccuracies. He includes a boat and water skiers. He has correctly positioned the skiers on the left up in the air and connected by lines to the boat. However, the photo has one person on the left, whereas Dave sketched a couple of people in the boat with two outside where only a single skier should be. Also, the skier in the photo is using a wake board instead of two water skis. Otherwise, Dave nailed this target. He wrote "tricks and careful execution," "water skier," "fun" and "people," noting that water skiing was a very prominent part of the target. He identified the water as a "lake" and at the same time an "ocean." "Ocean" is probably not correct; it is more likely a lake or a river. This serves as a reminder that even someone looking at a photo (with physical rather than psychic "eyes") there is information about the target we will not be able to confirm. That is why viewers sometimes "deserve a break," particularly when the rest of the transcript is so accurate.

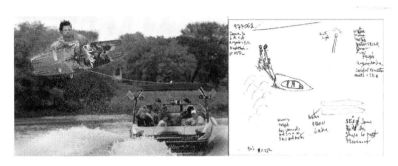

Figure 26.70 & 26.71 – Boat & Man on Air and RV Sketch by Dave Silverstein

Example of major distortion

In the following example, the target was a computer-generated image (CGI) on a white background. It was for a project testing viewers' ability to describe objects set in different backgrounds. If one looks closely, the brick pattern is incorporated, the curves and the columns are also depicted, but nothing is put together correctly. Some advise never to use computer-generated images in ARV (for example, the large lottery winner cited in our lottery chapter gives that advice).

Figure 26.72 & 26.73 – Castle and Sketch

More developed sketches and composite sketches

There is a psychological impact in remote viewing when a sketch is drawn artistically. The person evaluating such a sketch is more inclined to expect it to be correct, which, of course, is a problem if it is not! Highly polished renditions can be as "off" as much as extremely basic ones. The former will often include embellishments not present in the

567

feedback photo or location. If the viewer is an artist, they may try too hard to present a perfect sketch.

The late Paul Hennessy was a longtime remote viewer and professional artist. Everything he did looked great, even if he wasn't always exactly correct. In this example, there is a striking resemblance. Here he provides two sketches of microscopic entities – phages, which are viruses that attack bacteria. They can only be seen under an electron microscope, which – at least at the time of the session – could not take photographs of the phages themselves. For this reason, scientists had to rely on artistic renditions and visual models, and that is what comprises the feedback image in this instance.

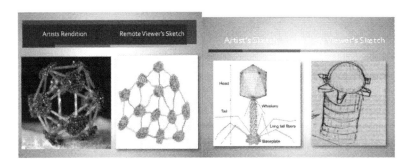

Figure 26.74 & 26.75 – Phage and Sketch

Keck Array detectors in the Pomerantz Observatory at Amundsen-Scott South Pole stations

This next example is from a project tasked by Angela T. Smith and the viewer is Joyce Wahlberg. Joyce clearly takes time with her RV sessions. This was from page 13 of her session. Earlier she had simpler sketches with many matching words. The first sketch portrays the Keck Array detector designed to study gravitational waves. She actually had the phrase "gravitational waves" in her session. The second photo below is the building where the Array detector is located. While this sketch is relatively simple and seems to represent only half the building, notice how the most unique and unusual part of the building is captured in Joyce's transcript – the basketlike object on top.

Figure 26.76 (left) Newspaper Article from Feedback and
Figure 26.77 (right) Joyce's Words on Final Page (13) of Her Transcript

Figure 26.78 (left) Photo of Metal Unit and
Figure 26.79 (right) Joyce's Sketch

Figure 26.80 (left) Photo of Building with Array Detector and
Figure 26.81 (right) Joyce's Sketch

From simple to complex (again)

Here two viewers were tasked with the same target for different projects. Carl McLelland, mentioned earlier, is an artist and a talented viewer. This session took only three minutes. It is one of his earlier ones, before he had much practice or training. The second session (by Debra) took about 30 minutes.

Like Carl, Debra initially had submitted just a shape – a simple curve. This was for IRVA's Focal Point target practice, for which viewers could upload their transcripts and look at each other's work before receiving the photo feedback. Debra uploaded her transcript, thinking she was done with her session. She skimmed through the other viewers' sessions and was surprised to see they all had a very similar curved shape. However, as newer viewers, none had attempted to explore the shape. One viewer labeled her session "AOL rainbow." Debra knew whatever this object was, it wasn't a rainbow. At this point, she didn't know if it was even manmade or natural. She removed her transcript and went back into session, retasking herself to get more information. After a few minutes, she had a sense of people with cameras on top of something and had an image of someone rappelling off the top to the ground below. She was quite sure they

569

were rappelling. Since people rappel off mountains (as opposed to office buildings), she felt she was on top of a mountain. She then did mapping exercises as part of the CRV methodology to flesh out that there was land on either side (as opposed to buildings).

This is an example of how a simple sketch can be developed further. Notice it is not that all the details came to her at once. It took conscious, directed, intentional moving and mapping around the location. Carl's three-minute drawing would have been more than enough to satisfy any ARV manager he was describing Uluru/Ayers Rock, unless the other photo had a similar shape in it. However, if someone wanted to know where they left their wallet, much more information would have been needed.

Figure 26.82 – Uluru/Ayers Rock Photo and Sketch by Carl M.

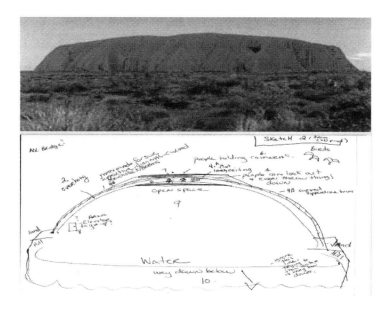

Figure 26.83 – Uluru/Ayers Rock Photo and Sketch by Debra

Here are further examples of projects involving satellite technology. The first is simple, the second complex. The first is by Australian viewer Simon Turnbull and is simple, straightforward and extremely representative of the target. In addition to the sketch, on a second sheet the viewer wrote the word "satellite."

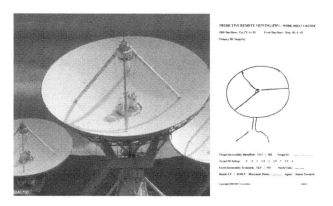

Figures 26.84 & 26.85 – Photo and Sketch of Satellite Dish by Simon T.

The second satellite example was developed over a period of time. This was done by Debra, who spent about an hour on this entire session, first drawing images of the round disk and then attempting to pull it all together. Sometimes critics of more intensive (and expensive) training methods will point out anyone can RV without training. As

571

some of the very simple drawings illustrate, that is definitely the case. Yet training can help develop a drawing of an entire scene, scenario or situation, which is rarely seen in sessions of those who are new or untrained.

Here Debra did not receive a single comprehensive image of the entire scene in her mind's eye. She had to pull together many individual elements, from a combination of relational aspects plus logic, intuition, guessing and the hope that however it was all being assembled, it would work out.

While there are mountains and very high/low aspects of the terrain, it's likely the spiky angular peaks are more about the angular aspects of technology in the target than the mountain shapes themselves, but these were integrated into the landscape. This is a common aspect of drawings in RV: shapes or aspects of one aspect of a target are integrated into the drawing, but not where they should be.

Also, she drew people with tall hats to denote they are military or governmental. We don't know whether workers at this location wear hats like these; the inclusion of them was analogical. This is often the case – suggestive concepts and literal aspects are integrated in the same drawing.

The entire sketch, while representative of the target, is still artistically very simple with little to no shading. It is more about the shapes and elements coming together.

Figure 26.86 – Photo of Dish and Sketch by Debra

Below is another by Debra, which she did as a test for a radio show host who bills himself as "Canada's most dangerous mind." He tasked her with a trip to Disneyland he took as a boy and specifically the outside of the Small World Exhibit. Notice this is more detailed than her other sketches in this chapter.

As noted, one of the reasons some of her sketches are more developed or expressive (i.e., adding shading or depth) is the amount of time and care taken. For this example, Debra devoted several hours to the session as she knew her work would appear on the radio host's site. Even so, when she had trouble drawing the overall placement and perspective, she gave up in the end and wrote: "This is a backward view because I can't draw from the other direction." In her written description, she indicated she was describing a theme park where families go for amusement rides, shows, restaurants and public bathrooms, but she did not know which one it was.

Figure 26.87, 26.88 & 26.89 – Photos and Sketch of Amusement Park by Debra

This next example was an outbounder exercise. The Sublime group project manager, Nancy Smith, was out taking a walk with her spouse and dog on what is a ski slope in the winter. Again, Debra is the viewer and here, too, she attempts to pull together the whole scene. Notations included "seems like a place where there are signs indicating it's colder weather" and "people walk their dogs there."

573

Figures 26.90 & 26.91 – Photos and Sketch of Nature Walk

The next session is from a professional-level applied RV project. The project manager was Angela T. Smith and the viewer was Russell Pickering. This was for a Gaia TV show special: "Evidence of Giants in Sardina," *Open Minds* with Regina Meredith. The correspondence is evident and very strong.

Figures 26.92 & 26.93 – Sketch and Photo of a Pyramid Ruin by Russel P.

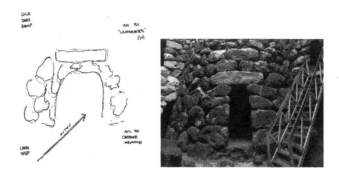

Figures 26.94 & 26.95 – Stone Entrance Sketch and Photo

The next set of sketches is by Joffre Perreault during a training session with Paul Smith. The target is the El Djem Amphitheatre, August 2019. The set shows the development from simple to highly developed information. Notice how he correctly depicts a destroyed part of the building and in the correct place.

Figure 26.96 – Colosseum Sketches by Joffre P.

Perreault had the words "Roman" and "colosseum" at this point in his session.

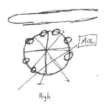

Figure 26.97

Figures 26.98 – 26.99 – RV Sketch by Joffre Perreault

Figures 26.100 – 26.101 – RV Sketch by Joffre Perreault

One question would be: If Perreault thought it was the Roman Colosseum and if he recalled images of it, is that why he drew the missing section quite well? Or did he not know of this gap but produced it spontaneously?

"Giant Crystal Cave" at Naica, Mexico

Applications Instructor: Angela Thompson-Smith
Remote Viewer: Joyce Wahlberg
The following target was tasked to Joyce during her class with Angela T. Smith, who provided this information about how she conducts her class:

All targets are assigned an individual alphanumeric coordinate. Students are always blind to the practice targets receiving only the alphanumeric. As part of the applications protocol, students do not receive in-session feedback. I remain partially-blind to the targets, only knowing the title and enough information to keep the student in structure. That way, telepathic overlay from the instructor can be minimized. Students receive feedback in the form of a video or article about the target. If they are in-person students, they receive the feedback following the session. If the class is virtual, they are emailed the link to their target.

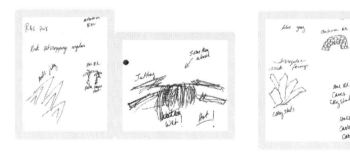

Figures 26.102 – 26.104 – RV Sketches 1, 2, 3 by Joyce Wahlberg

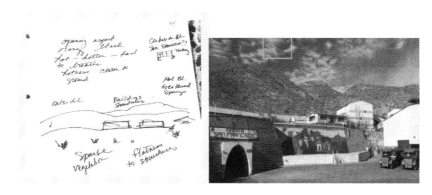

Figures 26.105 – 26.106 – Sketch 4 & Photo

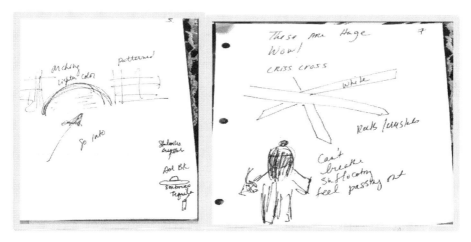

Figures 26.107 – 26.108 – 2 RV Sketches - 6 by Joyce Wahlberg

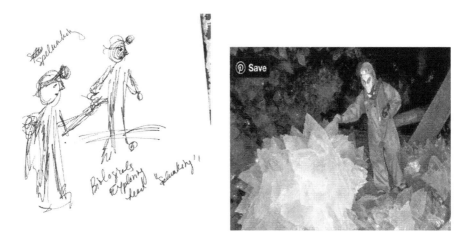

Figures 26.109 – 26.110 – RV Sketch 7 by Joyce Wahlberg & Photo

Final Sketch (Sketch 8)

A few things stand out about Joyce's 13-page session. First, one can see she was already honing in on important aspects of the target from the very first sketch, with "rock croppings" and a hot desert terrain. By Sketch 3, she has identified crystals and sketched them, noting they are "irregular." She also has "crystal caves" and "cavern," so she has identified the target. This would have been good enough for any ARV session at this point.

Let's look at what she continues to develop. She moves away from the inside and goes outside, describing two structures outdoors and hot dusty terrain. In her notations, she wrote "San Francisco" and then indicated she needed to take a "confusion break" as her analytical mind could tell the concept of San Francisco was not matching up with the data she already had. Now one might just assume she was just off on this point, but if you look at the photo next to Sketch 4, there are the words "San Francisco" in front of the opening to the cave!

By page 6, she is starting to get a sense of items related to Mexico – the cave entrance and a person facing huge rocks having a hard time breathing. By Sketch 7, she has described not just a person but a "spelunker" (a person who explores caves) and she has sketched the light on their head (which the people in the photos are wearing). Finally, on page 13 of her transcript, for the final sketch she has written the words "crystal caves in Mexico" and drawn shapes identical to those in the photo. She says "these are hot rocks" and it's suffocating. She is clearly having a visceral, somatic experience, which Ingo Swann called "aesthetic impact."

Nothing remains to describe that she has not already shared. It's likely this would be enough in a real-life applied case, let's say an archeological investigation. If someone wanted to know where a treasure was buried, she would have led the researcher to the correct cave and perhaps even the spot in the cave. This is a complete, detailed example of a transcript that demonstrates what a remote viewer can do. If you can do this or someday may be able to do a session on this level, why not try?

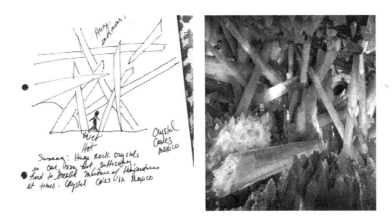

Figure 26.111 & 26.112 – RV Sketch 8 by Joyce Wahlberg & Photo

The need for sketches and learning to sketch

Some newer viewers create very small sketches to save paper or because they are not confident in their work. The problem with this is that judges may overlook tiny sketches – completely missing them, disregarding them or just not getting their full impact. My students have mostly been able to modify this with guidance.

It's important to recognize that sketching comes from the subconscious and is an intimate affair. A person should never be forced to sketch, just gently encouraged. Some people have intensive trauma based on childhood experiences with drawing and sketching where they were shamed or made to feel inadequate or embarrassed. On the other hand, many adults have suppressed the creativity they once had as children and, with some nudging, they can be helped to tap back into that. Debra worked with one student whose sessions often were very accurate; however, she noticed that other project managers did not show much interest in working with her. Her transcripts were devoid of sketches. During a one-on-one monitoring session with a target called "stairway to heaven," the student described seeing images of an escalator and steps. As she talked, she made sketching motions a few inches above her paper. She was sketching in the air! After encouragement, she finally put her images onto the paper. Her sketch was a fantastic match for the image in the feedback photo.

ARV projects – sketching matters

For ARV projects, sketching is essential. Recently a researcher sent along his data for an informal ARV project. The untrained viewer had done about two dozen sessions without a single sketch. Her words were often quite accurate but were filled with a lot of Analytic Overlay the raters had to translate. For example, one feedback photo was of a cluster of grapes. She described the photo as balloons tied together. Had she sketched this, her sketch of balloons would have probably been almost identical to the grapes. I began rejudging her sessions but felt exhausted after going through about five of them. When a viewer provides only words, the observer/rater has no recourse but to recreate his/her own mental image of what the viewer is describing in words, and then compare that mental image to the feedback photo.

Sketching is very important since it allows immediate recognition or rejection. This is particularly true when complex shapes are involved. Any time a viewer starts to verbally describe shapes, they should draw the shape on paper. Just imagine someone saying, "Well, there is a rectangle and above that a triangle, and above that a small rectangle, a pattern of little squares all over…" You get the idea. One picture is indeed worth 1,000 words.

Ingo Swann wrote in *Natural ESP*[520] that he worked for months as a research subject before he realized sketches could be highly correct even when his analytic mind didn't have the slightest idea what the target was. An upside-down can of soda caused his epiphany. When he first saw the target photo, he did not think he had done a good job. It wasn't until the can was turned right side up that he saw he had nailed it.

Figure 26.113 – Ingo Swann's Sketch and Feedback from APSR Experiment. From Natural ESP.

Bottom line, there is a reason why architects and engineers always draw visual plans. Remote viewers are essentially artists of the soul, architects of the subconscious.

How detailed do sketches need to be for ARV?

Opinions differ about how detailed sketches need to be for ARV projects. Some go for very simple transcripts and rely on quantity. For example, Lincoln Lounsbury (Chapter 24) states he only spends a couple of minutes creating a sketch. Samples of his work demonstrate that sometimes all he has is a single shape such as a square, and this works well for him. Greg K also made very quick sketches. Like Lincoln and Greg K, Mark White's method for many years was to do many short sessions and use the consensus to make a single prediction. All three relied on multiple sessions to be confident about risking money on a trade or outcome of a game. They felt many sessions – not just one or two – are necessary to achieve a good winning percentage and positive a ROI (Return on Investment). All three have been moderately or very successful using this approach.

Other ARVers who have had success do fewer and longer sessions (e.g., three pages of CRV or an offshoot) and they may take time to develop sketches. They, too, recognize that in ARV less data is required than is needed in a regular remote viewing session to distinguish the target from the decoy. Given the vagaries of ARV (particularly displacement) and the changing probabilities of events, they may do one or two sessions to feel confident about a prediction or use consensus from a team of viewers with established track records. Or, a third alternative, as with the CAS software, they may just wait out a large number of passes till they have a transcript with a c. 95% chance of success. For these folks, more detail and better sketches play a role in determining how and when they will be successful because the judge/scorer has more and better information to work with and can assign a reasonably accurate score to the transcript. These folks, too, have had successes. It is a different and equally valid approach.

As judges and viewers, we know how good transcripts can be in both ARV and RV. With the increasing lure of ARV apps and easy-to-use online target sites, "newbies" are becoming used to doing very quick rudimentary sessions. They may be missing out on what is possible not only in ARV, but also in regular remote viewing. This is one reason we have included the "How to" chapter (Chapter 25), which demonstrates ways to enhance transcripts by doing additional probing and taking time to develop sketches and verbal data. Of course, there is no shortage of instructional opportunities for both remote viewing and ARV.

All those who have kindly allowed us to use their sketches in this chapter have received some level of remote viewing training. Those with very developed sketches have had the benefit of extensive training, sometimes from several instructors. Additionally, many of these viewers have worked with project managers who are these same trainers. In this way, such viewers receive ongoing mentoring and support. While some might point out we start the chapter with examples from untrained people, we remind readers that these psi participants (such as Mary Craig and Warcollier's subjects) received tremendous personal attention from engaged researchers. The beneficial influence of that contact and connection should not be underestimated.

In closing, basic drawing skills are very useful for remote viewing. These include learning how to draw perspective, basic shading and cross-hatching techniques. Many remote viewers take art classes to develop these skills. According to Betty Edwards, author of *Drawing on the Right Side of the Brain,*[521] contrary to popular thought, drawing is a skill that can be learned just like anything else. Much like remote viewing, we'd add.

Unless otherwise stated, all images in this chapter and book that have been shared were actual targets in blind-tasked remote viewing sessions with evidence provided that the transcript was turned in to a third party prior to the feedback being known.

We hope this chapter has inspired you to try remote viewing yourself and that those who already are viewing but are just doing "quick and dirty" sessions without further probing will be inspired to put more time and effort into their sessions. Also, we hope to have demonstrated that remote viewing and ARV are alive and well in the year 2021, with more people than ever doing fascinating projects and working really well together.

CHAPTER 27

Ethics, Core Values, Common Practices and Related Considerations in Remote Viewing and ARV

In this chapter, we will discuss a variety of core values, ethical considerations, and related topics in remote viewing in general and ARV in particular. Our aim in discussing ethical issues is threefold.

First, we will highlight values and ethical issues as they relate to remote viewing as a profession. Unlike professions such as psychology or law, in remote viewing ethics and values are not articulated nor governed through licensing boards, societies, exams or courses. By contrast, values and ethics have been communicated by the originators of remote viewing and those who have emerged in the past 25 years as instructors, writers, speakers, researchers, project managers, viewers and organizers. These contributors have expressed these values across many platforms such as social media posts, discussion boards, email lists, books, classes, articles, videos, conferences and in organizational work.

Our second goal is to answer common questions students new to the topic of intuitive development and applied psi often ask and wrestle with.

Our third goal is to set forth standards or expectations that remote viewers can utilize as guideposts, to understand and protect their rights, whether or not they are being paid.

Last, but certainly not least, is our desire to help the reader understand their own belief system, which may affect not just the outcome of their remote viewing work, but their willingness to explore their own intuitive nature. When remote viewers or other aspiring intuitives face blocks to their practice, often they find their preconceptions and misconceptions arise from parental fears or religious teachings. These ideas were often not derived from direct personal experience but were taught to them by others who had little experience or understanding of such experiences.

Professional standards in remote viewing as a thought collective

So far, we've used the term *remote viewing community*. Another term that might be more appropriate is remote viewing as a *thought collective*. The expression *thought collective* was first introduced by Ludwik Fleck (1896–1961), a Polish-Jewish microbiologist. The term is not widely used today and, until this chapter, it has not been used in relation to remote viewing. Here is a description of what Fleck intended by *thought collectives*, as summarized by Sady:[522]

> Fleck claimed that cognition is a collective activity, since it is only possible on the basis of a certain body of knowledge acquired from other people. When people begin to exchange ideas, a thought collective arises, bonded by a specific mood, and as a result of a series of understandings and misunderstandings a peculiar thought style is developed. When a thought style becomes sufficiently sophisticated, the collective divides itself into an esoteric circle (professionals) and an exoteric circle (laymen). A thought style consists of the active elements, which shape ways in which members of the collective see and think about the world, and of the passive elements, the sum of which is perceived as an objective reality. What we call "facts" are social constructs: only what is true to culture is true to nature. Thought styles are often incommensurable: what is a fact to the members of a thought collective A sometimes does not exist to the members of a thought collective B, and a thought that is significant and true to the members of A may sometimes be false or meaningless for members of B (p. 1).

From this perspective, one might think of the remote viewing community as a thought collective.

The Worldwide Federation of Psi Collectives

From Debra: Most people don't realize how many different thought collectives there are when it comes to psi-based or psi-related communities, which encompass very different practices, approaches, beliefs, discourse, philosophies, values and norms. Many people in these collectives are not even aware other collectives exist. If they have heard of them, it's often from an outsider's perspective – perhaps they read something about them or had an encounter or two with a few members. They are then prone to make sweeping generalizations about the groups, which they then pass along to the members of their own groups.

While watching a *Star Wars* film in which the characters traveled from planet to planet, some being highly populated and some isolated in outer regions, I had an image of myself traveling between these different worlds of psi. On the one hand, I felt privileged to have been able to visit so many worlds, and on the other hand, it made me feel frustrated because I've been part of many collectives that don't have a clear understanding of all the other collectives or of the overall universe of collectives that exists. It's not that different from someone who has traveled the entire world and then encounters a person who has never left their own country yet doesn't hesitate to offer opinions about every place based on what they have seen on TV or the individuals they have met along the way.

My formal introduction to psi was through the *Theosophical Society* as a teen. That was one thought collective. Then I became very involved in clairvoyant programs originally created by Louis Bostwick – that was another – and remote viewing, a third thought collective. Along the way, I met spiritual mediums, who were part of their own collectives, as well as various groups of energy healers and formal religious groups, too. Meanwhile, Jon has been involved in many other collectives, both outside of psi and within the arena of remote viewing.

Traveling between collectives is not physically hard to do once you are aware they exist, though it does sometimes require one to adjust their own behavior, language and practices to open up to new teaching and to tolerate the biases held by each group. There's nothing quite like sitting around a table with people who say how much better their collective is than the group you just left and continue to identify with. Not only is one group often unaware of the importance and value of the other group's work, but they also don't understand the benefits one gains from combining practices and teachings.

Overall, I've found most people are not aware of the entire field or universe in which these collectives co-exist. With so much knowledge and variety when it comes to all things psi, one could easily have universities with entire departments devoted to a single collective. It's my hope someday we'll have just that.

One of the things I've observed is many of these psi collectives offer training and direct (professional) practices. Also, some groups have greater interaction with one another than with other groups, with members traveling between the two, while other groups have very little connection. Remote viewing, parapsychology and academia are linked in varying degrees. In the *World of Psi* chart (below), I try to demonstrate these links with "laser beams" between those that have the most interactions. (A lot of subjectivity is involved here, so this really should be considered the *World of Psi* according to Debra.)

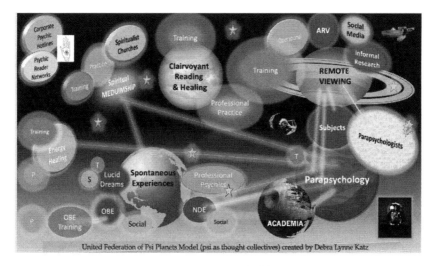

United Federation of Psi Planets Model (psi as thought collectives) created by Debra Lynne Katz

Figure 27.1 – World of Psi

For the rest of this chapter, we will discuss the thought collective of *remote viewing*.

When it comes to professional standards in remote viewing and all psychic work, we are still in the wild west. No single institution or person tells us how to behave. Still, pioneering individuals such as Ingo Swann certainly set the stage regarding values, principles and practices, which have become foundational for remote viewing.

As a psychic working alongside researchers at various labs, Ingo and his colleagues (Janet Mitchell and Karl Osis) coined the term "remote viewing." Ingo conceived of specific psi practices, and he molded steps already in progress at SRI into his Controlled Remote Viewing system. While continuing to work as a research subject, Ingo more than any other psychic aligned himself with scientific principles, values and practices that continue to be emulated today in the field. Many of these values are reflected in formal statements by groups, while others appear in personal statements in less formal venues, and all of which have carried over into the entire remote viewing community.

International Remote Viewing Association (IRVA)

Every remote viewer is not a member of IRVA, nor does IRVA have any governing power over any individual or group of remote viewers, but IRVA has been in existence for two decades now. Many of the members have been in the forefront of remote viewing activities, such as teaching, social media appearances, conference participation, communications, discussing norms and practices, and – to a much lesser extent – engaging in research.

IRVA is the sole international remote viewing organization. It is an organizing body that holds conferences, puts out a newsletter (*Aperture*, now a magazine), and most recently started a research endeavor called IRU (IRVA Research Unit). The addition of a new generation of board members offers hope that IRVA will have more success in completing the mission it undertook when founded. Hopeful indicators are the addition of a research unit (co-founded by Debra and Dale Graff), a new educational unit and initiation of an International Chapters program that is just getting under way in early 2021.

We will cite a lengthy quote from the IRVA website, which promulgates the values to which the founders and present-day Board of Directors and members have agreed to adhere. These are also, broadly speaking, the values upheld and demonstrated by the Applied Precognition Project (2,000 members) and by a variety of remote viewing instructors and practitioners (many of whom have served on the IRVA Board of Directors or have been longtime organizational members).

From the IRVA website

The International Remote Viewing Association (IRVA) was organized March 18, 1999, by selected scientists and practitioners meeting in Alamogordo, New Mexico, in conjunction with the first professional conference on remote viewing in Ruidoso, New Mexico. The concluded objective was to create an organization that would provide a mechanism for evaluating the discipline called "remote viewing," encourage scientifically sound research, propose ethical standards and provide overview educational information to the public. Below are the guidelines that have been in effect for the last decade. Some of these are in the process of slight modification.

Ethical Guidelines

The International Remote Viewing Association (IRVA) is the largest and most respected international organization promoting the responsible practice of, education and training in, and research into the art, science and phenomenon of remote viewing. We believe in and support the principles of verifiable truth, integrity, honesty, transparency and responsibility in dealing with clients, persons subject to remote viewing as targets, the scientific community, the news media, law enforcement and the general public. It is the purpose of these Ethical Guidelines to provide our members with a clear understanding of their responsibilities as active members of the Association and operational remote viewers. These Guidelines are also intended to protect the public and the Association from the unethical practice of remote viewing, wherever and in whatever nation remote viewers train, practice and operate worldwide.

Ethical Guidelines for remote viewers

A "client" shall be construed to include any individual person, group or legal entity, whether public or private, that solicits, engages or retains the services of one or more remote viewers or remote viewing organizations, whether on a free or payable-fee basis.

"Operational Remote Viewing" shall be construed to mean remote-viewing activity conducted toward any real-world target to accomplish some practical or pragmatic intentional objective, whether on a free or payable-fee basis. Such remote-viewing activity shall not be deemed to include any remote viewing conducted exclusively for one or more of the purposes of training, practice, general education or scientific research.

- A **remote viewer** shall adhere to all applicable laws, statutes and regulations of the state or province in which they are working, as well as of their nations of work and residence, in carrying out any operational or other remote-viewing activity on behalf of clients or themselves, and, in particular, concerning any living human person or persons as targets.
- A **remote viewer** shall provide honest, accurate, remote-viewing-based reports to clients to the best of his or her ability, using and acting in conformance with remote-viewing protocols generally accepted as facilitating the reception of truthful, reliable and accurate remote viewing-originated information.
- A **remote viewer** shall safeguard all confidential information provided to him or her by clients and exercise the utmost care to prevent any unauthorized disclosure of such information.
- A **remote viewer** shall maintain confidentiality with clients to protect the privacy interests of all persons involved in the remote-viewing activity, unless duly and properly authorized otherwise. The targeting of persons and the collection of personal information about them shall only be done for lawful purposes. And, except when in aid of a *bona fide* law-enforcement investigation, any personal information so collected shall not be disclosed to any third party without the knowing permission, secured beforehand, of the particular person or persons so targeted, identified, or about whom personal information has been collected. No remote viewer shall make a disclosure of information to any person not authorized by the client or by applicable laws, statutes or regulations.
- A **remote viewer** shall disclose to any client any conflict, whether legal, moral, or personal, that would prevent the remote viewer from performing an

objective, fair, accurate and scientifically sound remote-viewing session. When soliciting work, a remote viewer shall always conduct himself or herself in an ethical manner and shall refrain from misrepresenting the nature, character, accuracy potential or reliability potential of remote viewing and its various protocols and processes beyond what is verifiably known or reasonably posited by documented experience or reputable scientific research.

- Notes: (1) *In "conducting oneself in an ethical manner," a remote viewer should also undertake to refrain from misrepresenting or disparaging any other remote viewer in any public or media forum in order to obtain a work assignment or an unfair advantage while performing an active work assignment, or while carrying out the duties of the Association.* Notes: (2) *The term "reputable scientific research" is intended to mean peer-reviewed, published research performed according to generally accepted scientific methods. This provision seeks to set a cognizable standard to increase the credibility of proper remote-viewing activity, as distinguished from other, less rigorously performed forms of paranormally cognitive functioning.*

- A **remote viewer** shall, within the scope of his or her personal authority and to the best to his or her ability, act to ensure that all other persons associated with a remote-viewing assignment for a client adhere to these Ethical Guidelines while performing remote-viewing activities on behalf of the client. Such activities shall include, among others, targeting, tasking, remote viewing, session analysis and the operational management of the remote-viewing process. Note: *This provision lists the essential elements of standard remote-viewing practice, known to and accepted by those in the remote-viewing training and operational communities. It is intended to encourage the practice and self-regulation of ethical behavior according to norms embodied in these guidelines.*

- A **remote viewer** shall refrain from any conduct that would bring reproach by or negative attention from the general public, news media or law enforcement to the remote viewer acting as a remote viewer; the field of remote viewing in general; his or her client, if any; or the Association. Note: *This provision is not an enforcement tool, but rather seeks to encourage the practice of ethical behavior as it pertains to remote viewing, while practicing remote viewing, so as not to bring any undue negative publicity to the practice of remote viewing in general or to the individual remote viewer engaging in such activity.*

- A **remote viewer** shall never undertake a remote-viewing assignment that is or might reasonably be construed as being contrary to the protection of the national or internal security interests of that state, province or nation in which he or she is resident.

For comparison, here are the brief ethical guidelines of one of, if the not the largest remote viewing organization, the **Applied Precognition Project**. The APP states its mission is to "publicly explore, research and apply logic and intuition/emotion to predict future event outcomes, enabling participants to evolve personally while contributing to the elevation of global consciousness." APP explains that "evolve personally" means "A primary intent of APP is to be a catalyst to the individual spiritual evolution of each participant…Elevation of global consciousness: We believe that personal evolution is the primary mover behind the elevation of the world's consciousness, in the global shift now occurring."

Our (Debra and Jon's) understanding is that the goal of the above statements is not that every remote viewer must follow or accept every aspect of them or has sworn an oath to them. Rather, they are aspirational and useful in highlighting the overall values of these two large and influential groups.

"Nine Commandments" of Remote Viewing

The values articulated in the above statements, along with tenets we suggest, are important from our own experiences as remote viewers, project managers and RV instructors. You might think of the following as our Nine Commandments of RV.

- Maintain the definitions of remote viewing and the historical traditions
- Adhere to scientific principles and ethics
- Be honest
- Advance the field of RV before or in addition to advancing one's own interest
- Respect confidentiality but give proper credit where credit is due
- Respect and cooperate with other RV professionals
- Protect human rights, the safety and well-being of clients and viewers
- Be open to others' methods, approaches and ways of doing things
- Be contributors of knowledge, not just psi participants.

Some of these "commandments" are self-explanatory, while others are noted or alluded to in other sections of this book. Here we will highlight a few of them:

Maintaining the definitions, integrity of remote viewing and historical traditions and adhering to scientific principles and ethics

In 2020 and in years past, on various RV and ARV discussion boards, concern has been raised that people coming into the field do not understand the difference between intuitive work and remote viewing. To present the viewpoint of many established remote viewers in the field – essentially ideas that make up the core of the RV "thought collective"

– we've been granted permission to republish the following statement by John Herlosky and shared in the largest remote viewing Facebook group at the time of this writing.

John Herlosky is the author of the fascinating book *A Sorcerer's Apprentice: A Skeptic's Journey into the CIA's Project Stargate and Remote Viewing*.[523] He has worked for two large metropolitan police departments as a police officer and academy instructor and has been active as a professional remote viewer. He has trained extensively with former military remote viewer David Morehouse.

Remote Viewing: definition, protocol and corollaries, by John Herlosky

There is a lot of confusion out there over what constitutes remote viewing. People mix the terms psychic functioning (psi, ESP, anomalous cognition etc.), remote viewing protocols and remote viewing methodologies as a single term – remote viewing. People are saying they're natural "remote viewers." They've been doing remote viewing all their life, etc.

Nothing could be further from the truth.

You can't be "born a remote viewer." You can be born with psychic ability. In fact, you must have an innate psychic ability in order to become a remote viewer! But you can't have been "remote viewing all your life." Why? Because remote viewing isn't even psychic functioning! It uses your natural psychic ability. It's a totally artificial way of utilizing psychic functioning to enhance and keep it within scientific requirements. Prior to the '70s, the term didn't exist.

The term was created by the primary developer of Coordinate Remote Viewing – also known as Controlled Remote Viewing – Ingo Swann, to distinguish it from the previous terms for extra sensory perception, psi and clairvoyance. Why would he add another term seemingly for the same thing?

Because remote viewing isn't any of those things.

Remote viewing was created as a way to categorize and standardize innate psychic ability in order to facilitate training others to use their psychic ability in a militarily useful context. To do this, two things were done. One, Hal Puthoff and Russell Targ – laser physicists at the Stanford Research Institute – created the scientific protocols under which the remote viewing would be done, in order for it to conform to the scientific method. Two, Ingo Swann would be tasked with creating a methodology, a set of steps or instructions that would be followed to conform to the protocol, such that anyone could learn to use their natural psychic functioning in a standardized manner. This would facilitate its use in a military intelligence context, yet remain within scientific requirements.

Remote viewing is not something you are "born" with as it is an artifice that was required by scientific investigation. RV is NOT a phenomenon! That is psi. RV cannot increase your innate psi ability, and if you have nearly no innate ability, using either the protocol or its associated methodology will not make you psychic.

Therefore, a lesson in the terms…The protocols that define Remote Viewing are:

Planned and Aimed. The psychic session must be planned and done on purpose. If you get a "spontaneous insight" or have a dream, that is not remote viewing. RV is when you intend to collect information about a specific target.

Recorded. The remote viewing data is recorded in some format – written, audiotaped or videotaped.

Double-blind. In most experiments, if the person giving the answers does not know the question, it would be called "blind" or "single-blind." Remote viewing is required to be "double-blind." That means there are two (double) layers of "blinding." It means the psychic cannot know the target AND nobody else who is present with the psychic during the session (even by remote means such as webcam or phone) can know the target either. This is because even pheromones, voice-frequencies and many other "invisible" physiological senses can transfer information below the conscious level.[524]

Feedback. Although you can be psychic about anything (the future, for example), in order to "validate" the data IS psychic and not just a wild guess, it has to be at least partly correct. In order to know what is correct, we need the real info to "compare to" the session data. We call that info "feedback." Remote viewing results are considered experiential until validated by a resource OTHER than another remote viewer.

Now these are the scientific protocols. In operational work sometimes there was some information given regarding the target, called frontloading, that was given to facilitate the viewer to get on target. However, it was extremely neutral and gave very little about the actual unknown that was being investigated. Frontloading normally wouldn't even have been given until after the initial contact with the target by the viewer was made, at which time the monitor assisting might ask a passive and innocuous question like "What's inside a building?" Anything more would be in violation of the protocol. One of the reasons for getting trained is the necessity for being fully familiar with blind work before moving to operational work. Also, only someone fully versed in the use of remote viewing operational targets is qualified to choose or decide what might be given in specific circumstances as frontloading. That is not something for the novice. As you have surmised, the viewer should be blind to the target in experiments. However, as stated above, there are some limited exceptions in operational work.

Methodology. CRV, ERV, etc., are examples of methodology. Methodology is only important in that it must encompass the protocols. If it does, either for experimentation or operational work, then it is considered remote viewing. The methodology choice does not matter. It can be one of the initially created ones – CRV, ERV (known as Extended Remote Viewing) or WRV (known as Written Remote Viewing). A number of others have created their own methodology to better conform with their own preferences and opinions. As long as you have a set of standardized instructions that can be taught and conform to the protocol, you are doing remote viewing. They don't have to be elaborate. In the very beginning, it was nothing more than – Sit down, take pencil and paper, and use it to describe a target.

Then there are three corollaries that go along with remote viewing, as well. They are a result of the original reason for the development of remote viewing… which was intelligence work. They are:

1) Remote viewing is not a 100% accurate endeavor. It never was expected to be and probably never will be. Why? Because humans are terrible eyewitnesses! That's one of the reasons for training under a competent instructor. You aren't just learning remote viewing, you're also learning to become a trained observer.

Psychic functioning is a human ability. Its expression can change on a day-to-day or even hourly basis. Remote viewing is also like any other human skill. The more you practice, the more you learn, the better your skill at using your psychic ability will be, too.

Notice I said "the better your skill at using your psychic ability will be." I did not say your psychic ability will get greater. You have a given set of psychic ability that is determined, as far as we know, by genetics. It seems to follow a Poisson Distribution – the familiar bell-shaped curve. The majority will have some skill, with others having a high score and others a low score. We do not at present know how to increase the ability.

It has a range of success that can change for any number of reasons, at any time. How many basketball players who are normally excellent suddenly have their skills drop precipitously? It's the same for viewers. And for psychics. Remote viewing and psychic functioning success rates can vary by the hour. There is no guarantee how well any session will be.

2) Remote viewing was never designed to replace traditional forms of intelligence gathering. It was designed to complement them by the ability to gather information that was outside the capability of those traditional methods. The use of remote viewing outside of this consideration is outside its original purpose. Remote viewing is a PROCESS, NOT a singular event. Remote viewers never operate by themselves. Why? Because a remote viewer must remain blind to the target! Therefore, a remote viewer must always have a tasker. A tasker cannot remote view their own targets! You cannot make up your own target pool, then view it. That is in violation of protocol. Remember that remote viewing was designed for intelligence gathering, not personal exploration. This is why one cannot remote view "all their lives or be born a remote viewer."

3) The results of any remote viewing session outside of verifiable feedback (which does NOT include another remote viewer's session) is strictly experiential until such time as it is verified by other means. No remote viewing session stands on its own. The choice of how to use the data is up to the controlling agency or client. They must understand the above two corollaries to make an informed decision. It is unethical to present remote viewing results in the absence of verifiable feedback as a totally reliable result. The client must always be informed of the ephemeral nature of psychic functioning and remote viewing. Any attempt at sensationalism or falsely representing the accuracy potentials of remote viewing results is ethically and morally repugnant.

So again. To reiterate. Psi, psychic functioning, clairvoyance and ESP are synonymous with natural psychic ability. You are born with as much as you're going to get.

Remote viewing is a protocol and a methodology. It can't do any psychic functioning if you don't have the above. It's totally a manmade artificial creation. It has to be learned. If you have the psychic ability of a hammer, it won't matter how much or how well you learn the above, it won't make you psychic, and it won't make you a remote viewer either, since without the ability, the methodology and the protocols mean nothing.

And that is what is remote viewing.

(End of statement by John Herlosky)

When John posted this comment, dozens of remote viewers concurred. Then Katherine T. Hoppe joined in with her article "Remote Viewing isn't an Umbrella Term for Psychic Phenomenon."[525] Her statement is in alignment with Herlosky's when she writes:

I want to point out for the newbs, the term "remote viewing" isn't used in the remote viewing community as an "umbrella term" for all psychic phenomena. The use of the term "remote viewing" in this fashion is shunned in the remote viewing community and must be avoided at all costs so that the writer does not seem ridiculous.

Some may feel this wording is harsh. We prefer a less stringent approach, and as much as possible, a compassionate attitude toward our fellow psi enthusiasts. Also – a confession – we do sometimes in our personal lives use the term "remote viewing" loosely as a verb, such as, "Oh you can't find your keys, let me tune in for a second and see if I can remote view where they are." We've argued that many terms in the English language have several meanings, which develop organically over time, so there may be room for both usages. However, we seem to be in the minority on this point.

Scientific protocols: Blinding

While overall remote viewers value maintaining blinding protocols, as Herlosky alluded to, this is not always possible. In remote viewing, being "blind" to a target has different meanings. In the strict use of the word, the viewer receives no information up front about a target – just a random target number. A looser use of the word – when the viewer is told "the target is a location. Describe the location" – is still considered "blind to the target." Others would describe this as minimal frontloading – frontloading being when the viewer is told something about the target. For some, this minimal frontloading is very useful in operational/applied situations because it allows the viewer to home in on the target more quickly. That is, the information will help them focus their attention while remaining unaware of all other information about the target and specifically unaware of what it is the client wishes to know.

Still others (Stephan Schwartz) define blinding as simply being unaware of the specific information one is seeking. He points to drug trials for pharmaceutical studies in which being "blind" simply means participants don't know if they are receiving the drug or the placebo. Double-blind means the physicians delivering the drug also don't know which group a subject is in. However, in these scenarios, everyone is fully aware they are participating in a project for the purpose of studying the effects and efficacy of the drug. In ARV projects, the viewer usually knows some or all of the mechanics of the trial. They usually know the target will be a photograph, a video, a physical object, a color, sound or even an emotion. They may not know what the ARV is specifically being used for but trust the tasker or project manager.

This issue can be important because viewers have different techniques for different types of targets. If the task is to describe a two-dimensional photo, the viewer may have a different approach than for a three-dimensional object or location where one can attempt to smell, taste, touch or listen to the target's environment. Few want to view a target that is just a simple shape, unless they are practicing that skill. Also, a simple shape might take two to three minutes to perceive whereas a target that is a location might take 30 minutes to describe in detail. We point out again that most ARV sessions last under 15 minutes. Many last only three to five minutes because in ARV using photos, which is the most prevalent form, most viewers feel you only need enough data to differentiate between the two photos; you don't need the wealth of impressions you can get in non-ARV remote viewing.

Even if the viewer is not told any information beforehand, if a target pool consists of only one type of thing – such as photos of locations (the most common), a physical location, real objects set on white background or simple pictograms – after a few trials, the viewer will begin to get a sense of this sameness. However, this kind of uniformity of potential targets is fine and, in fact, is how most trials with pools are conducted and are understood by viewers. This is generally not considered frontloading.

Another challenge with blindness is that viewers in a group are never totally devoid of information. If they have no frontloading, they will still make assumptions that may or may not be correct. Just knowing the sort of projects, a project manager tends to work with can serve as frontloading – correct or not. For example, TransDimensional Systems tasked many sessions (Jon was a viewer in these) about the so-called "Big Event," an event which was supposed to occur in the near future (this was in 2001-02) and TDS members would have an important, very vital role to play. The event never happened. A second common target was anti-terror work for the FBI. TDS did scores of sessions on these and viewers began to sense that a terror attack or the "Big Event" might be the target for that day. The targets were "in the air."

Angela Thompson Smith[526] was asked about blinding protocols in a panel discussion that focused on RV and telepathy. She wrote:

> In my opinion, conditions such as "blind tasking" and "incognito viewers" have no place in applications work, where unknown information is sought. Such conditions might be necessary in experimental work and training but not in real-life projects. Doctors do not work "blind," neither do the police: it should not be expected of remote viewers. The goal in application work is not to test the viewer for their accuracy (that should already have been done) but to apply their skills to solving real-time problems. In such cases, telepathic overlay may

be the vital link to solving a problem, i.e. "seeing through the eyes" of a missing child to assess their location, circumstances and condition (p. 10).

Additionally, at any given time a viewer may have a number of internal anxieties that might be alleviated or exacerbated by how much information they are given up front. There is no situation in life we approach without some degree of information about what we are dealing with. Even if we are about to walk into a dark room and don't know what awaits us, we can logically rule out many things that can't be ruled out in most remote viewing situations. For instance, in the real-life example, we know we are entering the room through a door. If we don't know whose house it is (such as what an investigator might experience), we could be in a state of anxiety. This is how it can be for remote viewers and, in order to avoid that, some level of frontloading can help.

During a recent survey, remote viewers were asked[527] when doing applications work as opposed to RV practice or research targets: "Do you feel it is always possible or practical to use blinding protocols?" Out of 72 participants, 36 (43.37%) responded "Yes," 47 (56. 63%) responded "No."

A related question was, "When doing applications work, what level of frontloading do you prefer to work with?" Frontloading was defined as having some foreknowledge of what the target is. Over half of the respondents (45) responded "No frontloading" (53.57%). These responses largely corresponded with those who indicated they always work blind. Thirty-two (38.10%) answered, "Minimal, but I do prefer to know the nature of the project (for example, is it finding a missing person vs. for financial applications)." Four responded (4.76%): "I like to know more specifics (such as for a missing item, what exactly is the item – a ring, a wallet or treasure?) but still limited." Only three responded (3.57%): "I like to know as much as possible about a target and what is known before proceeding with using RV to find out what is not known."

Themes that emerged were respect for blinding, the difference between doing practice targets vs. applications work, the desire to focus on the correct information in an economical way, ways of trying to mitigate frontloading (e.g., if too much frontloading, do multiple sessions), and that being able to work with frontloading is a higher skill.

Regardless of one's preferences or beliefs regarding blinding protocols, feeling the need to discuss it is reflective of the importance remote viewers attach to blinding and their range of views about its definition and implementation. Blinding is important, too, in that it flies in the face of claims by skeptics and debunkers that all psychics operate by "cold reading," which consists of full frontloading, reading facial cues or being given information not only up front, but throughout the session.

Honesty

By honesty, what is meant is an accurate depiction of intentions, methods used and results. While one would expect and hope that honesty would be a core value of any human endeavor or practice, it is particularly important in psychic work. This is because so much doubt and skepticism already surround anything having to do with intuitive experiences, skills and practices.

Over the years, there have been instances of psychics, mediums and healers who were very carefully researched by highly credentialed investigators and found to be credible, and yet on one or more occasion, they embellished their results with trickery. If this happens even once in a thousand times, that is one time too many. It casts a permanent shadow on their other work and affects the entire field. The extrapolation of dishonesty to others may be absolutely unfair, but that is where we are in this day and age.

This should not make people nervous about making honest mistakes. For example, Debra was out taking photos at a "vortex" in Sedona and a striking anomaly appeared in front of the camera lens. It showed up in several photos. She yelled to her friend and both were amazed at the unusual, orange-colored object. They started exploring different angles and positions by snapping photos in different lighting conditions and with another camera to rule out all reasonable and logical explanations. They then realized it was not a UFO, orb, ghostly apparition, act of PK or anything unusual. It was simply…Debra's thumb. There was no attempt to be dishonest, just a good old-fashioned mistake and misinterpretation. When one makes such a mistake, it is important to set the record straight with anyone who might have been led to believe the phenomenon was something other than what it turned out to be.

Presentation of remote viewing sessions

Those who are publishing accounts of remote viewing, whether in books, papers, articles, presentations or websites, often share parts of a remote viewing session – half a page or an element or two (ourselves included here). Such practices may leave out much of the transcript, including part that contained inaccurate data. This is almost always done not to attempt to fool anyone but for simplicity of presentation. Most of the time, a person sharing such excerpts would gladly show the entire transcript if asked and may, in fact, offer the complete session on online. The edited presentation is done for brevity and to demonstrate a good match. It focuses the reader's attention on the single aspect being discussed.

This practice may give the mistaken impression that the transcript had only correct data or that a particular remote viewer gets only fantastic matches. While it's actually very common for viewers to have excellent matches to a target, especially in ARV, often there's a mixture of correct, incorrect and uncertain data. That is why using a CRV scale of 0-7 allows representation of the degree to which the session matches the target – 4s or 5s most of the time, an occasional 7 (which requires the transcript to have detailed descriptions and/or sketches with zero inaccuracies).

In his recent book,[528] *Signal and Noise: Advanced Psychic Training for Remote Viewing, Clairvoyance, and ESP*, Sean McNamara provides a good example of someone sharing full session data. One can get a great feel of all that is involved by looking at page after page of transcripts. Moreover, one sees right away there really is a lot of data to plow through. A reader has to be motivated to carefully read over each sketch and word to compare them with the feedback photo before a sense of the match emerges. It is not those matches aren't there – the sessions display many excellent matches to the photographic targets. It's just that the analyst has to sort through the great matches, the so-so matches and the incorrect or random data. Not all readers may be motivated to work that hard. Compare this with the impression if he had provided only pages featuring the best matches.

In this vein, leading viewer Joe McMoneagle has opined that others, including researchers and publishers, have taken his transcripts and edited them down to the most impressive pages and sketches. He feels this has given an inaccurate depiction of his own work, in that so many sources depict him as if he has 100 percent accuracy all the time, which is just not the case. That being said, we've seen him in action and there is no doubt he can come very close. At one APP conference, the target was a bus moving at a certain angle and he sketched a train moving at the same angle. The two looked quite similar. His sketches are also beautiful.

While more experienced viewers do seem to consistently achieve "site contact," with location-based targets (meaning the session displays enough data to ascertain the viewer has described the correct location), it's important for aspiring viewers to understand they can get wrong data, can misinterpret data, can miss data and sometimes miss the target entirely. We all have good days and bad days, good sessions and bad ones. A realistic depiction of what RV and ARV sessions are like is important for the sake of understanding human perception and human potential. Otherwise, aspiring remote viewers or intuitives are comparing themselves to an ideal that doesn't exist and will always feel let down.

In sum, we don't hold that it's always necessary to publish a full RV transcript, but we do suggest if just a portion is shared, then wording should be included to explain this isn't the full transcript and, if possible, provide links to the full transcript.

If it's not recorded, it doesn't count

Lots of impressions, images and thoughts pop into the minds of remote viewers while doing a session. The instruction given in the dominant methods (CRV and offshoots) is that all the impressions should be recorded (paper or audio) if they are to count. Until it is recorded, while it may be significant to the viewer, one can't really claim it as evidence of one's remote viewing skills.

A quite different viewpoint, which Joe McMoneagle holds, is that you should edit your impressions. In an instant, you will get (or he gets) a large amount of data flooding in, and the trick is to sort out what is relevant to the target and what is not. To the viewer, the impression they got may be all the proof they need, but in this second approach, learning to know which data to report and which data to leave out are key aspects of learning the craft.

These views affect how one treats a session transcript. One approach is to email the session to oneself or the tasker so it has a time/date stamp, which preserves an unmarked transcript. In ARV, many believe you should do a feedback session – that is, go over your transcript and highlight correct data. Some viewers like to mark up their transcripts, indicating correct and incorrect data. If you or the tasker is concerned about preserving the transcript in its original state, then scanning the session will preserve the original and the paper copy can be marked up.

It's recommended that remote viewers keep careful records of their transcripts and progress and stay as organized as possible. Creating a folder on your desktop with subfolders that contain your transcript and the feedback photo with notes about who tasked the project, any frontloading, etc., will come in handy later when you want to share your work or to just look back to examine your progress. If you would like to have solid proof of your work, we suggest emailing your work to yourself and to another person prior to anyone knowing the feedback.

Accurate representation to clients or investors of potential performance, outcome and fees

It is important that remote viewers accurately represent themselves to clients and investors. If a viewer has tracked their accuracy (highly recommended) and have a good record, a summation can be shared with the client to indicate proficiency. At the same time, a caveat should accompany the report noting that due to a variety of personal and objective factors, and the nature of remote viewing itself, "past performance may not be indicative of future results."

If a potential client questions a fee schedule, they should be told they are paying for the viewer's or team's time, their level of experience, training and success. This is on par with how other professionals present themselves. Doctors, attorneys and therapists charge clients for office visits regardless of whether they have uncovered the exact problem or delivered a satisfactory remedy. They can't promise to have a definitive answer or provide a cure or even be accurate in their assessments. Some of their patients may even die, but they still collect their fees.

It is the responsibility of the remote viewer or project manager supervising projects that involve investments to educate clients about the potential for monetary loss. Clients new to ARV should be presented with a balanced view of the history. (Give them a copy of this book!) Make them aware issues can arise that get in the way of results. ARV is in no way a fail-safe method. Honesty means you give an accurate depiction of the strengths and weaknesses of the ARV or RV setup that will be used and what you will deliver. Even a strong track record cannot ensure success. This needs to be communicated, and the client also needs to acknowledge they understand this going in.

Another aspect of being honest with clients is to be clear about fees up front. There should never be hidden fees. If money is involved, be clear about how and when payment will be made. The project manager or solo viewer can also offer a money-back guarantee if the client is not satisfied. This is what one of the very few remote viewers who makes a living at it – Daz Smith – offers.

The project manager or solo viewer may wish to offer more than one session as part of the package and, at the same time, set a limit to the number and timing of any additional sessions. The viewer should agree with the client/investor on what sort of feedback the client will give to the project manager or viewer. However, in the end, the client/investor may not share feedback since they feel it is proprietary. This is not a rare occurrence in client work.

Payment based on percentages of earnings: protection of both parties

In ARV projects, the possibility may arise to perform remote viewing tasks for the client in exchange for a cut of profits. Sometimes the client offers this or the viewer may suggest it. Agreement based on a percentage of profits should be spelled out in writing up front. If agreement is not reached, the viewer must decide if he will view for an agreed-upon sum.

A written agreement should hold the viewer harmless if things don't go well. Whatever agreement is formed, both participants should have the ability to opt out at a specified time, with the allowed reasons also specified. If a viewer becomes uncomfortable with a project and is forced to continue, their subconscious could revolt and the project might go awry. Of course, if the project is not going well, the investor should not have to risk further monies.

Teaching, training and ownership issues

To advance the field of remote viewing, information should be shared in ways that allow everyone to learn and participate, no matter their budget. This doesn't mean those capable of teaching should do so for free any more than a person teaching any other subject would be expected to do so. Most remote viewing instructors offer multiple ways for aspiring students to learn. These range from free resources on their websites – articles, videos, blog posts, inexpensive print and ebooks – to courses with tuition. Prices for remote viewing classes range from minimal amounts (some even free) to thousands of dollars per course.

From time to time, the practice of charging large amounts is questioned. Anyone who has taught a remote viewing course, however, understands the amount of time and work it takes to train others in what Joe McMoneagle[529] referred to as a "mental martial art." Not only is there lecture and discussion, but since practice is the most essential aspect of any RV training program, this requires continued management of every homework assignment. This includes setting up appropriate targets, tasking students, retasking them, analyzing their transcripts, commenting on the transcripts and helping them with any emotional issues (e.g., fear, lack of confidence, etc.). The more an instructor provides individual attention to a student, the fewer students they can manage and, therefore, the more expensive the class will be. Remote viewing classes can also serve students financially since those who excel and can demonstrate professionalism in remote viewing may be able to earn some money doing this, particularly if they are up for the task of managing projects for themselves or others. (However, to date, very few viewers are able to support themselves through their remote viewing work.)

Of note, even the more expensive CRV courses are on par with or less expensive than a single course at a community or state college. Still, it is important that motivated students who cannot afford training should be able to receive it in some form. No one has a right to receive free instruction from whomever they want, but if they are motivated enough, they can find free resources online to keep them busy for a lifetime. To learn more about these resources, visit https://remoteviewing.link (the updated version of Jon's 120+ web page).

Finally, some instructors offer reduced rates, full scholarships, work exchanges or payment plans in exchange for classes, as well as free retaking of the course.

Free sharing of information – for the advancement of humanity

From time to time, someone in the remote viewing community comes up with a method or technique and says others are not free to use it. This is way off base. Here we cite a document provided by the US Copyright Office Circular (https://www.copyright.gov/circs/circ31.pdf), explaining Section§ 102(b) of the Copyright Act of 1976 (1976 Act).[530] It reads:

Copyright law does not protect ideas, methods or systems. Copyright protection is therefore not available for ideas or procedures for doing, making or building things;

scientific or technical methods or discoveries; business operations or procedures; mathematical principles; formulas or algorithms; or any other concept, process or method of operation. Section 102 of the Copyright Act (title 17 of the U.S. Code) clearly expresses this principle: "In no case does copyright protection for an original work of authorship extend to any idea, procedure, process, system, method of operation, concept, principle or discovery, regardless of the form in which it is described, explained, illustrated or embodied in such work.…. Copyright protection extends to a description, explanation or illustration of an idea or system, assuming that the requirements of copyright law are met. Copyright in such a case protects the particular literary or pictorial expression chosen by the author. But it gives the copyright owner no exclusive rights in the idea, method or system involved.

The relevance to remote viewing is that no one owns a method of teaching, a method of scoring or a method of doing RV or ARV. Of course, this doesn't mean anyone is obligated to share their method, but if they do, anyone who hears about it is then free to use it themselves and even pass it along to others in any form they see fit – verbal or written, for profit or not.

For example, in Chapter 25 we provide instructions on how to do a remote viewing session. Anyone is free to take these methods (which are a combination of ones we have learned, developed or modified) and use them as they see fit. They can teach these to others and write about them. What they can't do is copy our words and pass them off as their own (plagiarism). They can cite short passages, put them in quotes and properly credit the authors. They can also reach out to us for permission to reprint longer passages. This is not us simply saying this – it's the law. We also feel strongly if someone learns a technique or approach from a particular instructor, researcher or remote viewer, the latter should be given proper credit. This not only shows respect for others in the field, but helps preserve the history of remote viewing, which, as noted above, is an important aspect of the remote viewing community ("thought collective").

According to law professor M. Samuelson,[531] the reason methods are not protected was articulated in a landmark court case in 1880. In *Baker v. Selden*, 101 U.S. 99, 100 (1880), the court ruled:

[t]he very object of publishing a book on science or the useful arts is to communicate to the world the useful knowledge which it contains. But this object would be frustrated if the knowledge could not be used without incurring the guilt of piracy of the book. The public domain status of this knowledge benefits users as well as subsequent authors. To ensure that authorial and user freedoms would

prevail insofar as the teachings of science and the rules and methods of useful art have their final end in application and use; and this application and use are what the public derive from the publication of a book which teaches them…(p. 13k).

Confidentiality vs. proper credit and respect

Remote viewers largely adopt the ethical guidelines of science and psychology. However, some practices established to protect subjects in scientific experiments serve to marginalize or discount the efforts of participants. One such practice is to decide the participants must remain anonymous and not receive recognition for participating in the study. This no doubt protects participants and is entirely appropriate in some circumstances. But for remote viewers who may contribute hours of their time per trial, over months or even years, it is unfair and disrespectful. Why should those who are managing the project receive credit and the benefits that come with publishing while the participants contributing the content are supposed to simply have the satisfaction of having been helpful in advancing knowledge?

We state this based on the several studies Debra has supervised. When polled, usually just one or two remote viewers express an interest in remaining anonymous. The majority state a strong preference for having their name published as a remote viewer who participated in the project. Those who do well like to have their name associated with their results – those who haven't done well often prefer/request their name be included as a participant but not in relation to their personal data.

We have found it is helpful to poll remote viewers on their preference at the time they sign their participatory agreements and to ask them again when the project is completed.

Following up with remote viewers on results, publications, etc.

While it takes extra time to do follow up with remote viewers, this is the ethical way to conduct projects. Bottom line, remote viewing is hard work. In fact, this is one of the main reasons many people decide it is not for them. It's not that they don't see results; it's just they don't have the patience or interest to spend the time and energy involved. (And it's another reason why ARV is so popular since the customary quick sessions require less stamina than other types of RV projects.) Since viewers have often put a lot into a project, it is important to get them feedback as quickly as possible so they can have closure ("close the feedback loop") and learn from their sessions.

When providing feedback is not possible or is delayed, the next best thing is "debriefing." Debriefing means the viewer gets to learn about the nature of the project, sometimes who the client was, the purpose, what the manager thought of the sessions, etc. Debriefing can take place via email, but viewers really appreciate it if you call so they can have a discussion and ask questions. Doing debriefings in this manner shows respect for viewers.

Additionally, just as it means a lot for a researcher or writer to see their work published, it is just as significant for a remote viewer to see their work was actually used. Keeping them apprised of the effects, benefits, outcomes, uses and especially any presentations or publications that discuss their work – whether soon after completion of the project or way out in the future – is most appreciated.

Debra has found one way to do this easily is to keep an ongoing subscription to a survey program like Survey Monkey. She creates participation agreements and collects viewer biographical data for every study, including contact information, which can easily be accessed at later dates to conduct follow up. Operating in this way improves the likelihood the viewers will want to work with the researcher in the future. It also helps to have a contact list that can be shared with other researchers, provided, of course, the viewers have indicated they would like to be referred for other projects (most state they would).

Meanwhile, Jon has participated in many RV and ARV activities and trials, as a viewer, project manager and/or trainer in TransDimensional Systems (2000-2003), the Aurora Remote Viewing Group (c. 2005-2009), the Applied Precognition Project (2010-2016) and in recent years working solo or with a small group. He has been an ARV group manager and APP membership coordinator), conducting numerous trials there, including CAS trials. None of this work has been research intended for publication, but was a practical effort with detailed records being kept (some of which resulted in articles in *Eight Martinis*). Proceeding in this way, he and many others have contributed to the field and also have something to refer back to.

Self-ownership of remote viewing sessions

The last thing remote viewers probably think about when they embark on a course of training or become involved in an ARV group is ownership of their remote viewing transcripts. Unfortunately, they need to address this. This issue has come up over the years in a variety of ways. When Debra was co-creating the website for the Applied Precognition Project, she wanted to share several of the successful remote viewing transcripts that had been turned in by various individuals to their group managers. Who could give permission – the managers or the viewers? None of this had been previously discussed between the participants.

Additionally, when we (Debra and Jon) studied the materials in Ingo Swann's archives, we got a very strong lesson about the concept of ownership of student and teaching materials. Ingo did several very interesting things as a contract instructor. First, he made it clear in multiple contract renewals with the government that he would maintain proprietary rights over his own materials and methods. Again, copyright law wouldn't actually allow him rights over the methods themselves, but it would over his materials.

Numerous memos between Ingo and SRI staff, as well as with government officials who had oversight of the projects, demonstrate Ingo's ongoing insistence that remote viewing could only excel under appropriate conditions for the viewers. The setup at Ft. Meade – where nothing other than remote viewing took place in the building, where viewers had no other work assignments, were able to take breaks and work at their own pace, and were supervised only by highly trained people – all can be traced to the Ingo's insistent memos that this was the only way remote viewing could work. Few people know this because the memos have not yet been made public.

Ingo had to argue for this approach because of different perspectives among the many people in government agencies who became enthusiastic (or not) about remote viewing. They eyed Ingo's training programs, his own work as a remote viewer and the collective work being done at SRI under the leadership of Hal Puthoff and Russell Targ. Agency supervisors wanted to be able to send their people to training for a few weeks and then have them return to the unit and make use of what they learned. Ingo vehemently opposed this – insisting those individuals would not be able to operate as remote viewers within an arena that did not understand the subject, particularly what was required for optimal functioning. In order to work blind targets, viewers are very dependent on analysts who understand their remote viewing process. Analysts also have to know how to interface with clients, which for the government "Star Gate" project was the lettered intelligence and military agencies (16 of them).

Documents in the archives show Ingo turned down offers for full- or part-time staff positions at SRI. He preferred the title and status of "independent contractor" because it meant he had the right to keep all his materials and the government did not. He did just that – and not just his own materials, but he also kept his students' materials. He kept his students' notes (which he required them to take) and he kept their remote viewing transcripts.

Under ordinary circumstances, keeping notes does not raise eyebrows, but there was nothing ordinary about Ingo. After every lesson, he had students write out the lessons they learned from him that day. Sometimes these consisted of just a single paragraph and at other times, they were several pages long. He would then look over the notes or have the student read them to him, then correct or improve them. Given this approach, he didn't have to produce his own written materials – the students did that for themselves. While we can't say for sure if his students (less than a dozen during the formal contract training period) found a way to make copies, we do know several ended training with their transcripts and all their notes in Ingo's possession and never saw them again. That is, until Ingo's archives were donated to the University of West Georgia.

We came across sessions by trainees like Paul Smith, Tom McNear, Bill Ray and Ed Dames. While Tom apparently later acquired many of his transcripts and notes from Ingo during the course of their friendship, most of the others did not. Paul Smith advised he had not seen them for 30 years and was quite thrilled to know they still existed.

While this practice may seem unfair (and we personally believe it was and do not recommend it), to Ingo's credit, he anticipated his archives would be preserved in a public place after his death. Without his foresight and the wherewithal to keep these materials together and in his possession, with legal backing, it's quite possible they would have been labeled as classified or taken by a government bureaucrat. Instead, they are now accessible to anyone who wishes to study them and can manage to make their way to the UWG library. (The full records are not available online at this time.)

Why should remote viewers retain their rights to their transcripts?

As remote viewers, we know remote viewing is as personal as recounting a dream. Our sessions are like our own signature and our own art. Some artists are trained, while others are born with an innate sense of perspective or color or technique. Regardless, art is always a reflection of one's being, body, emotions, the social and physical environment, the accumulation of who they are, where they have been and who they are becoming.

To have a viewer do a full remote viewing session under the guidance of an instructor, monitor or researcher and then take away ownership of it could be harmful to the viewer's inner being. If the effect was not immediate, in the future the viewer might want to access the transcript to gain more understanding of their remote viewing, which for some could mean understanding their journey in life.

Also, the transcript is important for viewers since it creates a record of their past performance, both in terms of establishing a track record as a viewer and also for later examination (as noted above). This isn't to say a copy can't be made by others and used for specific purposes, with the viewer's permission. In this digital age, the viewer can retain the original and send off the copy, but we highly recommend the viewer insist on keeping the rights to their own transcripts; they have the option of then leasing or lending these rights out per project or specific use.

For example, for this book we have received permission from a dozen viewers to include extracts from their transcripts, while others have written up their own work and sessions. However, we are not free to use either of these in another book, research project, website or class unless we obtain their specific permission.

An exception might be within a class or lecture while demonstrating some facet of remote viewing. Here one gets into the question of whether it is ethical to share their name. As a general rule, the viewer should be asked and their wish abided by. Seeking permission can be time-consuming and is not always easy, but the additional time it takes to work out these details tends to work in the best interests of all parties. We state this, having been on all sides of permission discussions as remote viewers, instructors, public speakers, writers and researchers.

Copyright of photographic and video targets

Ownership of targets that are images or videos is one of the things that newer RV or ARV project managers tend to not think about. This doesn't matter until one wishes to give a presentation of the results of a project or write them up in a publication. If you are going to show how well the viewer's transcript matched a photo, both the transcript and the photo will need to be shown. At that point, it may be quite late if you haven't already obtained permission to reprint the photo or session. Under these circumstances, getting permission can be quite cumbersome. Often it's not possible to find the original creator. Still, if one is in a bind, there are easy-to-use "reverse" photo search apps: the photo is uploaded and a search is done to trace the origin of the photo. These programs often work remarkably well.

Project managers can choose free photos from many photo-sharing websites online. They should also feel free to take their own photos and use them for targets. However, if they do this, their own personal experiences may come into play through the photo, which may or may not be helpful. For example, an independent judge might not understand the references if photos are of personal experiences.

Protection of human rights, safety, well-being of clients and viewers

Even though blind to the target, remote viewers frequently have "aesthetic impact" experiences, which involve not simply perceiving imagery but partial or full-blown sensory and visceral reactions to a target. Each viewer reacts differently. We've seen remote viewers become fearful when describing locations that involved height or other activities they find fearful in real life. We've been assigned targets involving human death, destruction and fear. At times, we've had little reaction and other times, have had intense emotional responses. Viewers may become traumatized by something they experience in a session. Therefore, assigning targets to a remote viewer is a great responsibility and should not be taken lightly.

In ARV projects, there is no need to choose a target that could trigger a highly negative reaction. Project managers, taskers and creators of photo pools need to show self-restraint. For example, we would not assign Auschwitz as a target or something else extremely painful or very frightening. Jon's teacher Pru assigned the Dresden bombing to trainees, explaining that a viewer should be able to view anything and everything. Jon does not recall being traumatized but he agrees with the others who criticized Pru for this target choice.

Of course, in non-ARV projects, the target may be something horrific, such as when viewers are recruited for a missing person case and the subject is dead. For these, you'd want to assign the target only to experienced viewers who understand the types of targets they might be tasked with. Further, viewers of all levels should be told in advance if the target contains potentially upsetting or emotionally triggering information.

Using protocols that keep the viewer blind creates endless complications in RV projects. For instance, you never know how a remote viewer will react. Debra once gave a student a beautiful basket of strawberries as a target. The student, who was at her home across the country and speaking on the phone, observed bright red circular objects in a green container. Everything was going great until Debra invited her to "smell the target." Suddenly the viewer became fearful, said the target was dangerous and became emotionally agitated. Debra immediately stopped the session and revealed the target. The student said she had an extreme allergy to strawberries. Of course, in a case like this, the tasker could not know the viewer would react so strongly to something that seemed benign. The example reflects the importance of staying in communication with viewers and being available to resolve problems. If the viewer experiences fear or trepidation about continuing and you are quite certain there is nothing frightening about the target, you can let them know that. This might be enough to help them to continue. However, if they continue to voice fear, it's best to stop the session and give them feedback as quickly as possible. This kind of experience could lodge in their memory until they have closure through feedback or debriefing.

The Big 3 – Common ethical considerations and dilemma that emerge with psi and ARV

The following discussion centers around the concerns, fears and ideas that often keep people from awareness of and connection to their own intuitive potential. These include making money with psi, privacy concerns and the relationship between precognition and influencing the future.

Some common questions regarding ARV and psi:

- Is it OK to make money with one's psi abilities?
- Aren't our gifts from God and shouldn't they be shared freely?
- Are we cheating in some way if we use our ability for wagering?
- Aren't our abilities spiritual, and isn't wagering bad and unspiritual?

- Is it ethical to wager?
- Is it legal to wager?
- Can someone get addicted to ARV?

Factors that can impact ARV results – Money and ethics

Bottom line, people place all sorts of value judgments on money. Some see money as the mother of invention and some see it as the mother of all evil. Debra's mother liked to say, "It's as easy to love a rich man as it is a poor man," to which Debra responds, "Why didn't you tell me that 30 years ago?!"

Many people have been taught by religious leaders that all intuitive abilities are the work of the devil unless they involve praying to a particular group of spirits, angels or saints sanctified by a church. Others have been taught that intuitive and healing abilities are gifts of the spirit and worry it may be abusing the gift or even committing a sin to charge money to help someone else with one's intuitive gifts.

Religious indoctrination aside, it makes little logical sense that some human skills to help others are gifts of God while others are not. It makes little logical sense why a remote viewer who has trained and practiced hard shouldn't make money from their work when an artist, an engineer, an athlete, musician or anyone else with a skill would never be expected to work for free. This is on par with the idea that those teaching about spiritual, personal or intuitive development are somehow taking advantage of people if they charge fees, while those who teach mathematics or science would never be expected to work for free (and instead receive salaries, health insurance and pensions).

Associative Remote Viewing was originally developed to make money through wagering or trading. It's been hypothesized over the years, based on anecdotal observations, that participants' feelings and attitudes about money could potentially have an impact on a project's success. CRV instructor Lori Williams, working on an ARV project years ago with Marty Rosenblatt, shared how she had an extraordinary hit rate until she was advised wagers were being made on the predictions. Then she had a sudden and ongoing decline (Rosenblatt, 2000).[532] While this decline could have been due to viewer fatigue, she suspected it had to do with her own attitude at the time toward money. As a teenager, she became deeply involved with a group of Christian missionaries who renounced earthly belongings and some of those beliefs still held sway over her at the start of her remote viewing career.

Regardless of whether one believes that acquiring money for oneself is a positive or negative goal, it appears that the higher the money stakes, the more performance anxiety remote viewers experience. In a recent interview with Debra, Stephan Schwartz confirmed his original ARV project lasting close to a year and yielding over $100,000 had to be shut down because of the level of attention and stress over the earnings. Even though proceeds were going toward funding future research, it was "destroying the lab." Russell Targ also attributed a focus on earnings and a change in the protocol by their investor as factors in the reversal of results from nine hits in a row to nine misses in a row.[533] That being said, was it simply the investor's attitude or did it have to do with the investor's demands that more trials be completed in a shorter time frame? Hard to say. What can be said is people's attitudes about money – whether positive or negative – place extra stress and pressure on an already delicate system.

Lack of control by viewers over wagering

Viewers in an ARV project may have no idea what those making wagers are doing behind the scenes. In addition, in some arrangements (made to reduce "feedback loops"), the viewer may not know who the wagerer is nor whether they can trust them. Often any communication about wagering occurs after the trial is completed.

For financial-based activities such as Forex trading – a highly technical process – remote viewers often do not understand how the system works.

Not only are viewers in the dark or perhaps confused, but it's not uncommon for those placing wagers or making trades to make a mistake. Sometimes the trading platform itself has technical glitches. Some viewers may appreciate focusing only on their role, but no one likes to think they are devoting time and effort if someone else's actions may not support the project's success.

On top of this, those placing a wager may not use the remote viewer's prediction or may even bet against it, such as when there are multiple viewers. What happens if multiple viewers' scores for the same trial are added together to come up with an aggregate prediction, and one viewer decides to bet against the group prediction? We have been on both sides of this scenario at conferences in Las Vegas (viewers betting against the consensus prediction). How this impacts the viewing (if there is a retroactive or other group effect) is unknown. Success rates at these conferences average about 60%, which is in the ballpark for groups conducting ARV over significant periods of time.

Further, the viewer has no way of knowing how much money the wagerer is really making. Are they making wagers on the side that others don't know about? Occasionally project managers provide information about the event and give the prediction to viewers, some of whom make use of this and place a wager themselves, while others pay no attention at all.

Unfortunately, out of all the variables that have been studied in RV and ARV trials, the one which has received the least amount of focus is the effect of wagering behaviors. In fact, ARV trial data often gets recorded with little to no mention of whether the viewers received the predictions and wagered themselves. Money, personal finances, and earning and spending behaviors continue to be a taboo topic in some quarters of American society. This seems to carry over into ARV projects. Unless a project establishes clear protocols for how all participants will handle wagers, finances, predictions, etc., these variables could end up fluctuating wildly and be the elephant in the room that goes on a rampage and destroys the entire operation. Because of this, a study of variables related to wagering behaviors is another frontier in ARV that needs to be explored.

Honesty and care in wagering

Viewers are not only intuitive about the target but are often aware when something is going wrong. Viewers might experience feelings of irritation, anxiety or general malaise when they go to do their session. For this reason, we recommend full understanding of wagering activities in a project, unless the viewer expressly asks not to be given this information. If someone changes things or if an error occurs – regardless of whose fault it was – it's important to communicate this to the viewers rather than to try to act like everything is fine when it's not. Keeping things from viewers can have the effect of driving them somewhat crazy. They may start acting out in various ways without understanding why themselves, such as having a sudden wave of displacement. Or they just may start forgetting to turn in sessions. Finally, wagers by all participants should be carefully documented, with notes as to whether they followed predictions or contradicted them. This is to preserve group harmony, which Targ felt was at issue in the two runs of silver futures predictions and which anecdotal evidence suggests is extremely important.

Legality of wagering

As discussed above, ARV trials frequently involve wagering. This can get complex since people in different countries and time zones may be working on projects. Some countries allow online wagering, some do not. Some countries permit their citizens to operate online wagering websites while others do not, but look the other way if citizens participate in sites operated by overseas companies. Then, too, some states and regions allow in-person wagering, while others do not. You cannot play the lottery in Nevada – the clout of the Vegas and Reno casinos prevents that option for residents and the millions of visitors to the state. It's highly recommended that anyone placing wagers familiarize themselves with local and national policies before proceeding.

Also, it's possible for those with a gambling addiction to get hooked on ARV – look out for this for yourself and others. To quote a wise man (Marty Rosenblatt), "If you are going to wager, wager wisely." Though ARV or RV can give you an advantage, making big bets is still brimming with risks.

When a field begins to gain credibility and grow…

Remote viewing and ARV in particular have finally caught fire. Membership in social media groups continues to grow and the re-established RV Reddit has surged to over 33K members at the time of this book's publication. Maybe the numbers don't look tremendous compared to other communities but they're remarkable for remote viewing. Given this rapidly spreading interest, all manner of individuals have come forward to claim what they do is remote viewing. Many well-intentioned people confuse remarkable spontaneous experiences – going out of body, receiving messages from dead relatives, encountering spectral entities – with remote viewing, which is, by definition, planned, done with intent, has feedback and other characteristics.

Other newcomers go further – they tell you what they do is remote viewing, and they will tell you how to do it. They're experts! Such people may offer a course but provide no evidence they are qualified to teach it. Sometimes they rely on the anonymity the internet affords to those trying to latch onto a good thing or make a buck. Even more concerning are those who try to engage people in trading, sports betting, lottery or other financially related activities who claim or even promise lots of money will be made. Relatively few people have demonstrated the ability to be successful using psi in any of these fields, so all such offers and promotions should be treated with extreme care.

Viewers should be careful whom they let task them. Unless a tasker is trained and knows what they are doing, the results are likely to be poor. Further unless you know the tasker, you don't know what purpose or motives the tasker has in mind. This is why some social media sites either don't allow taskings online or have strict guidelines. In addition, viewers and project managers should do "due diligence" about anyone they work with, whether paid or unpaid.

Also of note, especially for the future, are the hackers who plague the internet. Though not common, a few remote viewers' sites have been hacked, sometimes repeatedly. With the upsurge in interest in RV and ARV, more hackers are bound to try to steal data and methods or mess things up for those trying to bring RV and ARV into the mainstream. Some "thought communities" believe RV/ARV is evil. We surmise some governmental and other institutional bodies in the US and around the world may fear remote viewing will succeed and are taking steps to prevent that from happening. One group, TransDimensional Systems, was told in no uncertain terms to shut down – or else. It did. (Jon discusses this in his book *Remote Viewing from the Ground Up*.)[534]

Basic privacy

Newer remote viewing students often have questions about privacy. For example, does a remote viewer have the potential – or the right – to use their abilities to access information about another person? This concern is not paramount in a standard ARV-related project, but it can be with the unitary ARV protocol in which viewers seek to access the emotions of a coach, fan, player, bettor, etc. Most such experiments have tuned in to anonymous individuals. Was their privacy infringed upon?

Again, anything that causes anxiety or concern for one's sense of right or wrong could have an impact on their intuitive perceptions in terms of what they allow themselves to perceive. The following applies to "reading" people, but it has carryover for operational remote viewing when a person or persons is often what a client wants information about.

For 25 years, Debra has been working with clairvoyant students who have a desire to learn how to "read" people. However, when it comes time to do it, what emerges is a very long list of beliefs about what is too private to talk about. Often these include anything one might classify as negative, so even a sense that another person is feeling sad or lonely, angry or defeated is questioned as a perception that might be too personal or intimate. If you classify even simple feelings or states of being as too intimate, there is little hope of accessing much of the human condition or life. It's often useful to understand cultural norms about what people feel comfortable discussing and sharing.

Anyone who is dealing with the access of information through psi – whether through intentional acquisition such as via remote viewing or through unintentional experiences – must ask which is the higher value: keeping in alignment with what mainstream society finds acceptable, avoiding any possible intrusion on someone else's boundaries, or getting to the heart of the matter (whatever the goal is)?

What the higher aim and value is depends on the overall purpose of accessing information about another person. If the purpose is to help them understand themselves and their situation, then holding back the truth of what they are experiencing is not going to serve them. If the purpose is to access information for a missing person, to know if they are alive and safe, then we might want to bypass social norms about respect for privacy. However, people go missing for many reasons. If one's aims are of a selfish nature, to take advantage of someone or access information about them so a person could fulfill their own personal needs (such as a request by a stalker to remote view someone), then we'd all agree this is unethical.

Many other scenarios require the project manager to make a value-based decision (e.g., tracking people involved in a murder case or someone who has bilked a company). Viewers often have to trust their project manager's judgment, but the parameters of the work should be spelled out. In addition to those considerations, there will always be grey areas.

To complicate things further, psi-based perceptions certainly do not only happen within a bubble called a remote viewing session, as we alluded to earlier. As Upton Sinclair discovered and outlines in his book *Mental Radio*,[535] people are constantly receiving intuitive-based perceptions through thoughts, feelings and bodily sensations. In this case, it's not a matter of saying, "I'm going to spy on someone." It might be as quick and natural as wondering how someone is doing or how they might respond in a meeting the next day and then bam – suddenly the information lands in one's mind. Or feeling something in one's body that turns out to be a matching signal, like suddenly feeling a sharp pain in a tooth when wondering if someone will be on time tomorrow and then the following day finding out they just visited the dentist. If psi is really this easy and natural, this means to even think about someone else is a potential invasion of their privacy. Does this mean it's not okay to think about people? Do we have to guard against having a single thought about another person because we didn't first get their permission?

Consider how many people talk about others, analyze them or look them up on social media, all without their permission. Why would this be ethical but applying one's intuitive faculties toward them in an intentional way be unethical?

ARV and privacy

The great majority of ARV work uses photographs as targets. Sometimes people in the photos exhibit emotional states and behaviors. While there would be no reason to get deep into the person's psychology, history or future, observing someone in a target can inform a viewer about what is happening at the location. For example, a person standing at the top of a ski slope might be fearful but have a smile on her face. She may wear a heavy winter coat, snow pants and ski boots, with a tight cap around her long hair that seems to be blowing in the wind. Even without having perceived the environment around her, we can ascertain much from our description of her. We know she is in colder, windier weather. She is scared, not about being hurt by an attacker, but a fun kind of scared. We guess gender because of the longer hair (an assumption that very well could be wrong) and we also know the activity involves equipment she attaches to her feet and hands.

People can be indicators of what is occurring at a location and what the location itself is like. This is why we don't want to avoid people but to "make use of them" when describing physical locations or objects. Some viewers say they tend to miss people in photographs while others say they will almost always see the person first. Those in the latter category tend to be people who are used to applying their psi abilities in other contexts, such as doing intuitive-based readings or providing therapy. Those who say they tend to miss people at locations or in photos are often not ones most used to working with others in a deeply personal way. A psychotherapist would likely perceive the person before even buildings and objects, whereas an engineer might miss the people entirely.

We'll end this discussion regarding privacy with a quote from the recently re-published book *Penetration*.[536] It's one of our favorites:

If telepathy exists, then it would be of such overreaching and extraordinary importance that all Earth side institutions would have to be 'reorganized' in the face of it…it is one human faculty that has a most excellent chance of being summarily shot down before it has a chance to open and wink its all-seeing eye. The most visible explanation for this is that telepathy penetrates MINDS – and so its development is definitely cast into troubled waters where any format or element of mental secrecy might be involved…the funding agencies did sponsor the secret development work in remote viewing – somewhat on the grounds that it penetrates things, not minds. This is to say that remote viewing pertains to penetration of "physicals" not to penetration of "mentals"…This is to say that the principal reason why ALL formats of Psi research are marginalized, treated to energetic diminishment or suppressed altogether is that those formats do include potentials too near the hated and unwanted telepathic faculties (p. 144).

Are we changing the future when we attempt to predict it? Precognition vs. PK-Magic-Intention

Some people have a hard time understanding the difference between predicting the future and influencing it. When Debra and her twin sister were about 12 years old, her sister took a book on witchcraft out of the library. One day Debra's sister had a fight with a friend and decided to put a spell on the friend to make her sick. The next day the friend was very sick and had to go to the hospital. Debra's sister was absolutely horrified, totally convinced she had harmed her friend. As a result, for years she didn't even want to talk about these topics, afraid even thinking about them might lead to harming someone else.

It wasn't until many years later that she realized it was just as likely a case of precognition as it was influencing. In other words, it was very possible information about the friend's upcoming illness and hospitalization entered into her sister's mind on a semiconscious level. We say semiconscious because it didn't enter her mind in a way she could translate it properly as "my friend is about to get sick," but rather it commingled with her tumultuous feelings about her friend, leading to the idea of using the spells in the book to make her ill. In other words, her friend would have become sick regardless of any spells and had this not been the case, Debra's sister might never have had the idea to attempt a spell about illness (instead she would have tried to turn her into a frog). While we can't know for sure whether this was precognition or the effects of "magic," the precognitive theory is strengthened by the idea that she had never before attempted to make someone sick (and never would after).

This confusion over precognition is not unusual. Many people have dreams of the future and are concerned that dreaming about the future somehow means they caused it.

People are programmed by the educational system and science to believe cause and effect in terms of cognition and memory are unidirectional, rather than bidirectional (moving forward and backward). They have a hard time understanding an alternative theory of how consciousness and time works. The causality part might come about in this way: if we can intentionally act so as to create an effect in the future, we may be able to act in the future so as to create an effect in the past. So, we didn't cause the future accident we dream about; the accident in the future caused our thought about it in the past. In terms of remote viewing, we are reaching into the future to help us get information in the present. This viewpoint has many adherents in Associative Remote Viewing circles.

In truth, we just don't know much about remote viewing influences. Some evidence suggests when one looks at something clairvoyantly, they change it. For example, Ingo Swann was challenged at SRI to influence a quark detector/magnetometer buried beneath the floor.[537] Ingo also participated in a series of experiments with Gertrude Schmeidler[538] with the aim of changing the temperature of thermistors. According to Swann's notes, he had no idea how to consciously manipulate the devices when asked to do so. He just tried to peer within to get an impression of what they were like. It was only at the point when he began putting his attention into the internal structures of these complex objects that he achieved the intended effect. For the Schmeidler study, he was actually on lunch break – between trials but still hooked up to sensing devices – taking a peek for his own knowledge when the thermistor suddenly made a noticeable change.

Bottom line, is there anything wrong about trying to influence the future? If there was, we probably wouldn't go to school, have kids or exercise. All these activities are major attempts to influence the future. In fact, most people's lives are oriented minute by minute toward a purpose they hope will serve their future. All of work and most communications are oriented this way. So if that's the case, why should we be worried about what we may be able to do by looking psychically into the future?

We can add this illogical belief to the growing list of paradoxical assumptions:

- It's okay to influence all of life with every aspect of our behavior but not with anything psi-related;
- It's perfectly fine to make money from one's other God-given gifts but not from one's psi-based gifts;
- It's all right to think about another, to even obsess about them, to research them online and troll their social media pages, but it is not all right to take a psychic peek at them, remote view them without their permission; and finally,
- It's all right to reveal a researcher's identity who participated in a project, so the researchers may enjoy more power and prestige in his/her profession, but it's not all right to name a remote viewer who spends hours of their time doing session work.

We can definitely see who is getting the short end of the stick here.

The good news is these assumptions are not eternal truths; they are socially constructed value judgments and, therefore, ones each individual can choose to adopt or choose to swap out with ones that make more sense.

The Buzz of the Firefly & Other Forex/Stock Market Ventures

An Ethnographical Assessment of Project Firefly: A Yearlong Endeavor to Create Wealth by Predicting FOREX Currency Moves with Associative Remote Viewing

Debra Katz, Igor Grgić, T. W. Fendley

The following article was published in the Journal of Scientific Exploration. 32(1):21-54 · March 2018. DOI: 10.31275/2018/1141535

Abstract—More than 60 remote viewers contributed 177 intuitive-based associative remote viewing (ARV) predictions over a 14-month period. These viewers comprised pre-established, self-organized groups cooperating under the rubric of "Project Firefly" (PFF), and were supervised by experienced ARV group managers operating under the umbrella of the Applied Precognition Project (APP), a for-profit organization exploring precognition and leveraging ARV methodology as an investment enhancement tool. Based on predictions from the ARV sessions, PFF used the Kelly wagering strategy to guide trading on the Foreign Exchange (FOREX) currency market. Viewers performed under typical scientific protocols, including double-blind conditions, appropriate randomization, etc., using a variety of ARV application methodologies. Investors, many of whom were also participants (viewers and judges), pooled investment funds totaling $56,300 with the stated goal of "creating wealth aggressively." Rather than meeting that goal, however, most of the funds were lost over the course of the project. Beyond merely reporting on an extensive remote viewing experiment, the present study is an examination of what went wrong, providing lessons learned for further ARV research whether involving for-profit activities or basic research, as the principles are relevant to both. Associative remote viewing is a research paradigm that harkens back to early days in science where competent non-academic researchers can provide datapoints and breakthroughs in a field typically peopled solely by professional researchers. Adapting a form of ethnographic study, we refer not only to the statistical results produced by the PFF effort, but also employ a mixed-methods qualitative approach to exploit the information and insights contributed by numerous participants about what happened, what worked, and what didn't. This creates a reference we believe will be useful for those conducting future applied precognition projects involving multiple participants or groups. We feel that the insights gleaned from this study will both improve ARV experimental design and execution of research protocol, benefitting professional and amateur researchers alike in their future ARV experimentation.

Background

In October 2014, the Applied Precognition Project (APP) began Project Firefly (PFF), a yearlong effort to predict FOREX currency moves with ARV. APP serves as an umbrella for a variety of self-organized groups, which contribute predictions to an overall predictions list. According to the mission statement on its website, the APP's mission is "to publicly explore, research, and apply logic and intuition/emotion to predict future event outcomes, enabling participants to evolve personally while contributing to the elevation of global consciousness."

EXAMPLE 1: Applied Precognition Project. Long-time ARV enthusiast and former nuclear physicist Marty Rosenblatt founded APP in 2013, along with Tom Atwater and Chris Georges (since resigned). Prior to APP's creation, Rosenblatt operated P-I-A. APP serves as an umbrella for a variety of self-organized groups, which contribute predictions to an overall predictions list. APP groups are created by and overseen by volunteers who act as independent managers. They determine their own methodologies, recruit viewers, and choose which events to predict. Since APP's inception, Rosenblatt has overseen operations, kept data, managed active discussion lists, and planned yearly conferences, where he presents the overall group statistics. APP groups have primarily operated and communicated with each other via electronic technologies such as private, individual, or group emails, discussion email lists, and webinars. Some groups, such as the Winning Entanglements (WE) groups, use a web-based software program Rosenblatt designed.

WE members receive target numbers and tasking from their group manager, then can do self-judging and input their own predictions.

TABLE 1
ARV Hit Rate Summary from June 2013 to June 2014 (Prior to PFF)
Hit Rate = 62.4%, P-onetail = .000509, Znormal = 3.3; Odds vs. Chance = 1964-1

Group	Protocol	Hit Rate (%)	Hits	Misses	Passes
WebinarWorkshops	WE	100.0%	4	0	1
CAS-OAK A	CAS	100.0%	4	0	16
Vampires	1ARV	100.0%	1	0	1
PASR	PASR	80.0%	8	2	0
Solo	Binary	71.2%	52	21	30
Sublime	Binary	69.2%	9	4	7
Omega	WE	60.0%	6	4	7
Pegasus	WE	58.3%	7	5	9
WWCdinner	WE	58.3%	7	5	4
Financial	WE	53.8%	7	6	6
Croatorum	CAS	50.0%	1	1	6
Sage	WE	42.9%	3	4	13
First Groove	WE	27.3%	3	8	7
Poised	WE	14.3%	1	6	3
CAS-OAK C	CAS	0.0%	0	2	6
Totals		**62.4%**	**113**	**68**	**116**

Data shown by M. Rosenblatt at June 2014 APP conference in Henderson, Nevada.

Project Firefly used the Kelly wagering method to determine trade size—a probability-based system relying on a mathematical edge tied to past performance, used most often in sports betting with binary outcomes (Kelly 1956). The plan also implemented a majority vote (MV) procedure on every prediction made. For PFF to be successful, the Kelly wagering method required performance significantly above the 50% random rate. According to the "Assets Growth Simulation" APP completed prior to the project, the break-even point was a 55% hit rate. Before PFF began, APP founder Marty Rosenblatt had reported APP hit rates of 62% between June 2013 and June 2014 (Table 1).

Instead of holding steady or rising, however, Firefly's hit rate plunged to 48%. In December 2015, the project halted 14 months after it began with 177 predictions completed (Table 2), of which 152 were executed as trades. Of these 152 trades, only 72 (47.4%) were successful (Table 3). Only $4,114 remained of the $56,300 invested by 62 members.

TABLE 2
Firefly: 177 Daily Aggregate Predictions Oct. 20, 2014, to Dec. 18, 2015

Hits	Misses	Passes	Hit Rate
85	92	72	48.0%

trades. Of these 152 trades, only 72 (47.4%) were successful (Table 3). Only $4,114 remained of the $56,300 invested by 62 members.

TABLE 3
Firefly: 152 Actual Trades Taken on Daily Aggregate Predictions
October 20, 2014, to December 18, 2015

Hit	Miss	Pass	Hit Rate
72	80	97	47.4%

Following, the overall approach the authors used to report on the project and its scope are described. This includes a description of how PFF predictions were made and a discussion of what worked and what went wrong, with an emphasis on adjusting protocols for future projects.

At Firefly's completion, the managers made it clear they did not intend to do a formal writeup of the results, other than reporting to investors, stating it was an investment club and not a formal scientific research project. The authors and many contributors to this paper—all of whom participated in Project Firefly in various roles—felt otherwise.

There is scientific value in examining not just the actual numerical results, but also the lessons learned for the sociology of science in this 14-month project. Although not its expressed purpose, Firefly had all the underpinnings of an exploratory scientific experiment, in which there were repeated, blind trials conducted by experienced project managers, who replicated aspects of prior formal experiments. A project of this magnitude, carried out in a diligent manner on par with other exploratory research-based projects, should not merely disappear into the fog of history.

In search of an effective model, we, as a self-appointed "insiders" team, opted for a mixed-methods, qualitative-based approach, borrowed from the field of anthropology, known as "ethnography"—the study of social interactions, behaviors, and perceptions that occur within groups, organizations, and communities (Reeves, Kuper, & Hodges 2008). Whenever possible, direct quotes and data taken from written interviews, emails, presentations, documents, surveys, promotional materials, datasheets, etc., are provided. All contributors were given the opportunity to review earlier drafts of this paper and to provide input.

Metagroup Method: Project Firefly Begins

Carlos Mena, a Brazilian businessman and long-time remote viewing enthusiast, conceived Project Firefly. Together, he and Rosenblatt invited all APP members to attend an introductory webinar held in August 2014. Mena's PowerPoint slides summarized the proposed project: "Firefly is not a new group, it is a metagroup. That is, a group of groups. . . . It is aimed at creating wealth aggressively." The plan established a majority vote (MV) procedure for every prediction made by the private investment club.

Trading would take place on the Foreign Exchange market (FOREX) via Interactive Brokers, an online broker and trading platform. Although sports betting tended to be more popular within APP than financials, Project Firefly would use FOREX because— unlike sports betting—its legality in the United States is unquestioned. Also, FOREX has no limits on how many trades can be placed or when they can be placed.

Traders would define each Firefly trade prediction as an event with a binary outcome. Based on this, Firefly entities would use an ARV protocol to predict if a particular FOREX currency pair would move either "Up" or "Down" for a specific and predefined number of "pips," based on a predefined trade entry time. A pip is the smallest price move that an exchange rate makes for a given currency pair.

At the heart of the new project was the Kelly wagering method. This method is dependent on previous statistics, as it integrates an already established baseline into a formula to determine the optimal size of the wagers (Kelly 1956). APP had already demonstrated it could achieve a long-term hit rate of 62%, even with some groups performing at chance or even lower.

Encouraged by this high hit rate, Mena proposed an aggressive wagering strategy:

"We will be betting 20% of total assets in each trial in order to maximize our growth rate. If we reach a 60% total hit rate after 240 trials, we should expect $125,527 on our Excel sheet for each $1,000 invested . . . if we man-age to improve on our base hit rate and reach 65%, we may expect around $16,000,000 on our Excel sheet for each $1,000 invested after 240 trials."

The slides that followed included a disclaimer that "of course, the project could fail."

Under the proposed plan, all APP groups and all remote viewers were considered as equal contributors. Since no one was excluded, the project had plenty of viewers and groups providing predictions. While it would require considerable coordination and communication between group managers and Firefly Traders, the groups all maintained independence to set their own procedures related to photo selection, judging, rating, participants, and issuing of predictions (Appendix A and Appendix B describe the methodologies used).

To achieve the proposed 240 trades, each group had to contribute only one session a week. The Firefly trading team assigned each group manager a weekly event and date with a specified deadline for returning the prediction, which would then be entered into a shared predictions spreadsheet. Prior to the initiation of Project Firefly, many of APP's group managers were already submitting predictions to a shared "predictions list" that all paid APP members in good standing could make use of however they wanted. Now the difference would be that the Firefly Traders would use the predictions to place trades with money from investors. Each investor was required to participate in at least one group as a remote viewer.

Planning and Implementation

APP members and their personal contacts signed up as investors for Phase One between early August 2014 when the plan was introduced and early October 2014. Potential investors were counseled to only contribute monies they could afford to lose. The minimum investment amount was $100. Shares were based on $100 increments (e.g., a $100 investment was one unit of the total, for purposes of profit disbursement). Participants could not withdraw funds after the main phase began until the yearlong project was complete. Table 4 describes the number of investors and monies collected for each phase of the project.

APP co-founder Chris Georges set up the project as a legal financial business entity, according to U.S. tax law, and controls were established to ensure that no single person had access to the funds. Those placing trades via the FOREX system had authority to move money around within the system, but could not make withdrawals. As an additional safeguard, two Traders were to be involved in making every trade.

Only a few APP members understood how to place online trades in FOREX. Those who had the skill and time to devote to the project as unpaid volunteers—Mena, Rosenblatt, and another APP group manager,

TABLE 4
Financial Summary from Firefly Administrative Officer Chris Georges

	Phase 1	Phase 2	Total	Largest	Smallest	Average
Members	54	62	62			
Collected	$43,200	$18,000*	$61,200			
Invested	$38,500	$17,800	$56,300			
Retained			$4,114			
Investment amounts				$10,000	$100	$987

* Includes funds from 8 new investors and additional funds from Phase One members

Igor Grgić—comprised the Firefly trading team. Jon Knowles, a less-experienced Trader, stood in for Rosenblatt when he went on vacation at the start of Phase Two. Knowles also served as a consultant for the trading team.

Some Firefly members expressed concern about the proposed management structure, citing the need for an independent Oversight Committee that excluded members of the trading team. Also, no procedures were in place in the event of early losses. Not all APP members felt it was prudent to use under-performing viewers and groups, but that also remained an integral part of Firefly's initial design.

The Firefly Investors Manual was emailed to the APP Discussion Group on October 7, 2014, two weeks prior to the start of Phase One and after most of the investors had made their financial-contributions. The manual made no mention of what would happen if early losses occurred. It listed Oversight Committee members as Georges and trading team members Mena, Rosenblatt [Committee chairperson], and Knowles.

The manual gave the Oversight Committee power to adjust protocols as needed:

"At any point in time, Firefly may make adjustments for accepting predictions in order to strengthen our predictive capabilities. If made at all, these adjustments will be based on data gathered as the project advances and will be made by the Committee."

Per the manual, Traders were responsible for acting on each prediction, executing the trade in the market of choice, and following rules detailed internally for accepting the trade. An online document titled Firefly Tasking and Predictions tracked each trade decision. Traders were notified by SMS message via the Interactive Brokers platform for each executed trade (no matter who executed it). Before each Run, if existing rules or protocol changed, then the new rules were implemented.

Methodology

Overview of the ARV Process

As noted earlier, Firefly Traders executed FOREX trades based on predictions of whether the price would go up or down. The Trader for each trial would assign the event to one or more group managers who had previously indicated their group's availability to submit a prediction.

Each group manager handled all other aspects of the trial, which started with compiling a set of photos, one of which was designated for the "Up" outcome and the other for the "Down" outcome. The group manager assigned a target reference number (TRN), which represented the photo associated with the future winning outcome. The manager emailed the TRN to the group's remote viewer(s), along with "tasking" instructions. The tasking invited the remote viewers to use their intuitive abilities to tune into the feedback photo designated for the winning outcome, which they would receive after the trade was completed. During the remote viewing session, the viewer(s) recorded all intuitive impressions via words and sketches onto blank paper; afterward, they emailed this "transcript" to their manager.

Next came analysis and judging. Each group determined whether to use independent or self-judging, as well as what judging methodology to use. Some groups used a 7-point scale, some a 3-point scale, and others simply chose the best match. In each case, the remote viewer's transcript was compared to the two photos. Ideally, the transcript(s) would be a strong match for only one photo and a weak match for the other. If the transcript had no matches or weak ones, or if it matched both photos equally well, this indicated a breakdown in the process and the judge would call a "pass."

The group manager submitted the prediction to the Firefly Trader, who would use it to execute the trade. The Trader would trade in accordance with the group's prediction. When more than one group submitted for the same prediction, the Trader would apply the majority vote rule to come up with an aggregate prediction.

After completing the trade, the Trader communicated the outcome to the group manager(s) in a timely manner so he/she could provide the feedback photo associated with the actual outcome to the remote viewer(s). Most groups reported they received feedback within 48 to 72 hours. Remote viewers were encouraged to complete a "feedback session" by closely comparing their transcripts to the feedback photo to determine what matched. This completed what is referred to as a "feedback loop."

The trial's outcome would then be recorded in a shared spreadsheet maintained by the Firefly Traders.

Firefly Group Practices and Characteristics

To better understand specific methodologies used by the groups and characteristics of the group members, about a year after the project concluded the authors submitted an online survey to all Firefly group managers. Seven of the 8 group managers responded to the survey presented in Appendix B.

The 12 ARV groups that contributed sessions to Project Firefly had highly trained project managers with exposure to and training from ARV and RV founders. They were well-versed in the technical aspects, such as ensuring blind conditions, methodologies for judging, scoring, and making predictions, and target-pool creation. Prior to Project Firefly, they had worked hard for years to improve ARV statistics and learn from past performance. Collectively, they donated thousands of hours to this field.

Given ARV's goal of predicting an unknown future outcome, it would be impossible for viewers and group managers to be anything other than blind to the outcome itself. Based on knowledge of the group protocols (and self-reporting by all but one manager), the authors have high confidence that all remote viewers in Project Fire y were also blind to both target photo options prior to the judging phases. Some group managers were blind to both photo options, having used randomization procedures, while others were aware of the photo options, having personally chosen them without randomization. Following submissions of transcripts, some groups used self-judging (meaning the remote viewers would need to see both photos in order to determine which photo was the best match to their transcript) while others used independent judging (meaning the manager or a third party would judge the transcripts instead of the remote viewers. This would prevent them from seeing the unactualized photo).

Three of the Firefly groups had only one member, while the others averaged 7 members each. More viewers were in groups that used self-judging rather than independent judging, including 6 groups using the online Winning Entanglements (WE) computer system. Three groups also used CAS (Computer Assisted Scoring software), a system created by Ed May based on Fuzzy Set Theory. One used ARV Creator (scripted Excel spreadsheet) and one used ARV Studio software. While Binary ARV was the standard protocol, the target pools varied between groups, ranging from simple objects only, to include locations, activities, and lifeforms (see Appendix A).

Some groups (i.e. P7B and WE groups) included newer and experienced remote viewers, while others (i.e. Sublime, Sharp, Evans) had only experienced remote viewers. Viewers were trained in a variety of methodologies, including ARV, Controlled Remote Viewing (Smith 1985), Extended Remote Viewing, simple clairvoyance, and dowsing. Most reported using modified versions of these.

Further responses to the post-project survey are presented in Appendix B, which contains specific information related to judging, predictions, randomization, communications, and viewer experience level.

Results, Protocol Adjustments, and Wrap-Up

Phase One

Following a rigorous trading schedule, the PFF Traders wagered 20% of the investment in each of the 33 trades between October 20, 2014, and December 19, 2014. Funds were relatively stable and fluctuated around the initial investment figure until they dwindled in the last two weeks. The losses resulted from 3 misses and unrealized winnings of $4,000 on one prediction when a Trader was not able to enter the trade at the designated time. Phase One began with 54 investors and $43,200 collected. Of the $38,500 invested, $21,014 remained at the end of Phase One, which had an overall 54% hit rate, as shown in Table 5. The solo groups (those with only one viewer) had a 59% hit rate.

Investors could cash out at the end of Phase One or contribute more money, and managers could revise their plans, if necessary. Eight new investors joined Firefly for Phase Two and 7 added more funds, bringing the total funds available to $38,723.

Phase Two

After the Phase One losses, the Firefly trading team decided Phase Two would be organized into a series of short "runs" so adjustments to the protocol could be made, as needed. Chart 1 reflects the fluctuation of funds after each trade throughout the entire project. It also indicates the account balance after completing each run and outlines the different approaches taken and their results. At first, the Traders wagered 20% of the total Firefly account balance on each trade (full Kelly), but as the balance depleted they lowered the amount to 16% per trade and later to 10% per trade (half Kelly).

Run 1 began on January 26, 2015 (Week 11). Daily trades were based on a majority vote (MV) procedure using predictions from aggregate groups.

TABLE 5
Firefly Phase One, Run 1 – Hit Rate 54%

Firefly: Phase 1 Run 1 schedule: October 20, 2014, to December 19, 2014

Week 1-9	Monday	Tuesday	Wednesday	Thursday	Friday
Group 1	FIRST GROOVE	P7B	FINANCIAL	SAGE	PSICHISENSI SOLO
Group 2	MWHITE SOLO	CROATORUM	SUBLIME	PEGASUS	JFK SOLO
Group 3		OMEGA	TRANSCENDENT		SHARP SOLO
Week 9	Miss	Pass	Pass	Miss	Hit
Week 8	Pass	Miss	Hit	Hit	Miss
Week 7	Hit	Miss	Hit	Hit	Pass
Week 6	Hit	Miss	Hit	Pass	Miss
Week 5	Miss	Hit	Pass	Hit	Pass
Week 4	Hit	Hit	Hit	Miss	Pass
Week 3	Miss	Hit	Hit	Miss	Miss
Week 2	Pass	Pass	Miss	Pass	Pass
Week 1	Hit	Miss	Miss	Hit	Hit

Predictions based on majority vote—several entities/groups per day

Around this time, the Traders debated whether to tell the membership at large of the losses or even to disband the project. Each member of the trading team later indicated they were under a huge amount of stress as the money continued to dwindle and misses continued. During the last two weeks of Phase Two, Run 1, they made only simulated trades. Run 1 ended after 38 trades with a 36% hit rate.

In Run 1, a new precognitive tool that had shown a 64.7% hit rate in 25 trials prior to December 21, 2014, was added as a "group." Instead of remote viewing, the "Survey" method relied on a participant's instant response to a nonsensical pair of words, which was then associated with a particular undisclosed outcome. Mena sent the Survey weekly to all APP members until February, when he moved back to Brazil from Spain.

At that time, Mena told Rosenblatt he could no longer keep up with the day-to-day trading overview obligations because of the move and needed to find another setup. According to Mena, Rosenblatt suggested he could step down as Firefly General Manager, leaving Rosenblatt and Grgić in charge. Mena felt it was within Rosenblatt's right as APP founder to make such a request and therefore complied. When asked, Rosenblatt said he remembered it differently, as being a joint decision.

Mena announced the change at the next meeting, before Run 2 began. Some members who weren't present said they were unaware of the changes in the management structure until Firefly ended in the fall. According to Mena, he remained on the Oversight Committee throughout the project.

Run 2 began on March 30, 2015 (Week 19), with a new approach that relied on predictions by the four best viewers, who had hit rates of 70% to 75%. Trades on Mondays, Tuesdays, and Thursdays were based on predictions by a single entity made of 2 viewers selected from the best 4. Traders placed simulated trades based on aggregate predictions from the other groups on Wednesdays and Fridays. Run 2 ended after 12 trades with a 50% hit rate. Including the 13 simulated trades, the hit rate was 52%.

Run 3 began May 25, 2015 (Week 27), with a return to trading each weekday using the prediction provided by each group's manager. Trading was aborted in Week 30 due to 5 misses in a row. By the end of Run 3, the accuracy of the 4 best viewers had dropped to between 50% and 54%. Run 3 ended after only 8 trades with a 25% hit rate.

Run 4 began July 6, 2015 (Week 31), with one of 5 groups/entities—not the best individual viewers any more—providing predictions and with trading each weekday. This run showed the only increase in the hit rate, ending after 25 trades with a 60% hit rate. Previously, trades had preselected entry times and preselected currency pairs. During Run 4, however, neither a trade entry time nor preselected currency pair was used. Instead, when the Trader got the group manager's prediction, he searched FOREX charts of different currency pairs for the best trade opportunity. For instance, if the group manager submitted an "Up" prediction, the Trader searched (with intention) for the best "Up" move opportunity for a 1:1 risk-to-reward trade.

Run 5 began August 31, 2015 (Week 39), with one of 5 groups/entities providing predictions and with trading each weekday. Starting in Week 45, Traders used predictions from APPI entities (solo viewers with high hit rates). Run 5 ended after 48 trades with a 48% hit rate.

Wrapping Up Firefly

Once the end date arrived, Chris Georges hosted a webinar with Firefly investors. While some questioned what went wrong and suggested improvements for future projects, many expressed pride at having engaged in such a grand experiment.

During a January 2016 webinar, Grgić gave a breakdown of the phases with their various protocols, stats for all Firefly groups, and an explanation of decisions made.

In a subsequent presentation entitled "Proposal for Phase 3," Grgić suggested keeping any future endeavor simple, to eliminate complexity, focus on individual calls, and use groups of 2 to achieve the best psi efficiency. To help eliminate complexity, he suggested operating Firefly with only one tasker (for financials/FOREX) and Trader. If needed, the Trader could report to an oversight committee.

"I think that a team of two or three Firefly General Managers/Traders is not good for functioning of psi and psi efficiency," Grgić said. He suggested either using viewers from existing groups/solos with hit rates of 60% or greater, or creating several new entities/groups comprising two top viewers. To keep things simple, only one group would be active at a time. Runs would be short, with breaks between runs. A side would be called only if both viewers agreed; if one passed or if their predictions canceled each other out, the prediction would be a pass.

Rick D. was one Firefly's highest-contributing investors. Despite his losses, he continued to be enthusiastic, with an attitude of "let's understand what happened so we can make use of that knowledge and perhaps move on to Phase Three or a new large-group endeavor." He also performed some independent inquiries of the trading team, which left him satisfied that everyone had dealt with the monies and wagering in an ethical manner. While a few others also expressed interest in continuing on to Phase Three or a new project, no one volunteered to manage it, all citing a lack of time. In early January 2016, Georges mailed investors their remaining funds, along with a final report and tax documents, and Project Firefly closed as an official entity.

Discussion

What Went Wrong? This was an extremely complex project involving multiple groups of individuals producing predictions. These were aggregated to form meta-predictions, which were then wagered upon according to the Kelly wagering method, and finally input into a financial apparatus (FOREX). Ultimately, that complexity, more than any other factor, may be at the root of the problem.

As we will demonstrate below, decisions to initiate Project Firefly, as well as those involved in how to apply the Kelly wagering strategy, were based on preliminary performance statistics that were too "large-grained." The outcomes from earlier projects had been aggregated into a single statistic (the 62% hit rate), but those results included variables and individual group outcomes that were either unknown or unanalyzed prior to Firefly's initiation. The post-Firefly analysis of the earlier Zulutrade project is one such example.

Other factors examined below include the effect of Majority Vote, displacement within single groups and aggregate group predictions, the number of trials, and the judging method used.

Kelly wagering strategy. Project Firefly was based on the premise that the past is a strong predictor of the future. Mena initiated the project after he became aware that APP groups were achieving hit rates above 60%. In the field of parapsychology, success rates in precognitive-based trials tend to be around 53% (Bem 2011).

When invited to submit comments for this paper, Mena provided the following statement:

"The Kelly wagering system was simply chosen as the mathematical frame-work to optimize our betting strategy. It is not a controversial method; it is the optimal strategy. 'Aggressive creation of wealth' would be the natural consequence of using an optimal approach, as long as the groups were able to perform around the 60% level indicated by historical data."

Alexis Poquiz, an active APP member and Firefly investor, who posted the following to the Firefly Investment Club (FIC) Google page, echoed his sentiments:

"To blame our failure…to the adoption of the Kelly wagering strategy would be a mistake.…The bottom line is that our project was a disaster because we failed spectacularly to achieve our expected hit rate. Going forward, I would make two adjustments. The first adjustment would be to use a Kelly factor that is based on a lower hit rate than 60 percent. The second adjustment would be to change how the project ends. Originally we ended the project based on a set date. Instead of a set date, I would end the project based on a set number of wagerable predictions. This will alleviate the tension of having to produce a prediction week in and week out. I whole-heartedly believe that we can achieve success using the Kelly wagering strategy."

The chart Rosenblatt had shared at conferences and online showed the 62% hit rate was an aggregate of group statistics. Some groups predicted sporting events (i.e. the over/under scores of football, basketball, or baseball games), and others made financial predictions using the stock market or Zulutrade (FOREX).

Among APP groups that predicted sporting events, the methodologies and results varied widely. A closer look at the top-ranked APP groups showed one used a mixture of logic and remote viewing with self-judging, and another group viewed "live." Its members included some of the top viewers. Other high-performing groups based their predictions on dreams or tuning-in to emotions.

Although it wasn't known prior to Firefly, many groups making financial-related predictions were operating much closer to chance levels and, in some cases, below chance. This raises the question of whether measuring only groups mostly involved in financial predictions might have been a more accurate predictor of future performance than including higher-performing groups, many involved in other types of events.

Analysis of Zulutrade project. One way to assess ARV groups' future predictive behavior is to look at the most recent statistically significant historical data. Such data was collected by APP during the Zulutrade project, which lasted from April 28, 2014, to October 17, 2014.

Zulutrade is an online platform where one can execute FOREX trades without risk in a demo account and perform as a "FOREX signal provider." Other investors can follow these trades.

TABLE 6
Zulutrade Project – APP FOREX Groups April 28, 2014, to October 17, 2014

Group Name	Hit	Miss	Pass	ARV Protocol
Croatorum	2	5	16	Standard binary ARV
Financial	2	1	2	W.E.
FirstGroove	8	6	3	W.E.
LaurSolo	3	1	0	Standard binary ARV
Omega	2	5	10	W.E.
P7B	3	1	1	Standard binary ARV
Pegasus	7	7	4	W.E.
Sage	3	4	4	W.E.
Sublime	1	0	0	Standard binary ARV
TOTALS	**31**	**30**	**40**	

Results through 101 total Zulutrade trials: 51% Hit Rate

After Firefly ended, Grgić and APP member Mark Samuelson completed an assessment of APP data from that prior six-month project, which shared some similarities with Firefly. According to Grgić, 7 APP groups that participated in the Zulutrade project switched to Firefly, maintaining essentially the same structure in both projects (e.g., the remote viewers involved, protocol used, etc.). A technically identical ARV tasking was used to predict FOREX currency pair moves, and the trading team executed one trade per day / 5 trades per week. ARV groups were scheduled per trade day. Both projects had defined goals. In the earlier project, the goal—which wasn't achieved—was to rank among the top-performing Zulutraders; the project's 51% hit rate reflected 31 hits and 30 misses.

The data shows, to put it simply, that the Zulutrade ARV groups did not produce a combined hit rate above 60%, as needed for Firefly success (Table 6).

The majority of the Zulutrade groups used what is referred to as the Winning Entanglements (WE) protocol. These group statistics were easier to access than those for groups using other protocols because WE automatically collects the data, which viewers input into the online system. Predictions and outcomes are published to the APP "predictions email list" that full members can access, which allowed for easy assessment.

Most WE viewers did self-judging and didn't have to wait for independent judges to assess their sessions. That allowed more viewers to participate, and WE managers tasked more sessions (68 WE Zulutrade sessions vs. 33 by other groups). Additionally, APP often placed new viewers into WE groups, so more inexperienced viewers may have been in these groups.

TABLE 7
Comparison of WE Firefly Groups and Other Groups/Entities

	Winning Entanglements	Other
Hits	66	60
Misses	66	60
Passes	68	75
Total predictions	**200**	**195**

An assessment of Project Firefly's data showed many of these same WE groups went on to contribute slightly more predictions than other groups (Table 7) despite their lower hit rates during the earlier, pre-Firefly Zulutrade trials. Table 8 lists all the groups and protocols used in Project Firefly, with their hit rates.

Consequently, the commonly cited 62% pre-Firefly hit rate, while deemed an accurate statistic by Grgić and Samuelson, was not well enough defined nor understood to serve as a predictor of success, as mandated by the Kelly wagering method. Based on this analysis, a more conservative approach than investing 20% of all monies should have been applied at the start of the project.

Majority Vote: Single Group vs. Multiple Groups

Project Firefly had an aggressive wagering schedule driven by 5 predictions a week. At first, it was thought having input from 2 or more groups might lend strength to a prediction. That made it desirable to have more than one group of viewers contribute predictions each day so Traders could get trading direction based on majority vote (MV).

Mena told the authors:

"Project Firefly provides an important insight into the effect and inner workings of Majority Vote procedures applied to psi. Redundancy methods in general, and MV procedures in particular, are techniques designed to improve the reliability of psi to a level suitable for practical application. Redundancy provides the basis for the methods of increasing the accuracy of signals in normal communications systems, and many techniques pro-posed to enhance the reliability of psi follow this same path. The 'signal-enhancement' hypothesis holds that if a low-level psi effect occurs on the individual predictions, then Majority Vote procedures will be expected to increase the accuracy of psi to a high level. This did not happen in Firefly. In fact, the only run that reached a 60% hit rate was Phase Two, Run 4. Grgić partially attributed its success to having a prediction from only one entity per day. He also cited a new-to-APP trading protocol used only during Run 4 (described previously).

TABLE 8
Firefly Hit Rates by Group for October 20, 2014, to December 18, 2015

Group Name	Hit	Miss	Pass	Protocol	Group Type	Judging Type	Hit Rate
Mark S	2	0	0	Binary*	Solo	Indep.	100.0%
SuperSolos	1	0	0	Binary*	Group	Indep. & Self	100.0%
SHARP	9	4	3	Binary*/ ARV Creator	Solo	Self	69.2%
Sublime	9	4	12	CAS, live binary ARV	Group	Indep.	69.2%
Mwsolo	8	5	5	Binary*/ ARV Creator	Solo	Self	61.5%
FirstGroove	22	15	13	W.E.**	Group	Self	59.5%
PSI-SOLO	9	8	8	Other binary ARV	Solo	Self	52.9%
Financial	19	17	11	W.E.**	Group	Self	52.8%
P7B	9	9	15	Binary*/ ARV Studio	Group	Indep.	50.0%
APPI/other	5	6	1	Various	Group	Indep. & Self	45.5%
Sage	5	6	14	W.E.**	Group	Self	45.5%
Pegasus	12	15	21	W.E.**	Group	Self	44.4%
SURVEY	3	4	22	Survey	Group	Survey	42.9%
Omega	8	12	9	W.E.**	Group	Self	40.0%
JFK	4	10	3	Binary*	Solo	Indep.	28.6%
Transcendent	1	6	6	CAS (modified)	Group	Indep.	14.3%
Alpha Omega	0	2	0	Binary*	Group	Indep.	0.0%
Live	0	1	0	Binary*	Group	Indep.	0.0%
SuperBinary	0	1	0	Binary*	Group	Indep.	0.0%
SuperWE	0	1	0	W.E.**	Group	Self	0.0%
TOTAL:	126	126	143				50.0%

* Standard binary ARV
** Winning Entanglements

Displacement Affects Single Group and Aggregate Group Predictions

Another factor affecting Firefly's results was displacement, a common and troubling phenomenon where remote viewers accurately describe something other than the intended target. It occurs in ARV and other experimental parapsychology projects that use sets of photos as a judging method. Dr. Patrizio Tressoldi, a parapsychologist who has conducted extensive meta-analysis in areas such as the Ganzfeld body of research, advised in email correspondence with the authors that displacement is one of the most perplexing issues he and other researchers continue to witness. At face value, it makes it appear statistically that psi was not present, when in actuality psi may have been operating in full force but toward the wrong subject matter.

This happened 6 times between October 2014 and July 2015. In these 6 instances, all groups submitting predictions on a specific Firefly trade day were in agreement (no passes), but they predicted the unactualized side. After July 2015, the trading team abandoned the approach of having more than one group make a prediction for the same trade. Afterward, predictions from only one Firefly entity (group or solo) per trade day were used.

Additionally, other examples address possible displacement within a single group. Thirty-nine instances of strong consensus predictions occurred at the group level, resulting in a 48% hit rate. Strong consensus occurred when there was a 3-point spread difference or advantage for one side, such as 3 sessions predicting one side and 0 sessions for other side.

Number of trials. Jon Knowles, who served as an "Apprentice Trader" from October 2014 through March 2015, posted to the Firefly Investment Club Google page:

"The mandate to have 240 or so trades in the course of 15 months placed a heavy burden on the project in a variety of ways. Making so many trades means lots of taskings each week, lots of sessions, and lots of analysis."

In support of Knowles' observation, studies have shown that fewer trials seem to be more effective than too many close together. In 1984, Russell Targ and Keith Harary completed two ARV studies (Harary & Targ, 1985). The first, featured in The Wall Street Journal, yielded $120,000. On a second, unsuccessful attempt, they shortened the intervals between trials and viewers sometimes started a new trial before receiving feedback on an earlier one (Targ, 2012). In 1995, Targ repeated the study with the earlier protocol's less-frequent trials and results were highly significant (Targ, Kantra, Brown, & Wiegand 1995).

These researchers suggested too many trials in a short period of time may lead to both viewer and manager fatigue.

Judging

Outside of Project Firefly, fluctuations in judging have been observed in independent tests performed by Grgić, as well as those conducted by Poquiz, creator of the Dung Beetle Method of scoring (Poquiz 2013). While these exploratory trials did not include large sample sizes, their results demonstrated the need for further evaluation of differences in judging styles and predictive decision-making. Various factors can lead to misjudging: judging style and experience, taking into account AOLs (analytical overlays), or relying on late-session data. (Some argue that first impressions or the first gestalts are usually correct.) Accurate judging can also be impaired or derailed when photo targets are too similar to each other or when they differ in entropy or numinosity (May, Spottiswoode, & Faith 2000).

Grgić found instances where scores for both photo targets (whether actualized or unactualized) were high (each above 3.5 on the 7-point SRI/Targ scale) and when scores for both sides were too close, with less than two points of separation between them. Despite that, sometimes a judge made a call for one side when he should have passed because of a mixed signal, as evidenced by data in transcripts matching both sides.

Within Project Firefly, no quality control measures ensured the accuracy of group managers' judging or predictions. The Traders did not generally question the group managers' predictions, particularly in earlier runs when most of the losses were sustained.

Self-judging. In ARV projects where viewers are tasked with describing the feedback photo they will see after the outcome of the event is known, self-judging is controversial because it also exposes viewers to the unactualized photo. Over the years, on many remote viewing email lists and online forums, numerous APP members and others involved in ARV have repeatedly commented that self-judging derailed their sessions. However, Rosenblatt suggested this belief only serves as a self-fulfilling prophecy for some viewers, citing instances where viewers were able to overcome displacement with practice and self-discipline.

With so many other variables to consider, the effect of self-judging on the outcome of Project Firefly, if any, cannot be determined. As noted earlier, most, but not all, Winning Entanglements groups used self-judging.

WE groups use an online system Rosenblatt developed that automates the ARV process. Viewers see their coordinates in the system, upload their transcripts, and most self-judge them against the photo sets. The overall hit rates for those groups ranged from First Groove's 59.5% to Omega's 40%. At one point, a self-judging solo viewer had 9 hits in a row with only one pass. Non-WE groups that used independent judging had hit rates ranging from Sublime's 69.2% to Transcendent's 14.3%, as shown in Table 8.

Conclusions and Future Study

In summary, the consensus among this paper's authors, supported by the extensive contributions made by other Firefly key participants, are as follows:

First, predictions based on aggregate groups on a single trade day did not fare as well as single entities (groups or solos). Instead, the data generally support using the best viewers and teams, as per their hit rates listed in Table 8, and keeping the protocol simple. An exception to this was seen in Phase Two, Runs 2 and 3, when the top solo viewers' hit rates dropped from around 70% to roughly 50%. That data was not statistically significant, however, because no solo viewer did more than 11 non-passing predictions during those runs.

Second, the goal of having 240 trades in a single year placed a great deal of stress on the trading team. Of 249 predictions, 72 were passes. This may be an example of too many predictions in too short a time span, as seen in the Targ/Harary study (Targ 2012).

Third, an independent Oversight Committee could provide valuable support for the trading team by serving as a check and balance on trading activity, monitoring protocol, and implementing a process to make changes with greater transparency for the viewer/ investors. This could be critical if an aggressive wagering method is being used and early losses are incurred.

Fourth, the Kelly wagering method should only be used after verifying the hit rate for the specific viewers and a specific protocol. In this instance, subsequent examination of the pre-Firefly data showed many of the entities used in Firefly had hit rates below chance for similar financial predictions. In such cases, a more conservative approach than investing 20% of all monies should be applied. Further study on the hit rates of different protocols is needed.

Post-Firefly

Since the conclusion of Project Firefly, APP has continued to gain members and flourish. At APP's annual conference in June 2016, Rosenblatt included Firefly's hit rate in the charts shown, but he focused on APP's successes.

He often repeated two of his favorite sayings: "Wager wisely, if you wager," and "Get rich slowly." He also wrote, "What seems most important is to use what we believe we have learned to improve our personal ARV/RV skills and group applications."

When asked about Grgić's and Samuelson's study of pre-Firefly data, which showed the financial groups' overall hit rate was only 51%, Rosenblatt indicated he had never assessed the data in that way before. In a February 3, 2016, email response to the first draft of this paper, he stated: "I believe the FF [Firefly] low hit rate is due to internal money issues, plus the intensity/stress unwittingly placed on the project at the beginning." Mena said he believes other factors were at play:

"I disagree with any hypothesis that states that unconscious money issues related to this aggressive wealth approach are behind the group's inconsistent results. Historically, inconsistent psi effects were attributed to unconscious processes. It is time this meme is recognized and discarded as use-less. This approach has provided little explanatory or predictive value after 70 years of discussion and research. More specific hypotheses are needed."

In a February 5, 2015, post to the Firefly Investment Club Google list, Georges said:

"[The] project was not a financial success. In terms of organization and coordination involving many people throughout the world with varying tasks, it was a monumental achievement in the ARV community. Surely something to be proud and part of. The knowledge obtained and the experiences realized will continue leading us in paths of discovery."

In a similar vein, APP member Poquiz posted:

"Financial success is but a mere step in our journey of elevating global consciousness to the reality of precognition. We must not allow this temporary failure to weaken our resolve. Albert Einstein once said, "Failure is success in progress." And on that account, we have made very good progress toward success. We need only continue our efforts."

Acknowledgements

Slides, information, and contributions were made by Jon Knowles, Mark Samuelson, Chris Georges, Carlos Mena, Alexis Poquiz, and Marty Rosenblatt. The authors would like to offer a special thanks to JSE Reviewer Paul Smith and to JSE Associate Editor Roger Nelson for their suggestions and input.

Disclaimer

As with any project involving multiple "players," this paper reflects diverse viewpoints, opinions, interests, and concerns. We, the authors, have done our best to create a balanced picture by soliciting and including comments from those who were both longtime members of APP and most intimately involved with the project from start to finish. Earlier drafts of this paper underwent extensive peer review within and outside of the Applied Precognition Project. That being said, any opinions presented within this article should be read as reflective of the authors' own viewpoints (as both project participants and subsequent investigators) and/or of belonging to those specifically quoted within the article itself, rather than as representative of the former Firefly management team members or Applied Precognition Project's owners. It is our sincere hope that this paper will encourage further productive discussion for and between all those who were involved.

Appendix A

ARV Methodologies Used in Project Firefly Binary ARV. Binary ARV is the standard protocol within the ARV subculture. It has two possible outcomes, and a photo is attached to each outcome. The viewer does one session per trial with the intention of describing the feedback photo they will see after the event, which is the photo connected only to the winning outcome.

Binary ARV–"ARV Studio" software. During and following Project Firefly, Igor Grgić used the "ARV Studio" software he developed to manage the P7B group (Grgić 2015). The full-featured computer program automates and simplifies all phases of a standard binary

ARV trial. Those phases include: tasking, photo target selection and pairing, judging, and feedback.

The software features ARV task creation, random coordinate number (Task Reference Number) generation, automated task sending to remote viewers' emails, random and double-blind photo target selection, random and double-blind association of the outcomes to the photo targets, judging and scoring sheet, automated ARV prediction email sending, feedback photo email delivery and data-keeping. It can be used for both solo and group projects.

Built-in algorithms ensure dissimilarity of computer-selected photo targets from a large pool of photo fi les, and also ensure non-repetition of selected photo targets for a pre-defined number ARV trials (www.arv-studio.com).

Binary ARV–"ARV Creator" scripted spreadsheet. Two of Firefly's solo remote viewers, Gary Gholson and Mark White, used "ARV Creator." Over many years, White developed and refined this scripted Excel file, which enables a user to quickly and easily generate a standard binary ARV project.

ARV Creator automatically generates Target Reference Numbers (TRN), randomly selects two photographs by category from a very large photo set, and creates a project with the click of a button. The customizable spreadsheet can be used solo or by a team of viewers. The user interface and accompanying target set are very user-friendly.

Lively ARV ('Live' Binary ARV). "Lively" is a term Sublime's group manager borrowed from group manager Mark Samuelson to designate "live" viewing sessions. During Project Firefly, Sublime group members met online via webinar. They started by socializing, seeing each other on video, then turned off the video while their group manager led them through an opening meditation involving light running through the body. Then they completed their viewing sessions. It is unknown how many of Sublime's predictions for Project Firefly used the Lively method vs. the other reported methods.

Winning Entanglements (WE) software. Prior to and during Project Firefly, APP leader Marty Rosenblatt personally managed several groups that used his Winning Entanglements (WE) software. It has a varied photo pool of locations, activities, objects, etc., which allows for double-blind conditions, given that the project manager doesn't see the photo choices prior to the viewer completing the session. Most WE groups in Project Firefly used self-judging. Over the years Rosenblatt has conducted numerous, in-depth free webinars demonstrating WE. These videos are available on the APP website and can provide further insight into the general protocols and technology WE groups use (www.appliedprecognition.com). During Project Firefly, Rosenblatt exclusively used the WE software for the following groups: Omega, Financial, and Pegasus. Scott Williams used either WE or CAS (see below) for his Sage and Transcendent groups. A few individuals acting as a group of one used WE, with modifications. Those who used WE ranged from inexperienced through advanced viewers.

Computer Assisted Scoring (CAS) software. The Sublime Group, Transcendent, and Sage used the CAS software/protocols. "CAS" is the acronym APP group managers gave to the computer software system designed by Dr. Edwin May, who does not refer to it as "CAS." His system is based on Fuzzy Set Theory, and on the decades of research he and his colleagues performed at SRI aimed at overcoming errors and challenges in human judging and target selection (May 2006). One distinctive feature of this system is its use of a specific target pool comprising solely photos of locations collected from National Geographic archives and "cleansed" of people, animals, and transportation devices. This system was used by Bierman (2013) and by a few APP group managers for about one year prior to its use in Project Firefly.

CAS is designed to eliminate the need for a human judge to actually see the photo options. However, it does require an independent "rater" to look at the viewer's transcript and indicate on a scoresheet if a pre-determined set of descriptors are present. This information is input into a computer. According to APP group managers, informal trials using CAS prior to Firefly showed mixed results. Software glitches at times resulted in missed trials, and raters required a learning curve to understand the items they were scoring. The efficacy of the CAS method in Project Firefly cannot be determined because groups that used CAS also used other protocols.

A breakdown was not available of how many predictions were made using each protocol.

Survey. Carlos Mena devised a "Survey" based on parapsychological studies that suggest spontaneous occurrences of psi occur from quick, unconscious responses. Rather than pair photos with the direction of the FOREX moves, it used nonsensical word lists. The premise was to use the unconscious somatic responses of a viewer, who was advised to rapidly select the best word from a list of multiple-choice options. Because it took very little time to complete, Mena sent the Survey to all willing Firefly participants, not to one particular group.

APPENDIX B

Scoring, Prediction Criteria, Errors Related to Metagroup Communications, Table Hit Comparison, Additional Information (self-reported by group managers) [Tables created by Igor Grgić]

Manager Name	Group/Solo Name	Group Type	Number of Viewers	Judging Type	Manager's Other Roles
Gary Gholson	Sharp	Solo	1	Self-judge	No
John Kovacs	JFK	Solo	1	Independent judge	No
Russ Evans	Psichisensi	Solo	1	Self-judge	No
Igor Grgic	P7B	Group	7	Independent judge (group manager)	Trader, Firefly manager
Nancy Smith	Sublime	Group	7	Independent judge (group manager)	Judge for another group
Marty Rosenblatt	Omega, Financial, Pegasus, Firstgroove, APPI groups	Group	5 to 10 per group	Self-judge	Trader, Firefly manager
Scott Williams	Sage, Transcendent	Group	Several	Self-judge, independent judge	no

Manager/ Group	Descriptions of the ARV Protocol(s) Used	Target Pool Description	Target Selection and Randomization
Gary Gholson / Sharp	Binary ARV using *ARV Creator* (scripted Excel spreadsheet)	Locations and Activities, Simple Objects	*ARV creator* randomly picked target pairings blind
John Kovacs / JFK	Binary ARV	Simple Objects	Independent judge
Russ Evans / Psichisensi	High volume of data, sketch input direction, 3 advance visuals. Great data separation 20+ target direction. Advance image priority, regular sketches.	Locations and Activities	Solo/Viewer
Igor Grgic / P7B	Binary ARV using *ARV Studio* software. Software selects photo targets double blind.	Locations and Activities, All: lifeforms, structures, landscapes, activity	prepaired by indep. judge; ARV Studio randomly selects pairings blind
Nancy Smith / Sublime	CAS, Binary ARV - 'Lively' method where remote viewer's cooldown and do RV online live	Locations and Activities, Simple Objects, CAS (Ed May's Pool), Other types of targets	independent judge (group manager); viewers; CAS
Marty Rosenblatt / 5 groups	WE. Online system sends two blind coordinates to viewer's email. Viewer submits two sessions and selfjudges.	Locations and Activities, Simple Objects	WE system randomly selects prepaired target pairs
Scott Williams / 2 groups	WE (see above), CAS (Computer Assisted Scoring)	Locations and Activities, Simple Objects, CAS target pool	WE system randomly selects prepaired target pairs

Manager/ Group	Target Selection Guidelines	Viewer's Blindness to the Target at Viewing Time	Manager's Blindness to the Target Prior Viewing Time
Gary Gholson / Sharp	Random photosites using *ARV Creator*	Yes, at all times	Yes, at all times
John Kovacs / JFK	Private guidelines based off of 10 yrs of private signal line data	(not answered)	(not answered)
Russ Evans / Psichisensi	Divergent aspects	(not answered)	(not answered)
Igor Grgic / P7B	Dissimilar as possible in all aspects	Yes, at all times	Yes, at all times
Nancy Smith / Sublime	(not answered)	Yes, at all times	(not answered)
Marty Rosenblatt / 5 groups	(not answered)	Yes, at all times	Yes, at all times
Scott Williams / 2 groups	(not answered)	Yes, at all times	(not answered)

Manager/ Group	Predictions per Week	Prediction Communication to the Trader	Outcome Communication from the Trader	Feedback Sent to Viewers	Private Wagering
Gary Gholson / Sharp	1	Email	Email	Within 24 hours of *viewing* time	No
John Kovacs / JFK	1	Email	Email	Within 24 hours of *viewing* time	No
Russ Evans / Psichisensi	1	Traders / managers did as they chose regardless of input	Email	Within 24 hours of *viewing* time	No
Igor Grgic / P7B	1	Email	Trader - direct outcome access via trading platform	Within 48 hours of *viewing* time	Yes, GM and some viewers
Nancy Smith / Sublime	1	Email	Email. Sometimes made personal outcome decision.	Within few days of *viewing* time	Yes, GM
Marty Rosenblatt / 5 groups	1 per each of the groups	Email	Trader - direct outcome access via trading platform	Within few days of *viewing* time	Yes, some viewers
Scott Williams / 2 groups	1	Email	Email	(not answered)	(not answered)

Manager/Group	Experience Level of the Viewer(s)	RV Techniques Used by Viewer(s)	Acquaintance with Viewers	Firefly Participation — Impact on Group	Group Performance during Firefly
Gary Gholson / Sharp	10 years	Loose and simplified CRV	Very well	It was fun, but I quickly lost motivation when I was personally doing well, yet the group was not.	Stayed the same
John Kovacs / JFK	10+ years	CRV	Very well	I didn't like the energy of it and told it was doomed for failure, too many overlapping intentions . . .	Decreased
Russ Evans / Psichisensi	Plenty	ERV, dowsing, mental images	Very well	Not positively	Improved
Igor Grgic / P7B	Most 5-10 yrs of experience; 1 or 2 novices	Simple CRV, freestyle ARV	Very well	Performance was same as in our other projects	Stayed the same
Nancy Smith / Sublime	Experienced, advanced	(not answered)	Very well	It was a long project that encouraged a little boredom.	Don't know
Marty Rosenblatt / 5 groups	From novices to very experienced	Various RV techinques	(not answered)	(not answered)	(not answered)
Scott Williams / 2 groups	(not answered)	(not answered)	(not answered)	(not answered)	(not answered)

APPENDIX 2

Publications related to ARV

Year	Researcher(s)	ARV-related Publications – compiled by Debra Lynne Katz
1977-79	Schwartz, S.A.	Two Application-Oriented RV Experiments Employing a Submarine. Conference Presentations.
1984	Larson, E.	Did Psychic Powers Give Firm a Killing in the Silver Market? Wall Street Journal, Oct 22.
1984	Houck, J,	Associative Remote Viewing, Archaeus, (4), 31 -37.
1985	Putoff, H.E.	ARV Applications. Research in Parapsychology, Scarecrow Press, 121-122.
1985	Harary, K., & Targ, R.	A New Approach to Forecasting Commodity Futures. PSI Research, (4), 79-85.
1990	May, Utts, Luke, Frivold, Trask.	Advances in Remote Viewing Analysis. JP. Vol. 54, Sept.
1994	May, Spottiswoode	Shannon Entropy: A Possible Intrinsic Target Property. JP, Vol.58, Dec.
1994	May, Spottiswoode	Managing the Target Pool Bandwidth. JP. Vol 58., Sept.
1995	Targ, Kantra, Brown, Wiegand.	Viewing the future: A pilot study with an error-detecting protocol. JSE. 9, 3, 67-80.
2000	Rosenblatt, M.	Applications: AVM Precognition Project: Results for Protocol-1. Connections Through Time. (7).
2001	Rosenblatt, M.	Protocol 5 Report. Connections Through Time - Issue 13
2006-07	Atwater, T.	Results from Application of ARV to Predicting Winners of Horse Races.
2005	Brown, C.	Remote Viewing: The Science & Theory of Nonphysical Perception, Farsight Press, 2005.
2006	May, E.	Two Protocols for Data Collection and Analysis. L.F.R.

2007	Schwartz, S.A.	Opening to the Infinite: The Art & Science of Nonlocal Awareness. Nemoseen Media.
2009	Walker, D.	RV outcomes for fun and profit or How to be a Zen monk while in Las Vegas, July.
2010	Knowles, J.	Trailmarkers in the Forest: Results from Two Team ARV Trials, Eight Martinis. (4). August.
2011	Atwater, T.	My experience with the 12th National Handicapping Championship. Online.
2013	Katz & Bulgatz	Remote Viewers Correctly Predict the Outcome of the 2012 Presidential Election. Aperture.
2013	Poquiz, A.	The Dung Beetle System (AkA Alexis Poquiz Method of Scoring). Online.
2013	Bierman, D.	Can Psi Sponsor Itself? Simulations & Results of a Automated ARV-Casino Experiment. Online.
2014	Smith, Laham, & Moddell.	Stock Market Prediction Using ARV by Inexperienced Viewers. JSE., Vol.28., No. 1., pp 7-16, 2014.
2014	Schwartz, S.A.	The Origins of ARV, For the Record, Aperture, 2014 (fall/winter).
2015	Samuelson, M.	Year-long Lively Project. No write up.
2015	Kolodziejzyk, G.	13-Year Associative Remote Viewing Experiment Results. J.P., pp 349 to 368.
2015	Schwartz, S.A.	Through Time and Space. Damien Broderick and Ben Goertzel, McFarland Jefferson. pp. 204-209.
2015	Fendley, T.W.	WWC Group sets the pace in Associative Remote Viewing, Eight Martinis, Issue 14, April.
2016	Rosenblatt, Knowles, Poquiz.	Applied Precognition Project (APP) and a Summary of APP-2014, Connections Through Time, 38.
2017	Katz, Grgić, Fendley, T.W.	Project Firefly: An Ethnographic Reporting of a Large-Scale ARV Project. Vol. 32, No. 1, pp. 21–54.
2017	Katz & Bulgatz.	Predicting the 2016 Presidential Election with ARV. Eight Martinis.
2017	Mueller, M.& Wittmann, M.	RemoteViewing:Proof-of-Principle-Studie. Zeitschrift für Anomalistik Band 17, S. 83–104.
2017	Knowles, J.	Putting SUARV to the Test. Eight Martinis, (15)
2019	Atunrase, T.	Remote Viewing the FIFA 2018 World Cup. Eight Martinis, (17).

2019	Katz, D.L., Smith, N., Bulgatz, M., Graff, D & Lane, J.	The ARV Dreaming Experiment. JSPR, Vol. 83, No. 2, 65–84.
2020	Schwartz, S.A.	The Origins of ARV. Mindfield,12(1), 5-14.
Accepted	Katz, D.L., Grgić, I. Patrizio, T., Fendley, T.W.	The ARV Rejudging Project. Journal of Society for Psychical Research (2021).
Pending	Katz, D. L., Lane, J. & Bulgatz, M.	Effect of Background Condition on Remote Viewing Objects (pending peer review)
Pending	Katz, & Tressoldi.	RV Applications Historical Overview & Survey, (2021).
2021	Kruth, J.	Associative Remote Viewing for Profit: Evaluating the Importance of the Judge and the Investment Instrument. Journal of Scientific Exploration, Vol. 35, No. 1, pp. 13–35, 2021.

Endnotes

1. Swann, I. (2020). *Resurrecting the mysterious: Ingo Swann's 'great lost work'*. Swann-Ryder Productions.

2. Danziger, K. (1990). *Constructing the subject: Historical origins of psychological research*. Cambridge University Press.

3. Gieryn, T. (1983). Boundary-Work and the demarcation of science from non-science: Strains and interests in professional ideologies of scientists. *American Sociological Review, 48*(6), 781–795.

4. Bowler, P. J. (2009). *Science for all: The popularization of science in early twentieth-century Britain*. University of Chicago Press.

5. Franczak, K. (2016). The circulation of knowledge in public discourse –Between 'popularization' and 'populization'. *Polish Sociological Review, 193*(1), 19–32.

6. Knowles, J. (2017). *Remote viewing from the ground up*. CreateSpace, DBA of On-Demand Publishing.

7. The 120+ site has been moved to remoteviewing.links and is now maintained by moderators of the reddit Remote Viewing group, primarily Grin Spickett and Nyko Tar (reddit makes wide use of pseudonyms).

8. This flyer was not a reference to Teresa Fendley's website ARV4Fun, which was created later.

9. Utts, J. (1996). An assessment of the evidence for psychic functioning. *Journal of Scientific Exploration,10*(1), 3–30.

10. Katz, D. L. (2004). *You are psychic: The art of clairvoyant reading and healing*. Llewellyn Worldwide Publishing (1st printing). Living Dreams Press (2nd printing).

11. Honorton, C. (1975). Objective determination of information rate in psi tasks with pictorial stimuli. *Journal of the American Society for Psychical Research, 69*, 353 –359.

12. Schwartz, S. A. (2016). *The Alexandria Project*. Open Road Distribution.

13. Struck, T. (2016). *Divination and human nature: a cognitive history of intuition in classical antiquity*. Princeton University Press.

14. Swann, I. (1987). *Natural ESP: The ESP core and its raw characteristics*. Bantam Books.

15. Mitchell, J. L. (1987). *Out-of-body experiences: A handbook*. Aquarian. (Original work published in 1981).

16. Swann, I. (1996). *The real story*. The American Prophecy Project, p. 309.

17. Nelson, R. D., Dunne, B. J., Dobyns, Y. H., & Jahn, R. G. (1996). Precognitive remote perception: Replication of remote viewing. *Journal of Scientific Exploration, 10*(1):109–110.

18. Swann, Ingo (n.d.). SRI Files, Special Collections, Irvine Sullivan Ingram Library, University of West Georgia.

19. Swann letter to Hal Puthoff, February 15, 1985, Swann archives.

20. This statement is verified in Mitchell's book, in Swann's various writing and in archival materials. Further also in Information Transfer Under Conditions of Sensory Shielding (Targ & Puthoff 1974), when they refer to the Osis study they are referring to the work Osis did with Swann, Osis, Mitchell at the ASPR. Still, it appears that Targ and Puthoff would go on to refine this approach and it became the hallmark of the studies they performed with visitors to their lab to convince them psi was real by creating an opportunity for them to have their own experiences. Many of these visitors included government and military officials who would help secure funding for these programs spanning two decades.

21. Targ, R., Puthoff, H. E., & May, E. C. (1977). State-of-the-art in remote viewing studies at SRI. *Proceedings of the IEEE 1977 International Conference of Cybernetics and Society, Washington D.C., 1977, 7,* 519-529.

22. Targ, R., & Puthoff, H. E. (1974). Information transfer under conditions of sensory shielding, *Nature, 251,* 602–607.

23. Puthoff, H., & Targ, R. (1976). A perceptual channel for information over kilometer distances: historical perspective and recent research. *Proceedings of the IEEE, 64*(3), 329–354.

24. Puthoff, H., & Targ, R. (1976). A perceptual channel for information over kilometer distances.

25. Kress, K. A. (1977). Parapsychology in intelligence: A personal review and conclusions. *Studies in Intelligence, 21*(4), 7–17.

26. Swann, Ingo (n.d.). SRI Files, Special Collections, Irvine Sullivan Ingram Library, University of West Georgia.

27. Swann, Ingo (n.d.). SRI Files, Special Collections.

28. Kress, K. A. (1977). Parapsychology in intelligence: A personal review and conclusions. *Studies in Intelligence, 21*(4), 7–17.

29. Puthoff, H. E., May, E. C., Humphries, B. S., Lavelle, L.A. (1983). Project Grill Flame. Defense Intelligence Agency.

30. Puthoff, H. E. (1984). ARV applications. In R. White & J. Solfvin (Eds.), *Research in parapsychology* (pp. 121–122). Scarecrow Press.

31. Atwater, F. H. (2001). *Captain of my ship, master of my soul: Living with guidance.* Hampton Roads Publishing.

32. Atwater, F. H. (2014). *Nonlocal empathy* [Video]. Applied Precognition Project, Talk With Series.

33. Smith, P. H. (2015). *The essential guide to remote viewing: The secret military remote perception skill anyone can learn.* Intentional Press.

34. Swann, Ingo (n.d.). SRI Files, Special Collections, Irvine Sullivan Ingram Library, University of West Georgia.

35. Swann, Ingo (n.d.). SRI Files, Special Collections.

36. Defense Intelligence Agency. Directorate for Scientific and Technical Intelligence (1984). *Psycho-energetics research report.*

37. *Science Panel Report, SRI Studies, a program review,* dated March 1984. It can be found in the book *The Star Gate archives 1972-1995,* compiled and edited by May & Marwaha.

38. Swann, Ingo (n.d.). SRI Files, Special Collections, Irvine Sullivan Ingram Library, University of West Georgia.

39. Smith, P. H. (1998). *Coordinate remote viewing training manual.* Stanford Research Institute – International.

40. Smith, P. H. (2005). *Reading the enemy's mind: Inside Star Gate—America's psychic espionage program.* Tom Dougherty Associates.

41. Williams, L.L. (2016). 18 years of excitement: CRV stories from a professional remote viewer. *Eight Martinis,* #14, 13–18.

42. McMoneagle, J. (1998). *The ultimate time machine: A remote viewer's perception of time and predictions for the new millennium.* Hampton Roads.

43. Targ, R., Puthoff, H. E., & May, E. C. (1977). State-of-the-art in remote viewing studies at SRI. *Proceedings of the IEEE 1977 International Conference of Cybernetics and Society, Washington D.C., 1977, 7,* 519-529.

44. Targ, R., & Harary, K. (1984). *The mind race: Understanding and using psychic ability.* Villard.

45. May, E. C., & Marhawa, S. B. (2018, 2019). *The Star Gate Archives: Reports of the United States Government sponsored psi program, 1972-1995: Volumes 1-4.* McFarland & Company.

46. McMoneagle, J. (2014). *Mind trek.* Crossroad Press.

47. Morehouse, D. (1998). *Psychic warrior: The true story of America's foremost psychic spy and the cover-up of the CIA'S top secret Stargate Program.* St. Martin's Paperbacks.

48. Schnabel, J. (1997). *Remote viewers: The secret history of America's psychic* spies. Dell Books.

49. Buchanan, L. (2009). *The seventh sense. The secrets of remote viewing as told by a "psychic spy" for the U.S. military.* Pocket Books.

50. Schwartz, S. A. (2013). *Secret vaults of time: Psychic archaeology and the quest for man's beginnings.* Nemoseen Media.

51. Schwartz, S. A., Mattei, R. J. D., & Smith, R. C. (2019). The Caravel Project. The location, description, and reconstruction of marine sites through remote viewing, including comparison with aerial photography, geological coring, and electronic remote sensing. *Zeitschrift für Anomalistik, 19,* 113–139.

52. Graff, D. E. (1998). *Tracks in the psychic wilderness. An exploration of ESP, remote viewing, precognitive dreaming, and synchronicity.* Element Books.

53. Graff, D. E. (2000). *River dreams.* Element Books.

54. Knowles, J., & Katz, D. (2019, October). In the archives of a many-sided man – Ingo Swann, the 'Father of remote viewing'. *Eight Martinis, #17,* 29–32.

55. The history of IRVA is documented on the IRVA.org website. Thompson-Smith was on the IRVA Board of Directors, continues to serve on their board of directors and was interviewed about the history.

56. Atwater, F. H. (2002). A message from the President. *Aperture, 1*(3 & 4), 1–2.

57. Champion, A., Couval, M. E., & Tournier, A. (2019, June 26). *Intuition and remote viewing: Ten years of R&D and applications for public and private organizations* [abstracts of presented papers]. 62nd Annual Convention of the Parapsychological Association.

58. Coronado, P. (2018, June). Perceiving murder: Tales from a psychic detective. *Edge Science, 34,* 3–6.

59. Husick, G. (2018). Application of remote viewing in the medical field: Viewing twins with autism [Conference proceedings]. International Remote Viewing Association.

60. Calabrese, P. (2002). *Remote healing and diagnosis.* [DVD]. International Remote Viewing Association.

61. Klieman, M. (2004). *Remote viewing as part of healing by utilizing the whole human consciousness* [DVD]. International Remote Viewing Association.

62. Atunrase, T. (2013, November). In search for a cure for cancer. *Eight Martinis, #10,* 12–19.

63. Husick, G. (2017). CRV case file: Mother and child reunion [Conference Proceedings held online]. International Remote Viewing Association.

64. Liaros, C. A. (2004). Project Blind Awareness: A humanitarian application of remote viewing [Conference proceedings]. International Remote Viewing Association.

65. Smith, A. T. (2015). *Remote viewing in humanitarian aid work.* [DVD]. International Remote Viewing Conference Proceedings, New Orleans, LA.

66. Smith, A. T. (2014). The ring anomalies of Saturn –Frontloading, "high strangeness," and current feedback. *Eight Martinis, #11,* 19–23.

67. Katz, D. L., Bulgatz, M., & Fendley, T. W. (2013). Remote viewing the outcome of the 2012 Presidential Election: An expedition into the unexplored territory of remote viewing and rating human subjects as targets within a binary protocol. *Aperture,* 46–56.

68. Katz, D. L., Lane, J., & Bulgatz, M. (2016, September). *A double-blind study of the comparison of single objects as targets placed in natural vs. unnatural environments as perceived through nonlocal*

perception by experienced remote viewers. [Conference session]. International Remote Viewing Annual Conference, New Orleans.

69. Allgire, D., & Akamatsu, H. (2013, November). Remote viewing the God particle. *Eight Martinis,* #10, 5-11.

70. Morse, M. L., Beem, L., Schwartz, S. A., & Katz, D. L. (2011). The effects of consciousness at a distance on tomato plants [Conference proceedings]. 2011 Science of Consciousness Convention Stockholm, University of Arizona.

71. Katz, D. L., Beem, L., & Fendley, T.W. (2015). Explorations into remote viewing microscopic organisms – "The Phage." *Aperture 26*, 42–49.

72. Allgire, D. (2009). *Masking and entrainment. A case study* [DVD]. IRVA.

73. Smith, D. (2020). Mind to mind –What part does telepathy play within remote viewing? *Eight Martinis,* #17, 33–41.

74. Champion, A., Couval, M. E., & Tournier, A. (2019, June 26). *Intuition and remote viewing: Ten years of R&D and applications for public and private organizations* [abstracts of presented papers]. 62nd Annual Convention of the Parapsychological Association.

75. Smith, N., & Smith, S. (2014, 2016). *Music from the fringe (I & II)* [Conference Proceedings]. International Remote Viewing Association, University of Illinois, College of Idaho.

76. Mungia, L. (Director). (2019). *Third eye spies* [Film]. Conscious Universe Films.

77. Smith, A. T. (2014). The ring anomalies of Saturn –Frontloading, "high strangeness," and current feedback. *Eight Martinis,* #11, 19–23.

78. Brown, C. (2012). Remote viewing the future with a tasking temporal outbounder. *Journal of Scientific Exploration, 26*(1), 81–110.

79. McNear, T. (2020). *Mars through the eyes of remote viewing and science* [Conference Proceedings, held online]. Applied Precognition Project (APP Fest 2020).

80. Atwater, F. H. (2001). *Captain of my ship, master of my soul: Living with guidance.* Hampton Roads Publishing.

81. Smith, P. H. (2005). *Reading the enemy's mind: Inside Star Gate—America's psychic espionage program.* Tom Dougherty Associates.

82. Williams, L. L. (2019). *Boundless: Your how-to guide to practical remote viewing —Phase one.* Amazon Digital Services.

83. Atunrase, T. (2015, January). Remote viewing Japan Air Lines flight 1628 & a UFO encounter over Alaska. *Eight Martinis,* #12, 84–91.

84. Atunrase, T. (2015). *Remote viewing UFOs and the visitors: Where do they come from? What are they? Who are they? Why are they here?* CreateSpace.

85. Brown, C. (2020). *Celestial projects* [Videos]. Farsight Institute.

86. Marrs, J. (2007). *Psi spies: The true story of America's psychic warfare program.* Weiser.

87. Schnabel, J. (1997). *Remote viewers: The secret history of America's psychic* spies. Dell Books.

88. *Third Eye Spies,* co-produced by our friend Lance Mungia and directed by Russell Targ.

89. May, E. C., & Marhawa, S. B. (2018, 2019). *The Star Gate Archives: Reports of the United States Government sponsored psi program, 1972-1995: Volumes 1-4.* McFarland & Company.

90. Katz, D. L., Grgić, I., & Fendley, T. W. (2018). An ethnographical assessment of Project Firefly: A yearlong endeavor to create wealth by predicting FOREX currency moves with associative remote viewing. *Journal of Scientific Exploration. 32*(1), 21–54.

91. Katz, Grgić, & Fendley, An ethnographical assessment of Project Firefly, 21–54.

92. Müller, M., Müller, L., & Wittmann, M. (2019). Predicting the stock market: An associative remote viewing study. *Zeitschrift für Anomalistik 19*, 326–346.

93. Katz, D. L., Bulgatz, M., & Walter, N. (2017). Predicting the 2016 U.S. Presidential Election. *Eight Martinis,* #15.

94. Katz, D. L., Smith, N., Graff, D., Bulgatz, M., & Lane, J. (2019). The associative remote dreaming experiment: A novel approach to predicting future outcomes of sporting events. *Journal of the Society for Psychical Research, 83*(2), 65–84.

95. Katz, D. L., Grgić, I., & Fendley, T. W. (2018). An ethnographical assessment of Project Firefly: A yearlong endeavor to create wealth by predicting FOREX currency moves with associative remote viewing. *Journal of Scientific Exploration. 32*(1), 21–54.

96. Schwartz, S. A. (2020). Origins of ARV. *Mindfield - Bulletin of the Parapsychological Association, 12*(1), 5-15.

97. Regarding what the letters stand for, Schwartz wrote: "Associated (or Associative; originally 'Associational') Remote Viewing (ARV)" (p. 5). The most commonly used form now is "Associative Remote Viewing."

98. Schwartz, S. A. (1977). *Two application-oriented experiments employing a submarine involving a novel remote viewing protocol, one testing the ELF hypothesis.* Invited paper for The Philosophical Research Society Conference on Extraordinary Human Functioning (August 1977); Annual Meetings of the Southwestern Anthropology Association (March 1978); The Association for Transpersonal Anthropology (March 1978); Parapsychological Association Annual Meetings (1978); Proceedings of the American Society for Psychical Research (November1979).

99. E.C. May, private correspondence via email to Debra Katz, 2016.

100. Schwartz, S. A. (2016). *The Alexandria Project.* Open Road Distribution.

101. Targ, R. (2010). *Do you see what I see? Memoirs of a blind biker.* Hampton Roads Publishing.

102. May, E. C. (1984, December). *Data Base Management.* SRI International.

103. Central Intelligence Agency. (n.d.). *Mental communication investigation.* Ingo Swann Archives at Ingrim Library, University of West Georgia.

104. H.E. Puthoff, letter to McDonnell Corporation, February 18, 1981. UWG archives.

105. Puthoff, H.E. (1996). CIA-initiated Remote viewing experiments at Stanford Research Institute. Institute for Advanced Studies at Austin.

106. *The Secret History of US Remote Viewing Program,* October 21, 2014, [Video: 17:46].

107. Angela Ford email to Jon Knowles, April 8, 2021.

108. Humphries, B. S. (November 1985). *Psi communication experiments* [Defense Intelligence Agency White Paper]. SRI International.

109. Harary, K., & Targ, R. (1985). A new approach to forecasting commodity futures. *Psi Research, 4,* 79–85.

110. Puthoff, H. E. (1984). ARV applications. In R. White & J. Solfvin (Eds.), *Research in parapsychology* (pp. 121–122). Scarecrow Press.

111. Targ, R., Kantra, J., Brown, D., & Wiegand, W. (1995). Viewing the future: A pilot study with an error-detecting protocol. *Journal of Scientific Exploration, 9*(3), 367-380.

112. Rosenblatt, M. (2000). Applications: AVM Precognition Project: Summary of results for Protocol-1. *Connections Through Time, 7.*

113. Kolodziejzyk, G. (2015). Thirteen-year associative remote viewing results. *Journal of Parapsychology, 76*(2), 349–368.

114. Smith, C., Laham, D., & Moddel, G. (2014). Stock market prediction using associative remote viewing by inexperienced viewers. *Journal of Scientific Exploration, 28*(1), 7–16.

115. M. Samuelson, Yearlong "Lively" Project email to Debra Katz, 2016.

116. Ibid.

117. Katz, D. L., Grgić, I., & Fendley, T. W. (2018). An ethnographical assessment of Project Firefly: A yearlong endeavor to create wealth by predicting FOREX currency moves with associative remote viewing. *Journal of Scientific Exploration. 32*(1), 21–54.

118. Bem, D. J. (2000). Psi phenomena [prepared for inclusion in Atkinson, R. L., Atkinson, R. C., Smith, E. E., & Bem, D. J. (1990). Introduction to psychology (10th ed.). Harcourt Brace Jovanovich.

119. Wimberger, L. (2015). *Neurosculpting: A whole-brain approach to heal trauma, rewrite limiting beliefs, and find wholeness.* Sounds True.

120. Katz, D. L., Grgić, I., & Fendley, T. W. (2018). An ethnographical assessment of Project Firefly: A yearlong endeavor to create wealth by predicting FOREX currency moves with associative remote viewing. *Journal of Scientific Exploration. 32*(1), 21–54.

121. Targ, R. (2012). *The reality of ESP: A physicist's proof of psychic abilities.* Quest Books.

122. Katz, D. L., Smith, N., Graff, D., Bulgatz, M., & Lane, J. (2019). The associative remote dreaming experiment: A novel approach to predicting future outcomes of sporting events. *Journal of the Society for Psychical Research, 83*(2), 65–84.

123. Carington, W. (n.d.). *Experiments on the paranormal cognition of drawings, III steps in the development of a repeatable technique*. Kaiser Legacy Reprints.

124. Milton, J. (1997). A meta-analytic comparison of the sensitivity of direct hits and sums of ranks as outcomes measures for free-response studies. *The Journal of Parapsychology, 61*(3), 227–241.

125. Solfvin, G. F., Kelly, E. F., & Burdick, D. S. (1978). Some new methods of analysis for preferential-ranking data. *Journal of the American Society for Psychical Research, 72*, 93-114.

126. Katz, D. L., Lane, J., & Bulgatz, M. (2016, September). *A double-blind study of the comparison of single objects as targets placed in natural vs. unnatural environments as perceived through nonlocal perception by experienced remote viewers.* [Conference session]. International Remote Viewing Annual Conference, New Orleans.

127. Targ, R., Kantra, J., Brown, D., & Wiegand, W. (1995). Viewing the future: A pilot study with an error-detecting protocol. *Journal of Scientific Exploration, 9*(3), 367-380.

128. Katz, D. L., Smith, N., Graff, D., Bulgatz, M., & Lane, J. (2019). The associative remote dreaming experiment: A novel approach to predicting future outcomes of sporting events. *Journal of the Society for Psychical Research, 83*(2), 65–84.

129. Katz, Smith, Graff, Bulgatz, & Lane. The associative remote dreaming experiment, 65–84.

130. Grgić. I. (2019). *ARV Studio Software*. http://www.arv-studio.com/About/

131. Smith, C., Laham, D., & Moddel, G. (2014). Stock market prediction using associative remote viewing by inexperienced viewers. *Journal of Scientific Exploration, 28*(1), 7–16.

132. Katz, D. L., Grgić, I., Tressoldi, P., & Fendley, T.W. (2020). Associative Remote Viewing Projects: Assessing rater reliability and factors affecting successful predictions. *Journal of Society for Psychical Research*, Spring/Summer 2021.

133. Smith, D. (2005). *Open Source CRV Manual*.

134. Buchanan, L. (2003). *The seventh sense. The secrets of remote viewing as told by a psychic spy for the U. S. military*. Pocket Books.

135. Angela Thompson-Smith used the method in a CRV experiment described in IRVA's very first *Aperture* newsletter (2001).

136. Poquiz, A. (2013). Dung beetle scoring system (AKA Poquiz Rating Method). *Applied Precognition Project*.

137. Katz, D. L., Beem, L., & Fendley, T.W. (2015). Explorations into remote viewing microscopic   organisms – "The Phage." *Aperture 26*, 42–49.

138. Poquiz, A. (2013). Dung beetle scoring system (AKA Poquiz Rating Method). *Applied Precognition Project*.

139. Katz, D. L., Bulgatz, M., & Fendley, T. W. (2013). Remote viewing the outcome of the 2012 Presidential Election: An expedition into the unexplored territory of remote viewing and rating human subjects as targets within a binary protocol. *Aperture*, 46–56.

140. Katz, D. L., Lane, J., & Bulgatz, M. (2016, September). *A double-blind study of the comparison of single objects as targets placed in natural vs. unnatural environments as perceived through nonlocal perception by experienced remote viewers.* [Conference session]. International Remote Viewing Annual Conference, New Orleans.

141. Honorton, C. (1975). Objective determination of information rate in psi tasks with pictorial stimuli. *Journal of the American Society for Psychical Research, 69*, 353 –359.

142. May, E.C., Utts, J., Humphries, B., Luke, W., Frivold, T., & Trask, V. (1990). Advances in remote viewing analysis. *Journal of Parapsychology, 54*(3) 193-228.

143. Humphries, B. S., Trask, V. V., May, E. C., & Thomson, M. J. (1986). Remote viewing evaluation techniques. In E. C. May, & S. B. Marwaha (Eds.). *The Star Gate Archives: Reports of the first decade of remote viewing research and operations at SRI United States Government sponsored psi program, 1972-1995, Vol. 2: Remote viewing 1985-1995*. McFarland.

144. Humphries, B. S., May, E. C., & Utts, J. M. (1988). Fuzzy set technology in the analysis of remote viewing. *Proceedings of the 31st Annual Convention of the Parapsychological Association*, 378-394.

145. Jahn, R. G., Dunne, B. J., & Jahn, E. G. (1980). Analytical judging procedure for remote perception experiments. *Journal of Parapsychology. 44*, 207–231.

146. Targ, R., Puthoff, H. E., & May, E. C. (1977). State-of-the-art in remote viewing studies at SRI. *Proceedings of the IEEE 1977 International Conference of Cybernetics and Society, Washington D.C., 1977, 7,* 519-529.

147. May, E.C., Utts, J., Humphries, B., Luke, W., Frivold, T., & Trask, V. (1990). Advances in remote viewing analysis. *Journal of Parapsychology, 54*(3) 193-228.

148. Katz, D. L., Grgić, I., Tressoldi, P., & Fendley, T.W. (2020). Associative Remote Viewing Projects: Assessing rater reliability and factors affecting successful predictions. *Journal of Society for Psychical Research,* Spring/Summer 2021. This paper has been accepted to the JSPR – should be published in 2021.

149. Lange, R. T. (2011). Inter-rater reliability. In Kreutzer J. S., DeLuca J., Caplan B. (Eds.) *Encyclopedia of Clinical Neuropsychology.* Springer.

150. Replacing each photo that was seen by a viewer when you go to the next trial in the series is recommended by the group which has won the largest lottery to date using ARV. See Chapter 21 for details of their method.

151. Utts, J. (1991). Replication and meta-analysis in parapsychology. *Statistical Science 6*(4), 363-403.

152. Rhine, J. B., Pratt, J. G., Smith, B. M., Stuart, C. E., & Greenwood, J. A. (1940). *Extra-sensory perception after sixty years.* Holt.

153. Storm, L., Tressoldi, P. E., & Di Risio, L. (2010). Meta-analysis of free-response studies, 1992–2008: Assessing the noise reduction model in parapsychology. *Psychological Bulletin, 136*(4), 471–485. (Retraction published 2010, *Psychological Bulletin, 136*[5], 893).

154. Brown, C. (2005). *Remote Viewing: The science and theory of nonphysical perception.* Farsight Press.

155. Milton, J. (1986a). *Displacement effects, role of the agent, and mention categories in relation to ESP performance.* [Dissertation, University of Edinburgh], p. 350.

156. Definitions are from *A Glossary of Terms Used in Parapsychology* by Michael A. Thalbourne (republished by Puente Publications, Charlottesville, VA, USA, 2003).

157. McLuhan, R. (2019). *Paranormal, why they are wrong, and why it matters.* White Crow Books.

158. Sternberg, R. J., Roediger, H. L., & Halpern, D. F. (2007). *Critical thinking in psychology.* Cambridge University Press.

159. Katz, D. L., Lane, J., & Bulgatz, M. (2016, September). *A double-blind study of the comparison of single objects as targets placed in natural vs. unnatural environments as perceived through nonlocal perception by experienced remote viewers.* [Conference session]. International Remote Viewing Annual Conference, New Orleans.

160. Alvarado, C. (2008). Note on Charles Richet's "La Suggestion Mentale et le Calcul des Probabilités" (1884). *Journal of Scientific Exploration, 22*(4), 543–548.

161. Bruck, C. (1925). *Experimentelle telepathie,* Proceedings 35, pp. 466–9. BR/PSI-X/telepath/fran

162. Murphy, G., & Dale, L. A., (1943). Concentration versus relaxation in relation to telepathy. *Journal of the ASPR, 37,* 19ff.

163. Sinclair, U. (1930). *Mental radio.* Hampton Roads Publishing.

164. Haynes, R. (1982). *The Society for Psychical Research, 1882–1982: A history.* Macdonald.

165. Carington, W. (n.d. - c. 1925). *Telepathy: An Outline of its Facts, Theories, and Implications* (2nd. ed.). Methuen & Company, p. 31.

166. Walker, K. (2019). *The extrasensory mind.* Sun Wise Books. (Original work published in 1961), p.85.

167. Carington, W. (1940–41). Experiments on the cognition of drawings [Conference proceedings]. *Journal for Society of Psychical Research, XLVI,* containing parts 35-151, 161-165.

168. Carington, Experiments on the cognition of drawings.

169. Milton, J. (1986a). *Displacement effects, role of the agent, and mentation categories in relation to ESP performance.* [Dissertation, University of Edinburgh], p. 48.

170. Rhine, J. B. (1950). Psi phenomena and psychiatry. *Proceedings of the Royal Society of Medicine, 1950; 43*(11): 804-814.

171. Rhine, Psi phenomena and psychiatry, 804-814.

172. Rhine, L. (1962a). Psychological processes in ESP experiences: Part I. Waking experiences. *Journal of Parapsychology, 26*, 88-111.

173. Rhine, L. (1962b). Psychological processes in ESP experiences: Part II. Dreams. *Journal of Parapsychology, 26*, 172-199.

174. Rhine, Psychological processes in ESP experiences: Part I, 88-111.

175. Rhine, Psychological processes in ESP experiences: Part II, 172-199.

176. Puthoff, H., & Targ, R. (1976). A perceptual channel for information over kilometer distances: historical perspective and recent research. *Proceedings of the IEEE, 64*(3), 329–354.

177. Hastings, A., & Hurt, D. (1976). A confirmatory remote viewing experiment in a group setting. *Proceedings IEEE, 64*(10), 1544–1545.

178. Tart, C. T. (1980). Are we interested in making ESP function strongly and reliably? A reply to J. E. Kennedy. *Journal of the American Society for Psychical Research, 74*(2), 210-222.

179. Crandall, J. E., & Hite, D. D. (1983). Psi-missing and displacement: Evidence for improperly focused psi? *Journal of the American Society for Psychical Research, 77*(3), 209–228.

180. Milton, J. (1986). *Displacement effects, role of the agent, and mention categories in relation to ESP performance.* [Dissertation, University of Edinburgh].

181. Milton, *Displacement effects.*

182. Ibid.

183. Harary, K., & Targ, R. (1985). A new approach to forecasting commodity futures. *Psi Research, 4*, 79–85.

184. Psi Research. (1985, Sep-Dec). (Reprint of *Wall Street Journal* article), p. 84.

185. Targ, R., Kantra, J., Brown, D., & Wiegand, W. (1995). Viewing the future: A pilot study with an error-detecting protocol. *Journal of Scientific Exploration, 9*(3), 367-380.

186. Houck, J. (1986). Associative remote viewing. *Archaeus, 4*, 31–37.

187. Butzer, B. (2019). Bias in the evaluation of psychology studies: A comparison of parapsychology versus neuroscience, *EXPLORE, 16*(6), 382-391.

188. Debra developed many of the ideas in this chapter, so we utilize first person.

189. Broadbent, D. E. (1952). Listening to one of two synchronous messages. *Journal of Experimental Psychology, 44* (1): 51–55.

190. Broadbent, D. E. (1972). *Decision and stress.* Academic Press.

191. Titchener, E. B. (1908). *Lectures on the elementary psychology of feeling and attention.* Macmillan.

192. James, W. (1890). *The principles of psychology, Volume 1.* Henry Holt and Company.

193. Merleau-Ponty, M. (1962). *Phenomenology of perception.* Routledge Classics. (Original work published 1945).

194. Ribot, T. (2012). *The psychology of attention.* Forgotten Books. (Original work published in 1903).

195. Wertheimer, M. (1923). *Laws of organization in perceptual forms.* Classics in the History of Psychology.

196. Warcolier, R. (2010). *Experiments in telepathy.* Kessinger Publishing. (Original work published in 1948).

197. Wagemans, J., Elder, J. H., Kubovy, M., Palmer, S. E., Peterson, M. A., Singh, M., & von der Heydt, R. (2012). A century of Gestalt psychology in visual perception: I. Perceptual grouping and figure–ground organization. *Psychological Bulletin 138*(6), 1172–1217.

198. Rutledge, R. (2020). Gestalt principles of perception – 4: Common fate [Personal website].

199. Warcollier, R. (2001). *Mind to mind* (Josephine B. Gridley, Trans.). Hampton Roads Publishing. (Original work published in 1948).

200. Perls, F. (1973). *The gestalt approach and eyewitness to therapy.* Science and Behavior Books.

201. Rutledge, R. (2020). Gestalt principles of perception – 4: Common fate [Personal website].

202. In Chapter 19 (on Election Predictions), readers will find a description of a study by Frank (2020). However, there may have been some methodological problems to his study, so his findings are in no way definitively showing that period of time from targeting or viewing to feedback makes a difference; it only serves to demonstrate that researchers are exploring this issue.

203. Brown, C. (2005). *Remote Viewing: The science and theory of nonphysical perception*. Farsight Press.

204. Hull, C.L. (1952). *A behavior system*. Yale University Press.

205. Cherry, K. (2019). *Drive-reduction theory and human behavior*. VeryWellMind, p.1

206. We make a point of the distinction because one well-known remote viewing instructor (Pru Calabrese) held that women experience sexual arousal while remote viewing.

207. Noble, J. (2018). *Natural remote viewing: A practical guide to the mental martial art of self-discovery* (2nd ed.). Intentional Press.

208. Honorton, C. (1978, February 18). *Replicability, experimenter influence, and parapsychology: An empirical context for the study of mind*. [Paper presentation]. Annual Meeting of the American Association for the Advancement of Science, Washington, D.C.

209. Braude, S. (2002). *ESP and psychokinesis: A philosophical examination*. Universal Publishers.

210. Etzel, C., Palmer, J., Marcusson-Clavertz, D. (Eds.). (2015). *Parapsychology: a handbook for the 21st century*. McFarland & Co 204. Wiseman, & Schlitz, 1998).

211. Wiseman, R., & Schlitz, M. (1997). Experimenter effects and the remote detection of staring. *Journal of Parapsychology, 61*(3), 197–208.

212. Mind-Matter Mapping Project Round Table Series Colloquium #2. (2013, June). The tip of the iceberg: placebo, experimenter expectation and interference phenomena in subconscious information flow. *The Journal of Nonlocality, II* (1).

213. Bengston, W. F., & Moga, M. M. (2007). Resonance, placebo effects, and type II errors: Some implications from healing research for experimental methods. *Journal of Alternative and Complementary Medicine, 13*(3), 317–327.

214. Warcolier, R. (2010). *Experiments in telepathy*. Kessinger Publishing. (Original work published in 1948).

215. Ullman, M., Krippner, S., & Vaughan, A. (1973). *Dream telepathy*. Macmillan; p. 30-42, 89-92, 116-124, 210-213.

216. Mishlove, J. (1975). *The roots of consciousness*. Random House.

217. Kennedy, J. E., & Taddonio, T. (1976). Experimenter effects in parapsychological research. *Journal of Parapsychology, 40*, 1–33.

218. Radin, D. (1997). Unconscious perception of future emotions: An experiment in presentiment. *Journal of Scientific Exploration,11*(2), 163–180.

219. Radin D. (2006). *Entangled minds*. Paraview Pocket Books.

220. Buchanan, L. (2003). *The seventh sense. The secrets of remote viewing as told by a psychic spy for the U. S. military*. Pocket Books.

221. McMoneagle, J. (2015, June 22-25). *Scoring considerations*. [Conference talk]. Applied Precognition Project, New Orleans, LA. USA.

222. Müller, M., Müller, L., & Wittmann, M. (2019). Predicting the stock market: An associative remote viewing study. *Zeitschrift für Anomalistik 19*, 326–346.

223. Müller, M., Müller, L., & Wittmann, M. (2019). Predicting the stock market.

224. A 1993 thought-provoking film directed by Harold Ramis, starring Bill Murray, a TV weatherman sent to report on Groundhog Day in a small town, where he finds himself repeating the same day over and over. Each time he learns from his mistakes, particularly in relation to how to seduce his producer (Andie Macdowell).

225. Pratt, J. G., Smith, B. M., Rhine, J. B., Stuart, C. E., & Greenwood, J. A. (1940). *Extra-sensory perception after sixty years: A critical appraisal of the research in extra-sensory perception*. Henry Holt and Company.

226. McNamara, S. (2020). *Signal and noise: Advanced psychic training for remote viewing, clairvoyance, and ESP*. [Independently published].

227. Katz, D. L., Grgić, I., Tressoldi, P., & Fendley, T.W. (2020). Associative Remote Viewing Projects: Assessing rater reliability and factors affecting successful predictions. *Journal of Society for Psychical Research*, Spring/Summer 2021.

228. A long series of trials in APP called 1ARV had an analogous setup with one event, two teams, two photos, with separate scoring for each photo. No significant improvement was noted with this dual setup.

229. Thouless, R. H., Brier, R. M. (1970). The stacking effect & methods of correcting for it. *Journal of Parapsychology, 34*(2), 124-128.

230. McMoneagle, J. (1998). *The ultimate time machine: A remote viewer's perception of time and predictions for the new millennium.* Hampton Roads.

231. Swann, I., *Superpowers of the Human Biomind,* [database] p. 426.

232. Mossbridge, J. (2019). Review of Eric Wargo´s *Time loops: Precognition, retrocausation and the unconscious. JSE, 33*(2), 288-295.

233. Wolchover, N. (2020, April 7). Does time really flow: New clues come from a century-old approach to math. *Quanta Magazine.*

234. Smolin, L. (2013). *Time reborn: From the crisis in physics to the future of the universe.* Mariner Books.

235. Hawking, S., & Mlodinow, L. (2010). *The grand design.* Bantam Books.

236. McMoneagle, J. (1998). *The ultimate time machine: A remote viewer's perception of time and predictions for the new millennium.* Hampton Roads.

237. McMoneagle, J. (2000). *Remote viewing secrets: A handbook.* Crossroad Press.

238. "Finding a Missing Person Using Remote Viewing" at APP-2016 conference Learn ARV, Seminars. Presented by Joe McMoneagle APP-2016 in Vegas: "Consciousness is FUNdamental Precognition: Health-Wealth-Wisdom." June 13th-16th

239. McCrae, M. (2017, March 11). *Physicists Find That as Clocks Get More Precise, Time Gets More Fuzzy.* Science Alert.

240. Swann to May, April 1987.

241. Purcell, C. (2018, April 30). *A quantum physicist reveals why time is not as simple as it seems.* In Books, et al. V. Thomson (Ed.). AAAS (*Science* magazine).

242. MEST is an oft-repeated concept in "Scio" ("I know" in Latin).

243. *Superpowers of the Human Biomind* [Database].

244. Mossbridge, J. (2021). How to avoid the time wars. *Medium.*

245. Swann, I. (1997). *Your Nostradamus factor. Accessing your innate ability to see into the future.* Fireside.

246. Bem, D. J. (2011). Feeling the future: Experimental evidence for anomalous retroactive influences on cognition and affect. *Journal of Personality and Social Psychology, 100*(3):407-25. doi: 10.1037/a0021524.

247. Swann, I. (2018). *Psychic literacy & the coming psychic renaissance.* Swann Ryder Productions.

248. Cheung, T., & Mossbridge, J. A. (2018). *The premonition code.* Watkins Publishing.

249. May, E.C. & Marhawa, S. B. (2016). Precognition: The Only Form of Psi? *Journal of Consciousness Studies, 23*(3–4), 76–100.

250. Julie Mossbridge presents her views on this in *Time and the unconscious mind: A brief commentary* (2015).

251. Julia Mossbridge explained that through personal correspondence.

252. Penrose, R. (2012). *Cycles of time: An extraordinary new view of the universe.* Vintage Books.

253. Unpublished as of 2020.

254. Swann, I. (1997). *Your Nostradamus factor. Accessing your innate ability to see into the future.* Fireside.

255. Mossbridge, J. (2019). Review of Eric Wargo, *Time Loops: Precognition, retrocausation and the unconscious, Journal of Scientific Exploration, 33*(2), 288-295.

256. Central Intelligence Agency. (1987, October 15). Sun Streak Report – Third Quarter CY 87.

257 Tart, C. T. (1983). Information acquisition rates in forced-choice ESP experiments: Precognition does not work as well as present-time ESP. *Journal of the American Society for Psychical Research, 77*(4), 293–310.

258. McMoneagle, J. (2000). *Remote viewing secrets: A handbook.* Crossroad Press; p. 157.

259. Mossbridge, J. A., Tressoldi, P., Utts, J., Ives, J. A., Radin, D., & Jonas, W. B. (2014). Predicting the unpredictable: Critical analysis and practical implications of predictive anticipatory activity. *Frontiers in Human Neuroscience, 8,* p. 146.

260. Vallee, J. (1979). *Messengers of deception: UFO contacts and cults.* And/Or Books.

261. Sheldrake, R. (2009). *A new science of life / Morphic resonance.* Park Street Press.

262. Wolff, R. (2001). *Original wisdom: Stories of an ancient way of knowing.* Inner Traditions.

263. Honorton, C. (1975). Objective determination of information rate in psi tasks with pictorial stimuli. *Journal of the American Society for Psychical Research, 69,* 353 –359.

264. Rhine, J. B., Pratt, J. G., Smith, B. M., Stuart, C. E., & Greenwood, J. A. (1940). *Extrasensory perception after sixty years.* Holt.

265. Swann, I. (1987). *Natural ESP: The ESP core and its raw characteristics.* Bantam Books.

266. Sinclair, U. (1930). *Mental radio.* Hampton Roads Publishing.

267. Warcolier, R. (2010). *Experiments in telepathy.* Kessinger Publishing. (Original work published in 1948).

268. Warcollier, R. (2001). *Mind to mind* (Josephine B. Gridley, Trans.). Hampton Roads Publishing. (Original work published in 1948).

269. Warcollier, R. (2001). *Mind to mind* (Josephine B. Gridley, Trans.). Hampton Roads Publishing. (Original work published in 1948).

270. Krippner, S., & Zeichner, S. (1973). Telepathy and dreams: A descriptive analysis of art prints telepathically transmitted during sleep. *A.R.E. Journal, 8,* 197–201.

271. Honorton, C., & Schechter, E. I. (1987). Ganzfeld target retrieval with an automated testing system: A model for initial ganzfeld success. In D. H. Weiner & R. D. Nelson (Eds.), *Research in parapsychology 1986* (pp. 36–39). Scarecrow.

272. Honorton, C., Berger, R. E., Varvoglis, M. P., Quant, M., Derr, P., Schechter, E. I., & Ferrari, D.C. (1990). Psi communication in the ganzfeld: Experiments with an automated testing system and a comparison with a meta-analysis of earlier studies. *Journal of Parapsychology, 54,* 99–139.

273. Watt, C. A. (1989). Characteristics of successful free-response targets: Theoretical considerations. In L. A. Henkel & R. E. Berger (Eds.), *Research in parapsychology 1988,* (95-99). Scarecrow.

274. Morris, R. L. (1977). The Airport Project: A survey of the techniques for psychic development   advocated by popular books. In J. D. Morris, W. G. Roll, & R. L. Morris (Eds.), *Research in Parapsychology 1976* (pp. 54–56). Scarecrow Press.

275. Delanoy, D. L. (1989). Characteristics of successful free-response targets: Experimental findings and observations. In L. A. Henkel & R. E. Berger (Eds.). *Research in parapsychology 1988,* 92–95. Scarecrow.

276. Krippner, S., & Zeichner, S. (1973). Telepathy and dreams: A descriptive analysis of art prints telepathically transmitted during sleep. *A.R.E. Journal, 8,* 197–201.

277. Swann, I. (2002, October 13). Presentation at UFO Conference, Bordentown, New Jersey. Transcript by R.J. Durant.

278. Mitchell, J. L. (1987). *Out-of-body experiences: A handbook.* Aquarian. (Original work published in 1981).

279. Targ, R., & Puthoff, H. E. (1974). Information transfer under conditions of sensory shielding, *Nature, 251,* 602–607.

280. Swann, Ingo (n.d.). SRI Files, Special Collections, Irvine Sullivan Ingram Library, University of West Georgia.

281. Vallee, J. (1988). Remote viewing and computer communications –an experiment. *Journal of Scientific Exploration, 2*(1), 13–27.

282. Smith, P. H. (1998). *Coordinate remote viewing training manual.* Stanford Research Institute – International.

283. Smith, P. H. (2015). *The essential guide to remote viewing: The secret military remote perception skill anyone can learn.* Intentional Press.

284. Schwartz, S. A. (2016). *The Alexandria Project.* Open Road Distribution.

285. Knowles, J., & Katz, D. (2019, October). In the archives of a many-sided man – Ingo Swann, the 'Father of remote viewing'. *Eight Martinis,* #17, 29–32.

286. Targ, R., & Puthoff, H. E. (1974). Information transfer under conditions of sensory shielding, *Nature, 251,* 602–607.

287. Targ, & Puthoff. Information transfer under conditions of sensory shielding, p. 604.

288. Hansel, C. E. M. (1980). *ESP and parapsychology: A critical reevaluation*. Prometheus Books.

289. Tart, C., Puthoff, H. E., & Targ, R. (1980). Information transmission in remote viewing experiments. *Nature 284*, 191.

290. Puthoff, H., & Targ, R. (1976). A perceptual channel for information over kilometer distances: historical perspective and recent research. *Proceedings of the IEEE, 64*(3), 329–354.

291. Targ, R., & Puthoff, H. E. (2005). *Mind-reach: Scientists look at psychic abilities*. Hampton Roads Publishing.

292. Targ, R. (2004). *Limitless mind, a guide to remote viewing and transformation of consciousness*. New World Library.

293. Puthoff, H.E. (1996). CIA-initiated remote viewing experiments at Stanford Research Institute. Institute for Advanced Studies at Austin.

294. Puthoff, CIA initiated remote viewing experiments at SRI.

295. Puthoff, H. E., Targ, R., May E. C., Humphries, B. S., & Langford, G. (1980, June). *3-Year Joint Services Psychoenergetics Program*. Menlo Park, CA: SRI International.

296. Targ, R., Puthoff, H. E., & May, E. C. (1977). State-of-the-art in remote viewing studies at SRI. *Proceedings of the IEEE 1977 International Conference of Cybernetics and Society, Washington D.C., 1977, 7*, 519-529.

297. Targ, R., Kantra, J., Brown, D., & Wiegand, W. (1995). Viewing the future: A pilot study with an error-detecting protocol. *Journal of Scientific Exploration, 9*(3), 367-380.

298. Puthoff, H. E. (1984). ARV applications. In R. White & J. Solfvin (Eds.), *Research in parapsychology* (pp. 121–122). Scarecrow Press.

299. McMoneagle, J. (2014). *Mind trek*. Crossroad Press.

300. Morehouse, D. (1998). *Psychic warrior: The true story of America's foremost psychic spy and the cover-up of the CIA'S top secret Stargate Program*. St. Martin's Paperbacks.

301. Buchanan, L. (2009). *The seventh sense. The secrets of remote viewing as told by a "psychic spy" for the U.S. military*. Pocket Books.

302. Watt, C. A. (1989). Characteristics of successful free-response targets: Theoretical considerations. In L. A. Henkel & R. E. Berger (Eds.), *Research in parapsychology 1988,* (95-99). Scarecrow.

303. May, E.C., Utts, J., Humphries, B., Luke, W., Frivold, T., & Trask, V. (1990). Advances in remote viewing analysis. *Journal of Parapsychology, 54*(3) 193-228.

304. May, E. C., Spottiswoode, S. J. P., & Faith, L. V. (2000). Correlation of the gradient of Shannon entropy and anomalous cognition: Toward an AC sensory system. *Journal of Scientific Exploration, 14*(1), 53-72.

305. May, E. C. (2007). Advances in anomalous cognition analysis: A judge-free and accurate confidence-calling technique. In E. C. May & S. B. Marwaha (Eds.), *Anomalous cognition: Remote viewing research and theory* (pp. 80-88). McFarland.

306. May, E. C., Spottiswoode, S. J. P., & James, C. L. (1994a). Shannon entropy: A possible intrinsic target property. *Journal of Parapsychology, 58*, 384–401.

307. May, E. C., Spottiswoode, S. J. P, & James, C. L. (1994b). Managing the target-pool bandwidth: Possible noise reduction for anomalous cognition experiments. *Journal of Parapsychology, 58*, 303–313.

308. Lantz, N. D., Luke, W. L. W., & May, E. C. (1994). Target and sender dependencies in anomalous cognition experiments. *Journal of Parapsychology, 58*, 285–302.

309. Longmore, K. (2019, June). *Shannon entropy: A genius gambler's guide to market randomness*. Robot Wealth. Applying Science to Trading.

310. May, E.C., Utts, J., Humphries, B., Luke, W., Frivold, T., & Trask, V. (1990). Advances in remote viewing analysis. *Journal of Parapsychology, 54*(3) 193-228.

311. Warcollier's books were originally in French. They were translated by Gardner Murphy of Harper publishers in 1938. There have been several editions published since then. In both *Mind to mind* (1948) and *Experiments in telepathy* (1948) he mentions gestalt principles of psychology and perception.

312. Delanoy, D. L. (1989). Characteristics of successful free-response targets: Experimental findings and observations. In L. A. Henkel & R. E. Berger (Eds.). *Research in parapsychology 1988*, 92–95. Scarecrow.

313. May, E. C., & Marwaha, S. B. (2018, 2019). *The Star Gate Archives: Reports of the United States Government sponsored psi program, 1972-1995: Volumes 1-4.* McFarland & Company.

314. Lantz, N. D., Luke, W. L. W., & May, E. C. (1994). Target and sender dependencies in anomalous cognition experiments. *Journal of Parapsychology, 58*, 285–302.

315. May, E. C., Spottiswoode, S. J. P., & James, C. L. (1994a). Shannon entropy: A possible intrinsic target property. *Journal of Parapsychology, 58*, 384–401.

316. May, E. C., Spottiswoode, S. J. P, & James, C. L. (1994b). Managing the target-pool bandwidth: Possible noise reduction for anomalous cognition experiments. *Journal of Parapsychology, 58*, 303–313.

317. Lantz, N. D., Luke, W. L. W., & May, E. C. (1994). Target and sender dependencies in anomalous cognition experiments. *Journal of Parapsychology, 58*, 285–302.

318. Rhine, J. B., Pratt, J. G., Smith, B. M., Stuart, C. E., & Greenwood, J. A. (1940). *Extra-sensory perception after sixty years.* Holt.

319. Puthoff, H., & Targ, R. (1976). A perceptual channel for information over kilometer distances: historical perspective and recent research. *Proceedings of the IEEE, 64*(3), 329–354.

320. May, E. C., Spottiswoode, S. J. P., & James, C. L. (1994a). Shannon entropy: A possible intrinsic target property. *Journal of Parapsychology, 58*, 384–401.

321. May, E. C., Spottiswoode, S. J. P, & James, C. L. (1994b). Managing the target-pool bandwidth: Possible noise reduction for anomalous cognition experiments. *Journal of Parapsychology, 58*, 303–313.

322. The paper based on this proposal is still undergoing peer review with a formal journal.

323. Crandall, J. E., & Hite, D. D. (1983). Psi-missing and displacement: Evidence for improperly focused psi? *Journal of the American Society for Psychical Research, 77*(3), 209–228.

324. Hastings, A., & Hurt, D. (1976). A confirmatory remote viewing experiment in a group setting. *Proceedings IEEE, 64*(10), 1544–1545.

325. Katz, D. L., Bulgatz, M., McLaughlin-Walter (2018). Predicting the 2016 U.S. Presidential Election using a double blind associative remote viewing protocol. *Eight Martinis*, #18, 4–15.

326. Milton, J. (1997). A meta-analytic comparison of the sensitivity of direct hits and sums of ranks as outcomes measures for free-response studies. *The Journal of Parapsychology, 61*(3), 227–241.

327. Tart, C. T., & Hastings, A. C. (1976). *Apparent displacement effect in a remote viewing experiment: A methodological note.* Parapsychological Research Group report, Palo Alto.

328. Tart, C. T. (1980). Are we interested in making ESP function strongly and reliably? A reply to J. E. Kennedy. *Journal of the American Society for Psychical Research, 74*(2), 210-222.

329. Katz, D. L., Grgić, I., & Fendley, T. W. (2018). An ethnographical assessment of Project Firefly: A yearlong endeavor to create wealth by predicting FOREX currency moves with associative remote viewing. *Journal of Scientific Exploration. 32*(1), 21–54.

330. Rosenblatt, M. (2000). Applications: AVM Precognition Project: Summary of results for Protocol-1. *Connections Through Time, 7.*

331. Rosenblatt, R., Knowles, J., & Poquiz, A. (2015). Applied Precognition Project (APP) and a Summary of APP-2014. *Connections Through Time, 38.*

332. Williams, L. V., & Siegel, D. S. (2014). *The Oxford handbook of the economics of gambling.* OUP.

333. Delanoy, D. L. (1989). Characteristics of successful free-response targets: Experimental findings and observations. In L. A. Henkel & R. E. Berger (Eds.). *Research in parapsychology 1988*, 92–95. Scarecrow.

334. Barenholtz, E. (2013). Quantifying the role of context in visual object recognition. *Visual Cognition, 22*(1), 30–56.

335. Ribot, T. (2012). *The psychology of attention.* Forgotten Books. (Original work published in 1903).

336. In Chapter 13 we present other computer-based programs for use on desktop, laptop, tablet and smart phone.

337. May, E.C. (2013, June). *Improving Precognitive ARV: How Good Can It Get?* APP Conference.

338. Humphries, B., et al., (1987). *Fuzzy set applications in remote viewing analysis.* SRI Final Report.

339. May, E. C., Spottiswoode, S. J. P., & Faith, L. V. (2000). Correlation of the gradient of Shannon entropy and anomalous cognition: Toward an AC sensory system. *Journal of Scientific Exploration, 14*(1), 53-72.

340. May, E. C., Faith, L. V., Blackman, M., Bourgeois, B., Kerr, N., & Woods, L. (1999, August). *A target pool and database for anomalous cognition experiments.* Paper presented at the Parapsychological Association, Stanford, California.

341. May, E.C. (2013, June). Improving Precognitive ARV: How Good Can It Get? APP Conference.

342. Humphries, B., et al., (1987). *Fuzzy set applications in remote viewing analysis.* SRI Final Report.

343. May, E. C., Spottiswoode, S. J. P., & Faith, L. V. (2000). Correlation of the gradient of Shannon entropy and anomalous cognition: Toward an AC sensory system. *Journal of Scientific Exploration, 14*(1), 53-72.

344. May, E.C., Utts, J., Humphries, B., Luke, W., Frivold, T., & Trask, V. (1990). Advances in remote viewing analysis. *Journal of Parapsychology, 54*(3) 193-228.

345. Humphries, B., et al., (1987). *Fuzzy set applications in remote viewing analysis.* SRI Final Report.

346. May, E. C., Spottiswoode, S. J. P., & Faith, L. V. (2000). Correlation of the gradient of Shannon entropy and anomalous cognition: Toward an AC sensory system. *Journal of Scientific Exploration, 14*(1), 53-72.

347. May, Spottiswoode, & Faith, Correlation of the gradient of Shannon entropy and anomalous cognition, 53-72.

348. We recommend reading the series of articles (and May 2010 in particular), for details about the setup and results as well as topics we have not covered in this skeletal summary such as constructing target pools so they have comparable numbers of elements, dynamic vs. static targets, particulars of Shannon Entropy measurements and the mathematical formulas utilized in the experiments.

349. Wittgenstein, L. (1953). *Philosophical Investigations.* Blackwell, p. 66.

350. Lakoff, G. (1987). *Women, fire and dangerous things: What categories reveal about the mind.* University of Chicago Press.

351. Joe McMoneagle claimed at an APP Conference in 2013 that he took part in a CAS series with an investor in which a net of $3 million was raised. This figures to be the largest amount realized using ARV thus far.

352. Targ, R., & Puthoff, H.E. (1974). Remote viewing of natural targets. Standford Research Institute.

353. Targ & Puthoff, H. E. (1977). *Mind Reach.* Delacorte Press.

354. Targ, R., & Harary, K. (1984). *Mindrace: Understanding and using psychic ability.* Villard.

355. Targ, R., Kantra, J., Brown, D., & Wiegand, W. (1995). Viewing the future: A pilot study with an error-detecting protocol. *Journal of Scientific Exploration,* 9:367–380.

356. Targ, R. (2010). *Do you see what I see? Memoirs of a Blind Biker.* Hampton Roads Publishing.

357. Targ, R. (2012). *The reality of ESP: A physicist's proof of psychic abilities.* Quest Books.

358. Structure in Rome, Italy, that was originally the mausoleum of the Roman emperor Hadrian and became the burial place of the Antonine emperors until Caracalla. It was built in AD 135–139 and converted into a fortress in the 5th century. It stands on the right bank of the Tiber River and guards the Ponte Sant'Angelo, one of the principal ancient Roman bridges. In plan, the fort is a circle surrounded by a square; each corner of the square is protected by an individually designed barbican, or outwork, while the central circle is a lofty cylinder containing halls, chapels, apartments, courtyard, and prison cells.

359. Katz, D. L., Beem, L., & Fendley, T.W. (2015). Explorations into remote viewing microscopic organisms – "The Phage." *Aperture 26,* 42–49.

360. Katz, D. L., Bulgatz, M., & Fendley, T. W. (2013). Remote viewing the outcome of the 2012 Presidential Election: An expedition into the unexplored territory of remote viewing and rating human subjects as targets within a binary protocol. *Aperture*, 46–56.

361. Fendley, T. W. (2016). WWC Group sets the pace in Associate Remote Viewing. *Eight Martinis*, #14.

362. Fendley, WWC Group sets the pace.

363. From an email by Sumner to WWCD viewers.

364. Spottiswoode, J. P. (1997). Apparent association between effect size in free response anomalous cognition experiments and local sidereal time. *Journal of Scientific Exploration, 11*(2), 109-122.

365. Spottiswoode, Apparent association between effect size in free response anomalous cognition experiments and local sidereal time, 109-122.

366. Spottiswoode email to Marty Rosenblatt, post on APP Discussions, March 29, 2021.

367. Katz, D. L., Grgić, I., & Fendley, T. W. (2018). An ethnographical assessment of Project Firefly: A yearlong endeavor to create wealth by predicting FOREX currency moves with associative remote viewing. *Journal of Scientific Exploration. 32*(1), 21–54.

368. Personal communication.

369. Note by Debra and Jon: These six "gestalts" are found as ideograms in Gary Langford's training proposal and other trainers have taught them as well (such as Pru Calabrese in TransDimensional Systems and John Vivanco in Right Hemispheric.)

370. James Spottiswoode reported in 2021 that an even larger data set showed there was indeed a positive Effect Size at 13:00 LST, + or - 1 hour. (Marty Rosenblatt citing email by Spottiswoode on APP Discussions, March 29, 2021.)

371. A member of P-I-A, Mr. A.R., suggested an ARV method using just one dice throw to choose one photo. He wrote a protocol but did not test it, as far as is known. The method involves three people: one viewer (person 1), two tasker-judges (persons 2 and 3). The viewer gets coordinates from Person 2, who also selects a game, e.g., NBA Over/under. Person 2 assigns six paired types of events (land vs. water, outside vs. inside), one pair to each face of a single die, and rolls the die. Person 2 then flips a coin having assigned heads and tails to each member in the six pairs (e.g., head = land, tail = water). One member of each pair has also been preassigned Over or Under. If the coin comes up with the number that corresponds to land-water, then (by predetermination) water is the Over. Person 3 waits till the end of the game. After the game Person 3 searches for a photo of water if the total match points were Over. He mails the photo to the viewer. Person 3 does not get the prediction. "By this system no one can disrupt the future."

The method proposed throwing the dice and correlating the numbers with six categories (e.g., land vs. water) allowing for a binary choice. It seems doubtful using a single category would be enough to make a successful choice most of the time.

372. In this chapter and elsewhere, the letters TRN or the word "Tag" will sometimes be substituted for the word Target Reference Number or identifier. They all refer to an arbitrary numeric or alphanumeric reference to the intended target (e.g., 6734 8229 or GL34 TR21).

373. Issue 2, July 2009.

374. The Aurora Remote Viewing Group (2006-2010) was the first multi-method group that tried to establish itself as a company. After a prolonged period building the group, it did do a little operational work. However, the group was not successful and eventually dissolved. The reasons include not having a core in one place (it was based on three continents with different time zones); the differing methods the viewers used and the difficulties that posed; and inability of the management team and viewers to sustain the business. "Ken" was the name Jon Knowles used in the group.

375. Issue 4.

376. Marty is referring to the public 1ARV effort as opposed to the intra-PIA trials.

377. A viewer in Sublime.

378. Knowles, J. (2019, April). Putting SUARV to the test. *Eight Martinis*, #15.

379. The article refers to "two targets," which is a very common usage. However, in this book, on Debra's initiative, we refer to the actual target and other potential targets, also known as "decoys" in some experiments.

380. Kolodziejzyk, G. (2015). Thirteen-year associative remote viewing results. *Journal of Parapsychology, 76*(2), 349–368.

381. From video by Greg Kolodziejzyk, *Thought Power* (2013).

382. Kolodziejzyk, G. (2015). Thirteen-year associative remote viewing results. *Journal of Parapsychology, 76*(2), 349–368.

383. From the *Managed Futures Podcast*.

384. Kolodziejzyk, G. (2015). Thirteen-year associative remote viewing results. *Journal of Parapsychology, 76*(2), 349–368.

385. Harary, K., & Targ, R. (1985). A new approach to forecasting commodity futures. *Psi Research, 4*, 79–85.

386. Targ, R., Kantra, J., Brown, D., & Wiegand, W. (1995). Viewing the future: A pilot study with an error-detecting protocol. *Journal of Scientific Exploration, 9*(3), 367-380.

387. Smith, C., Laham, D., & Moddel, G. (2014). Stock market prediction using associative remote viewing by inexperienced viewers. *Journal of Scientific Exploration, 28*(1), 7–16.

388. Müller, M., Müller, L., & Wittmann, M. (2019). Predicting the stock market: An associative remote viewing study. *Zeitschrift für Anomalistik 19*, 326–346.

389. Morehouse, D. (2011). *Remote viewing: The complete user's manual for coordinate remote viewing*. Sounds True Publishing.

390. End of material prepared by Mr. Laurino.

391. Top psychics Laurie Campbell and Pam Coronado team up with police to bring fresh insight to unsolved murder mysteries.

392. Personal communication, March 19, 2021.

393. Coindesk, March 19, 2021.

394. Cheung, T., & Mossbridge, J. A. (2018). *The premonition code*. Watkins Publishing.

395. Vivanco, J. (2016). *The time before the secret words: On the path of remote viewing, high strangeness, and Zen*. Amazon Digital Services.

396. Julia Mossbridge, personal communication.

397. Igor Grgić, post on Remote Viewing reddit, Jan. 16, 2021.

398. Igor Grgić, Remote Viewing group subreddit, Feb. 10, 2021.

399. Graff, D. E. (1998). *Tracks in the psychic wilderness. An exploration of ESP, remote viewing, precognitive dreaming, and synchronicity*. Element Books.

400. Graff, D. E. (2000). *River dreams*. Element Books.

401. Graff, D. E. (2007). Explorations in precognitive dreaming. *Journal of Scientific Exploration,* 21(4), 707–722.

402. Graff, D. E., & Cyrus, P. S. (2017). Perceiving the future news: Evidence for retro-causation. *AIP Conference Proceedings, 1841*(1): 030001.

403. Targ, R., Kantra, J., Brown, D., & Wiegand, W. (1995). Viewing the future: A pilot study with an error-detecting protocol. *Journal of Scientific Exploration, 9*(3), 367-380.

404. Katz, D. L., Smith, N., Graff, D., Bulgatz, M., & Lane, J. (2019). The associative remote dreaming experiment: A novel approach to predicting future outcomes of sporting events. *Journal of the Society for Psychical Research, 83*(2), 65–84.

405. Targ, R., Kantra, J., Brown, D., & Wiegand, W. (1995). Viewing the future: A pilot study with an error-detecting protocol. *Journal of Scientific Exploration, 9*(3), 367-380.

406. Targ, R., Kantra, J., Brown, D., & Wiegand, W. (1995). Viewing the future: A pilot study with an error-detecting protocol. *Journal of Scientific Exploration, 9*(3), 367-380.

407. Katz, D. L., Smith, N., Graff, D., Bulgatz, M., & Lane, J. (2019). The associative remote dreaming experiment: A novel approach to predicting future outcomes of sporting events. *Journal of the Society for Psychical Research, 83*(2), 65–84.

408. Katz, D. L., Smith, N., Graff, D., Bulgatz, M., & Lane, J. (2019). The associative remote dreaming experiment: A novel approach to predicting future outcomes of sporting events. *Journal of the Society for Psychical Research, 83*(2), 65–84.

409. Acknowledgements. We'd like to thank *JSPR* for allowing us to reprint limited portions of the original study. Please consider joining the *JSPR* to read the full article. (For links, please visit http://www.arvbook.com)

We'd like to thank John Kruth and Rhine Research Institute for performing a proposal review prior to our embarking on this Study. We'd like to thank Jon Knowles, Alexis Poquiz, and Igor Grgić for their help with selecting the photographic targets and creating the sets, which is always an arduous task. Also thank you to Dr. James Lane for doing our statistics free of charge.

This project was completed with a $100, demonstrating that research, even that involving wagering, doesn't require much funding, just plenty of enthusiasm, patience, perseverance, appreciation and respect for one another, and in our case, cooperation from passionate individuals within a variety of remote viewing and parapsychological communities.

410. From "Facebook RV Community discusses effectiveness," 2014.

411. Ingo Swann's background in Scientology was provided to us by remote viewer Russell Pickering.

412. *Advance!* (1973). An interview with Ingo Swann. *Advance! 21,* October/November.

413. Wikipedia entry for Ingo Swann.

414. All of the above Star Gate participants left Scientology over the years (Ingo around 1982). Pat Price died in 1975.

415. "Scio" is the short form used in Scientology.

416. End of comments by Russell Pickering.

417. Schwartz, S. A. (2020). Origins of ARV. *Mindfield –Bulletin of the Parapsychological Association, 12*(1), 5-15.

418. Targ, R., & Harary, K. (1984). *The mind race: Understanding and using psychic ability.* Villard; p. 98.

419. Targ, R. (2012). *The reality of ESP: A physicist's proof of psychic abilities.* Quest Books, p. 137.

420. Tom Atwater notes that the word was invented by the French and therefore this is the correct spelling. However, we note virtually every source spells the word "parimutuel."

421. Foster, L. (2012, August 31). *Does the US presidential election impact the stock market?* Enterprising Investor.

422. Stevenson, P. W. (2016, May 12). This professor has predicted every presidential election since 1984. He's still trying to figure out 2016. *The Washington Post.*

423. Silver, N. (2016, November 5). *Election update: The campaign is almost over, and here's where we stand.* FiveThirtyEight.

424. Silver, N. (2016, November 5). *Election update: The campaign is almost over, and here's where we stand.* FiveThirtyEight.

425. Katz, D. L., Bulgatz, M., & Fendley, T. W. (2013). Remote viewing the outcome of the 2012 Presidential Election: An expedition into the unexplored territory of remote viewing and rating human subjects as targets within a binary protocol. *Aperture*, 46–56.

426. Katz, D. L., Bulgatz, M., McLaughlin-Walter (2018). Predicting the 2016 U.S. Presidential Election using a double blind associative remote viewing protocol. *Eight Martinis*, #18, 4–15.

427. Katz, Bulgatz, & McLaughlin-Walter, Predicting the 2016 U.S. Presidential Election, 4–15.

428. Katz, D. L., Bulgatz, M., & Fendley, T. W. (2013). Remote viewing the outcome of the 2012 Presidential Election: An expedition into the unexplored territory of remote viewing and rating human subjects as targets within a binary protocol. *Aperture*, 46–56.

429. Ed May personal communication to Jon.

430. The author can be contacted via reddit at www.reddit.com/u/FrankandFriends or by email at FrankandFriends@protonmail.com

431. In March 2019, Daz ran this public ARV project through his Facebook group. He was the tasker and project manager. The seven viewers knew nothing about the project.

432. Note by authors: Biden was duly inaugurated on January 20, 2021.

433. McNamara, S. (2017). *Defy your limits: The telekinesis training method.* Lightning Source.

434. McNamara, S. (2020). *Signal and noise: Advanced psychic training for remote viewing, clairvoyance, and ESP.* Mind Possible.

435. Note by Debra and Jon: As noted above, Jan. 20 came and went and Joe Biden was inaugurated president, so the majority of the viewers had it right.

436. Jon adds: It is a fact, however, that TransDimensional Systems used no frontloading in operational work. Just the tag (Target Reference Number) or a made-up word. This did take more

tasker, viewer and analyst time. But TDS was the most successful early commercial remote viewing company. Viewers were well-paid from client work and in part thanks to a wealthy patron/investor. Keeping the viewers fully blind worked – and worked very well. (I was a TDS viewer, Intern and became the Training Coordinator.)

437. Rhine, J. B. (1934). *Extra-sensory perception*. Boston Society for Psychic Research.

438. Pratt, J. G., Smith, B. M., Rhine, J. B., Stuart, C. E., & Greenwood, J. A. (1940). *Extra-sensory perception after sixty years: A critical appraisal of the research in extra-sensory perception*. Henry Holt and Company.

439. Swann, Ingo (n.d.). SRI Files, Special Collections, Irvine Sullivan Ingram Library, University of West Georgia.

440. Targ, R., & Puthoff, H. E. (2005). *Mind-reach: Scientists look at psychic abilities*. Hampton Roads Publishing.

441. Kress, K. A. (1977). Parapsychology in intelligence: A personal review and conclusions. *Studies in Intelligence, 21*(4), 7–17.

442. Tom McNear email to Jon, April 18, 2021.

443. Schnabel, J. (1997). *Remote viewers: The secret history of America's psychic* spies. Dell Books.

444. Joe McMoneagle email correspondence with Michelle Bulgatz, Oct. 8, 2016.

445. May, E. C., & Marwaha, S. B. (2018, 2019). *The Star Gate Archives: Reports of the United States Government sponsored psi program, 1972-1995: Volumes 1-4*. McFarland & Company.

446. May & Marwaha, *The Star Gate Archives*, p. 161-62.

447. Analytics was a hypothesized Stage 8 of CRV (at one time referred to as Stage 7), according to Tom McNear. However, it was never formally developed. (Video: Controlled Remote Viewing Training with Ingo Swann, and the Origin of the CRV Manuals (Youtube), April 6, 2021.)

448. Tom McNear personal communication, April 18, 2021.

449. We discuss Ingo's views on time in Chapter 7.

450. In Chapter 7, Jon describes the related concepts of "lumps" which Swann came up with from his experiences in these experiments.

451. Video: Controlled Remote Viewing Training with Ingo Swann, and the Origin of the CRV Manuals (Youtube), April 6, 2021.

452. May, E. C., & Marwaha, S. B. (2018, 2019). *The Star Gate Archives: Reports of the United States Government sponsored psi program, 1972-1995: Volumes 1-4*. McFarland & Company.

453. May & Marwaha, 2018, *The Star Gate Archives*, p. 280.

454. Targ, R., & Puthoff, H. E. (2005). *Mind-reach: Scientists look at psychic abilities*. Hampton Roads Publishing.

455. Ireland, R. (2011). *Your psychic potential: A guide to psychic development*. North Atlantic Books.

456. Miller, G. A. (1956). The magical number seven, plus or minus two: Some limits on our capacity for processing information. *Psychological Review, 101*(2), 343-352.

457. Personal communication to Debra, Oct. 4, 2016.

458. Treffert, D.A. (2009). The savant syndrome: an extraordinary condition. A synopsis: past, present, future. Philosophical transactions of the Royal Society of London. Series B, Biological sciences, *364*(1522), 1351–1357.

459. It tells the story of selfish yuppie Charlie Babbit (Tom Cruise), who discovers his father left a fortune to his savant brother Raymond (Dustin Hoffman) and a pittance to Charlie; both brothers travel together cross-country.

460. Peek, F. (1996). *The real Rain Man: Kim Peek*. Harkness Publishing Consultants.

461. Powell, D.H. (2008). *The ESP enigma: The scientific case for psychic phenomena*. Walker & Co.

462. Powell, D.H. (2015). Autistics, savants, and psi: A radical theory of mind. *EdgeScience* 23, 12-18.

463. Rimland, B. (1978). *Savant capabilities of autistic children and their cognitive implications*. In G. Serban (Ed.), *Cognitive defects in the development of mental illness* (p. 43–65). Brunner/Mazel.

464. Powell, D.H. (2015). Autistics, savants, and psi: A radical theory of mind. *EdgeScience* 23, 12-18.

465. Selfe, L. (1977). *Nadia: A case of extraordinary drawing ability in an autistic child*. Academic Press.

466. Top psychics Laurie Campbell and Pam Coronado team up with police to bring fresh insight to unsolved murder mysteries.

467. Email to Debra Katz, April 2, 2021.

468. Aurora Remote Viewing Group, *Remote Viewing the Nina Reiser Case*, July 2008.

469. Freedman, M., Leach, L., Kaplan, E., Winocur, G., Shulman, K., Dells, D. C. (1994). *Clock drawing: A neuropsychological analysis*. Oxford University Press.

470. Swann, I. (2002, October 13). Presentation at UFO Conference, Bordentown, New Jersey. Transcript by R.J. Durant.

471. California's Fantasy 5 is a lotto number handicapper's dream because it has a same-game history going back to 1992 with more than 5,600 drawings. You choose five numbers from 1 to 39. The odds of winning the first prize jackpot are one in 575,575 compared with the jackpot odds of one in 41 million of winning the California Super Lotto. The top Fantasy 5 prize starts at $50,000 and can top $500,000 depending on how often the top prize has recently rolled over (not been won).

472. Schwartz, S. A. (2007). *Opening to the infinite: The art and science of nonlocal awareness*. Nemoseen Media.

473. Ryzl, M. (2007). *Voyage to the rainbow: Reminiscences of a parapsychologist*. Trafford.

474. Humphries, B. S. (November 1985). *Psi communication experiments* [Defense Intelligence Agency White Paper]. SRI International.

475. Martin Wszolek. Post about Milan Ryzl's methods. June 30, 2020 (on Facebook).

476. Mishlove, J. (1975). *The roots of consciousness*. Random House; p. 244.

477. Martin Wszolek, Facebook Lottery Group, June 30, 2020.

478. Adapted from Knowles, 2017, p. 262-63.

479. In 2012 I was doing several sessions for each sports event and one session turned out to be so strong for the next tasking that I felt confident to bet it – and it was a win. I recognized it was a displacement in time, as opposed to displacing to the "wrong" photo in a binary pair.

480. A wheel. If you psychically get the numbers 1, 2, 3, 5 for a Pick 3, a wheel for a box bet would be to bet 123, 125, 135, 235. You have four numbers but since you need three for each pick; you end up with four bets. When you five or six numbers, wheels usually generate too many choices to make money betting them. I have used wheels with Pick 4's and sometimes have had success.

481. Roberts, J. (1979). *The afterdeath journal of an American philosopher: The world view of William James*. Prentice Hall.

482. Mark Zilberman, The Public Testing of Artificial Intuition Device Using Pick 3 Lottery, 2009 (white paper).

483. McNamara, S. (2020). *Signal and noise: Advanced psychic training for remote viewing, clairvoyance, and ESP.* [Independently Published].

484. Swann archives, n.d.

485. Atunrase, *Remote Viewing UFOs and the Visitors: Where Do They Come From? What Are They? Who Are They? Why Are They Here?*

486. Mark White email to Jon Knowles, April 9, 2012.

487. We address time displacement in the chapter on the lottery, Chapter 21.

488. We include more examples by Mark White in Chapter 26 (Pictograms).

489. Atunrase, T. (2019, October). Remote viewing the FIFA 2018 World Cup: An experiment in using associative remote viewing (ARV) to successfully predict a global event 2 months into the future. *Eight Martinis*, #17.

490. From Tunde Atunrase's Presentation at RV Meetup in 2019.

491. Atunrase, T. (2019, October). Remote viewing the FIFA 2018 World Cup: An experiment in using associative remote viewing (ARV) to successfully predict a global event 2 months into the future. *Eight Martinis*, #17.

492. Joe McMoneagle both at SRI and Ft. Meade often viewed with an experienced monitor or interviewer assisting him in accessing in-depth information about a target. However, in later years McMoneagle viewed solo. Gary Langford reported he could produce many pages of data (by himself) from a single short session, based on extensive self-training.

493. Those familiar with CRV structure will note that what I am calling "secondary probing" is found in stage 6 of CRV methodology.

494. Knowles, J. (2017). *Remote viewing from the ground up*. CreateSpace, DBA of On-Demand Publishing.

495. Katz, D. L., & Tressoldi, P. (2021). Remote viewing applications: An historical overview and a new survey. Presented at the joint online conference sponsored by the Parapsychology Association and Society for Scientific Exploration in July, 2021. Paper presently under formal peer review.

496. Smith, P. H. (2015). *The essential guide to remote viewing: The secret military remote perception skill anyone can learn*. Intentional Press.

497. Noble, J. (2018). *Natural remote viewing: A practical guide to the mental martial art of self-discovery* (2nd ed.). Intentional Press.

498. Smith, D. (2014a). *CRV—Controlled Remote Viewing: Manuals, collected papers & information to help you learn this intuitive art*. Amazon Digital Services.

499. Morehouse, D. (2011). *Remote viewing: The complete user's manual for coordinate remote viewing*. Sounds True Publishing.

500. Williams, L. L. (2019). *Boundless: Your how-to guide to practical remote viewing —Phase one*. Amazon Digital Services.

501. Smith, P. H. (1998). *Coordinate remote viewing training manual*. Stanford Research Institute – International.

502. David Morehouse, *Remote Viewing*, p 24. "Cognitron: This word was invented by the Remote Viewing community circa 1984, a combination of 'cognition' and the acronym TRON, to connote a combination of neural patterns. A cognitron is an assemblage of neurons, linked together by interconnecting synapses, which, when stimulated by the mind's recall system, produce a composite concept of their various subparts. Each neuron is charged with an element of the overall concept, which, when combined with the elements of its fellow neurons, produces the final concept, represented by the cognitron. As a human learns new facts, skills, or behaviors, neurons are connecting to form new cognitrons, the connecting synapses of which are more and more reinforced with use."

503. Source for all three: statements made at conferences or on social media.

504. Katz, D. L. (2004). *You are psychic: The art of clairvoyant reading and healing*. Llewellyn Worldwide Publishing (1st printing). Living Dreams Press (2nd printing).

505. Katz, *You are psychic*.

506. Knowles, J. (2017). *Remote viewing from the ground up*. CreateSpace, DBA of On-Demand Publishing.

507. This is the *only* public demonstration team that any remote viewing organization, group or company has had. However, individuals such as Joe McMoneagle, Daz Smith and Dick Allgire have done live demonstrations for media or on video.

508. "Dazchat" on Facebook, March 5, 2021.

509. See Swann, I. (1987). *Natural ESP: The ESP Core and Its Raw Characteristics*. New York, NY: Bantam Books.

510. Knowles, J. (2013, November). The "pictolanguage" of psi sketches. *Eight Martinis*, #10.

511. Videos of two PowerPoints on the subject are available to members of APP who have a paid subscription.

512. Swann, I. (1987). *Natural ESP: The ESP core and its raw characteristics*. Bantam Books.

513. Swann, *Natural ESP*, p. 187.

514. In remote viewing the term "subconscious" is often used. However, "unconscious" is the term preferred in the professional psychological literature. We will use both terms in reference to the processes underlying remote viewing.

515. Warcolier, R. (2010). *Experiments in telepathy*. Kessinger Publishing. (Original work published in 1948).

516. Sinclair, U. (1930). *Mental radio*. Hampton Roads Publishing.

517. Swann, I. (1987). *Natural ESP: The ESP core and its raw characteristics*. Bantam Books.

518. Swann, *Natural ESP*, p. 139.

519. Personal communication, March 18, 2021.

520. Swann, I. (1987). *Natural ESP: The ESP core and its raw characteristics*. Bantam Books.

521. Edwards, B. (2012). *Drawing on the right side of the brain*. TarcherPerigee.

522. Sady, W. (2016). Ludwik Fleck. In E. N. Zalta (ed.), *Online Stanford Encyclopedia of Philosophy*.

523. Herlosky, J. (2015). *A sorcerer's apprentice: A skeptic's journey into the CIA's Project Stargate and remote viewing*. Trine Day Publishing.

524. See Ingo Swann's article: Your 17 Senses (1994).

525. Hoppe, K. T. (2019). *'Remote Viewing' isn't an umbrella term for psychic phenomenon*. Medium.

526. In an International Remote Viewing Conference that took place in 2013 in Las Vegas.

527. Katz, D. L., & Tressoldi, P. (2021). Remote viewing applications: An historical overview and a new survey. Presented at the joint online conference sponsored by the Parapsychology Association and Society for Scientific Exploration in July, 2021. Paper presently under formal peer review.

528. McNamara, S. (2020). *Signal and noise: Advanced psychic training for remote viewing, clairvoyance, and ESP*. [Independently Published].

529. McMoneagle, J. (2014). *Mind trek*. Crossroad Press.

530. The Copyright Act of 1976, which provides the basic framework for the current copyright law, was enacted on October 19, 1976, as Pub. L. No. 94-553, 90 Stat. 2541. You can find it on the US Copyright office website.

531. Samuelson, P. (2007). Why copyright law excludes systems and processes from the scope of its protection. *Texas Law Review, 85*(1), 2007, UC Berkeley Public Law Research Paper No. 1002666.

532. Lori Williams reported a streak of 100 hits in a row in a trinary ARV setup (Animal-Vegetable-Mineral). We were unable to confirm this figure. Marty Rosenblatt indicates the most hits in a row in APP is 14, by a viewer named JFK (personal communication to Jon, April 9, 2021).

533. Targ, R. (2012). *The reality of ESP: A physicist's proof of psychic abilities*. Quest Books.

534. Knowles, J. (2017). *Remote viewing from the ground up*. CreateSpace, DBA of On-Demand Publishing.

535. Sinclair, U. (1930). *Mental radio*. Hampton Roads Publishing.

536. Swann, I. (2019). *Penetration: Special Edition: The question of extraterrestrial and human telepathy*. Swann-Ryder Productions.

537. Targ, R., & Puthoff, H. E. (2005). *Mind-reach: Scientists look at psychic abilities*. Hampton Roads Publishing.

538. Schmeidler, G. (1984). Further analysis of PK with continuous temperature recordings. *Journal of the American Society for Psychical Research, 78*, 355–362.

Bibliography

Allgire, D., & Akamatsu, H. (2013, November). Remote viewing the God particle. *Eight Martini's Remote Viewing Magazine, 10,* 5-11.

Alvarado, C. (2008). Note on Charles Richet's "La Suggestion Mentale et le Calcul des Probabilités" (1884). *Journal of Scientific Exploration, 22*(4), 543–548.

Atunrase, T. (2013, November). In search for a cure for cancer. *Eight Martinis,* (10), 12–19.

Atunrase, T. (2015, January). Remote viewing Japan Air Lines flight 1628 & a UFO encounter over Alaska. *Eight Martinis,* #12, 84–91.

Atunrase, T. (2015). *Remote viewing UFOs and the visitors: Where do they come from? What are they? Who are they? Why are they here?* CreateSpace.

Atunrase, T. (2019, October). Remote viewing the FIFA 2018 World Cup: An experiment in using associative remote viewing (ARV) to successfully predict a global event 2 months. *Eight Martinis,* #17.

Atwater, F. H. (2001). *Captain of my ship, master of my soul: Living with guidance.* Hampton Roads Publishing.

Atwater, F. H. (2002). A message from the President. *Aperture, 1*(3 & 4), 1–2.

Barenholtz, E. (2013). Quantifying the role of context in visual object recognition. *Visual Cognition, 22*(1), 30–56.

Bem, D. J. (2000). Psi phenomena [prepared for inclusion in Atkinson, R. L., Atkinson, R. C., Smith, E. E., & Bem, D. J. (1990). Introduction to psychology (10th ed.)]. Harcourt Brace Jovanovich.

Bem, D. J. (2011). Feeling the future: Experimental evidence for anomalous retroactive influences on cognition and affect. *Journal of Personality and Social Psychology, 100*(3), 407–425.

Bengston, W. F., & Moga, M. M. (2007). Resonance, placebo effects, and type II errors: Some implications from healing research for experimental methods. *Journal of Alternative and Complementary Medicine, 13*(3), 317–327

Bierman, D., & Rabeyron, T. (2013, August). *Can psi sponsor itself? Simulations and results of an automated ARV-casino experiment.* Presented at the 56th Parapsychological Association Convention in Viterbo, Italy.

Bowler, P. J. (2009). *Science for all: The popularization of science in early twentieth-century Britain.* University of Chicago Press.

Braude, S. (2002). *ESP and psychokinesis: A philosophical examination.* Universal Publishers.

Brier, B. (1967). A Correspondence ESP experiment with high I.Q. subjects. *Journal of Parapsychology, 31*(2), 143–48.

Broadbent, D. E. (1952). Listening to one of two synchronous messages. *Journal of Experimental Psychology, 44*(1): 51–55.

Broadbent, D. E. (1972). *Decision and stress.* Academic Press.

Brown, C. (2005). *Remote Viewing: The science and theory of nonphysical perception.* Farsight Press.

Brown, C. (2012). Remote viewing the future with a tasking temporal outbounder. *Journal of Scientific Exploration, 26*(1), 81–110.

Bruck, C. (1925). *Experimentelle telepathie*, Proceedings 35, pp. 466–9. BR/PSI-X/telepath/fran

Buchanan, L. (2003). *The seventh sense. The secrets of remote viewing as told by a psychic spy for the U. S. military.* Pocket Books.

Buhlman, W. (1996) *Adventures beyond the body: How to experience out-of-body travel.* HarperOne Publishers.

Bundesen, C., & Habekost, T. (2008). *Principles of visual attention: Linking mind and brain.* Oxford Psychology Series.

Butzer, B. (2019). Bias in the evaluation of psychology studies: A comparison of parapsychology versus neuroscience, *EXPLORE, 16*(6), 382–391.

Carington, W. (n.d. - c. 1925). *Telepathy: An Outline of its Facts, Theories, and Implications* (2nd. ed.). Methuen & Company.

Carington, W. (1940–41). Experiments on the cognition of drawings [Conference proceedings]. *Journal for Society of Psychical Research, XLVI,* containing parts 35–151, 161–165.

Carington, W. (n.d.). *Experiments on the paranormal cognition of drawings, III steps in the development of a repeatable technique.* Kaiser Legacy Reprints.

Carpenter, J. C. (1977). Intrasubject and subject-agent effects in ESP experiments. In B. B. Wolman (Ed.). *Handbook of parapsychology.* Van Nostrand Reinhold.

Central Intelligence Agency. (1987, October 15). Sun Streak Report – Third Quarter CY 87.

Central Intelligence Agency. (n.d.). *Mental communication investigation.* Ingo Swann Archives at Ingram Library, University of West Georgia.

Cham, J., & Whiteson, D. (2017). *We have no idea: A guide to the unknown universe*. Riverhead Books.

Champion, A., Couval, M. E., & Tournier, A. (2019, June 26). *Intuition and remote viewing: Ten years of R&D and applications for public and private organizations* [abstracts of presented papers]. 62nd Annual Convention of the Parapsychological Association.

Cherry, K. (2019). *Drive-reduction theory and human behavior*. VeryWellMind.

Cheung, T., & Mossbridge, J. A. (2018). *The premonition code*. Watkins Publishing.

Child, I. L. (1985). Psychology and anomalous observations: The question of ESP in dreams. *American Psychologist, 40*(11), 1219–1230.

Chollet, F., & Allaire, J. J. (2018). *Deep learning with R*. Manning Publications.

Coronado, P. (2018, June). Perceiving murder: Tales from a psychic detective. *Edge Science, 34*, 3–6.

Crandall, J. E., & Hite, D. D. (1983). Psi-missing and displacement: Evidence for improperly focused psi? *Journal of the American Society for Psychical Research*, 77(3), 209–228.

Dames, E., & Newman, J. H. (2010). *Tell me what you see: Remote viewing cases from the world's premier psychic spy*. (S. Bowlby, Narr.) [Audiobook]. Audible Studios.

Danziger, K. (1990). *Constructing the subject: Historical origins of psychological research*. Cambridge University Press.

Defense Intelligence Agency. Directorate for Scientific and Technical Intelligence (1984). *Psycho-energetics research report*.

Delanoy, D. L. (1989). Characteristics of successful free-response targets: Experimental findings and observations. In L. A. Henkel & R. E. Berger (Eds.). *Research in parapsychology 1988*, 92–95. Scarecrow.

Edwards, B. (2012). *Drawing on the right side of the brain*. TarcherPerigee.

Etzel, C., Palmer, J., Marcusson-Clavertz, D. (Eds.). (2015). *Parapsychology: a handbook for the 21st century*. McFarland & Co.

Fendley, T. W. (2016). WWC Group sets the pace in Associate Remote Viewing. *Eight Martinis*, #14.

Fontenrose, J. E. (1978). *The Delphic Oracle, its responses and operations, with a catalogue of responses*. University of California Press.

Foster, L. (2012, August 31). *Does the US presidential election impact the stock market?* Enterprising Investor.

Franczak, K. (2016). The circulation of knowledge in public discourse –Between 'popularization' and 'populization'. *Polish Sociological Review*, *193*(1), 19–32.

Freedman, M., Leach, L., Kaplan, E., Winocur, G., Shulman, K., Dells, D. C. (1994). *Clock drawing: A neuropsychological analysis*. Oxford University Press

Garfield, A., Drwecki, B., Moore, C., Kortenkamp, K., & Gracz, M. (2014). The Oneness

Beliefs Scale: Connecting spirituality with pro-environmental behavior. *Journal for the Scientific Study of Religion, 53*(2), 356–372.

Gieryn, T. (1983). Boundary-Work and the demarcation of science from non-science: Strains and interests in professional ideologies of scientists. *American Sociological Review, 48*(6), 781–795.

Graff, D. E. (1998). *Tracks in the psychic wilderness. An exploration of ESP, remote viewing, precognitive dreaming, and synchronicity.* Element Books.

Graff, D. E. (2000). *River dreams.* Element Books.

Graff, D. E. (2007). Explorations in precognitive dreaming. *Journal of Scientific Exploration, 21*(4), 707–722.

Graff, D. E., & Cyrus, P. S. (2017). Perceiving the future news: Evidence for retro-causation. *AIP Conference Proceedings, 1841*(1): 030001.

Grgić. I. (2019). *ARV Studio Software.*

Hansel, C. E. M. (1980). *ESP and parapsychology: A critical reevaluation.* Prometheus Books.

Harary, K., & Targ, R. (1985). A new approach to forecasting commodity futures. *Psi Research, 4,* 79–85.

Hastings, A., & Hurt, D. (1976). A confirmatory remote viewing experiment in a group setting. *Proceedings IEEE, 64*(10), 1544–1545.

Hawking, S., & Mlodinow, L. (2010). *The grand design.* Bantam Books.

Haynes, R. (1982). *The Society for Psychical Research, 1882–1982: A history.* Macdonald.

Herlosky, J. (2015). *A sorcerer's apprentice: A skeptic's journey into the CIA's Project Stargate and remote viewing.* Trine Day Publishing.

Hoffman, D. (2000). *Visual intelligence: How we create what we see.* W.W. Norton & Company.

Honorton, C. (1975). Objective determination of information rate in psi tasks with pictorial stimuli. *Journal of the American Society for Psychical Research, 69,* 353–359.

Honorton, C. (1978, February 18). *Replicability, experimenter influence, and parapsychology: An empirical context for the study of mind.* [Paper presentation]. Annual Meeting of the American Association for the Advancement of Science, Washington, D.C.

Honorton, C. (1985). Meta-analysis of psi ganzfeld research: A response to Hyman. *Journal of Parapsychology, 49*(1), 51–91.

Honorton, C., Berger, R. E., Varvoglis, M. P., Quant, M., Derr, P., Schechter, E. I., & Ferrari, D.C. (1990). Psi communication in the ganzfeld: Experiments with an automated ttesting system and a comparison with a meta-analysis of earlier studies. *Journal of Parapsychology, 54,* 99–139.

Honorton, C., & Schechter, E. I. (1987). Ganzfeld target retrieval with an automated testing system: A model for initial ganzfeld success. In D. H. Weiner & R. D. Nelson (Eds.), *Research in parapsychology 1986* (pp. 36–39). Scarecrow.

Hoppe, K. T. (2019). *'Remote Viewing' isn't an umbrella term for psychic phenomenon.* Medium.

Houck, J. (1986). Associative remote viewing. *Archaeus, 4,* 31–37.

Hull, C.L. (1952). *A behavior system.* Yale University Press.

Humphries, B. S., May, E. C., & Utts, J. M. (1988). Fuzzy set technology in the analysis of remote viewing. *Proceedings of the 31st Annual Convention of the Parapsychological Association,* 378–394.

Humphries, B. S., Trask, V. V., May, E. C., & Thomson, M. J. (1986). Remote viewing evaluation techniques. In E. C. May, & S. B. Marwaha (Eds.). *The Star Gate Archives: Reports of the first decade of remote viewing research and operations at SRI United States Government sponsored psi program, 1972-1995, Vol. 2: Remote viewing 1985-1995.* McFarland.

Humphries, B. S. (November 1985). *Psi communication experiments* [Defense Intelligence Agency White Paper]. SRI International.

Husick, G. (2017). CRV case file: Mother and child reunion [Conference Proceedings held online]. International Remote Viewing Association.

Husick, G. (2018). Application of remote viewing in the medical field: Viewing twins with autism [Conference proceedings]. International Remote Viewing Association.

Husserl, E. (2019). *The phenomenology of internal time-consciousness.* (James S. Churchill, Trans.). Indiana University Press. (Original work published 1928).

Hyman, R. (1996). Evaluation of program on anomalous mental phenomena. *Journal of Scientific Exploration, 10*(1), 31–58.

Ireland, R. (2011). *Your psychic potential: A guide to psychic development.* North Atlantic Books.

Jahn, R. G., Dunne, B. J., & Jahn, E. G. (1980). Analytical judging procedure for remote perception experiments. *Journal of Parapsychology. 44,* 207–231.

James, W. (1884). What is an emotion? [First published in *Mind, 9,* 188-205.] Republished by Christopher Green as an internet resource in Classics in the History of Psychology,

James, W. (1890). *The principles of psychology, Volume 1.* Henry Holt and Company.

Johnston, K. B. (2018, November 26*). The Pythia at Delphi: A topic of continued intrigue.* Women in Antiquity.

Katz, D. L. (2004). *You are psychic: The art of clairvoyant reading and healing.* Llewellyn Worldwide Publishing (1st printing). Living Dreams Press (2nd printing).

Katz, D. L. (2018). Remote viewing training survey. *Eight Martinis,* #16, 4–5.

Katz, D. L. (2019). Training interview with Debra Lynn Katz. *Eight Martinis,* #16, 33–45.

Katz, D. L. (2020). Multi-dimensional remote viewing. Theories and practice. [Conference Proceedings held online]. International Remote Viewing Association.

Katz, D. L., Beem, L., & Fendley, T.W. (2015). Explorations into remote viewing microscopic organisms – "The Phage." *Aperture 26,* 42–49.

Katz, D. L., Bulgatz, M., & Fendley, T. W. (2013). Remote viewing the outcome of the 2012 Presidential Election: An expedition into the unexplored territory of remote viewing and rating human subjects as targets within a binary protocol. *Aperture*, 46–56.

Katz, D. L., Bulgatz, M., McLaughlin-Walter (2018). Predicting the 2016 U.S. Presidential Election using a double blind associative remote viewing protocol. *Eight Martinis*, #18, 4–15.

Katz, D. L., Bulgatz, M., & Walter, N. (2017). Predicting the 2016 U.S. Presidential Election. *Eight Martinis*, #15.

Katz, D. L., Grgić, I., & Fendley, T. W. (2018). An ethnographical assessment of Project Firefly: A yearlong endeavor to create wealth by predicting FOREX currency moves with associative remote viewing. *Journal of Scientific Exploration. 32*(1), 21–54.

Katz, D. L., Grgić, I., Tressoldi, P. (2018). The ARV rejudging project. *Conference Proceedings of the Journal of Parapsychology, 82*(2), 118–119.

Katz, D. L., Grgić, I., Tressoldi, P., & Fendley, T.W. (2020). Associative Remote Viewing Projects: Assessing rater reliability and factors affecting successful predictions. *Journal of Society for Psychical Research*, Spring/Summer 2021.

Katz, D. L., Lane, J., & Bulgatz, M. (2016, September). *A double-blind study of the comparison of single objects as targets placed in natural vs. unnatural environments as perceived through nonlocal perception by experienced remote viewers.* [Conference session]. International Remote Viewing Annual Conference, New Orleans.

Katz, D. L., Smith, N., Graff, D., Bulgatz, M., & Lane, J. (2019). The associative remote dreaming experiment: A novel approach to predicting future outcomes of sporting events. *Journal of the Society for Psychical Research, 83*(2), 65–84.

Katz, D. L., & Tressoldi, P. (2020). Remote viewing applications: An historical overview and a new survey. Presented at the joint online conference sponsored by the Parapsychology Association and Society for Scientific Exploration in July, 2021. Paper presently under formal peer review.

Kelly, J. L. (1956). A new interpretation of information rate. *System Technical Journal, 35*(4), 917–926.

Kennedy, J. E., & Taddonio, T. (1976). Experimenter effects in parapsychological research. *Journal of Parapsychology, 40*, 1–33. https://jeksite.org/psi/jp76.htm

Knowles, J. (2009, March and July). Remote viewing from the perspective of "embodied mind." *Eight Martinis*, #1 and #2.

Knowles, J. (2010, August). Trailmarkers in the forest: Results from two team ARV trials. *Eight Martinis, #4.*

Knowles, J. (2012, May). What was that thing in the sky over Oakland, California in November 1896? *Eight Martinis, #7.*

Knowles, J. (2013, November). The "pictolanguage" of psi sketches. *Eight Martinis*, #10.

Knowles, J. (2014, June). Remote viewing meets the mystery of Oak Island. *Eight Martinis*, #11.

Knowles, J. (2017). *Remote viewing from the ground up.* CreateSpace, DBA of On-Demand Publishing.

Knowles, J. (2019, April). Putting SUARV to the test. *Eight Martinis*, #15.

Knowles, J., & Katz, D. (2019, October). In the archives of a many-sided man – Ingo Swann, the 'Father of remote viewing'. *Eight Martinis*, #17, 29–32.

Kolodziejzyk, G. (2015). Thirteen-year associative remote viewing results. *Journal of Parapsychology*, *76*(2), 349–368.

Kress, K. A. (1977). Parapsychology in intelligence: A personal review and conclusions. *Studies in Intelligence*, *21*(4), 7–17.

Krippner, S. (1993). The Maimonides ESP-dream studies. *Journal of Parapsychology*, *57*(1), 39–54.

Krippner, S., & Zeichner, S. (1973). Telepathy and dreams: A descriptive analysis of art prints telepathically transmitted during sleep. *A.R.E. Journal*, *8*, 197–201.

Lakoff, G. (1987). *Women, fire and dangerous things: What categories reveal about the mind.* University of Chicago Press.

Lange, R. T. (2011). Inter-rater reliability. In Kreutzer J. S., DeLuca J., Caplan B. (Eds.) *Encyclopedia of Clinical Neuropsychology*. Springer.

Lantz, N. D., Luke, W. L. W., & May, E. C. (1994). Target and sender dependencies in anomalous cognition experiments. *Journal of Parapsychology*, *58*, 285–302.

Larson, E. (1984, October). Did psychic powers give firm a killing in the silver market? – And did greed ruin it all? Californians switch over to an extrasensory switch. *Wall Street Journal*, p. 1.

Levine, R. V. (2020). Time and culture. In R. Biswas-Diener & E. Diener (Eds.), *Noba textbook series: Psychology*. DEF publishers. http://noba.to/g6hu2axd

Levine, R. V., & Norenzayan, A. (1999). The pace of life in 31 countries. *Journal of Cross-Cultural Psychology*, *30*(2), 178–205.

Liaros, C. A. (2004). Project Blind Awareness: A humanitarian application of remote viewing [Conference proceedings]. International Remote Viewing Association.

Lidwell, W., Holden, K., & Butler, J. (2003). *Universal principles of design*. Rockport Publishers.

Longmore, K. (2019, June). *Shannon entropy: A genius gambler's guide to market randomness.* Robot Wealth. Applying Science to Trading.

Marrs, J. (2007). *Psi spies: The true story of America's psychic warfare program.* Weiser.

May, E. C. (December 1984). *Data Base Management.* SRI International.

May, E. C. (2006). *Two protocols for data collection and analysis.* Laboratories for Fundamental Research, Palo Alto, CA.

May, E. C. (2007). Advances in anomalous cognition analysis: A judge-free and accurate confidence-calling technique. In E. C. May & S. B. Marwaha (Eds.), *Anomalous cognition: Remote viewing research and theory* (pp. 80–88). McFarland.

May, E. C., Faith, L. V., Blackman, M., Bourgeois, B., Kerr, N., & Woods, L. (1999, August). *A target pool and database for anomalous cognition experiments.* Paper presented at the Parapsychological Association, Stanford, California.

May, E.C., & Lantz, N. (2010). Anomalous cognition technical trials: Inspiration for the target entropy concept. *Journal of the Society for Psychical Research, 74.4*(901), 225–243.

May, E.C. & Marwaha, S. B. (2016). Precognition: The Only Form of Psi? *Journal of Consciousness Studies, 23*(3–4), 76–100.

May, E. C., & Marwaha, S. B. (2018, 2019). *The Star Gate Archives: Reports of the United States Government sponsored psi program, 1972-1995: Volumes 1-4.* McFarland & Company.

May, E. C., Spottiswoode, S. J. P., & Faith, L. V. (2000). Correlation of the gradient of Shannon entropy and anomalous cognition: Toward an AC sensory system. *Journal of Scientific Exploration, 14*(1), 53–72.

May, E. C., Spottiswoode, S. J. P., & James, C. L. (1994a). Shannon entropy: A possible intrinsic target property. *Journal of Parapsychology, 58*, 384–401.

May, E. C., Spottiswoode, S. J. P, & James, C. L. (1994b). Managing the target-pool bandwidth: Possible noise reduction for anomalous cognition experiments. *Journal of Parapsychology, 58*, 303–313.

May, E.C., Utts, J., Humphries, B., Luke, W., Frivold, T., & Trask, V. (1990). Advances in remote viewing analysis. *Journal of Parapsychology, 54*(3) 193–228.

McCrae, M. (2017, March 11). *Physicists Find That as Clocks Get More Precise, Time Gets More Fuzzy.* Science Alert.

McLuhan, R. (2019). *Paranormal, why they are wrong, and why it matters.* White Crow Books.

McMoneagle, J. (1998). *The ultimate time machine: A remote viewer's perception of time and predictions for the new millennium.* Hampton Roads Publishing Company, Inc.

McMoneagle, J. (2000). *Remote viewing secrets: A handbook.* Crossroad Press.

McMoneagle, J. (2002). *The Stargate chronicles: Memoirs of a psychic spy.* Hampton Roads.

McMoneagle, J. (2014). *Mind trek.* Crossroad Press.

McMoneagle, J. (2015, June 22-25). *Scoring considerations.* [Conference talk]. Applied Precognition Project, New Orleans, LA. USA.

McMoneagle, J. (2016, June 13-16). *Finding a missing person using remote viewing.* APP-2016 conference Learn ARV Seminars. Presented by Joe McMoneagle APP-2016 in Vegas: "Consciousness is FUNdamental Precognition: Health-Wealth-Wisdom."

McNamara, S. (2017). *Defy your limits: The telekinesis training method*. Lightning Source.

McNamara, S. (2020). *Signal and noise: Advanced psychic training for remote viewing, clairvoyance, and ESP.* [Independently published].

McNear, T. (2020). *Mars through the eyes of remote viewing and science* [Conference Proceedings, held online]. Applied Precognition Project (APP Fest 2020).

Mercer, A., Deane, C., & Mcgeeney, K. (2016, November 9). *Why 2016 election polls missed their mark.* Pew Research Center.

Merleau-Ponty, M. (1962). *Phenomenology of perception*. Routledge Classics. (Original work published 1945).

Miller, G. A. (1956). The magical number seven, plus or minus two: Some limits on our capacity for processing information. *Psychological Review, 101*(2), 343–352.

Milton, J. (1986a). *Displacement effects, role of the agent, and mention categories in relation to ESP performance*. [Dissertation, University of Edinburgh].

Milton, J. (1986b). Thesis on Displacement. *Proceedings of the Society for Psychical Research 1928-1952* – notes with mentions of displacement. *Journal of Society for Psychical Research* 1884–1952. https://www.cia.gov/library/readingroom/docs/CIA-RDP96-00792R000701040005-1.pdf

Milton, J. (1997). A meta-analytic comparison of the sensitivity of direct hits and sums of ranks as outcomes measures for free-response studies. *The Journal of Parapsychology, 61*(3), 227–241.

Mishlove, J. (1975). *The roots of consciousness*. Random House.

Mitchell, J. L. (1987). *Out-of-body experiences: A handbook*. Aquarian. (Original work published in 1981).

Monroe, R. A. (1977). *Journeys out of the body*. Anchor Press.

Morehouse, D. (1998). *Psychic warrior: The true story of America's foremost psychic spy and the cover-up of the CIA'S top secret Stargate Program*. St. Martin's Paperbacks.

Morehouse, D. (2011). *Remote viewing: The complete user's manual for coordinate remote viewing*. Sounds True Publishing.

Morris, R. L. (1977). The Airport Project: A survey of the techniques for psychic development advocated by popular books. In J. D. Morris, W. G. Roll, & R. L. Morris (Eds.), *Research in Parapsychology 1976* (pp. 54–56). Scarecrow Press.

Morse M. L., Beem, L., Schwartz, S. A., & Katz, D. L. (2011). The effects of consciousness at a distance on tomato plants [Conference proceedings]. 2011 Science of Consciousness Convention Stockholm, University of Arizona.

Mossbridge, J. A. (2015). Time and the unconscious mind: A brief commentary. Institute of Noetic Sciences.

Mossbridge, J. (2019). Review of Eric Wargo's *Time Loops: Precognition, Retrocausation and the Unconscious, Journal of Scientific Exploration, 33*(2), 288–295.

Mossbridge, J. (2021). How to avoid the time wars. *Medium.*

Mossbridge, J. A., Tressoldi, P., Utts, J., Ives, J. A., Radin, D., & Jonas, W. B. (2014). Predicting the unpredictable: Critical analysis and practical implications of predictive anticipatory activity. *Frontiers in Human Neuroscience, 8,*146.

Müller, M., Müller, L., & Wittmann, M. (2019). Predicting the stock market: An associative remote viewing study. *Zeitschrift für Anomalistik 19*, 326–346.

Murphy, G., & Dale, L. A., (1943). Concentration versus relaxation in relation to telepathy. *Journal of the ASPR, 37,* 19ff.

Nelson, R. D. (2017). Princeton Engineering Anomalies Research (PEAR). *Psi Encyclopedia.* London: The Society for Psychical Research. Retrieved 28 November 2019.

Nelson, R. D., Dunne, B. J., Dobyns, Y. H., & Jahn, R. G. (1996). Precognitive remote perception: Replication of remote viewing. *Journal of Scientific Exploration*, *10*(1):109–110.

Noble, J. (2018). *Natural remote viewing: A practical guide to the mental martial art of self-discovery* (2nd ed.). Intentional Press.

Open Science Collaboration (2015). Estimating the reproducibility of psychological science. *Science, 349*(6251), aac4716.

Palmer, J., Bogart, D. N., Jones, S. M., & Tart, C. T. (1977). Scoring patterns in an ESP ganzfeld experiment. *Journal of the American Society for Psychical Research*, *71*, 121–145

Peek, F. (1996). *The real Rain Man: Kim Peek.* Harkness Publishing Consultants.

Penrose, R. (2012). *Cycles of time: An extraordinary new view of the universe.* Vintage Books.

Perls, F. (1973). *The gestalt approach and eyewitness to therapy.* Science and Behavior Books.

Popkin, J. (2015, November 12). Meet the former pentagon scientists who says psychics can help American spies. *Newsweek Magazine.*

Poquiz, A. (2013). Dung beetle scoring system (AKA Poquiz Rating Method). *Applied Precognition Project.*

Powell, D.H. (2008). *The ESP enigma: The scientific case for psychic phenomena.* Walker & Co.

Powell, D.H. (2015). Autistics, savants, and psi: A radical theory of mind. *EdgeScience* 23, 12–18.

Pratt, J. G., Smith, B. M., Rhine, J. B., Stuart, C. E., & Greenwood, J. A. (1940). Extra-sensory perception after sixty years: A critical appraisal of the research in extra-sensory perception. Henry Holt and Company.

Psi Research. (1985, Sep-Dec). (Reprint of *Wall Street Journal* article), p. 84.

Purcell, C. (2018, April 30). *A quantum physicist reveals why time is not as simple as it seems.* In Books, et al. V. Thomson (Ed.). AAAS.

Puthoff, H. E. (1984). ARV applications. In R. White & J. Solfvin (Eds.), *Research in parapsychology* (pp. 121–122). Scarecrow Press.

Puthoff, H.E. (1996). CIA-initiated remote viewing experiments at Stanford Research Institute. Institute for Advanced Studies at Austin.

Puthoff, H. E., & Targ, R. (1975). Perceptual augmentation techniques, SRI Progress Report No. 3 (31 Oct. 1974) and Final Report to the CIA, covering the period January 1974 through February 1975, the second year of the program.

Puthoff, H., & Targ, R. (1976). A perceptual channel for information over kilometer distances: historical perspective and recent research. *Proceedings of the IEEE, 64*(3), 329–354.

Puthoff, H. E., Targ, R., May E. C., Humphries, B. S., & Langford, G. (1980, June). *3-Year Joint Services Psychoenergetics Program*. Menlo Park, CA: SRI International.

Puthoff, H. E., Targ, T., & May, E. C. (1981). Experimental psi research: Implications for physics. In R. G. Jahn (Ed.), *The role of consciousness in the physical world*. Westview Press.

Puthoff, H. E., May, E. C., Humphries, B. S., Lavelle, L.A. (1983). Project Grill Flame. Defense Intelligence Agency.

Radin, D. (1997). Unconscious perception of future emotions: An experiment in presentiment. *Journal of Scientific Exploration, 11*(2), 163–180.

Radin D. (2006). *Entangled minds*. Paraview Pocket Books.

Rhine, J. B. (1934). *Extra-sensory perception*. Boston Society for Psychic Research.

Rhine, J. B. (1950). Psi phenomena and psychiatry. *Proceedings of the Royal Society of Medicine, 1950; 43*(11): 804–814.

Rhine, J. B., Pratt, J. G., Smith, B. M., Stuart, C. E., & Greenwood, J. A. (1940). *Extra-sensory perception after sixty years*. Holt.

Rhine, L. (1962a). Psychological processes in ESP experiences: Part I. Waking experiences. *Journal of Parapsychology, 26*, 88–111.

Rhine, L. (1962b). Psychological processes in ESP experiences: Part II. Dreams. *Journal of Parapsychology, 26*, 172–199.

Ribot, T. (2012). *The psychology of attention*. Forgotten Books. (Original work published in 1903).

Rimland, B. (1978). *Savant capabilities of autistic children and their cognitive implications*. In G. Serban (Ed.), *Cognitive defects in the development of mental illness* (p. 43–65). Brunner/Mazel.

Rogers, K. (2014). *The attention complex: Media, archeology, method*. Palgrave Macmillan.

Rosenblatt, M. (2000). Applications: AVM Precognition Project: Summary of results for Protocol-1. *Connections Through Time*, 7. http://p-i-a.com/Magazine/Issue7/Applications_7.htm

Rosenblatt, R., Knowles, J., & Poquiz, A. (2015). Applied Precognition Project (APP) and a Summary of APP-2014. *Connections Through Time, 38*. http://p-i-a.com/Magazine/Issue38/Connections_38.html

Ryzl, M. (2007). *Voyage to the rainbow: Reminiscences of a parapsychologist.* Trafford.

Sady, W. (2016). Ludwik Fleck. In E. N. Zalta (ed.), *Online Stanford Encyclopedia of Philosophy*.

Samuelson, P. (2007). Why copyright law excludes systems and processes from the scope of its protection. *Texas Law Review, 85*(1), 2007, UC Berkeley Public Law Research Paper No.1002666.

Schmeidler, G. (1958). Analysis and evaluation of proxy sessions with Mrs. Caroline Chapman. *Journal of Parapsychology, 22*, 137–155.

Schmeidler, G. (1984). Further analysis of PK with continuous temperature recordings. *Journal of the American Society for Psychical Research, 78*, 355–362.

Schmidt, S. (2009). Shall we really do it again? The powerful concept of replication is neglected in the social sciences. *Review of General Psychology 13*(2), 90–100.

Schnabel, J. (1997). *Remote viewers: The secret history of America's psychic* spies. Dell Books.

Schwartz, S. A. (1977). *Two application-oriented experiments employing a submarine involving a novel remote viewing protocol, one testing the ELF hypothesis.* Invited paper for The Philosophical Research Society Conference on Extraordinary Human Functioning (August 1977); Annual Meetings of the Southwestern Anthropology Association (March 1978); The Association for Transpersonal Anthropology (March 1978); Parapsychological Association Annual Meetings (1978); Proceedings of the American Society for Psychical Research (November 1979).

Schwartz, S. A. (2007). *Opening to the infinite: The art and science of nonlocal awareness.* Nemoseen Media.

Schwartz, S. A. (2013). *Secret vaults of time: Psychic archaeology and the quest for man's beginnings.* Nemoseen Media.

Schwartz, S. A. (2016). *The Alexandria Project.* Open Road Distribution.

Schwartz, S. A. (2019). The location and reconstruction of a Byzantine structure in Marea, . Egypt, including a comparison of electronic remote sensing and remote viewing. *Journal of Scientific Exploration, 33*(3), 451–480.

Schwartz, S. A., Mattei, R. J. D., & Smith, R. C. (2019). The Caravel Project. The location, description, and reconstruction of marine sites through remote viewing, including comparison with aerial photography, geological coring, and electronic remote sensing. *Zeitschrift für Anomalistik, 19*, 113–139.

Schwartz, S. A. (2020). Origins of ARV. *Mindfield - Bulletin of the Parapsychological Association, 12*(1), 5–15.

Science Panel Report, SRI studies in remote viewing – A program review. (1984). In *The Star Gate Archives. Reports of the United States government sponsored psi program, 1972–1995*. Pp. 345–346 compiled and edited by May and Marwaha.

Selfe, L. (1977). *Nadia: A case of extraordinary drawing ability in an autistic child*. Academic Press.

Sheldrake, R. (2009). *A new science of life / Morphic resonance*. Park Street Press.

Sherwood, S., & Roe, C. A. (2003). A review of dream ESP studies conducted since the Maimonides dream ESP programme. *Journal of Consciousness Studies, 10*(6–7), 85–109.

Silver, N. (2016, November 5). *Election update: The campaign is almost over, and here's where we stand*. FiveThirtyEight.

Sinclair, U. (1930). *Mental radio*. Hampton Roads Publishing.

Smith, A. T. (2001). Research news: Remote viewing study: Summary and preliminary group data. *Aperture, 1(1)*.

Smith, A. T. (2014). The ring anomalies of Saturn – Frontloading, "high strangeness," and current feedback. *Eight Martinis*, #11, 19–23.

Smith, C., Laham, D., & Moddel, G. (2014). Stock market prediction using associative remote viewing by inexperienced viewers. *Journal of Scientific Exploration, 28*(1), 7–16.

Smith, D. (2014a). *CRV—Controlled Remote Viewing: Manuals, collected papers & information to help you learn this intuitive art*. Amazon Digital Services.

Smith, D. (2014b). *Remote viewing dialogues: Psychic spy veterans from the 23-year, U.S. military and intelligence remote viewing programs share their experiences and expertise*. CreateSpace.

Smith, D. (2020). Mind to mind – What part does telepathy play within remote viewing? *Eight Martinis*, #17, 33–41.

Smith, N., & Smith, S. (2014, 2016). *Music from the fringe (I & II)* [Conference Proceedings]. International Remote Viewing Association, University of Illinois, College of Idaho.

Smith, P. H. (1998). *Coordinate remote viewing training manual*. Stanford Research Institute – International.

Smith, P. H. (2005). *Reading the enemy's mind: Inside Star Gate—America's psychic espionage program*. Tom Dougherty Associates.

Smith, P. H. (2015). *The essential guide to remote viewing: The secret military remote perception skill anyone can learn*. Intentional Press.

Smolin, L. (2013). *Time reborn: From the crisis in physics to the future of the universe*. Mariner Books.

Solfvin, G. F., Kelly, E. F., & Burdick, D. S. (1978). Some new methods of analysis for preferential-ranking data. *Journal of the American Society for Psychical Research, 72*, 93-114.

Soto, C. J. (2019). How replicable are links between personality traits and consequential life outcomes? The life outcomes of Personality Replication Project. *Psychological Science, 30*(5), 711–727.

Spickett, G. (2020, August 30). *A deep dive into remote viewing tournament, the psychic competition app with real prizes*. Remote Viewing Community Magazine (on Medium).

Spottiswoode, J. P. (1997). Apparent association between effect size in free response anomalous cognition experiments and local sidereal time. *Journal of Scientific Exploration, 11*(2), 109-122.

Spottiswoode, J. P. & Faith, L.V. (2000). A correlation of the gradient of Shannon entropy and anomalous cognition: Toward an AC sensory system. *Journal of Scientific Exploration, 14*(1), 53–72.

Sternberg, R. J., Roediger, H. L., & Halpern, D. F. (2007). *Critical thinking in psychology.* Cambridge University Press.

Stevenson, P. W. (2016, May 12). This professor has predicted every presidential election since 1984. He's still trying to figure out 2016. *The Washington Post.*

Storm, L., Tressoldi, P. E., & Di Risio, L. (2010). Meta-analysis of free-response studies, 1992–2008: Assessing the noise reduction model in parapsychology. *Psychological Bulletin, 136*(4), 471–485. (Retraction published 2010, *Psychological Bulletin, 136*[5], 893).

Storm, L., Sherwood, S., Roe, C., Tressoldi, P., Rock, A. J., & Di Risio, L. (2017). On the correspondence between dream content and target material under laboratory conditions: A meta-analysis of dream-ESP studies, 1966–2016. *International Journal of Dream Research, 10*(2), 120–140.

Struck, T. (2016). *Divination and human nature: a cognitive history of intuition in classical antiquity.* Princeton University Press.

Styles, E. (2006). *The psychology of attention* (2nd. ed.). Psychology Press.

Swann, I. (1973). Experimental prophetic correlations. Experiment No. 70, SRI files. Ingo Swann papers found in Annie Belle Weaver Special Collections, Irvine Sullivan Ingram Library, University of West Georgia.

Swann, I. (1980–1984). Swann SRI employment contracts. Retrieved from Ingo Swann papers found in Annie Belle Weaver Special Collections, Irvine Sullivan Ingram Library, University of West Georgia. https://www.westga.edu/special/index.php

Swann, I. (1987). *Natural ESP: The ESP core and its raw characteristics.* Bantam Books.

Swann, I. (1993). On remote viewing, UFO's, and extraterrestrials. *Fate Magazine, 46*(9), 73–82.

Swann, I. (1996). *The real story.* The American Prophecy Project.

Swann, I. (1997). *Your Nostradamus factor. Accessing your innate ability to see into the future.* Fireside.

Swann, I. (2002, October 13). Presentation at UFO Conference, Bordentown, New Jersey. Transcript by R.J. Durant.

Swann, I. (2003). Remote viewing – The real story. In *Superpowers of the human mind*. [Database].

Swann, I. (2018). *Psychic literacy & the coming psychic renaissance*. Swann-Ryder Productions.

Swann, I. (2019). *Penetration: Special Edition: The question of extraterrestrial and human telepathy*. Swann-Ryder Productions.

Swann, I. (2020). *Resurrecting the mysterious: Ingo Swann's 'great lost work'*. Swann-Ryder Productions.

Swann, Ingo (n.d.). SRI Files, Special Collections, Irvine Sullivan Ingram Library, University of West Georgia.

Taddonio, J. L. (1976). The relationship of experimenter expectancy to performance on ESP tasks. *Journal of Parapsychology*, *40*, 107–114.

Targ, R. (2004). *Limitless mind, a guide to remote viewing and transformation of consciousness*. New World Library.

Targ, R. (2010). *Do you see what I see? Memoirs of a blind biker*. Hampton Roads Publishing.

Targ, R. (2012). *The reality of ESP: A physicist's proof of psychic abilities*. Quest Books.

Targ, R., & Harary, K. (1984). *The mind race: Understanding and using psychic ability*. Villard.

Targ, R., Kantra, J., Brown, D., & Wiegand, W. (1995). Viewing the future: A pilot study with an error-detecting protocol. *Journal of Scientific Exploration*, *9*(3), 367–380.

Targ, R., Puthoff, H. E., & May, E. C. (1977). State-of-the-art in remote viewing studies at SRI. *Proceedings of the IEEE 1977 International Conference of Cybernetics and Society, Washington D.C., 1977*, *7*, 519–529.

Targ, R., & Puthoff, H. E. (1974). Information transfer under conditions of sensory shielding, *Nature*, *251*, 602–607.

Targ, R., & Puthoff, H. E. (2005). *Mind-reach: Scientists look at psychic abilities*. Hampton Roads Publishing.

Tart, C. T. (1963). Physiological correlates of psi cognition. *International Journal of Parapsychology*, *5*, 375–386.

Tart, C. T. (1977). Toward conscious control of psi through immediate feedback training: Some considerations of internal processes. *Journal of the American Society for Psychical Research*, *71*, 375–407.

Tart, C. T., & Hastings, A. C. (1976). *Apparent displacement effect in a remote viewing experiment: A methodological note*. Parapsychological Research Group report, Palo Alto.

Tart, C. T. (1980). Are we interested in making ESP function strongly and reliably? A reply to J. E. Kennedy. *Journal of the American Society for Psychical Research*, *74*(2), 210–222.

Tart, C. T. (1983). Information acquisition rates in forced-choice ESP experiments: Precognition does not work as well as present-time ESP. *Journal of the American Society for Psychical Research, 77*(4), 293–310.

Tart, C., Puthoff, H. E., & Targ, R. (1980). Information transmission in remote viewing experiments. Nature 284, 191 (1980).

Thouless, R. H., Brier, R. M. (1970). The stacking effect & methods of correcting for it. *Journal of Parapsychology, 34*(2), 124-128.

Titchener, E. B. (1908). *Lectures on the elementary psychology of feeling and attention.* Macmillan.

Treffert, D.A. (2009). The savant syndrome: an extraordinary condition. A synopsis: past, present, future. Philosophical transactions of the Royal Society of London. Series B, Biological sciences, *364*(1522), 1351–1357.

Tsakiris, A. (Host). (2017, February 28). *Ex-Stargate head, Ed May, unyielding re materialism, slams Dean Radin.* [Audio podcast]. Skeptiko – Science at the Tipping Point.

Ullman, M., Krippner, S., & Vaughan, A. (1973). *Dream telepathy.* Macmillan.

Utts, J. (1991). Replication and meta-analysis in parapsychology. *Statistical Science 6*(4), 363–403.

Utts, J. (1996). An assessment of the evidence for psychic functioning. *Journal of Scientific Exploration, 10*(1), 3–30.

Vallee, J. (1979). *Messengers of deception: UFO contacts and cults.* And/Or Press.

Vallee, J. (1988). Remote viewing and computer communications –an experiment. *Journal of Scientific Exploration, 2*(1), 13–27.

Vivanco, J. (2016). *The time before the secret words: On the path of remote viewing, high strangeness, and Zen.* Amazon Digital Services.

Wagemans, J., Elder, J. H., Kubovy, M., Palmer, S. E., Peterson, M. A., Singh, M., & von der Heydt, R. (2012). A century of Gestalt psychology in visual perception: I. Perceptual grouping and figure–ground organization. *Psychological Bulletin 138*(6), 1172–1217.

Walker, D. (2009, July). Remote viewing outcomes for fun and profit, or how to be a Zen monk while in Las Vegas, *Eight Martinis, #2.*

Walker, K. (2019). *The extrasensory mind.* Sun Wise Books. (Original work published in 1961).

Warcolier, R. (2010). *Experiments in telepathy.* Kessinger Publishing. (Original work published in 1948).

Warcollier, R. (2001). *Mind to mind* (Josephine B. Gridley, Trans.). Hampton Roads Publishing. (Original work published in 1948).

Watt, C. A. (1989). Characteristics of successful free-response targets: Theoretical considerations. In L. A. Henkel & R. E. Berger (Eds.), *Research in parapsychology 1988,*

(95-99). Scarecrow.

Watzl, S. (2017). *Structuring mind. The nature of attention and how it shapes consciousness.* Oxford University Press.

Wertheimer, M. (1923). *Laws of organization in perceptual forms.* Classics in the History of Psychology.

White, R. A. (1964). A comparison of old and new methods of response to targets in ESP experiments. *Journal of the American Society for Psychical Research 58,* 21–56.

Williams, L. V., & Siegel, D. S. (2014). *The Oxford handbook of the economics of gambling.* OUP.

Williams, L.L. (2016). 18 years of excitement: CRV stories from a professional remote viewer. *Eight Martinis,* #14, 13–18.

Williams, L.L. (2017). *Monitoring: A guide for remote viewing & professional intuitive teams.* Amazon Digital Services.

Williams, L. L. (2019). *Boundless: Your how-to guide to practical remote viewing —Phase one.* Amazon Digital Services.

Williams, L., & Smith, D. (2019). Training interview with Lori Williams. *Eight Martinis,* #16, 66–77.

Wimberger, L. (2015). *Neurosculpting: A whole-brain approach to heal trauma, rewrite limiting beliefs, and find wholeness.* Sounds True.

Wiseman, R., & Schlitz, M. (1997). Experimenter effects and the remote detection of staring. *Journal of Parapsychology, 61*(3), 197–208.

Wittgenstein, L. (1953). *Philosophical Investigations.* Blackwell.

Wolchover, N. (2020, April 7). Does time really flow: New clues come from a century-old approach to math. *Quanta Magazine.*

Zimbardo, P., & Boyd, J. (2008). *The "time" paradox.* Simon & Schuster.

Name Index

Subject Index

JON KNOWLES has been active in the remote viewing community for 20 years. He was Training Coordinator and viewer with *TransDimensional Systems* (2000-2003); a viewer, project manager and administrator with the Aurora RV Group (2005-2008); and a group manager, viewer and Membership Coordinator in the *Applied Precognition Project* (2010-2016). He is the author of *Remote Viewing from the Ground Up* (2017)

and numerous articles about remote viewing published in Eight Martinis magazine. Jon is also a retired teacher, medical transcriptionist, developer of the ABCZ typing abbreviation system, and a former political activist. He has a B.A. in philosophy from Harvard University and an M.A. in English from U.C. Berkeley. He lives in Northern California.

DEBRA LYNNE KATZ resides has been conducting remote viewing and parapsychological research for a number of years and worked/studied in the Ingo Swann remote viewing archives for three years at the University of West Georgia. She is also the Director of the International School of Clairvoyance, one of the first schools of its kind to offer successful distance training programs via teleseminar and webinar. She is the

author of the landmark books *You Are Psychic: The Art of Clairvoyant Reading and Healing*, *Extraordinary Psychic: Proven Techniques to Master Your Natural Abilities* and *Freeing the Genie Within*, which has recently been released into the trilogy, *The Complete Clairvoyant*. She is an accomplished clairvoyant, remote viewer, medium and energy healer who works for business owners, manufacturers, stockbrokers, and other investors. Her celebrity clientele includes pop stars, actors, and musicians. Several of her students from the past two decades have gone on to start successful professional intuitive-related businesses. Her website is http://www.debrakatz.com. She has a Ph.D. in Psychology and a Masters Degree in Social Work. She is a former U.S. Probation Officer and legal victims advocate. She presently resides in Mapleton, Oregon.

REMOTE VIEWING

FROM THE GROUND UP

JON KNOWLES

The Complete Clairvoyant

You Are Psychic
Extraordinary Psychic
Freeing the Genie

A Trilogy

DEBRA LYNNE KATZ